MW00426726

INDUSTRIAL MOTOR CONTROL FUNDAMENTALS

Fourth Edition

Robert L. McIntyre
Late Assistant Director
National Joint Apprenticeship and
Training Committee for the Electrical Industry

Rex Losee
Formerly Instructor
Utah Valley Community College

McGRAW-HILL PUBLISHING COMPANY

New York Atlanta Dallas St. Louis San Francisco Auckland Bogotá Caracas
Hamburg Lisbon London Madrid Mexico Milan Montreal New Delhi Paris
San Juan São Paulo Singapore Sydney Tokyo Toronto

Sponsoring Editor: Paul R. Sobel
Editing Supervisor: Elizabeth R. Curione
Design and Art Supervisors: Nancy Axelrod and Meri Shardin
Production Supervisors: Catherine Bokman and Kathyrn Porzio

Text Designer: Suzanne Bennett & Associates
Cover Designers: Meri Shardin and Peri Zules
Cover Photographer: James Nazz
Technical Studio: Fine Line Illustrations, Inc.

Library of Congress Cataloging-In-Publication Data

McIntyre, R. L. (Robert L.)
Industrial motor control fundamentals / Robert L. McIntyre, Rex Losee.—4th ed.
p. cm.
Rev. ed. of: A-C motor control fundamentals / R.L. McIntyre. 3rd ed. 1974.
ISBN 0-07-045110-9 :
1. Electric controllers. 2. Electric motors, Alternating current. 3. Automatic control. I. Losee, Rex. II. McIntyre, R. L. A-C motor control fundamentals. III. Title.
TK2851.M25 1990 90-5423
621.46—dc20 CIP

INDUSTRIAL MOTOR CONTROL FUNDAMENTALS, Fourth Edition

1 2 3 4 5 6 7 8 9 0 SEMBKP 9 8 7 6 5 4 3 2 1 0

ISBN 0-07-045110-9

Contents

Preface

Industrial Motor Control Fundamentals, Fourth Edition, is an introduction to electric motors and their controls. It is written for students training to be industrial electricians or technicians. The text assumes that the reader has a working knowledge of electricity and is able to solve mathematical problems by using simple formulas. The text also provides the reader with the practical foundations of the subject, together with any additional electrical or mathematical theories required to understand the discussions.

Although the text discusses the historical bases of machines and methods, its emphasis is on modern equipment and practices. Wherever appropriate, reference is made to actual equipment likely to be found in the modern industrial environment. Therefore, attention is given to the continuing widespread use of magnetically operated (relay) controls as well as to some of the older solid-state equipment.

A systems approach is used to illustrate the close relationship between rotating equipment and control devices. Process systems are similarly utilized to integrate machines and their controls. Although integrated circuits and microprocessors are rapidly dominating modern controls, their application and operation are closely related to the older relay and contractor equipment and circuits. The use of the microprocessor in programmable controllers has had a dramatic impact on controls of all types. One chapter focuses on this subject and uses a number of well-known commercial P.C.s as examples.

No text on controls would be complete without a discussion of digital logic and boolean algebra. The application of these two subjects is as pertinent to relay and contactor circuits as it is to solid-state electronic circuits. In a carefully sequenced manner, the reader is introduced to the key elements, symbols, and operations of both subjects. Schematic, wiring, ladder, and logic diagrams are used to instruct the reader to progress, step by step, from the written specifications to the final circuit diagram.

The role of an industrial electrician and technician is largely one of "hands on." To aid such a person, this text includes code references applicable to the installation of new control systems and motors, as well as information on systematic maintenance and troubleshooting techniques. Most material in this text has been reviewed and tested in the classroom by both teachers and students, and in the field by electricians and technicians. To everyone involved goes my deepest appreciation.

It is fitting here to acknowledge the groundbreaking work of the originator of this text, the late Robert L. McIntyre. Bob was a master electrician, craftsman, instructor, and author in the field of motor controls. Much is owed him for his pioneering and tireless work and enthusiasm in the field.

Rex L. Losee

1
Fundamentals of Control

This chapter is concerned with the fundamentals and history of industrial motor control. The meaning of control is discussed. Open-loop, closed-loop, and multiloop control systems are presented. The basic concepts and methods of providing semiautomatic and automatic control are also discussed. Several block diagrams of control systems are illustrated.

In the modern plant many machines are made for fully automatic operation. The operator merely sets up the original process, and most or all operations are carried out automatically. The automatic operation of a machine is wholly dependent upon motor and machine control. Sometimes this control is entirely electrical or electronic, and sometimes a combination of electrical and mechanical control is used. The same basic principles apply, however.

1-1 MEANING OF CONTROL

A modern machine consists of three separate divisions which need to be considered. First is the machine itself, which is designed to do a specific job or type of job. Second is the motor, which is selected according to the requirements of the machine, as to load, duty cycle, and type of operation. Third, and of chief concern in this book, is the control system. The control system design is dictated by the operating requirements of the motor and the machine. If the machine needs only to start, run for some time, and stop, then the only control needed would be a simple toggle switch. If, however, the machine needs to start, perform several automatic operations, stop for a few seconds, and then repeat the cycle, it will require several integrated units of control.

The word *control* means to govern or regulate, so it must follow that when we speak of motor or machine control, we are talking about governing or regulating the functions of a motor or a machine. Applied to motors, controls perform several functions, such as starting, accelerating, regulating speed, protecting power, reversing, and stopping.

Any piece of equipment used to regulate or govern the functions of a machine or motor is called a *control component*. Each type of control component will be discussed in a separate section of this book.

1-2 MANUAL CONTROL

A manual controller is one having its operations controlled or performed by hand at the location of the controller (Fig. 1-1). Perhaps the single most popular type in this category is the manual full-voltage motor starter for lower-horsepower motors. This starter is

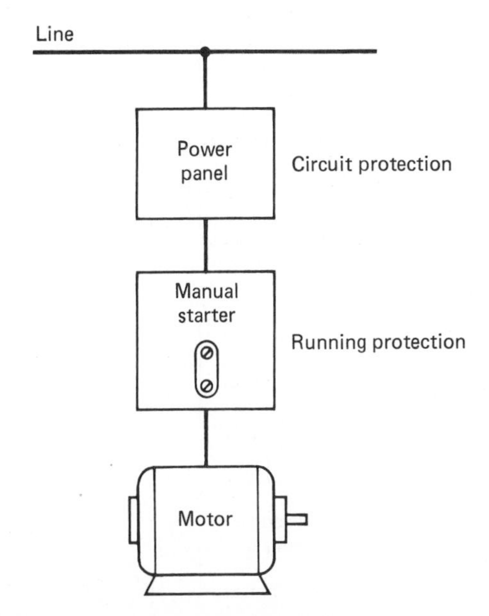

Fig. 1-1 Manual control of a motor.

frequently used where the only control function needed is to start and stop the motor. Probably the chief reason for the popularity of this unit is that its cost is only about one-half that of an equivalent magnetic starter. The manual starter generally gives overload protection and low-voltage release, but does not give low-voltage protection.

Manual control which provides the same functions as those achieved by the manual full-voltage motor starter can be had by the use of a switch with fusing of the delayed-action type, which will provide overload protection for the motor.

Examples of this type of control are very common in small metalworking and woodworking shops, which use small drill presses and lathes and pipe-threading machines. Another example is found in the exhaust fan generally used in machine shops and other industrial areas. In this installation the operator generally presses the START button for the fan in the morning when the plant opens, and the fan continues to run throughout the day. In the evening, or when the plant is shut down, the operator then pushes the STOP button, and the fan shuts down until needed again. The starter for welding machines of the motor-generator type is a very common application of this kind of control.

The compensator, or manual reduced-voltage starter, is used extensively to control polyphase squirrel-cage motors where reduced-voltage starting is required and the only control functions required are start and stop. The compensator gives overload protection, low-voltage release, and low-voltage protection. The compensator-type starter is quite frequently used in conjunction with a drum controller or wound-rotor motors (Fig. 1-2). This combination gives full manual control of start, stop, speed, and direction of rotation.

The compensator, being a reduced-voltage starter, is generally found only on the larger-horsepower motors. A very common use of the compensator and the drum controller combination is in the operation of many centrifugal-type air-conditioning compressors. The reduced-voltage feature is used to enable the motor to overcome the inertia of the compressor during starting without undue current loads on the system. The drum controller, through its ability to regulate the speed of a wound-rotor motor, provides a means of varying the capacity of the air-conditioning system, thus giving a flexibility that would not be possible with a constant-speed, full-voltage installation.

These are just a few examples of manual controllers, but you should have little trouble classifying any unit of this type because it will have no automatic control function. A manual controller requires the operator to move a switch or press a button to initiate any change in the operation of the machine or equipment.

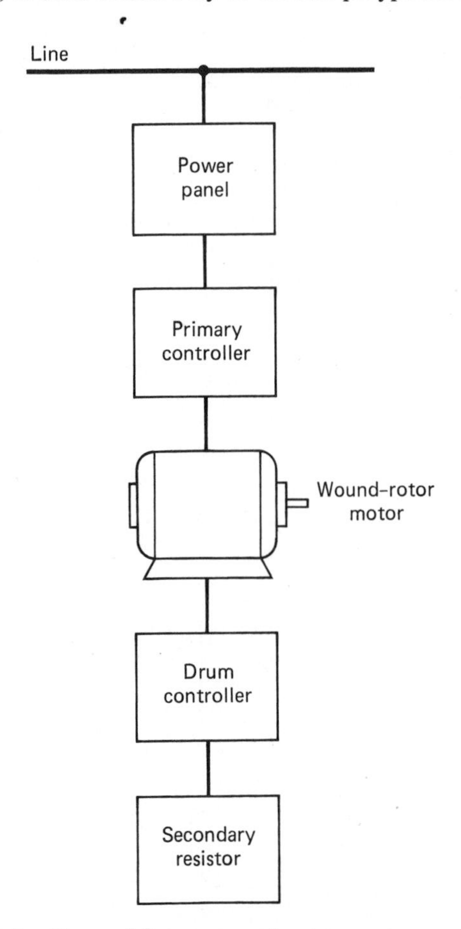

Fig. 1-2 Control for a wound-rotor motor.

1-3 SEMIAUTOMATIC CONTROL

Semiautomatic controllers use a magnetic starter and one or more manual pilot devices, such as pushbuttons, toggle switches, drum switches, and similar equipment (Fig. 1-3). Probably the most used of these is the push-

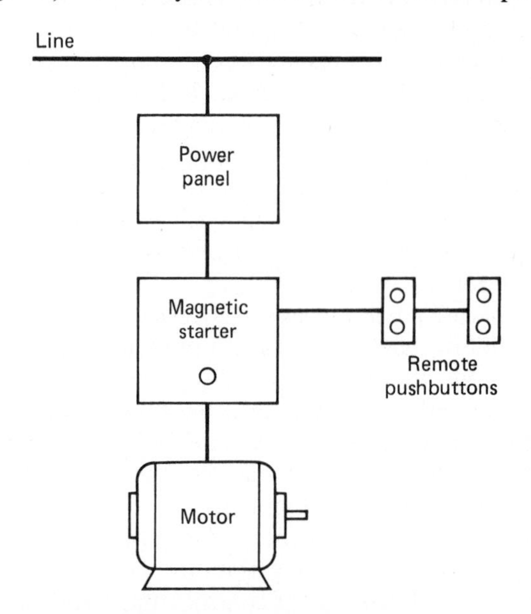

Fig. 1-3 Semiautomatic control for a motor.

button station because it is a compact and relatively inexpensive unit. Semiautomatic control is used mainly to give flexibility to the placement of the control in installations where manual control would otherwise be impractical.

A control system whose pilot devices are manually operated and whose motor starter is magnetic is classified as a semiautomatic control system. There are probably more machines operated by semiautomatic control than by either manual or automatic. This type of control requires the operator to initiate any change in the attitude or operating condition of the machine. Through the use of the magnetic starter, however, this change may be initiated from any convenient location, whereas manual control requires that the control point be at the starter.

1-4 AUTOMATIC CONTROL

In its simplest form, an automatic controller is a magnetic starter or contactor whose functions are controlled by one or more automatic pilot devices (Fig. 1-4). A pilot device provides a source of information to a control system. As an example, a thermostat, level switch, pressure switch, or flow switch can be used to provide information concerning the process variables that are being controlled. Temperature, liquid level, pressure, and flow are examples of process variables that are automatically controlled by magnetic starters, pilot devices, and motors.

Motor control can be classified as open-loop or closed-loop control. A switch that requires manual adjustment to start and stop a motor is an example of open-loop control. Open-loop control is used in manual control systems where an operator is always present to make decisions such as when to start and stop a motor.

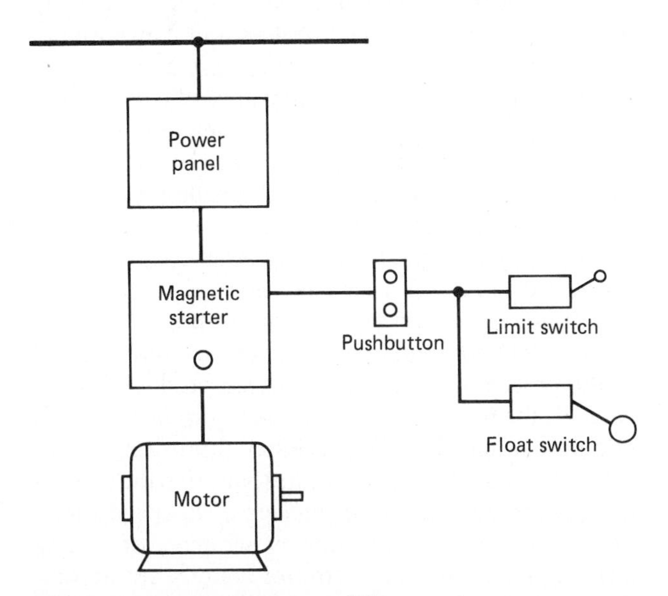

Fig. 1-4 Automatic control for a motor.

Closed-loop control is illustrated by the level-switch control used to regulate the level of liquid in a tank. Block diagrams for closed-loop control systems may appear as in Figs. 1-5 and 1-6.

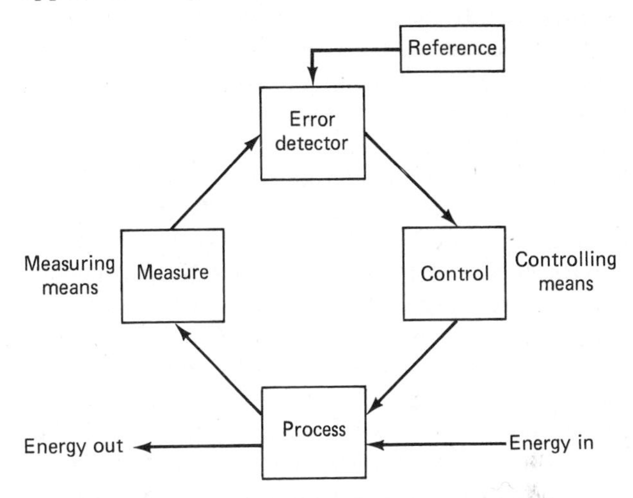

Fig. 1-5 Closed-loop control block diagram.

In a closed-loop system, as illustrated by the block diagrams in Fig. 1-5, it can be seen that by the feedback of information concerning a process parameter (level, for example) and the comparison of that information with a reference (desired level), a decision can be made by the error detector as to when the controlling means (starter and motor) should start and raise the level in the tank. In a closed-loop control system, it is obvious that the process controls itself by the feedback of its condition.

In the block diagram for the closed-loop control system of a refrigerator (Fig. 1-6), notice that the following control action would occur. The thermostat, or temperature-actuated switch, would be adjusted to the desired temperature. The thermostat responds to the actual temperature inside the box. The reference signal (R) is compared with the feedback signal (B). Any error or difference between the reference signal (R) and the feedback signal will produce an actuating signal (E). If the refrigerator temperature is too high, signal E will turn on the control system (motor and compressor)

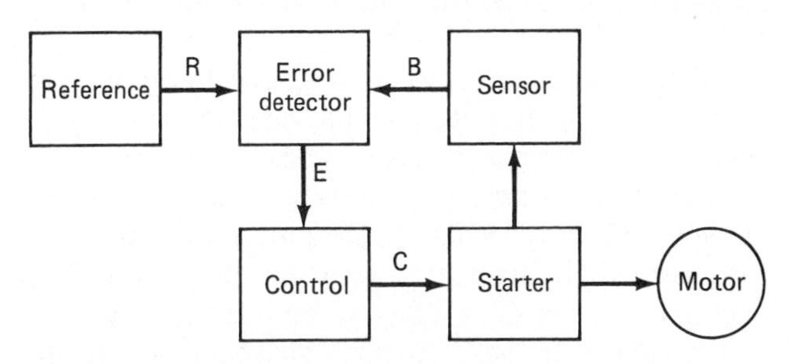

Fig. 1-6 Closed-loop control block diagram for a refrigerator.

to bring the refrigerator to the desired temperature. If the temperature is too low, the signal E causes the control system elements to turn off in order to restore C to the desired value. The blocks in Fig. 1-6 are continuously providing outputs that are fed around and around the loop, making the system a closed-loop control system. Once the start is initiated, the process controls itself in this type of control system.

Regulators are used in many industrial applications to control the voltage output of generators, speed of dc motors, and positions of loads. Regulators are essentially examples of closed-loop systems. In a later chapter we will see how solid-state regulators are used to provide speed control for dc and ac motors. Regulators may be defined as a closed-loop systems that maintain a steady level or value for some quantity, such as voltage, current, speed, temperature, pressure, and the like.

In industry there are control systems responsible for the positioning of large objects. Examples include the positioning of the rolls in a steel-rolling mill and the turning of a rudder on a large ship. The control system that is used for this type of control is called a *servomechanism,* or *servo system* for short. A servomechanism is a closed-loop system that moves or changes the position of the controlled object so that it will follow or coincide with the position of a control device or director. The servomechanism is often located at a location remote from the controlled object. A servo system requires motors to cause mechanical movement. Servos are discussed in greater detail in later chapters.

Automatic control is also illustrated by the system shown in the block diagram of Fig. 1-7.

The information section of the block diagram of Fig. 1-7 will contain pilot or information devices or sensors to initiate control action. The power control block will consist of motor starters and controllers or any other control device that provides power to the action devices. The action devices of the process are motors, generators, solenoid valves, heaters, etc. The block we jumped over, the decision-making block, is what this book is all about. This block is where all the control of the electric motor originates.

In the early automatic control systems decisions in control action were made by relays. The relay soon became unable to keep up with all the demands for high speed, long life, and high reliability that was placed on it. Static switching devices were developed in the 1950s that replaced relays and provided the same logical decisions that relays make. The transistor became the workhorse of the logic and static switching devices of 1960s and early 1970s. Logic control using transistors, printed circuit boards, resistors, diodes, and capacitors were found in many of the various electronic logic systems. In the 1970s integrated circuits, with all the logic of many transistors found on one chip, made great strides in decision-making control systems.

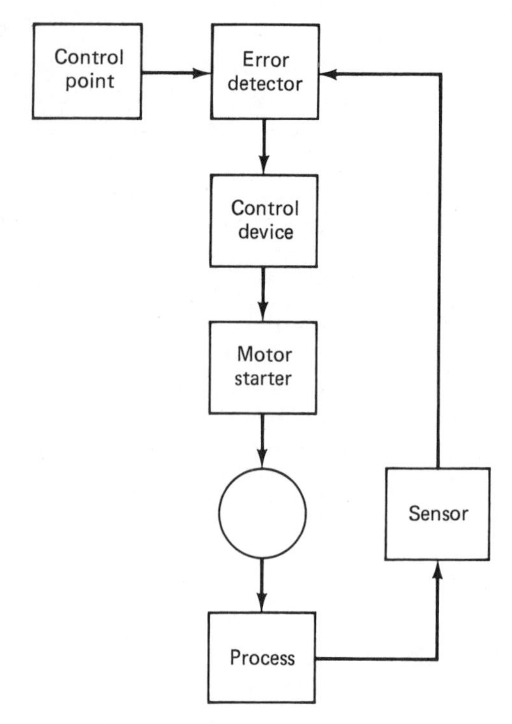

Fig. 1-7 Block diagram of an automatic control system.

In the late 1970s and the 1980s the major advances in solid-state devices for dedicated control systems have been the creation of the microprocessor and programmable controllers. When a control system is designed for a specific application, it is called a *dedicated control system.* The programmable controller by the middle 1980s was the workhorse of industry as far as providing dedicated and programmed control. Programmable controllers became the most popular decision-making devices in the automatic control of electric motors.

With all the changes through the years in automatic control, it is interesting that electric motors have really not changed significantly. Motors and conventional motor control are essentially the same as they have been for years. Automatic control has been through many changes in just a few years' time. A computerized integrated control system produced by General Electric Company will be used to illustrate the elements of automatic control (Fig. 1-8).

In the block diagram of the industrial control system illustrated by Fig. 1-8, each block is given a name. The raw materials enter the system in the lower-right corner. The equipment that converts the raw material to the processed material, or finished product, is indicated by the block labeled "Process Equipment." This equipment may be mills, machines, fabricating equipment, and the like. The motors that drive the process equipment are called *prime movers.* The speed, direction of rotation, time on, time off, sequence of starting and stopping, and other control actions are accomplished by the input to the block labeled Adjustable

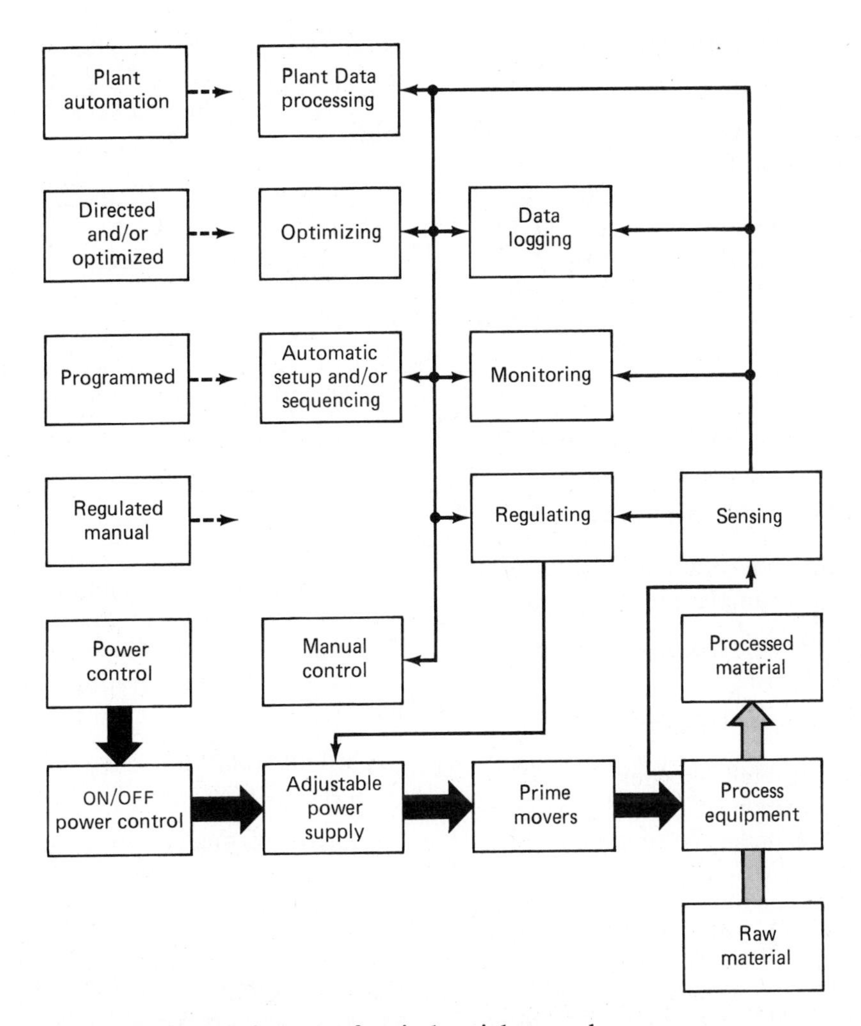

Fig. 1-8 Block diagram of an industrial control system.

Power Supply. All the remaining blocks of the control system in Fig. 1-8 contain many control loops. This type of control system is often referred to as a *multiloop control system.* By looking at only the motor control for large industrial control systems it can be seen that many loops are used to control the machines in a manufacturing process. Multiloop control has many applications in machine control. The following will summarize the function of the blocks in Fig. 1-8:

PROCESS CONTROL

Manual Control. Accepts manual commands and translates them into appropriate signal commands for the other functional blocks and receives control status signals and converts them to manageable information. Examples are

- Pushbutton
- Selector switch
- Rheostat
- Indicating light
- Digital display
- Horn

Regulating. Accepts a reference signal, compares it to a feedback signal, and provides an output to cause the feedback to equal the reference. Examples are

- Voltage regulator
- Position regulator
- Flow regulator

Sensing. Converts a sensible, physical quantity (or quantities) into a usable control signal proportional to a process parameter. Examples are

- Tachometer (converts rate of rotation to voltage)
- Current transformer (converts large current to signal-level current)
- X-ray thickness gage (converts penetration of x-ray to signal proportional to thickness)

Monitoring. Continually measures conditions within the system and communicates any departure of these conditions from predetermined limits. Examples are

- Steam-turbine monitoring system (checking that temperature and pressures have not exceeded limits)

Fault-monitoring system (checking order and occurrence of faults)

Automatic Setup and/or Sequencing. Provides predetermined reference signals for setup and command signals for sequencing in a prescribed order. An advance from one step of output to the next may be initiated by time or by some feedback from the process (such as operation completed). The predetermined program or sequence may be stored on manually operated switches, punched cards, memory, etc. Examples are

Start-up sequence control for a conveyor system

Card program control for a reversing hot mill

Changing control for a batch chemical process

Data Logging. Provides a record of conditions existing within the process. Examples are

Strip chart record of process speed, temperature, position, etc.

Typewriter log of performance data in a power plant

Optimizing. Provides overall direction for the process, taking into consideration what is to be made and how it can best be made under the conditions within the process at that instant in time. Thus, feedback for actual process operating conditions is utilized to determine the optimal manner in which the process is to be operated. Examples are

Adaptive schedule calculation for a reversing hot mill

Optimizing the relationship between temperature and catalyst feed rate in a chemical process

Plant Data Processing. The determination of the schedule of operations for each unit of process equipment in the overall plant complex. In general, the input to plant data processing is the customers' orders and the output to each process equipment control system is what to make, how much to make, and how fast to make it. Examples are

Manual scheduling of a small manufacturing operation

Computer scheduling of a large manufacturing operation

POWER CONTROL

On-Off Power Control. Controls, in a completely on or off manner, the flow of power from the main source to the adjustable power supply or prime movers, or both. Examples are

Motor starter

Valve

Adjustable Power Supply. Converts a constant amount of energy to an adjustable amount of energy such that the prime mover can be given the proper speed, torque, etc. Examples are

Generator

Valve

Static armature supply

Prime Movers. Equipment which converts some form of energy (such as electric) to mechanical energy usable by the process equipment. Examples are

Electric motor

Heat exchanger

Gas turbine

Process Equipment. Equipment which processes or acts on the material in the manufacturing process. The processing may include operations such as shaping, cutting, moving, heating, and distilling. Examples are

Distilling column

Reversing hot mill

Material-sorting conveyor

1-5 BASIC CONCEPTS OF AUTOMATIC MOTOR CONTROL

Automatic control had its beginning with the development of the relay. The design of relays lends itself to automatic control. Relays can perform many functions required by basic automatic control. Their contacts can be normally open or normally closed, or a combination of both. They can provide such timed operation to their contacts as time opening or time closing. They can be designed to provide stepping, sequenced, or latching operations. Many variations of contact arrangements are available to provide automatic motor control. Relays can provide such automatic control as reversing, braking, accelerating, decelerating, plugging, jogging, and the like.

For many years the relay provided adequate control for the automatic machinery of industry. As automatic control became more demanding, the relay was strained to its limits. High-production industries required liberation from the slow-acting, ever-failing, and short-lived relay. While the relay is still adequate for some automatic control systems, it cannot provide the reliability and dependability needed for high-speed, high-production industries. With the development of static and logic control, the relay panels for the conventional automatic control system have been rapidly replaced with this new control system. Static and logic control provides the dependable reliable control for high-production automatic control systems.

The three main divisions of an automatic control system are the information section, action section, and decision-making section.

The information section of the control systems contains such pilot devices as pushbuttons, limit switches, and sensing devices that are used to indicate such variables as pressure, temperature, flow, voltage, current, light, humidity, conductivity, or some other quantity. The information devices found in modern control systems may use solid-state devices to accomplish the measurement or detection of the different process control variables requiring control action.

The action section of automatic control systems contains the motors, starters, solenoids, heaters, coolers, and power devices that provide the regulating action in the control system. The action section may contain individual starters or motor control centers that control many motors at the same time.

With static and logic control devices replacing relays, logic components have become the decision-making devices in this generation of control devices. Static and logic control has opened a vast new system of automatic control for the rapid, fully automated machines and processes of industry.

STATIC CONTROL AND LOGIC CONTROL

Devices found in the earlier Westinghouse Cypak (logic in a package) and General Electric Company static control systems both used magnetic amplifiers in their logic units. Transistor logic systems followed the magnetic amplifier generation; solid-state integrated circuits (ICs) are used in the latest generation of logic and static control. Logic systems using transistors and ICs will be covered in later chapters.

Logic and reasoning are usually associated with the human mental process. However, circuit logic does not imply that circuits possess any mental or intellectual ability. Logic circuits are designed to give a logical response when presented with certain input conditions. Logic devices use two-state (on-off) devices to perform decision-making functions in the control of machines. The essence of a logic control circuit is in recognizing that a certain condition exists and making an appropriate response. The simplest form of logical response is "yes" or "no," or "on" or "off."

Logic circuits need control signals from sensing devices to give them the required information to trigger the logic response. Signals are the basis for all industrial control. They determine the number of operations that must be performed and how they should be performed to produce the required end results. Signals may be amplified, transmitted, or switched.

Signals fall into two general classifications: analog and digital. A digital signal is a sharp, discrete signal of definite short duration. A pulse is an example of a digital signal. An analog signal is a continuous signal that represents some quantity or value over a period of time. A signal from a tachometer generator is an example of an analog signal. The variations in speed indicated by the tachometer take place over an infinite number of steps. Logic functions are primarily concerned with digital signals.

A logic function is a means of expressing a definite condition. Three logic functions are used to express a condition, a set of movements, or a specific arrangement of information. These logic functions are AND, OR, and NOT.

In the AND logic function there is an output only when all of the input signals are present. In the OR logic function there is an output when any one of the inputs is present. In the NOT logic function there is an output when there is no input present. The memory function will give a continuous output after a momentary ON input. The output will cease after a momentary OFF input. Other logic devices, such as DELAY units, NANDS, NORS, amplifiers, signal converters, etc., will be included in logic control systems.

Logic elements are different than relays in several ways. Unlike relays, logic elements have no moving parts. Because they have no moving parts, they are often called *static switching devices.* Static control and static switching are terms that pertain to devices that switch without moving parts. Because static control has no moving parts, it offers a much higher degree of reliability and much longer life than conventional relays. Static switching devices are usually sealed in such a way that atmospheric environment will not have any effect on their operation. They are usually encapsulated in a durable compound that is unaffected by dirt, dust, oils, acid, alkalies, and solvents. Greater system reliability is ensured with the use of static components. Another advantage of static and logic switching is that it provides much higher speeds of operation than relays. Static switching differs from relays also in the number of inputs and outputs. A relay has only one input but several outputs. Static switching devices have several inputs but only one output. This requires a different type of circuitry than conventional relay control.

Binary logic consists of only two states or conditions. The two states can be represented by the numbers 0 and 1. The number 0 can be used to represent "off," "no," "minus," or "down" and the number 1 can be used to represent "on," "yes," "plus," or "up." Binary logic can be used to control basic circuit operations, whose electrical signals can have only one of two states. The principles apply equally well to pneumatic, hydraulic, electrical, electronic, mechanical, and even optical control.

As mentioned earlier, the three main divisions of control systems are information, decision, and action. Pushbuttons, limit switches, pilot devices, and sensing devices that sense temperature, pressure, voltage, light, humidity, rate of liquid or air flow, or some other quantity are information devices. The above devices produce and send the information signals to the logic or static control systems.

The signals from the information devices furnish the information which is used to establish the requirements to be fulfilled by the control. Relays and static switching devices are the logic elements which make decisions based on the information received from the information devices. Contactors and power-actuated switches perform the action called for by the decision-making or logic elements.

Automatic control for motors is continuously changing. New electronic devices are being developed to perform the decision making in automatic control systems. The continuous development of new solid-state and electronic devices combined with modern electric motor controllers presents constant challenges for electricians and technicians. Mills, factories, and manufacturing facilities, which use a large number of motors and controllers, are finding that the new automatic control systems are easier to use and are in many applications more reliable and dependable than their predecessors.

Both relay and solid-state logic automatic control systems have one very important disadvantage. Once a control system is developed and wired, it must be rewired if the sequential control of a machine or the program for an industrial process is changed.

In recent years the development of the programmable controller (P.C.) has eliminated the wiring problems of relay and hard-wired logic systems. With the P.C. changing the operational sequence or program for an industrial control system is very simple. When first introduced in automatic control systems, P.C.s made a strong impact on the direction of controls in the future. The ease with which they can be programmed and reprogrammed make them very impressive control devices. All an electrician has to do to change the program in a P.C. is to enter any desired change on a keyboard. The new logic sequence is stored in the memory of the P.C. By contrast, a relay or logic control panel would require rewiring in order to change the logic of the control system.

A P.C. is a solid-state control system which has a user-programmable memory. The memory is used for storage and to implement specific functions such as inputs and outputs controlling the operations of a system of electromechanical devices. Programmable controllers can perform the same functions as relay panels, electromechanical timers, electromechanical counters, drum timers, and wired solid-state logic control systems.

A programmer provides a means of giving instructions to the P.C. so that the P.C. controls motors and other power devices in the sequence desired. A electrician or technician pressing the keys on the programmer's keyboard can input a predetermined program in the P.C.'s memory. Any automatic control system can be programmed or reprogrammed if necessary to give the control desired. The programmer can be used to display program instructions as well as to analyze, monitor, and edit the program. The programmer also can be used to locate problems and to troubleshoot the program stored in memory. To direct the controller system the P.C. contains electronic logic circuitry that converts the instructions from the programmer into a sequence of control signals. Many of these electronic circuits will be discussed later in the sections on programmable controllers.

Other features of P.C.s have contributed to their widespread use in industrial control. Such factors as their low cost, reliability, ease of maintenance, and ruggedness have made them very popular with plant electricians and technicians. Another feature of P.C.s that makes them popular in industry is their ability to work in environments with temperatures from 0 to 60°C and relative humidities from about 0 to 95 percent. The P.C. is truly the workhorse of industry.

COMPUTER CONTROL

The same five basic logic functions that appear in static control systems are also contained in digital computers. Therefore, digital computers can be used as the decision-making elements to control electrical circuits. Digital computers can be used to control all or part of the machines in a plant that will respond to digital signals, such as the pulse of current from the contacts of a relay or a switch. The control information required for the computer to control the machines in their proper sequence is provided by a programmer.

Some machines operate on analog signals, which can be continuously variable throughout their range. For example, the control voltage in a thermostat varies in an infinite number of steps, in proportion to variations in temperature. The closed-loop or feedback control system described earlier is another example of an analog control system. If the output of analog devices is used as input to digital computers, the analog signal must be converted to a digital signal by a coder. It may be necessary to convert some of the output of a digital computer to an analog signal. This can be accomplished by the use of a decoder.

A typical computer system consists of a memory, a central processing unit (CPU), and input-output (I/O) ports. The memory serves as a place to store instructions. The instructions are coded bits of information that direct the activities of the computer. The CPU reads each instruction from memory and processes it in a logical manner to determine the operational sequence. It processes the program to introduce the actions that provide automatic control to machines and motors. We will concern ourselves more with computer control in later chapters.

The CPU is a special IC called a *microprocessor.* Microprocessor-based control systems are finding many applications in automatic control. Microprocessors can be programmed to perform many of the complex control applications found in modern industry. They are finding many applications in the measurement and control of process variables and in numerical control systems. They can perform sequential control. Microprocessors are used in control systems where high-production machines require high speed and high efficiencies and where very demanding and sophisticated control techniques are required.

SUMMARY

The basic difference among manual, semiautomatic, and automatic control systems lies in the degree of flexibility. With manual control the operator must go to the physical location of the starter to make any change in the operation of the machine. With semiautomatic control the operator can place the control point at any convenient location. With automatic control the operator does not need to initiate the changes for each automatic operation included in the control system.

In the modern automated industrial plant the operator merely sets up the operation and initiates a start, and the operations of the machine are accomplished automatically. Automatic operations of a machine are dependent upon the manufacturing process or the functions the machine is to accomplish or both. An automatic clothes washing machine and a steel rolling mill are both applications of automatic control. The rolling mill, which may be computer controlled, is obviously going to require a more complex control system than the washing machine.

Automatic control systems in industry require integrated control components. An industrial automatic control system may contain electrical, electronic, magnetic, mechanical, fluidic, hydraulic, and pneumatic devices. An integrated process control system may utilize a wide variety of analog and digital elements and modules to meet a specific set of functional requirements.

REVIEW QUESTIONS

1. What does the word control mean?
2. Define a control component.
3. What is the advantage of manual control?
4. What is the application of a compensator?
5. When is semiautomatic control used?
6. Describe a pilot device.
7. What is the difference between open-loop and closed-loop control?
8. Give an example of closed-loop control.
9. What is an application for a servomechanism?
10. Where are static switching devices used?
11. Describe a dedicated control system.
12. Where would multiloop control systems be used?
13. What are some of the operations of relays?
14. What are the three main divisions of an automatic control system?
15. What type of control devices used magnetic amplifiers?
16. What is the essence of a logic control circuit?
17. How important are signals to control systems?
18. What is the difference between a digital and an analog signal?
19. Define a logic function.
20. How do static switching control systems differ from relay control systems?
21. What are some of the advantages offered by programmable controllers?
22. What are some of the functions that can be accomplished by the programmer or terminal for programmable controllers?
23. What are some of the advantages offered by computer control systems?
24. How can an analog signal be converted to a digital control signal?
25. What is a CPU and what does it do in a computerized control system?
26. When are microprocessors used in control systems?
27. What is meant by an automatic control system with integrated control components?

2
A Brief Review of Motors

There are many different types of motors. The application of a motor dictates its type and size. How a motor is used will also determine its required speed as well as its voltage and horsepower rating. The machine a motor drives will have much to do with the appearance and mechanical construction of the motor. Some motors are connected to mechanical loads by V belts or by direct mechanical coupling. Others are connected to loads by gear reducers or clutches. In many cases a motor becomes part of the machine it drives.

Motors are classified according to the type of electric current required for their operation, that is, either alternating current (ac) or direct current (dc). Motors that run on alternating current are further classified as single-phase or polyphase (usually three-phase).

2-1 CHOOSING MOTORS

Alternating current (ac) motors are used for considerably more applications than direct current (dc) motors. They are widely used because ac current is much more economical to generate and distribute than dc current, and is thus more readily available. Alternating current is generated commercially at voltages as high as 138 kilovolts (kV), is fed into transmissions systems at as much as 1 megavolt (MV), and is delivered to users at voltages as high as 320 kV.

Direct current generation is limited to much lower voltages, though long-distance transmission sometimes delivers direct current at hundreds of kilovolts. Direct current, however, cannot compete economically with alternating current. The simpler construction of ac motors also makes them more desirable than dc motors. The dc motor, however, offers many advantages in the operation of industrial machines. If multispeed operation of a machine is required, then the drive motor will often be a dc motor. Industry also uses dc motors for machines that require wide ranges of torque limit, braking, and simple directional control.

The economics of using dc motors had been a problem for industry for years. The older methods of generating dc power required very expensive equipment. Some industries used large motor-generator units, while others used large rectifier banks. These methods of generating direct current made the cost of using dc motors considerably greater than the cost of using ac motors. Thanks to modern technology a low-cost, reliable, and relatively maintenance-free source of direct current is now available.

The development of such solid-state devices as the silicon-controlled rectifier (SCR) is making it possible to operate dc motors without dc generators or elaborate rectifier systems. In these motors alternating current is connected directly to the controller. The solid-state controller converts the alternating current to direct current as well as controls all the dc motor functions. Some of its typical control functions are acceleration, deceleration, speed control, reversing, dynamic braking, soft starts, current limit, etc. These are all performed electronically by the solid-state controller.

With these new dc solid-state drives, dc motors are becoming much more popular. Lack of dc power is no longer a problem and thus the use of dc motors in modern industrial applications is increasing considerably. Solid-state motor control is covered in depth in a later chapter.

2-2 THE STRUCTURE OF AC MOTORS

Because ac motors are used more than dc motors we will start our review with ac motors. The majority of ac single-phase motors are constructed in fractional-horsepower sizes for 120- to 240-V, 60-hertz (Hz, or cycles per second) power sources. Fractional-horsepower motors have a continuous rating of less than 1 horsepower (hp) when operating at speeds near 1800 rpm. The different needs and applications of fractional horsepower motors for industrial machines have dictated several different motor design. Some require very high starting torque, others require operating torques that vary. Some applications require a motor that has quiet running characteristics; other applications may require motors that can be reversed easily; and still others may require dynamic braking or plugging. To meet these various needs, ac single-phase motors are available with a wide range of characteristics.

Large-horsepower ac motors are generally polyphase motors. Almost all modern polyphase motors are three-phase. At one time two-phase motors were widely used in industry. Although two-phase power is no longer generated commercially, it is produced electronically for such special applications as the small motors in some recorders and measuring instruments.

The rotating part of an ac motor is referred to as the *rotor.* A squirrel-cage rotor consists of a laminated steel core with conductors cast into the rotor slots. The conductors are generally aluminum because of its low weight and good conductivity. End rings are used to short all of the conductors, while fins emanating from the rings serve as a fan to cool the motor. Figure 2-1 illustrates a typical single-phase capacitor motor.

Fig. 2-1 Cut-away of a single-phase capacitor motor illustrating squirrel-cage rotor construction. *(Courtesy of General Electric)*

The *stator* is the name given to the stationary part of the ac electric motor. The stator is formed from special sheet-steel laminations stacked and fastened together. The laminated core of the stator aids in preventing eddy currents that would circulate in the iron core, generating heat and causing power loss. The laminations contain slots in which are placed the slot insulation and coils that become the stator circuits. In single-phase motors, the slots may contain two or more sizes of wires, which are used in the starting and running winding.

Fig. 2-2 Three-phase motor stator and rotor. *(Courtesy of General Electric)*

The stator coils for the polyphase motors may be wye- or delta-connected. Figures 2-2 and 2-3 (on the next page) illustrate stator construction.

When three-phase current is supplied to the stator of a three-phase motor, a rotating magnetic field is produced. The speed of the rotating field depends upon the frequency of the source of current. The standard frequency for industrial three-phase power is 60 Hz. The number of poles also affects the speed of a three-phase motor. The number of poles in a motor is determined by the stator coil wiring.

The rotating magnetic field set up by current flowing in the stator will induce a voltage in the squirrel-cage rotor conductors. Current will flow in the squirrel-cage conductors, producing magnetic fields. The attractions and repulsion resulting from the interaction of the stator magnetic field and the rotor magnetic field cause the rotor to rotate. The rotor cannot rotate at exactly the same speed as the rotating magnetic field of the stator. If it did, there would be no voltage induced in the rotor conductors. The magnetic field of the stator must cut the rotor conductors or no voltage will be induced in them. For this reason the actual speed of the motor (rotor speed) and the synchronous speed (speed of rotating field in the stator) will be different. The difference between the rotor speed and the synchronous speed is called *slip.* All induction motors, whether they are single-phase or polyphase, must operate with some slip. Slip accounts for the difference between actual motor speed and calculated motor speed. The synchronous speed of a motor can be found by the following equations:

$$\text{Synchronous rpm} = \frac{60 \times \text{frequency}}{\text{Pairs of poles}}$$

$$\text{Synchronous rpm} = \frac{120\ \text{Hz}}{\text{Number of poles in stator winding}}$$

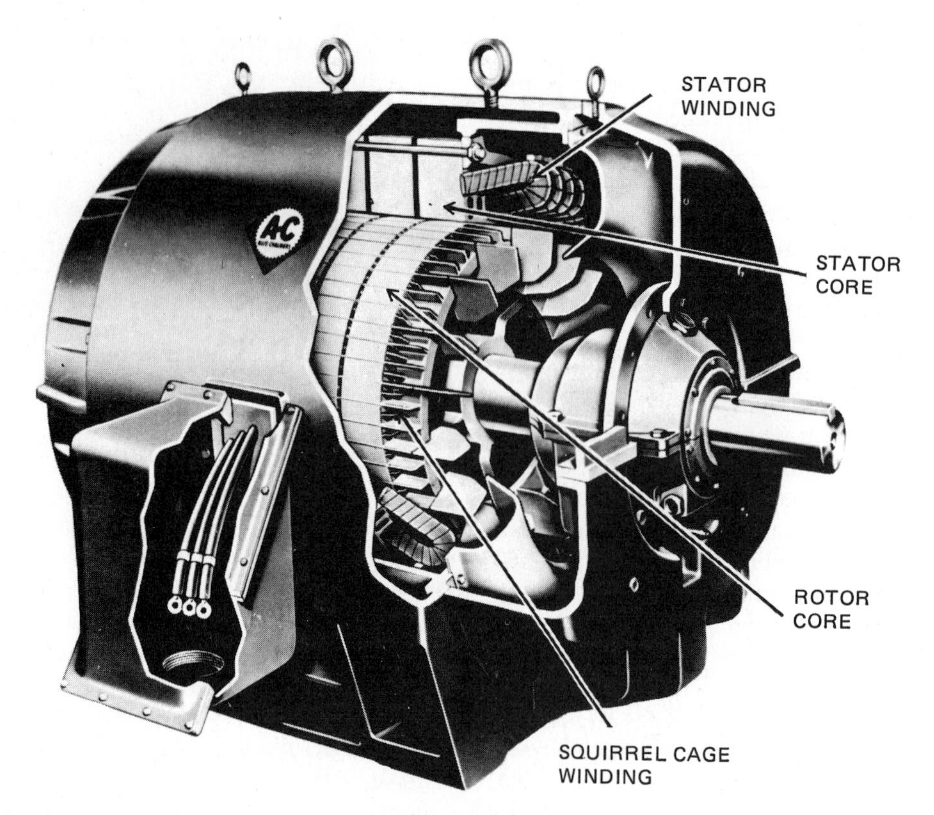

Fig. 2-3 Cutaway of a three-phase dual-voltage induction motor illustrating stator windings. *(Courtesy of General Electric)*

2-3 ELECTRICAL AND MECHANICAL CHARACTERISTICS OF MOTORS

The end bells and bearings housing support the motor shaft and bearings. They must be rigid enough to maintain alignment of the rotor under load conditions. They must be strong enough to support all the weight mounted on the shaft or hanging from the shaft, such as that of chain and sprocket or belt and sheave drives. These types of "overhung" loads determine the size of the shaft bearing and shape of the end bell.

The end bells are also responsible for maintaining the air gap between the stationary and rotating coils at a constant value. An uneven air gap around a rotor will cause the magnetic flux to be distributed unevenly in the fields. Bearing wear in large motors can cause uneven air gap. It can be detected by measuring the air gap between the rotor and stator. A tapered, long-blade feeler gauge can be used to measure air gap. Uneven air gap on the top, bottom, and sides of the motor will indicate bearing wear or improper shaft bearing alignment.

Motors of less than 5 hp have end bells made of an aluminum alloy. Larger-horsepower motors require strong and rigid support of the rotor and use end bells made of high-grade cast iron.

The most common type of motor mounting is the rigid mount. In this type of mount the frame of the motor is welded right to the base.

Resilient mounting is common for fractional-horsepower motors. Resilient motor mounting uses rubber rings to support the motor to the cylindrical end brackets attached to the motor base. Resilient bases have the advantage of isolating vibration or noise, and operate quietly. It also provides a "torque cushion" when starting or stopping the motor.

Many motors designed for fan operation use stud mounting for the motor. The bolts that extend through the motor end bells are long enough to secure the motor to the fan housing.

Flange-mounted motors have a special end bell with a machined flat surface. Four or more holes, which may be taped or not, are used to secure the motors to the driven machine. Typical flanges are the oil burner or the "NEMA" (National Electrical Manufacturers Association) flanges.

Fractional-horsepower motors generally use bushing or sleeve bearings. Some special applications may require antifriction ball bearings. Large-horsepower, three-phase motors commonly use antifriction ball bearings, or sleeve and babbit bearings.

Sleeve bearings or bushings are made from tubes of brass, steel-backed babbitt, or a self-lubricating material such as oilite. (See Fig. 2-4.) Oil grooves are

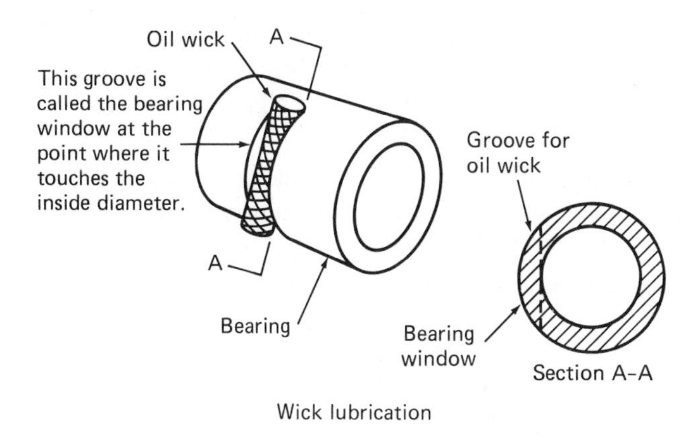

Fig. 2-4 Sleeve bearing construction.

"bearing windows" cut into the bearing. Bearing windows provide lubrication for the motor shaft through a wick sitting in the groove and extending into an oil reservoir. When they are lubricated properly at regular intervals, these bearings will provide years of service.

To be properly lubricated, motors with standard sleeve bearings should be mounted with their shafts horizontal. Oil will leak out of the bearing if this type of motor is mounted vertically. The oil wick must sit in the oil reservoir in such a way as to allow the oil to be drawn up the wick by capillary action. Motors that are designed to provide vertical service have vertical sleeve bearings.

Motors with sleeve bearings that are connected by belts or chains to loads often tend to pull the shaft to one side, creating special mounting problems. To prevent improper lubrication, be sure the belt does not pull the shaft in the direction of the oil hole (Fig. 2-5).

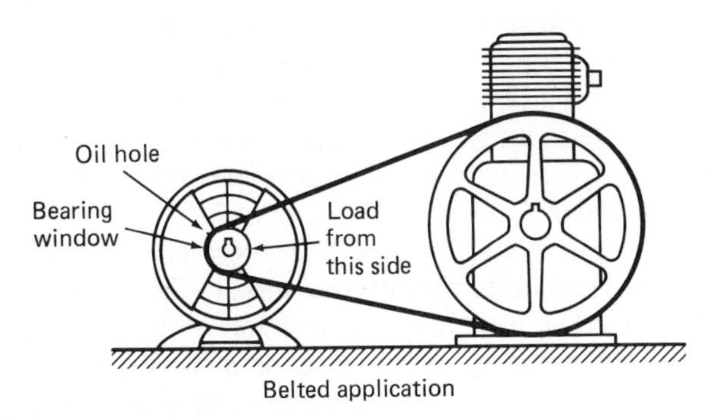

Fig. 2-5 Proper belt pull.

Antifriction ball bearings consist of a number of hard steel balls placed between the inner and outer races of the bearing. Most three-phase motors of 5 to 200 hp use antifriction bearings. The selection of sleeve bearing or antifriction ball or roller bearings for large motors is dictated by the proposed application of the motor.

With proper lubrication methods a motor with antifriction bearings should give years of service. Remember that if a little grease does a little bit of good, a whole lot of grease does not necessarily do a whole lot of good. More motors are lost from bearings being excessively lubricated than from not being lubricated. Follow the manufacturer's instructions for proper bearing lubrication.

Modern motors with new insulation materials are designed to run at higher temperatures. Insulation is grouped into "temperature tolerance" classes. At present five classes of insulation are used for wires used to wind motor coils. These classes are as follows: class O, which has a temperature limitation of 90°C; class A, which has a temperature limitation of 105°C; class B, which has a temperature limitation of 130°C; class F, which has a temperature limitation of 155°C; and class H, which has a temperature limitation of 180°C. A motor's temperature rating is determined by the motor's magnet wire insulation.

In a single-phase motor there are at least two coils in a stator slot. The insulation used to separate the two coils is called the *phase* or *coil separator*. When all the coils are in the slot, a fish paper wedge is used to close off the slot. The coils are then connected together to form proper pole groups. The ends of the coils that are to be brought out of the motor are placed in tubing. The entire assembly is dipped into an insulating varnish and baked. When the stator assembly is dry, it can then be pressed into or bolted to the motor frame. The leads from the stator assembly are brought out and connected to a terminal block. When the motor is ready for operation, power leads will be connected to the terminal block to provide the motor with its power source.

Some motors are constructed with insulating materials that are new developments of the plastic industry. The insulation materials are in the epoxy and silicone elastomer families. Epoxy cannot be dissolved and thus is not affected by many corrosive chemicals. High temperatures cannot affect silicone. These new materials are used in motors used in special high-temperature applications or environments with extreme conditions.

Some motors are classified by their type of enclosure. The terms used for different types of motor enclosures are: open; drip-proof; splash-proof; weather-protected; open, externally ventilated; totally enclosed; totally enclosed, fan-cooled; totally enclosed, water-cooled; explosion-proof; dust-ignition–proof; and water-proof machine.

A motor nameplate is illustrated in Fig. 2-6 on the next page. Nameplates provide electricians with a lot of important information about the use of motors. The numbered blocks in Fig. 2-6 represent the following basic electrical and mechanical characteristics of the motor that are given on the nameplate (this is sometimes called nameplate call-out):

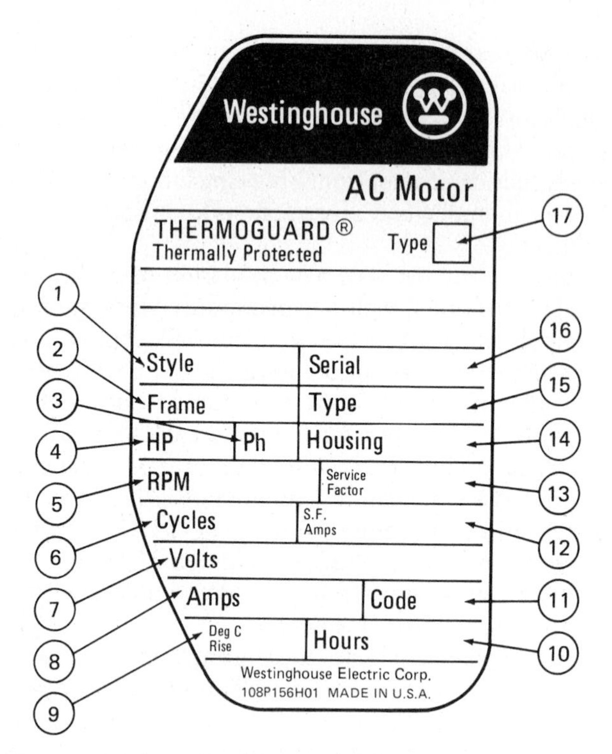

Fig. 2-6 Motor nameplate lists the basic mechanical and electrical characteristics of the motor. *(Courtesy of Westinghouse Electric Corp.)*

1. **Style.** This gives the manufacturer's specifications; code numbers indicate drawings, plans, and specifications for the motor.
2. **Frame.** This is defined by NEMA. Common frame-size numbers for small motors would be 42, 48, and 56 (dc, single, 2- or 3-phase ac).
3. **Ph.** This indicates the type of power that the motor was designed for.
4. **HP.** This is the horsepower rating of the motor at rated speed.
5. **RPM.** This indicates the speed of the motor in revolutions per minute. This speed is running, not synchronous, speed.
6. **Cycles.** This indicates the frequency of the power that the motor operates at.
7. **Volts.** This indicates the voltage that the motor is rated to operate at.
8. **Amps.** This indicates the full-load current at rated load, rated voltage, and rated frequency.
9. **Deg.C/Rise.** This indicates the permissible temperature in degrees centigrade that the motor temperature can rise above ambient temperature when operating at rated load and speed.
10. **Hours.** This indicates that period of time the motor may operate without overheating. Small motors are generally rated "continuously."
11. **Code.** This is a NEMA code letter designating the locked rotor kilovolt ampere (kVA) per horsepower. (Example: Letter *M* allows from 10 to 11.2 kVA per horsepower.)
12. **S.F./Amps.** This indicates what the starting full load current the motor will draw when it is loaded to its full-service factor.
13. **Service Factor.** This indicates that when the motor is operated at rated voltage and frequency, it may be safe to overload the motor to the horsepower obtained by multiplying the service factor by the horsepower rating of the motor.
14. **Housing.** This indicates the motor enclosure or housing.
15. **Type.** This is a code letter indicating something about the construction of the motor and power it runs on, such as split-phase, capacitor start, polyphase power, etc.
16. **Serial.** This indicates the manufacturer's code number or serial numbers for the motor.
17. **Type.** This indictes the type of thermoguard protector installed in the motor. *A* indicates automatic reset, U.L.-approved. *C* indicates locked rotor only. *M* indicates manual reset, UL-approved. *T* indicates automatic or manual reset, not UL-approved. UL refers to Underwriters Laboratory.

2-4 APPLICATIONS OF SINGLE-PHASE MOTORS

In general practice, it is usually more economical to buy a new fractional-horsepower motor than to repair a burned-out one. However, with today's cost of motors, repairing a motor is often justifiable. A knowledge of the types and structure of single-phase ac motors is therefore important to anyone involved with motor repair.

The single-phase motor with the highest torque rating is the ac series motor, or universal motor. The universal motor runs with about the same degree of success when connected to an ac or dc power source. It generally operates at high speed and includes design features that minimize commutation- and armature-reaction difficulties. Commutation- and armature-reaction problems will cause severe sparking at the brushes if not corrected. Using high-resistance brushes, using compensating windings, and skewing the armature slots so that the armature conductors are not parallel with the shaft are all methods used to improve the operating characteristics of the universal motor.

The universal motor is constructed with the field coils wound on the stator. They are connected in series with the armature windings passing through brushes and commutator. The connection for the motor leads is as follows. A motor field lead is connected to one line from an ac 60-Hz power source. The other field lead is connected to one of the armature leads. The remaining armature lead connects to the other ac power line. The motor is ready to go except for a direction of rotation check.

The direction of rotation of the universal motor is changed by interchanging the field terminals with respect to the armature. Some universal motors are wound with two separate fields. One of the fields provides the forward direction of rotation; the other, the reverse direction. The motor will have three external leads. A power source connected to the armature and one field will give one direction of rotation. When the other field is connected to the armature, it will give the opposite direction of rotation, as shown in Fig. 2-7.

There are several methods used to control speed in a universal motor. One method is to insert a rheostat in the line. This causes a voltage drop that results in a reduced voltage across the motor terminals. This provides for a rather widely controllable speed range. Another method of providing a wide range of speed adjustments is to use a tapped field. The motor in Fig. 2-8 illustrates a tapped winding that provides four operating speeds.

The universal (ac series) motor may reach exceedingly high speeds if operated unloaded. A popular method to prevent this from happening is to use a governor-controlled speed adjustment. The governor prevents the speed of the motor from becoming excessively high (Fig. 2-9 on the next page). A spring-tension setting on the contacts will determine speed of motor.

The high starting torque of universal motors and their ability to adjust to widely varying loads make them ideal for such electrical tools as saws, drills, sanders, etc. The universal motor is ideal for use on many small appliances, such as vacuum cleaners, mixers, blenders, and other portable household appliances.

The single-phase induction motor, or split-phase motor, as it is commonly called, uses a start and run winding in its stator. The two windings are also called the *main* and the *auxiliary windings*. Both windings are wound in the slots of the stator. The size of wire in the starting winding is smaller in size than the running windings. This is to ensure a phase displacement of the current flow in the starting winding that is relative to the running winding.

The coils that form the auxiliary windings (starting winding) are positioned in pairs directly opposite each other in the stator, and between the main windings (running windings). When looking at the end of the stator of a split-phase motor, you will see alternate running and starting windings (Fig. 2-10 on the next page).

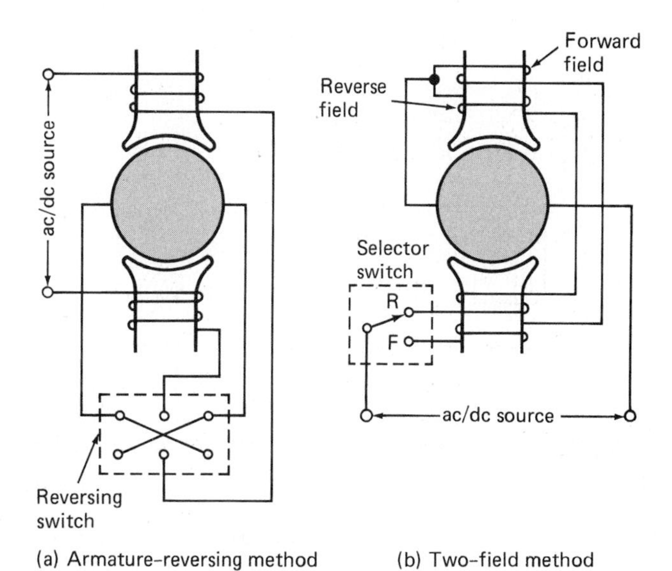

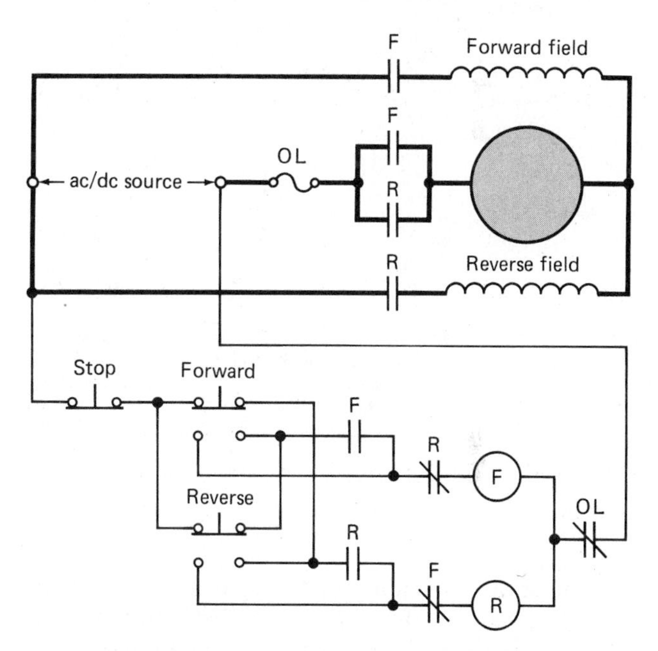

(c) Control circuit connected to a two-field reversing universal motor.

Fig. 2-7 Methods for reversing universal motors.

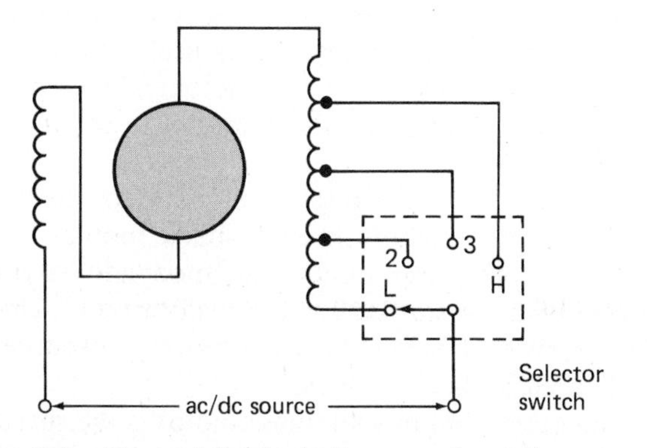

Fig. 2-8 Tapped-field winding speed—adjustment method for a universal motor.

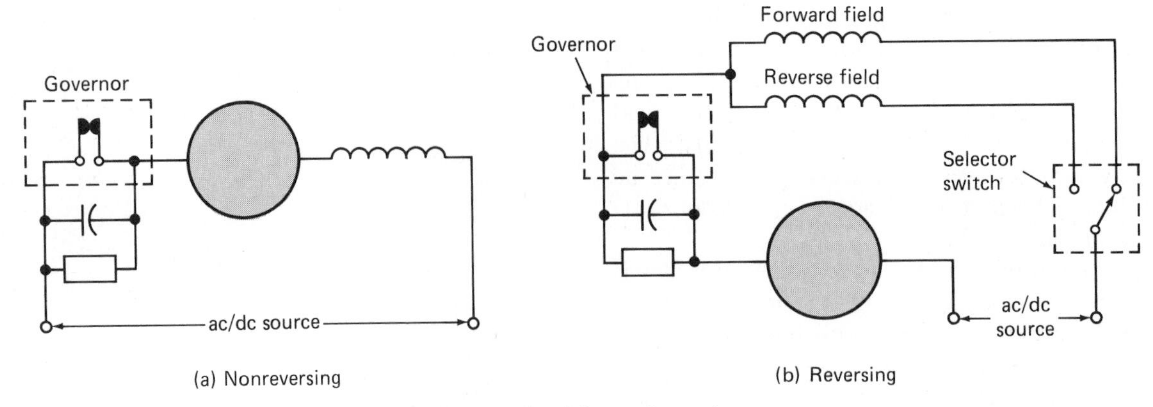

Fig. 2-9 Governor-controlled speed-adjusting method for universal motor.

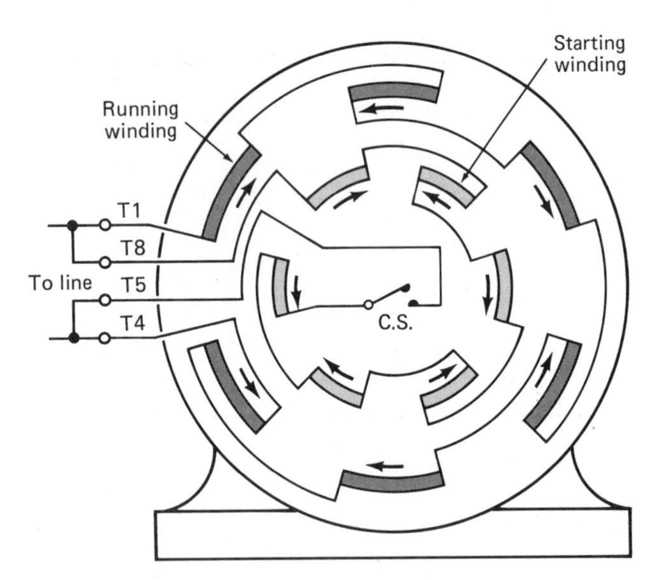

Fig. 2-10 Connections for a six-pole, split-phase motor.

The starting winding of the split-phase motor is opened by a centrifugal switch as the motor reaches about 75 percent of full speed. The circuit of the centrifugal switch and the starting winding will remain open as long as the motor is in operation.

Two general types of split-phase motors are manufactured to provide different starting torque, to minimize noise, and to improve mechanical performance characteristics. The split-phase motor and the capacitor split-phase motor are the names of these types of motors.

There are two types of split-phase motors, the standard and the special-purpose split-phase motors. The standard split-phase motor has moderate starting torque, with low starting and full-load current. These motors are ideal for fans, furnaces, blowers, oil burners, appliances, and unit heaters.

The other kind of split-phase motor is the special-purpose split-phase motor, which has high starting torque and high starting current. This motor is particularly applicable to devices that are hard to start but are started only infrequently. It is ideal for washing machines, evaporator coolers, dishwashers, etc.

Figure 2-11 illustrates the three types of capacitor motors. The capacitor motor is classified as (1) the capacitor start, (2) the permanent-split capacitor, and (3) the two-value capacitor motor. All these motors are wired similarly, with the capacitor placed in series with the starting winding and the power source. A centrifugal switch is placed in series with the capacitor and starting windings on the capacitor start motor. The switch will cut out (open) the capacitor and the starting winding as the motor reaches full speed. It should be understood that during starting the torque is higher and the inrush current is lower than in the standard split-phase motor. Once started, the capacitor start motor has the same running characteristics as the standard split-phase motor. If the starting period of the capacitor start motor is short in duration, electrolytic capacitors can be used with the motor. Electrolytic capacitors are designed for short-duty service only. Table 2-1 indicates the typical capacitor sizes for motors of different horsepower and speed.

A permanent-split capacitor motor has a low starting torque and should be used in applications that are generally shaft-mounted. The two windings of the permanent-split capacitor are identical. A continuous-duty capacitor is connected between the ends of the two motor windings. This makes it so the motor can be easily reversed. Figure 2-12 on page 18 shows how a permanent-split capacitor motor can be reversed. The selector switch in one position places the capacitor in series with one winding. When the switch is placed in its other position, the capacitor is in series with the other winding. This arrangement makes an easy reversing method for the motor.

The speed torque characteristics of this motor shift considerably with changes in supply voltage. Advantage is taken of this motor's speed torque characteristics by supplying adjustable voltages to the motor to obtain speed changes from the motor. Later we will see that a

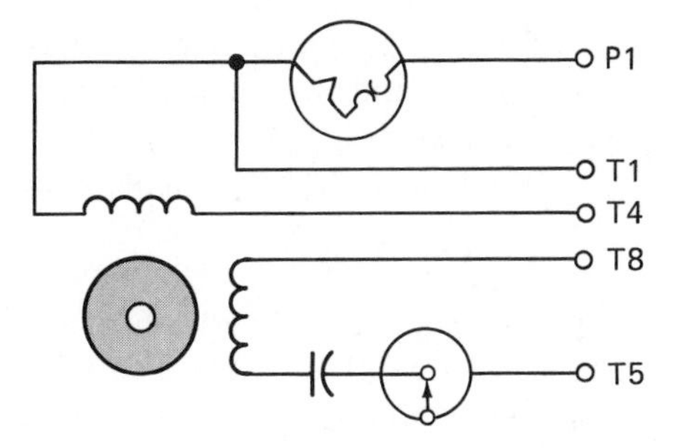

Rotation	L1	L2	JOIN
CCW	P1	T4-T5	T1-T8
CW	P1	T4-T8	T1-T5

Permanent-split capacitor-start motor and connection chart

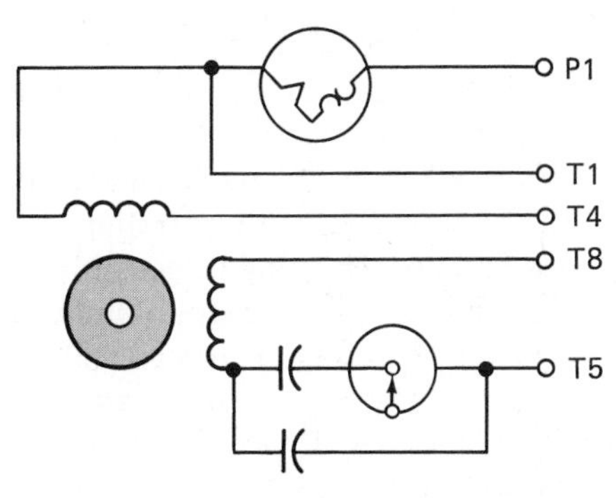

Rotation	L1	L2	JOIN
CCW	P1	T4-T5	T1-T8
CW	P1	T4-T8	T1-T5

Two-value capacitor motor with connection chart

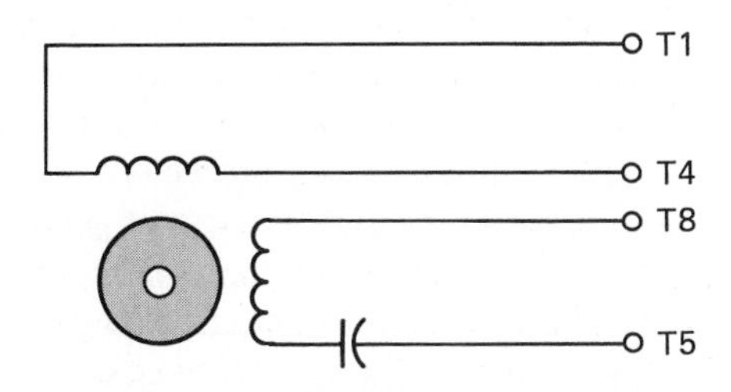

Rotation	L1	L2
CCW	T1-T8	T4-T5
CW	T1-T5	T4-T8

Permanent-split capacitor motor with connection chart

Fig. 2-11 Three types of capacitor motors.

TABLE 2-1
Typical Capacitance Values for 60-Hz Capacitor-type Split-phase Motors

Horsepower	Poles	Revolutions per Minute	Approximate Capacitor Values μF	Approximate Percent Starting Torque
	2	3450	135	350–400
¼	4	1750	135	400–450
	6	1140	135	275–400
	2	3450	175	350–400
⅓	4	1750	175	400–450
	6	1140	175	275–400
	2	3450	250	350–400
½	4	1750	250	400–450
	6	1140	250	275–400
	2	3450	350	350–400
¾	4	1750	350	400–450
	6	1140	350	275–400

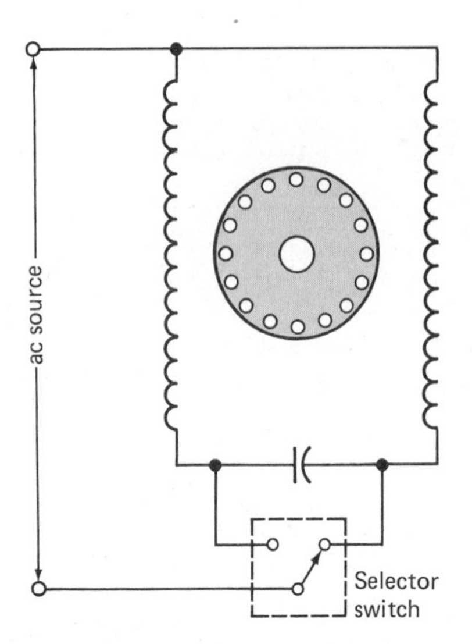

Fig. 2-12 Permanent split-capacitor motor connected for reversing service.

solid-state motor controller will give a wide range of speed control for this type of motor.

The two-value capacitor motor has the advantage that it develops extremely high locked-rotor torque with one value of capacitance in series with the auxiliary motor winding. When the centrifugal switch cuts out the starting capacitance, quiet running performance of the motor can be obtained with another value of capacitance in series with the start winding. The starting capacitance is about 10 to 15 times greater than the running capacitance.

The starting current of standard split-phase and capacitor motors is 3 or 4 times their running current. The capacitor motor's starting current is normally about 1½ times its running current.

Standard rotation for single-phase induction motors is counterclockwise. Rotation is observed when facing the end opposite the drive shaft. Figure 2-13 illustrates the standard connection diagrams for the single-phase induction motors discussed thus far.

The shaded-pole motor is the cheapest and simplest to manufacture. In the shaded-pole motor, a loop of heavy copper strap or wire replaces the starting winding. This loop of copper is mounted in the face of the steel laminations, as shown in Fig. 2-14 on page 20.

The magnetic field produced by the pole piece with a shader is first away from and then toward the loop of copper. The action of the shading coil is to produce a weak rotating field that shifts across the face of the poles. This rotating field gives the motor a very weak starting torque but is sufficient to make continuous rotation of the motor possible.

Shaded-pole motors can be reversed mechanically by turning the stator housing and shaded poles end for end. Reversing service can be provided electrically by two sets of shading coils that are normally open-circuited and are located at each end of the pole pieces. A switch that selects the set of shaded coils used and leaves the other set open-circuited will determine the direction of rotation.

Speed control of the shaded-pole motor can be accomplished in several ways. The three most common methods, which use a selector switch to determine desired speed, are the autotransformer method, the reactance method, and the tapped-winding method. Figure 2-15 on page 20 illustrates these methods.

Applications for shaded-pole motors would include fans and other similar loads that do not require high starting torques.

When the starting service is both severe and substained, it is generally necessary to use a repulsion-start motor. The starting of the repulsion-induction motor is based on one principle and its operation on another. The rotor coils during starting are not shorted by the mechanical shorting device at the commutator. As it reaches full speed, the motor changes its mode of electrical operation. The repulsion motor has a wound-rotor and vertical commutator. A shorting ring will short the commutator at full speed. This changes the motor's operation from a repulsion start to an induction run. This is why the motor is often refered to as a *repulsion-induction motor.*

The high starting torque of the repulsion-induction motor is produced by the repulsion between magnetic poles of the armature and the magnetic poles of the stator. The repulsion force is controlled and changed so that the speed of the motor increases rapidly. The motor is prevented from reaching excessively high speeds by the speed-actuated mechanical switch, which short-circuits the armature conductors. When short-circuited, the armature becomes the equivalent of a squirrel-cage rotor.

Direction of rotation in the repulsion motor is changed by shifting the brushes. On a straight repulsion motor such as might be found on a tugger or a hoist, the speed and direction of rotation are both controlled by the lever that shifts the brush holders.

Some repulsion-induction motors are equipped with a mechanical-actuated governor to give speed regulation to the motor. On these motors, once the governor is adjusted to the desired motor speed, the motor becomes a constant-speed motor.

Repulsion-induction motors can start very heavy loads without drawing excessive starting current. They range from ½ to 20 hp. They are useful also in locations where low voltage is not available. They are used on such applications as mills, grinders, air compressors, refrigeration systems, and hoists. The repulsion motor is expensive and requires a lot of maintenance. It is often replaced by a larger-sized split-phase capacitor motor, which is less expensive and relatively maintenance-free.

Single-phase synchronous motors operate at synchronous speeds for all load values. Single-phase syn-

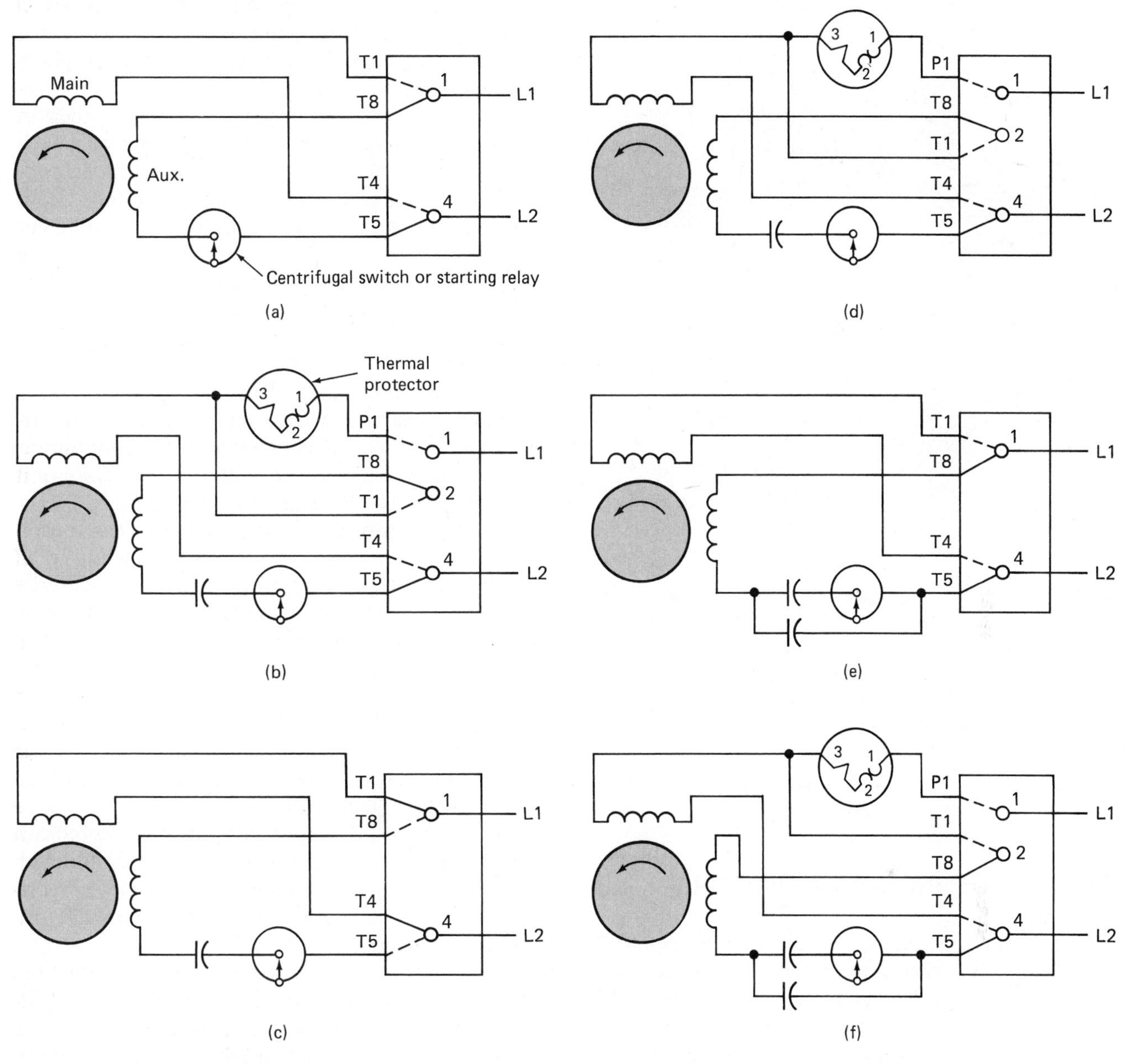

Single-phase motors connected for counterclockwise rotation—(end opposite drive shaft)

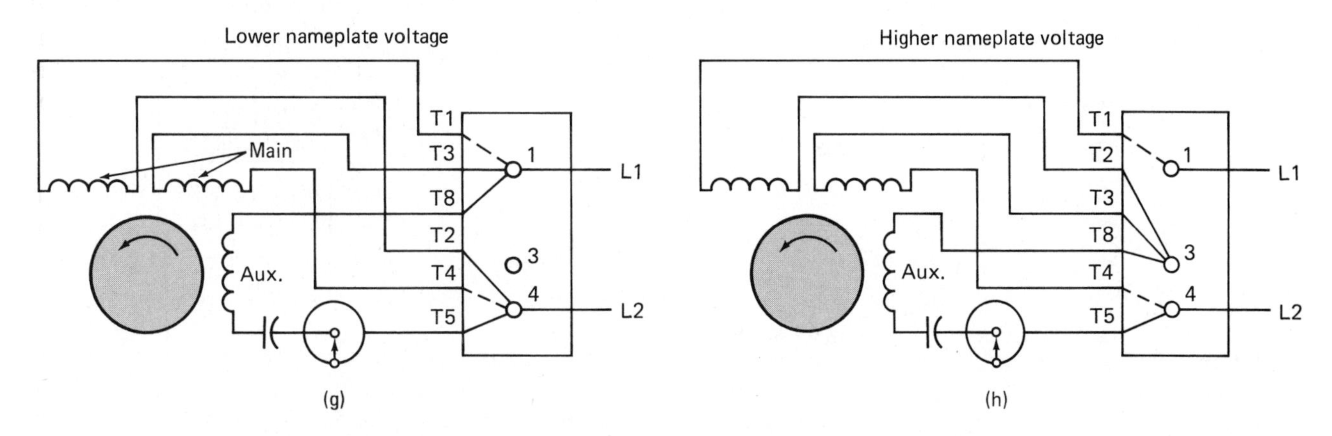

Dual-voltage, single-phase, split-phase capacitor motor connected for counterclockwise rotation.

Fig. 2-13 Split-phase and capacitor/split-phase motors connected for counterclockwise rotation.

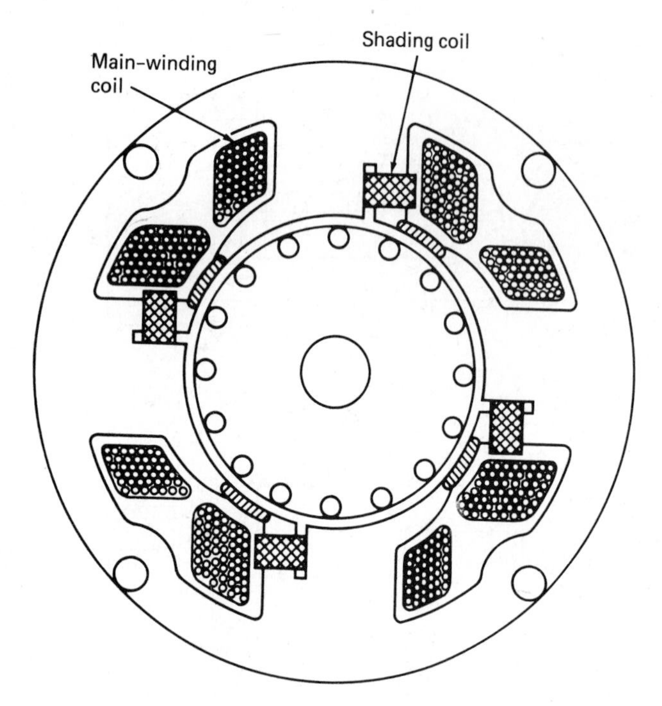

Fig. 2-14 Typical four-pole, shaded-pole motor.

chronous motors are manufactured in small sizes. They are used in timers, turntables, clocks, etc. The names given these motors are *reluctance motors, hysteresis motors,* and subsynchronous motors.

2-5 POLYPHASE INDUCTION MOTORS

The polyphase induction motor is the workhorse of industry. The three-phase motor is the most widely used polyphase induction motor. This motor has three sets of identical coils wound on the stator. The stator is a cylindrical structure that consists of a stack of laminations that are about 14 mils in thickness for motors operating on 60-Hz power. The coils are placed in slots in the laminations. The coils are terminated at a junction box on the external shell of the motor, where they are connected to a three-phase power source.

Polyphase induction motors have names such as *squirrel-cage motors* or *wound-rotor motors.* The names are related to and descriptive of their rotor construction. The two types of rotor structures used in polyphase induction motors are considerably different from each other. The rotor generally used in most polyphase induction motors is the squirrel cage. The other type of three-phase induction motor uses a wound rotor with slip rings. (It is sometimes called a *slip-ring motor.*)

The rotor of polyphase squirrel induction motors is similar in construction to the squirrel-cage rotor of single-phase motors. It consists of several aluminum conductors or bars cast into the rotor structure and uninsulated from the laminated core.

The conductors are connected to end rings at each end of the rotor. This provides the conducting paths for the rotor current. The rotor current is a function of magnetic induction. The source of power for the rotor is an induced voltage that is produced by the stator magnetic field sweeping past the rotor conductors. The end rings connect the conductors together and have fan blades for cooling. The rotor design will affect the starting and running characteristics of the motor. The classes of the motor (A, B, C, D, and F) are determined by the rotor design.

The squirrel-cage rotor bars are not parallel to the shaft. They are skewed slightly to create a condition that will give quiet operation as well as prevent cogging. Cogging is a condition that tends to make the motor run at subsynchronous speeds.

The stator windings of a three-phase induction motor can be star-connected or delta-connected. The windings are usually designed for two voltages, such as 220- and 440-V three-phase power. Figure 2-16 illustrates the delta and Fig. 2-17 illustrates the star (also called *wye*) dual-voltage connections.

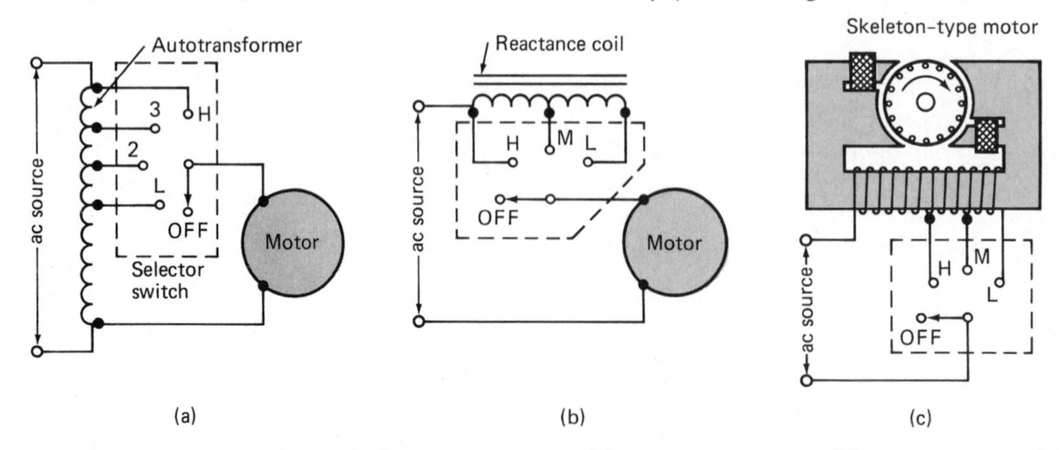

Fig. 2-15 Speed control methods for shaded-pole motors. *(a)* Autotransformer; *(b)* reactance coil; *(c)* tapped winding.

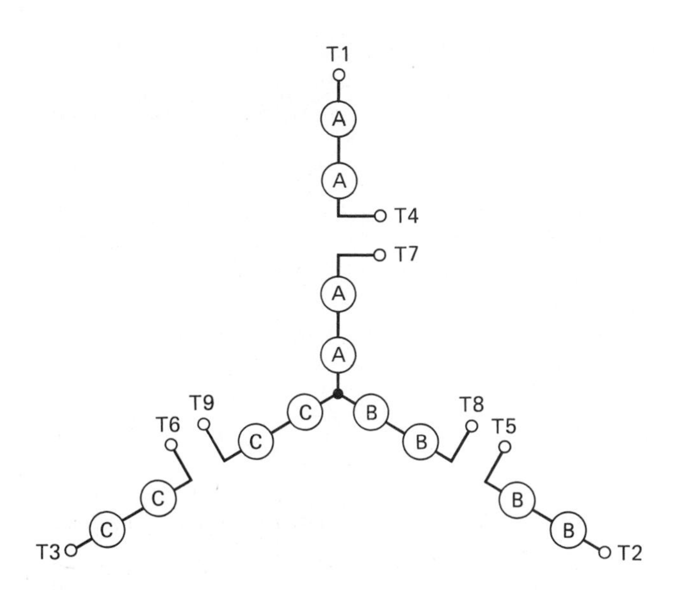

Dual-voltage star (wye) winding for a four-pole, three-phase motor.

Dual-voltage star (wye) connections

Voltage	L1	L2	L3	Tie Together
Low	1-7	2-8	3-9	4-5-6
High	1	2	3	4-7, 5-8, 6-9

Connection table for dual-voltage star-connected motor.

Fig. 2-16 Dual-voltage connections for delta connected three-phase motors.

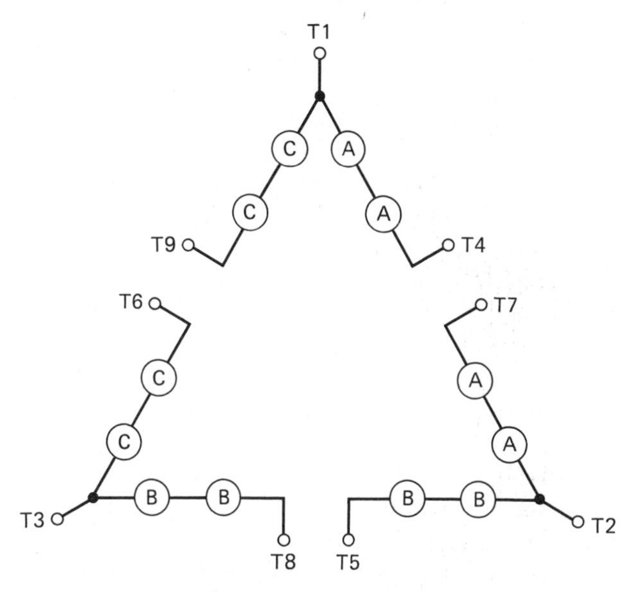

Dual-voltage delta winding for a four-pole, three-phase motor.

Dual-voltage delta connections

Voltage	L1	L2	L3	Tie Together
Low	1-6-7	2-4-8	3-5-9	- - - - -
High	1	2	3	4-7, 5-8, 6-9

Connection table for dual-voltage delta-connected motor.

Fig. 2-17 Dual-voltage connections for star (wye) connected motors.

2-6 REVERSING POLYPHASE MOTOR

No matter how the three-phase induction motor is connected, star (wye) or delta, the direction of rotation can be changed by reversing any two of the three-phase power connections to the motor.

In some industrial applications, when a three-phase induction motor is connected to a driven machine or load, rotation in the wrong direction may cause serious damage to the driven equipment. For these applications, the motor must be connected to obtain correct rotation before power is applied to the motor. A motor rotation indicator can be used to predict the rotation of the motor without applying voltage to the motor.

The first step in connecting a motor for proper direction of rotation is to determine the phase sequence of the three-phase power source with a phase-sequence tester. This is illustrated in Fig. 2-18.

Interchange the sequencer leads until the voltage source for the motor indicates ABC on the phase-sequence tester. Mark the leads with the corresponding ABC. When the power leads are marked ABC, connect the motor rotation indicator to the motor leads as indicated in Fig. 2-18. Hold the ON button closed and rotate the motor shaft one-quarter turn in the di-

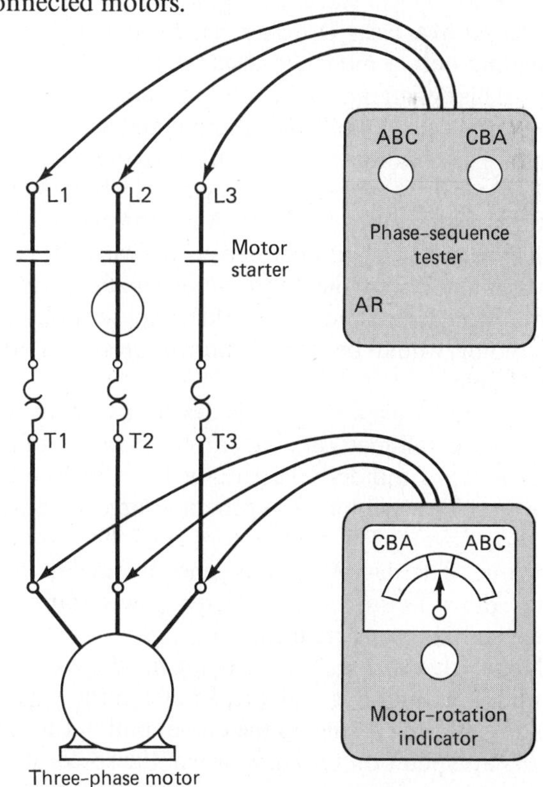

Fig. 2-18 Phase-sequence tester checking phase of supply voltage; motor rotation indicator indicating direction of rotation.

rection the motor is required to rotate. Observe the dial on the rotation indicator to determine if the direction of rotation will be correct.

2-7 DIFFERENT CLASSES OF POLYPHASE MOTORS

Applications for the class A and B three-phase induction motors cover an extremely wide area. Drives for pumps, fans, mills, lathes, grinders, drill presses, machine tools, conveyors, compressors, etc., are typical examples of where the class A and B motors can be used.

Class C motors have a double-cage rotor construction. These motors can be started across the line. Rotors designed with the "double deck" construction start with the rotor current crowded to the upper cage. When the motor reaches full speed, the fluxes penetrate deeper into the rotor, making the inner cage parallel with the upper cage. This condition in the rotor determines the running characteristics of the motor.

The class C motors are not adapted to high-inertia loads, which prevent rapid acceleration. They should be used with loads that require high starting torques but are of a static nature. The NEMA-design C also provides relatively low starting current and low slip.

The NEMA class D motor has a rotor with slots extending only a moderate distance into the punchings. This design makes the rotor resistance comparatively high during both the running and starting periods. These motors are generally large in size (25 hp and up). They are ideal for applications where energy is stored as the rotor speeds up. Mechanical machines with flywheels that give up stored energy when the peak load comes on are typical applications that the class D motor would be suited for. Other applications for the motor would be shears, punch presses, stamp presses, etc.

The NEMA class F motor is designed to draw less starting and full-load running current than any of the other types of motors. The motor has a high-resistance double-cage rotor. It is generally large in size (25 hp and above). The starting torque of this motor is very low. The class F motor is generally used in locations with a limited and restricted power source or where starting loads are light.

Notice that in Fig. 2-19 a point on the curves is identified as pull-out and breakdown torque. If a motor is starting (going up the curve) pull-out torque occurs at a point on the curve where the speed of the motor produces maximum torque. If the motor is operating at full speed and additional load is added, the speed of the motor will decrease. Breakdown torque occurs at the same point as pull-out torque. Breakdown torque is the maximum torque or load the motor will carry without an abrupt drop in speed. When the breakdown point is exceeded, the motor will stall.

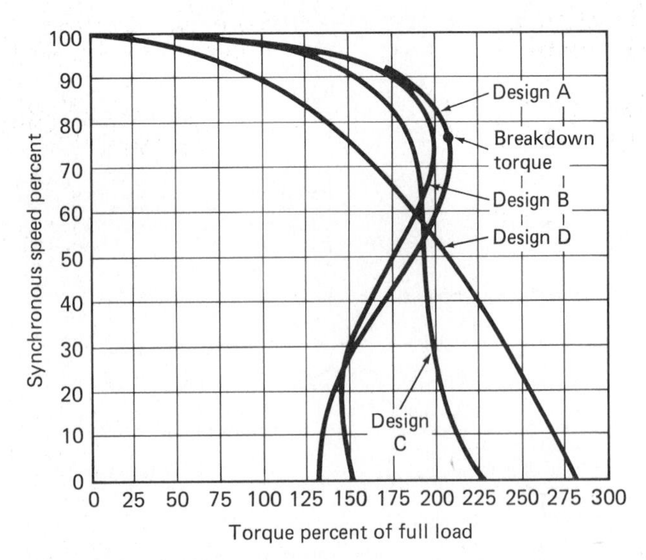

Fig. 2-19 Typical speed-torque curves for different designs of polyphase induction motors.

The three-phase induction motor is essentially a constant-speed energy converter. The speed will vary only from 2 to 8 percent between no-load speed and full-load speed. The speed of the induction motor is tied to the frequency of the power source and the number of poles. To obtain speed control for the squirrel induction motors that we have studied thus far, it is necessary to have a means of providing a change in frequency or a change in poles. A variable-frequency drive will provide adjustable speeds where a pole change will only provide multiple speeds. Variable-speed drives will be covered in a later chapter.

2-8 MULTISPEED POLYPHASE MOTORS

Motors can be wound with two separate windings with a different number of poles in each winding to obtain two-speed motors. These motors are more expensive and have other problems, such as heating, a low power factor, and low efficiency, which make them undesirable for industrial applications.

Another method that is commonly used to get two speeds from a single-winding motor is the consequent-pole method. (With this method a single motor is connected conventionally for the high speed; it is operated with the consequent pole for low speed.) In the consequent-pole connection, the paths for current flow produce poles of the same instantaneous polarity. At a given instant, all the windings would produce, say, a north pole. South poles are formed magnetically as consequence of the field formed by the north pole windings.

With the conventional motor connection, a four-pole motor would produce in its stator four poles (north-south, north south). A north pole cannot be produced without a south pole forming somewhere. If the consequent-pole motor coil groups are connected so they all have the same magnetic polarity, then opposite magnetic poles will exist in consequence of the wound poles. In the above example, at a given instant if all the poles were north in the consequent-pole connection, south poles would be formed because of the wound north poles. If we had four north poles, we would have four south poles in consequence of the four poles produced by the current flow in the coils. Two speeds then can be obtained from a consequent-pole motor.

The high speed is obtained when the motor has conventional connections, producing a given number of wound north and south poles. When the motor is connected for consequent-pole operation, the speed of the motor will be half the high-speed value because the number of poles is doubled.

2-9 APPLICATIONS AND CONNECTIONS FOR TWO-SPEED MOTORS

The National Electrical Manufacturers Association has standardized the general types of two-speed motors. The names and characteristics for the motors are as follows. The constant torque motor develops the same full-load torque at both high and low speed. The constant-horsepower motor develops a full-load torque that is inversely proportional to the motor's speed. The variable-torque motor develops full-load torque that is directly proportional to the speed. Remember that the motors are connected consequent-pole for low speed and conventional-pole for high speed. The phase connections (series-delta, two parallel star, and series-star) for two-speed motors depend upon the motors' classification.

Each of the two-speed motors has certain characteristics that make it particularly suitable for certain industrial applications. Constant-torque motors are used to drive compressors, conveyors, machine tools, printing presses, and textile, baking, and laundry machines. Constant-horsepower motors are used to drive lathes, shapers, grinders, boring mills, and any other machine where considerable torque must be developed at low speed and much less torque is required at high speed. Practical applications for the variable-torque motors are fan drives, pumps, blowers, or any loads in which the required horsepower increases approximately as the cube of the speed.

Figure 2-20 on the next page indicates the terminal markings and connections for single-winding, two-speed induction motors.

2-10 WOUND-ROTOR MOTORS

The wound-rotor, or slip-ring, motor differs from the three-phase induction motor in that it has a wound rotor in place of the squirrel-cage rotor. The rotor winding is wound in slots in the laminated core and terminates on the collector rings, or slip rings, as they are also called. Brushes that ride on the slip rings are connected to external resistors or adjustable rheostats for speed and torque control See Fig. 2-21 on the next page for motor diagram.

The stator winding of the wound-rotor motor is known as the *primary winding* of the motor. The wound rotor acts as the secondary winding in the motor. The voltage in the rotor winding is induced from the primary or stator winding by transformer action. Controlling the rotor current by external resistance affects the load slip and changes the motor's speed. Increasing the rotor's external resistance causes the speed to decrease. With no resistance in series with the slip rings (the rotor is short-circuited), the motor would rotate at its maximum speed. The maximum speed would be determined by the number of stator poles and the frequency of the applied power.

Besides supplying a means of adjusting speed, the external resistors can also be used to control torque. To compensate for different torque requirements for starting loads under a variety of load conditions requires only the adjustment of the external resistance in series with the wound-rotor motor. Maximum torque developed by a wound-rotor motor is fixed by the design constants of the motor but can be made to vary from the maximum value to satisfy any load by adjusting the resistance. Figure 2-21 is an illustration of a wound-rotor motor connected to external resistors for speed or torque control.

It is generally impractical to reduce the full-load speed of a wound-rotor motor below 50 percent of the rated speed of the motor because of the large values required of secondary resistors. Operating a wound-rotor motor at speeds below 50 percent–rated speed also causes extremely low motor efficiency.

Another method of obtaining adjustable speeds (either above or below synchronous speeds) from a wound-rotor motor is to inject a voltage into the slip rings of the motor. If the injected voltage acts to aid that generated in the rotor, the speed of the motor will be greater than its synchronous speed. If the injected voltage opposes the generated voltages, the motor's speed will be less than synchronous speed.

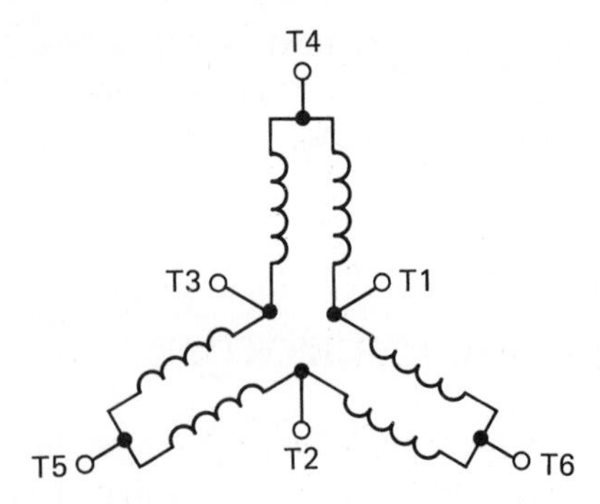

(a) Constant-torque connections

Speed	Line wires	Tie together	Winding polarity	Winding connection
Low	T1-T2-T3	----	Conseq.-pole	Series Δ
High	T4-T5-T6	T1-T2-T3	Conventional	2 Parallel Y

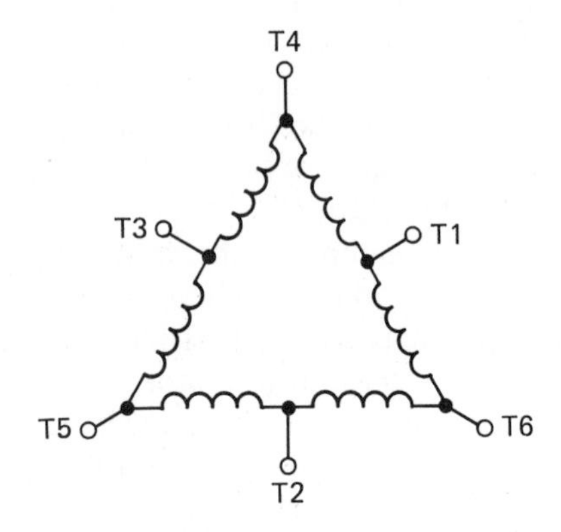

(b) Constant-horsepower connections

Speed	Line wires	Tie together	Winding polarity	Winding connection
Low	T1-T2-T3	T4-T5-T6	Conseq.-pole	2 Parallel Y
High	T4-T5-T6	----	Conventional	Series Δ

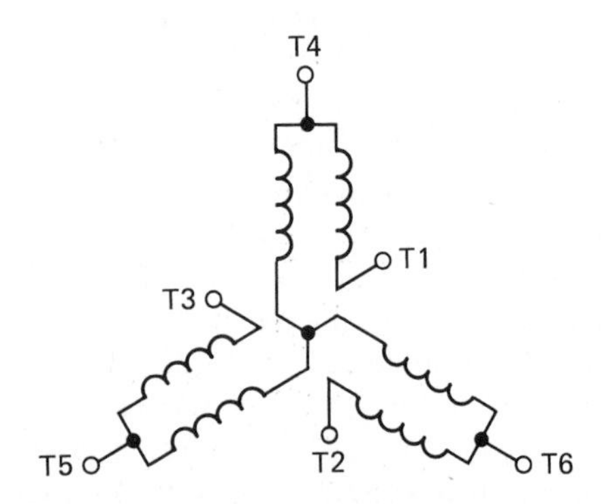

(c) Variable-torque connections

Speed	Line wires	Tie together	Winding polarity	Winding connection
Low	T1-T2-T3	----	Conseq.-pole	Series Y
High	T4-T5-T6	T1-T2-T3	Conventional	2 Parallel Y

Fig. 2-20 Terminal markings and connections for single-winding, two-speed induction motors.

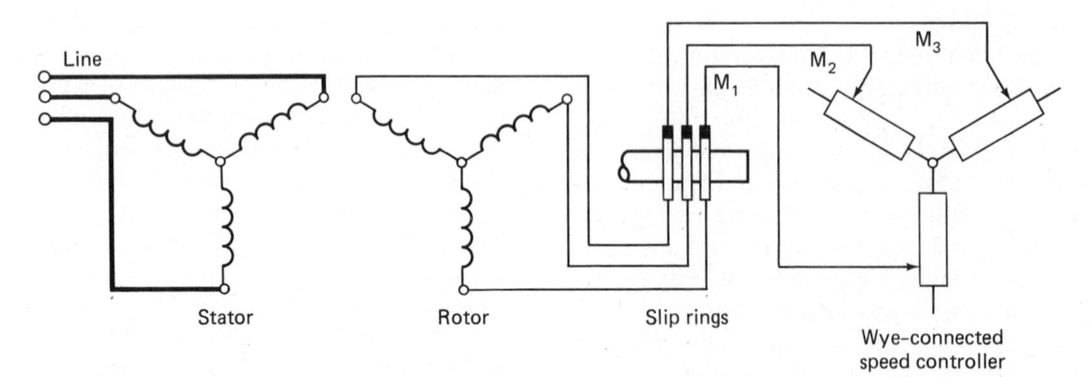

Fig. 2-21 Diagram of connections of a wound-rotor induction motor connected to a speed controller.

The voltage-injection method of controlling the speed of wound-rotor motors eliminates power losses and contributes to the efficiency of the motor. Voltage injection for speed control of wound-rotor motors is used only on larger motors because it requires costly control equipment and voltage-injection devices.

As compared with a squirrel-cage induction motor, the wound-rotor motor would be the choice for applications that require starting extremely high-inertia loads.

2-11 SYNCHRONOUS MOTORS

Synchronous three-phase motors operate at a constant speed. Their stator contains windings similar to those in the regular induction motors (Fig. 2-22*a*). The rotor contains coils that are of the salient-pole type (Fig. 2-22*b*). The rotor also contains a squirrel-cage or amortisseur winding used during starting the motor. The field coils are connected to slip rings that receive power from brushes that are connected to the dc power source. The stator power source is a three-phase voltage. The rotor voltage is direct current. The dc voltage is usually obtained from an exciter or generator. The motor is started as a squirrel-cage motor. At about 94 to 97 percent of synchronous speed, dc voltage is applied to the fields. The motor synchronizes and continues to run at constant speed.

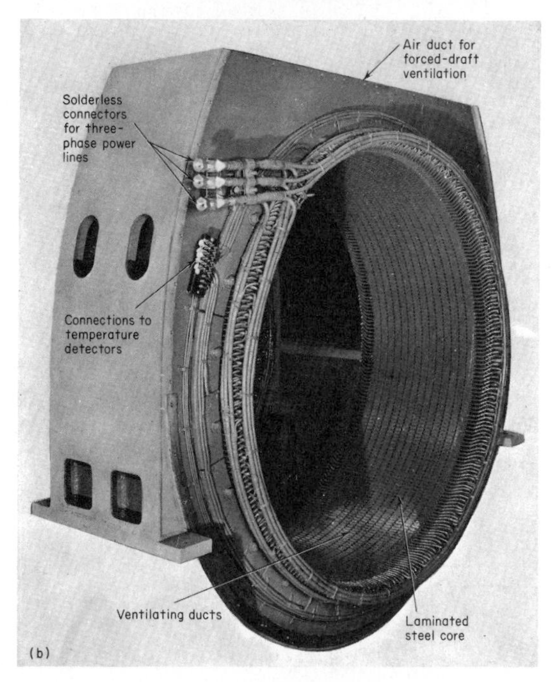

Fig. 2-22 Synchronous motor construction. *(a)* Rotor; *(b)* stator.

The synchronous motor has two important characteristics. One is that the motor operates at a definite constant speed. The other is that the synchronous motor can be used for power factor correction. By controlling the value of the dc excitation applied to the motor, the synchronous motor can be made to operate at unity or leading power factor. When connected to the ac power source of a mill or factory, the synchronous motor can improve the overall power factor of the entire system. When a synchronous motor is running by itself with no mechanical load attached to it, it is being used for power factor correction only. When the synchronous motor is used for power factor correction only, it is sometimes called a *synchronous condenser* or *synchronous capacitor.*

A synchronous motor is reversed the same way as any other three-phase induction motor. Reversing any two stator leads will reverse the motor.

2-12 DIRECT-CURRENT MOTORS

One of the advantages offered by dc motors is their ability to provide a wide range of adjustable speeds. Many machines found in industry require motors with a wide range of speed control. Overhead cranes, for example, could not perform their job if they were not equipped with adjustable-speed motors. Direct-current motors are used on the majority of big industrial cranes because of their speed control characteristics. Other features make dc motors desirable for certain applications in industry, such as ease of reversal, dynamic braking, and jogging, or inching, operations.

Requirements for a wide variety of industrial drives often can be met only by dc motors. Their performance and ease of control are basic necessities for many processing lines and other similarly complex systems. Applications for which dc motor types are most suitable may require one or more of the following features:

1. Wide speed range
2. Small incremental speed changes
3. Speed matching

4. Smoothness, or a specific rate, of acceleration or deceleration
5. Limited or controlled torque or tension

Direct-current motors can be used where basic ac drives are often unacceptable. Alternating-current motors are inherently constant-speed machines. Although multispeed and adjustable-speed operations are possible with ac motors, they require a very complex design of the motors and their associated control equipment to obtain adjustable-speed control. By comparison, dc motors are inherently adjustable-speed machines, and only a small change in design and slight changes in control equipment are necessary to take advantage of this important feature of dc motors.

With respect to application characteristics, dc motors are classified according to their type of field winding. The operating characteristics of the three types—shunt-wound, series-wound, and compound-wound motors—are illustrated in Figs. 2-23 to 2-26.

Regardless of the name of the dc motor (series, shunt, or compound), the rotating member of the motor is called the *armature* (Fig. 2-27). The armature of a dc motor consists of coils inserted in a laminated core. The windings are generally wound in a lap or wave configuration. Lap windings are used for low-voltage, high-current machines and wave windings are used in high-voltage, low-current applications. The coil ends are connected to the risers of the commutator bars.

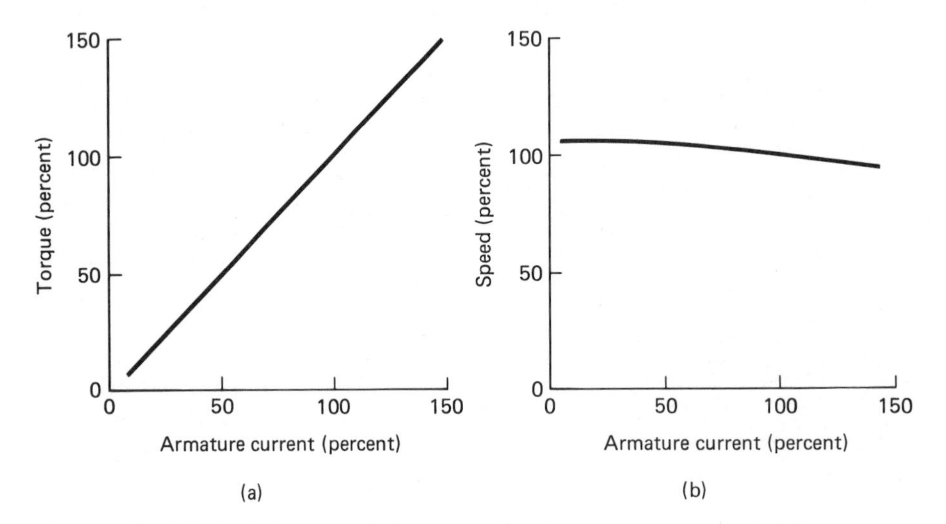

Fig. 2-23 Typical characteristic curves for dc shunt motors. *(A)* Current-torque curve; *(b)* speed-load curve.

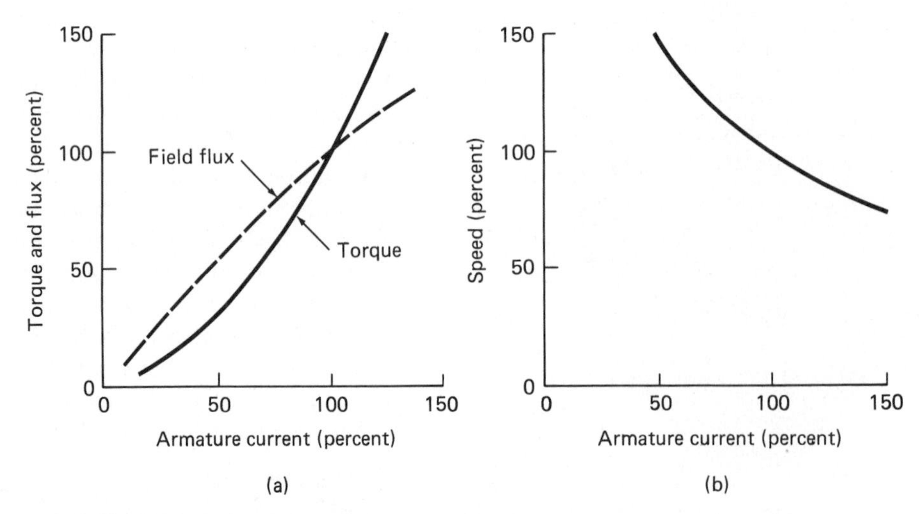

Fig. 2-24 Typical characteristic curves for series motors. *(A)* Current-torque and current-field strength curves; *(b)* speed-load curve.

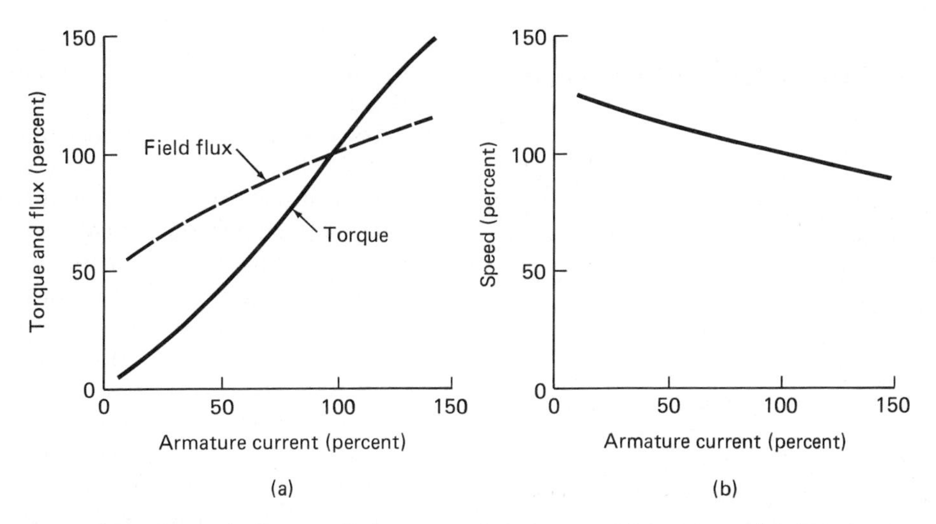

Fig. 2-25 Typical characteristic curves for compound motors. *(a)* Current-torque and current-field strength curves; *(b)* speed-load curve.

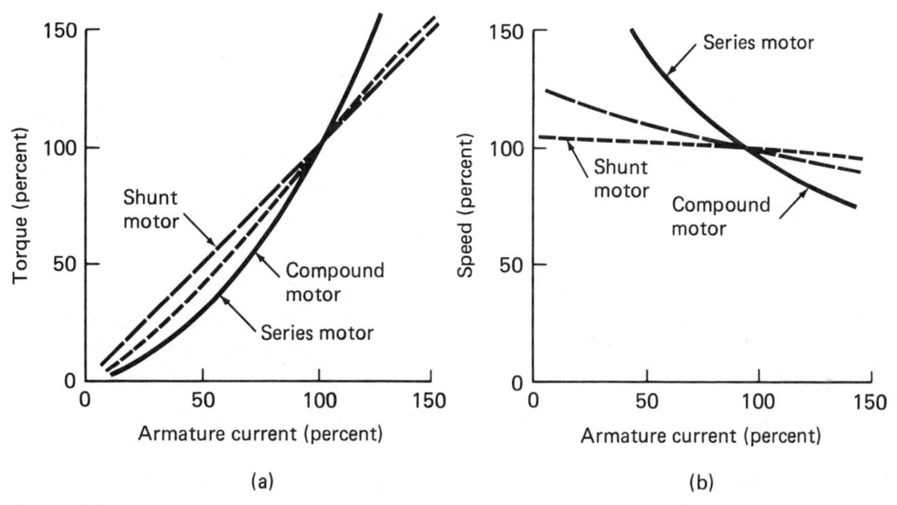

Fig. 2-26 Comparison of characteristics. *(a)* Current-torque curves; *(b)* speed-load curves.

Fig. 2-27 Direct-current armatures.

The commutator consists of copper bars separated by sheets of mica. The mica provides the insulation between the copper bars. Mica cones and rings insulate the copper bars from the iron core and shaft of the armature. The mica between the copper commutator bars is under-cut to prevent interfering with the electrical contact between the brushes and the commutator surface.

2-13 WORKING FIELDS AND CORRECTIVE FIELDS

The differences among series, shunt, and compound motors are determined by the field structures on the stationary part of each type of motor. The way the fields are connected in relation to the armature determines the name of the motor. If the field coils are connected in series with the armature, the motor is known as a *series motor* and is said to be *series-excited.* In the shunt motor, the field coils are connected in parallel with the armature and the motor is said to be *shunt-excited.* If a motor has a series and a shunt field, the excitation is produced by the sum of the magnetic fluxes from each field. The motor is known as a *compound motor.* When the series field aids the shunt field in producing the field excitation, the motor is called *cumulative-compounded.* When the series field opposes the shunt field, the motor is said to be *differentially compounded.*

The fields that produce the stationary magnetic fields that react with the armature field to cause motor rotation are called *working fields.* The dc motor also has some corrective fields that prevent arcing and sparking between the brushes and commutator. These corrective fields minimize the effects of armature reaction which cause poor commutation, motor instability at high speeds, and commutator flash-over conditions of suddenly applied heavy overloads.

The most widely used corrective field is the interpole, or commutating pole, as it is sometimes called. The interpoles are small poles located between the main poles. The purpose of the interpoles is to neutralize the inductive effect of the armature coils, which could cause poor commutation, producing arcing and sparking at the brushes and commutator.

Compensating windings, or pole face windings, are embedded in the surface of the main-field pole faces. The purpose of the compensating windings is to prevent armature reaction. Armature reaction in a motor shifts the flux across the pole face in a direction opposite to armature rotation. The application of sudden, heavy overloads can produce enough armature reaction to shift the magnetic field of the stationary field until flash-over can occur. This can destroy the commutator surface. The effect of the compensating field is to provide a magnetic field that will eliminate or minimize the effects of armature reaction.

When the shunt field of a motor is made weaker, the speed of the armature increases. The field can be made weak enough on a shunt motor that the motor will race to destruction. To prevent a weak shunt field from causing runaway or unstable operation, a stabilizing field is added to the shunt motor. Direct-current shunt motors designed to operate at high speeds with field weaking are called *stabilized shunt motors.* The stabilized field is a light series field placed directly over the shunt field to prevent runaway caused by field weakening.

Speed control of dc motors can be provided in a number of ways. The base speed of a dc motor is the speed at which the motor runs at rated load with full voltage on the field and armature. Adding resistance to the shunt field circuit will give speeds above base speed. Adding resistance in series with the armature gives motor speeds below base speed. In a later chapter we will see how solid-state drives provide speed control for dc motors.

To reverse a dc motor, interchange all the field connections or reverse the armature connections (Fig. 2-28). Reversing all the field leads is not recommended because of the ease of interchanging the armature connections.

2-14 DYNAMIC BRAKING

If the electrical source of power were removed, a motor would stop as a result of its own friction and the energy drain of the load. Such a stop would not be quick, and could be variable, depending upon the type of load. In many instances, a coast-to-stop response is satisfactory.

Another method of stopping a motor drive is friction braking with either a magnetic-, mechanical-, or hydraulic-actuating means. Because of the wear on the wheel and shoes of the brake, especially if frequent stops are required, friction braking is not always satisfactory.

Dynamic braking is a useful method for retarding a motor. This method uses one of the operational characteristics of a motor and requires equipment that is comparatively simple to provide.

When a motor is running, it is also internally acting as a generator. The voltage so produced has a polarity opposite to that of the power supply and hence is called the *countervoltage* or *counter-emf.* If the motor is disconnected from the power source, the armature still generates a voltage with the same polarity as before—opposite to that of the power supply.

Now, if a resistor is connected across the armature, as in Fig. 2-29, the motor acts as a generator, driven by the kinetic energy in the motor and load system, and forces current through the resistor. However, because of the polarity of the generated voltage, current flows through the armature in the direction opposite to the flow during normal running conditions. Torque created by this current opposes the rotation of the armature. Thus, a retarding force is developed and motor speed is reduced.

As motor speed is reduced, generated voltage, current, and retarding torque are also reduced. Conse-

Shunt motor connections

Counterclockwise		Clockwise	
Line 1 F1–A1	Line 2 F2–A2	Line 1 F1–A2	Line 2 F2–A1
F1 F2 L1 A1 (A) A2 L2		F1 F2 L1 A2 (A) A1 L2	

Standard shunt motor connections for counterclockwise and clockwise rotation

Series motor connections

Counterclockwise			Clockwise		
Line 1 A1	Tie A2–S1	Line 2 S2	Line 1 A2	Tie A1–S1	Line 2 S2
L1 A1 (A) A2 S1 S2 L2			L1 A1 (A) A2 S1 S2 L2		

Standard dc series motor connections for counterclockwise and clockwise rotation.

Compound motor connections

Counterclockwise			Clockwise		
Line 1 F1–A1	Tie A2–S1	Line 2 F2–S2	Line F1–A2	Tie A1–S1	Line 2 S2–F2
F1 F2 L1 A1 (A) A2 S1 S2 L2			F1 F2 L1 A2 (A) A1 S1 S2 L2		

Standard dc compound (cumulative) motor connections for counterclockwise and clockwise rotation.
For differential compound connection reverse S1 and S2

Fig. 2-28 Connection diagram for reversing dc motors. (*Courtesy of Adams/Rockmaker,* Industrial Electricity: Principles and Practices)

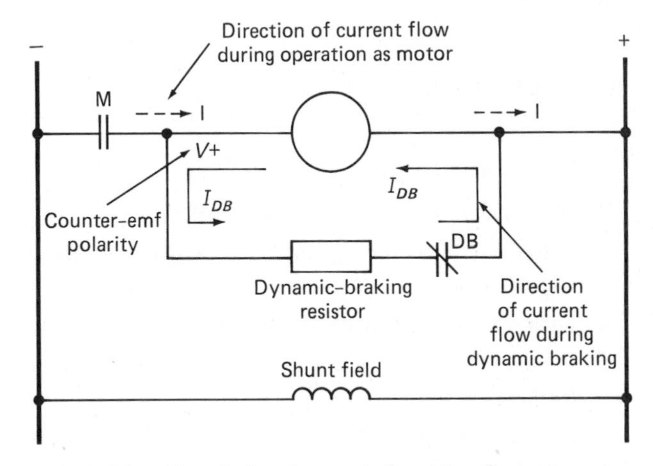

Fig. 2-29 Circuit for dynamic braking for a dc motor. Contacts M and DB are shown in condition for braking action.

quently, as the motor slows down, it is acted upon by a steadily diminishing, retarding torque. At comparatively low speeds, little braking force remains. When dynamic braking is used, the motor stops as a result of friction rather than electrical retardation.

Additional control equipment can be provided to reduce the value of the dynamic-braking resistor in one or more steps as the motor slows down. Thus, average retarding current and torque are kept at the higher level. This, in turn, provides more effective electrical- or dynamic-braking action. Although this method is used occasionally where faster deceleration is required, it is usually not necessary.

Dynamic braking is most conveniently applied to the shunt or lightly compound wound motor since there is little or no field flux change during the operation. Although it can be applied to the series-wound motor, dynamic braking requires extensive switching; the results, however, are not too satisfactory because of continuous reduction of field flux.

This is a summary of the characteristics of the different motor types presented in this chapter:

Alternating-current Squirrel-cage Motor; Single- or Three-Phase. Single-phase requires starting winding; three-phase, self-starting; stationary stator winding, no electrical connection to short-circuited rotor torque produced from magnetic reaction of stator and rotor fields; speed and function of supply frequency and number of electrical poles wound on stator; considered as constant speed even though speed decreases slightly with increased load; good starting torque; high inrush current during starting on full voltage; rugged construction; easily serviced and maintained; high efficiency; good running power factor when delivering full load; requires motor control only for stator windings.

Alternating-current Wound-rotor Induction Motor. Similar to squirrel-cage motor stationary stator winding; rotor windings terminate on slip rings; external addition of resistance to rotor circuit for speed control; good starting torque; high inrush current during starting on full voltage; low efficiency when resistor inserted in rotor windings; good running power factor requires motor controls for both stator and rotor circuits.

Alternating-current Synchronous Motor. Stationary ac stator winding; rotating dc field winding; no starting torque unless motor has starting winding; generally poor starting torque; constant speed when motor up to speed and dc field winding energized; can provide power factor correction with proper dc field excitation; requires special motor control for both ac and dc windings to prevent the dc field winding from being energized until a specified percentage of running speed has been obtained.

Direct-current Commercial and Industrial Motors. Control functions for constant horsepower; constant- or variable-torque machine loads; constant-torque metalworking machinery; variable-torque and -horsepower fans and blowers; constant-power heating pumps and motors for all types of applications may include speed regulation; good for cycling or infrequent starting and stopping.

Direct-current Shunt Motor. Main-field winding designed for parallel connection to armature; stationary-field rotating armature with commutator; has a no-load speed; full speed at full load less than no-load speed; torque increases directly with load.

Direct-current Series Motor. Main-field winding designed for series connection to armature stationary field; rotating armature with commutator does not have a no-load speed; requires solid direction connection to load to prevent runaway at no load; speed decreases rapidly with increase in load; torque increases as square of armature current; main motor for crane hoists; excellent starting torque.

Direct-current Compound Motor. Main field both shunt (parallel) and series; stationary fields; rotating armature with commutator; combination shunt and series fields produce characteristics between straight shunt or series dc motor; good starting torque; main motor for dc-driven machinery, mills, presses, etc.

SUMMARY

Motors are classified according to their voltage requirements (ac or dc) and the manner in which they are wound. Most industrial motors are of the ac type primarily because most of the industrial electrical systems are alternating current. Small ac motors generally operate on single phase, while all large ac motors are built to operate on polyphase (usually three-phase) systems. Since most electrical systems are ac, the use of dc motors requires converting the ac to dc. In the past this meant using an ac motor to turn a dc generator. Another method used electron tube rectifiers. Neither method was particularly efficient, and in the case of the motor-generator set, the usual maintenance, repair, and control problems associated with rotating machines were present. Modern solid-state electronic rectifiers with high current capacity have, in many ways, made the use of dc motors of all sizes more feasible. Direct current motors have always had one distinct advantage over ac motors, and that is the comparative ease with which they can be accurately controlled in terms of changing speed, reversing the direction of rotation, starting, and stopping. Clearly, modern solid-state devices have increased the scope of ac motor control, and thus for many applications where control is an important criterion for motor specification, ac motors are replacing dc motors.

REVIEW QUESTIONS

1. Why are ac motors used more than dc motors?
2. What advantages do dc motors offer?
3. What type power is used by large-horsepower motors in industry?
4. Describe the construction of a squirrel-cage rotor.
5. What is the function of the stator windings in an ac motor?
6. What determines the running speed of a polyphase ac motor?
7. Define slip.
8. What is the synchronous speed of a four-pole ac motor operating at 60 Hz?
9. Does an uneven air gap around the rotor of a motor indicate anything?
10. Describe a sleeve bearing and what motors use sleeve bearings.

11. What is the rule concerning the amount of grease an antifriction bearing should receive if it can be lubricated?
12. How many classes of insulation are there and what is the temperature rating of each?
13. How many different types of motor enclosures are there? What are they?
14. What is meant by nameplate call-out?
15. What determines the direction of rotation of a universal motor? How is it reversed?
16. Why are governors used with some universal motors?
17. Describe the construction of a split-phase induction motor.
18. What is the purpose of a centrifugal switch in a split-phase motor?
19. What are the different classifications for split-phase motors?
20. What size of capacitor is required for a ½-hp, 1750-rpm split-phase capacitor motor?
21. How do the inrush current and starting torque of a capacitor start split-phase motor and a standard split-phase motor compare?
22. When can an electrolytic capacitor be used for a capacitor split-phase motor?
23. How is a permanent-split capacitor motor reversed?
24. How are speed changes obtained with a permanent-split capacitor motor?
25. What is considered the standard rotation of a single-phase induction motor?
26. How are shaded-pole motors reversed?
27. What type motor would be used where the starting service is both severe and sustained?
28. How is the direction of rotation of a repulsion motor changed?
29. Why are split-phase capacitor motors being used as replacements for repulsion motors?
30. Where does the rotor obtain its power source in an induction motor?
31. What is cogging and how is it minimized?
32. What effect does the external resistance have on the operation of a wound-rotor induction motor?
33. How are polyphase induction motors reversed?
34. A three-phase induction motor must rotate in the correct direction the first time it is connected. How would you accomplish this?
35. What determines whether a three-phase induction motor can be started across the line?
36. What is the difference between pull-out torque and breakdown torque?
37. What is a consequent-pole connection used for on an induction motor?
38. What windings are known as the primary and secondary in a wound-rotor motor?
39. What effect does increasing the external rotor resistance have on a wound-rotor motor?
40. How can a wound-rotor motor be made to operate above synchronous speeds?
41. When is the dc field excitation applied to the field of synchronous motors?
42. How can a synchronous motor be used for power factor correction?
43. How are synchronous motors reversed?
44. Name three types of dc motors.
45. What is the difference between a working field and a corrective field in a dc motor?
46. What is the difference between an accumulative compound motor and a differential compound motor?
47. What is the function of interpoles in a dc motor?
48. What are compensating windings and what are they used for in dc motors?
49. Define base speed. How are speeds above and below base speed obtained in a dc motor?
50. What is a stabilizing field used for? What type motor has a stabilizing field?
51. What is the rule for reversing dc motors?
52. Describe dynamic braking and how it is accomplished in dc motors.

3
Control of Motor Starting

Control circuits and equipment can perform varied functions. These can be grouped into 11 general types according to the effect they have on the motor to be controlled. Each of the general types can be broken down into endless variations, but they each stem from a few basic principles which, if understood, are the key to control work. This chapter presents these principles using nonmathematical and familiar terminology as much as possible.

3-1 MOTOR STARTING

There are several general factors to be considered in the selection of motor-starting equipment. The most obvious of these are the current, voltage, and frequency of the motor and control circuits. Motors require protection according to the type of service, the type of motor, and the control functions that will be needed.

Whether to use a full-voltage starter or a reduced-voltage starter may depend on the current-carrying capacity of the plant wiring and the power company lines as well as power company rates. Other factors, such as the need for jogging, or inching, or acceleration control, or the type of motor to be used, will also affect this selection.

FULL-VOLTAGE STARTING

This type of starting simply requires that the motor leads and line leads be connected (Fig. 3-1). This could be accomplished merely by using a knife-blade switch, but this method would provide no protection for the motor except its circuit fusing.

For small fractional-horsepower motors on low-amperage circuits, the simple switch may be satisfactory and is used frequently. Many appliances use nothing more than the cord and plug as the disconnecting means, with a small toggle switch to start and stop the motor. Because the motor is not disconnected from the line on power failure, this type of starting control can be an advantage for fans and other devices that otherwise would have to be restarted.

For motors up to 10 hp and not over 600 V, the manual across-the-line motor starter may be used to give manual control. Most of these units give overload protection and undervoltage release.

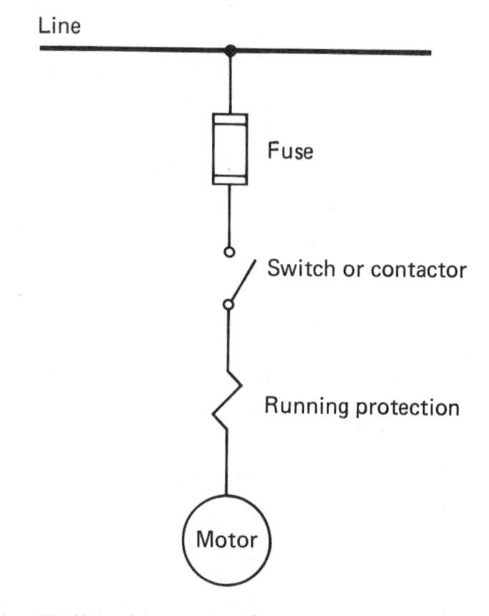

Fig. 3-1 Full-voltage starting.

The most popular starter for motors up to 600 hp and 600 V or less is the magnetic across-the-line starter. This starter, combined with pilot devices, can give full protection to the motor and fully automatic operation.

The vast majority of motors today are built to withstand the surge of current that occurs when they are suddenly thrown across the line. Not all our plant circuits can stand this surge, however, nor can all the power company equipment. When a large motor starts at full voltage, it may cause a voltage drop large enough to drop out other control equipment. Should the voltage drop be serious enough, it might even cause a dimming of lights in other buildings.

Utility companies penalize most industrial installations, in the form of higher rates, for surges of excessive current on the line. Using a demand meter, the maximum average demand of power on the line for a given period of time, generally a 15-minute period, can be measured. This factor should always be taken into consideration when the method of starting large motors is being decided. The extra cost of power, because of the excessive demand for starting large motors across the line, may well exceed the cost of reduced voltage starting, which would reduce the demand reading.

When considering using full-voltage starting, always check the building wiring and distribution system capacity. Should the wiring be inadequate, either it must be increased in capacity or reduced-voltage starting must be used.

REDUCED-VOLTAGE STARTING

Whenever the starting of a motor at full voltage would cause serious voltage dips on the power company lines or the plant wiring, reduced-voltage starting becomes almost a necessity (Fig. 3-2). There are, however, other reasons for using this type of control. The effect on the equipment must also be taken into account in the selection of motor starters. When a large motor is started across the line, it puts a tremendous strain or shock on such things as gears, fan blades, pulleys, and couplings. Where the load is heavy and it is hard to bring it up to speed, reduced-voltage starting may be

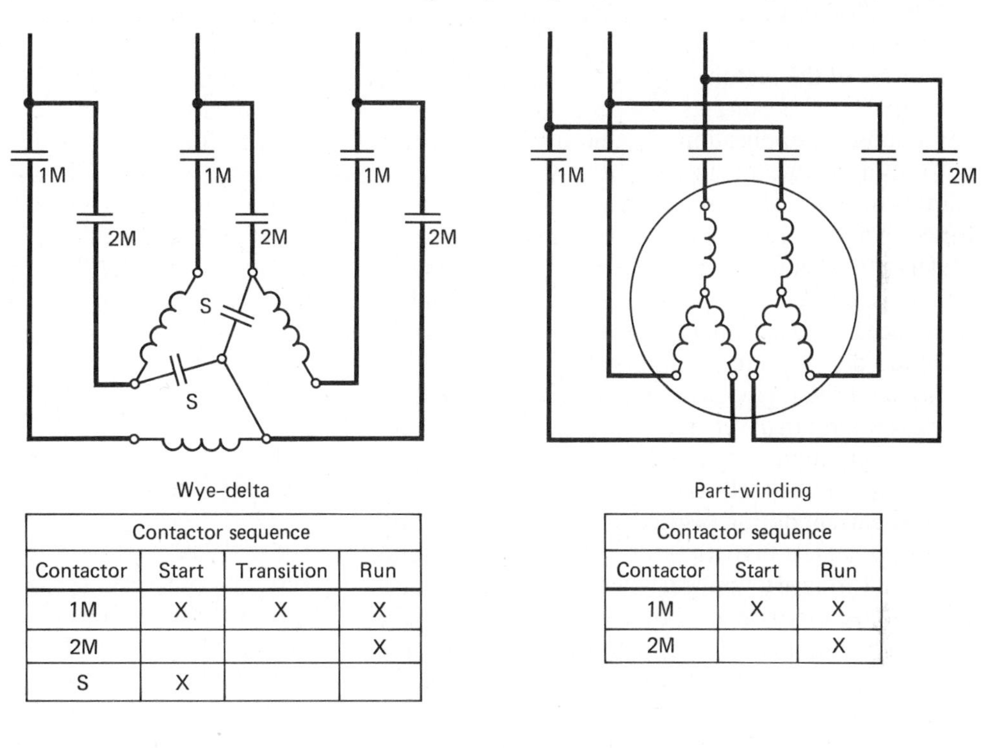

Wye-delta

Contactor sequence			
Contactor	Start	Transition	Run
1M	X	X	X
2M			X
S	X		

Part-winding

Contactor sequence		
Contactor	Start	Run
1M	X	X
2M		X

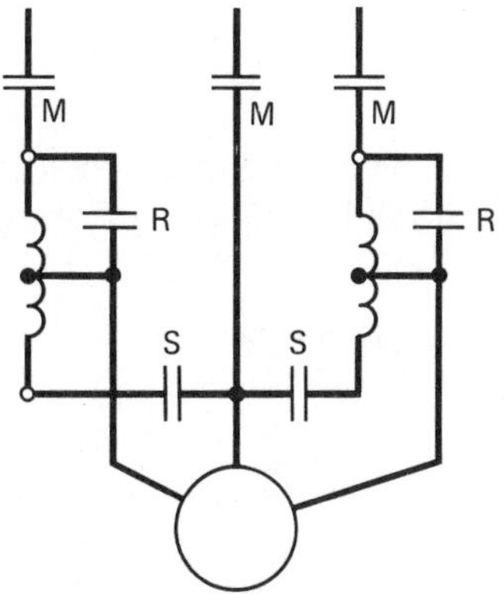

Two-coil autotransformer

Contactor sequence			
Contactor	Start	Transition	Run
M	X	X	X
S	X		
R			X

Fig. 3-2 Reduced-voltage starting.

necessary. Belt drives on heavy loads are apt to have excessive slippage unless the torque is applied slowly and evenly until full speed is reached.

Reduced-voltage starting is accomplished by the use of resistors, autotransformers, or reactors. Many electronic devices are also used to reduce starting current and starting torque. Regardless of the means of reducing the voltage, it must be designed to match the particular motor to be started. It is not within the scope of this book to go into the design of reduced-voltage starters, but rather to point out the need for starter selection that conforms to the specifications furnished by the motor manufacturer.

Regardless of the method used to provide reduced-voltage starting, it must be kept in mind that the starting torque of the motor is also reduced. If a motor is not capable of starting its load under across-the-line conditions, the application of reduced-voltage starting will only aggravate the situation because of the reduced starting torque. The torque of an induction motor is a function of the square of the rotor current, or approximately the square of the line current. If the starting voltage is reduced by 50 percent, the motor current will be reduced to 50 percent of normal, but the torque will be reduced to 25 percent of normal.

Some of the methods of obtaining reduced-voltage starting will result in very little or no acceleration under starting conditions. This requires that the total acceleration occur after full voltage has been applied. The starting current during reduced-voltage conditions will be somewhat less than across-the-line starting current. When full voltage is applied, however, the starting current will be approximately the same as it would have been if the motor had been placed across the line in the beginning. This type of starting is generally referred to as *increment starting* and is used generally to spread the rate of change of current demanded from the line over a longer period of time. Part-winding squirrel-cage and wye-delta squirrel-cage motor starting (Fig. 3-2) generally fall into the increment-start category.

JOGGING AND INCHING

Printing presses, cranes, hoists, and similar equipment require that the motor be started repeatedly for short periods of time in order to bring some part of the machine into a given position. This process is known as *jogging* or *inching*. Even though these terms are often used interchangeably, there is a slight difference in their meaning. If the motor is started with full power in short jabs, it is jogging. If the motor is started at reduced speed so as to let the machine creep to the desired spot, then it is inching.

When jogging service is required, the starter must be derated. For instance, a size 3 starter rated at 30 hp, 230 V three-phase normal duty should be derated to 20 hp for jogging duty. The manufacturer's literature should be consulted for ratings of starters in jogging service. The nonjogging and jogging ratings shown in Tables 3-1 and 3-2 were established by NEMA.

3-2 ACCELERATION CONTROL

Squirrel-cage motors do not generally lend themselves very well to speed or acceleration control. There are special types of squirrel-cage motors designed for two-, three-, or four-speed applications. These types of multispeed squirrel-cage motors do not have a true variable speed but rather several definite speeds which may be

TABLE 3-1
Ratings for Polyphase Single-speed Full-voltage Magnetic Controllers for Nonplugging and Nonjogging Duty

Size of Controller	Continuous Current Rating A	Three-Phase Horsepower at 200 V	230 V	460/475 V	Service-limit Current Rating A
00	9	1½	1½	2	11
0	18	3	3	5	21
1	27	7½	7½	10	32
2	45	10	15	25	52
3	90	25	30	50	104
4	135	40	50	100	156
5	270	75	100	200	311
6	540	150	200	400	621
7	810	—	300	600	932
8	1215	—	450	900	1400
9	2250	—	800	1600	2590

TABLE 3-2
Ratings for Polyphase Single-speed Full-voltage Magnetic Controllers for Plug-Stop, Plug-Reverse, or Jogging Duty

Size of Controller	Continuous Current Rating A	Three-Phase Horsepower at 200 V	230 V	460/475 V	Service-limit Current Rating A
0	18	1½	1½	2	21
1	27	3	3	5	32
2	45	7½	10	15	52
3	90	15	20	30	104
4	135	25	30	60	156
5	270	60	75	150	311
6	540	125	150	300	621

used as is or in steps. When adjustable speed is required, the wound-rotor motor with secondary control or the adjustable-speed ac commutator motor is probably the most appropriate.

Manual control of acceleration or speed may be accomplished with multispeed squirrel-cage motors by having the operator close the proper contactor as determined by the load or speed requirements. With wound-rotor motors, the secondary or drum controller is moved as needed to give the desired speed.

Automatic control of acceleration may be accomplished by several methods. Probably the simplest is *definite-time control.* With this method, a time-delay relay is used for each step or speed. When the motor is started in its lowest speed, the first time-delay relay is energized. When this relay times out, it energizes the second contactor, increasing the speed to its second step. This process may be carried through as many steps as necessary to give the speed and acceleration desired. The chief disadvantage of this method is that it is not affected by the conditions of the machine, its load, or the motor current.

For machines or equipment that cannot stand full torque until the load has reached a given speed, the system called *current-limit control* should be used to accelerate the motor. In this system each step is brought in by a current relay which will not close the circuit for its speed until the current has dropped to a safe value. The current, of course, will not drop until the motor and the load are running at nearly the same speed. This system is very well suited for belt or gear drives with high-inertia loads. This method of control must be designed according to application, and the relays must be set for the machine used, and its requirements. For this reason, it is not available as a standard stock controller.

Another system of acceleration control is *slip-frequency control.* This system is used on wound-rotor motors and is also used to energize the field of synchronous motors.

Because the secondary voltage and frequency of wound-rotor motors are inversely proportional to the speed, a frequency-sensitive relay may be used to energize each progressive step or speed. One disadvantage of slip-frequency control is that it must be started on the first, or lowest, speed.

Several features must be kept in mind when selecting a method of acceleration control. Manual control is sensitive only to the operator's reaction. Definite-time control is sensitive only to the lapse of time. Current-limit control is sensitive to the load on the motor. Slip-frequency control is sensitive to the speed of the motor.

It is quite possible that with certain specific applications a combination of any two or more of the above-mentioned control system functions might be needed. A controller combining such functions could be custom-built.

Single-speed squirrel-cage motors are generally started with across-the-line magnetic starters. A multispeed squirrel-cage motor, however, requires a controller that is built for its particular windings.

Two-speed motors may either have two separate stator windings or, when of the consequent-pole type (Fig. 3-3 on the next page), have only one stator winding. The two speeds are obtained with the consequent-pole motor by regrouping the coils to give a different number of effective poles in the stator. This type of motor gives a 2 to 1 speed ratio. Three-speed motors usually have one winding for one speed and a second winding that is regrouped to give the other two speeds. Four-speed motors usually are wound with two windings which are regrouped to give two speeds each. The number of contactors, the order in which they close, and the number and types of overload units required depend on the method of obtaining the various speeds.

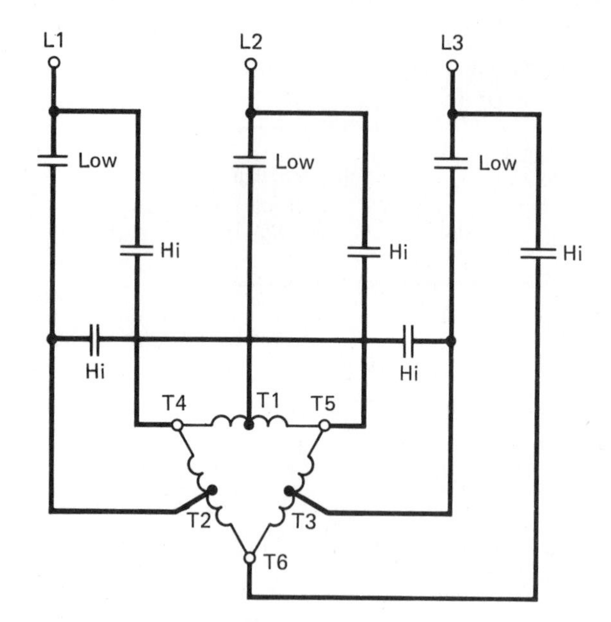

Fig. 3-3 Two-speed squirrel-cage motor and controller. High-speed, four-pole; low-speed, eight-pole.

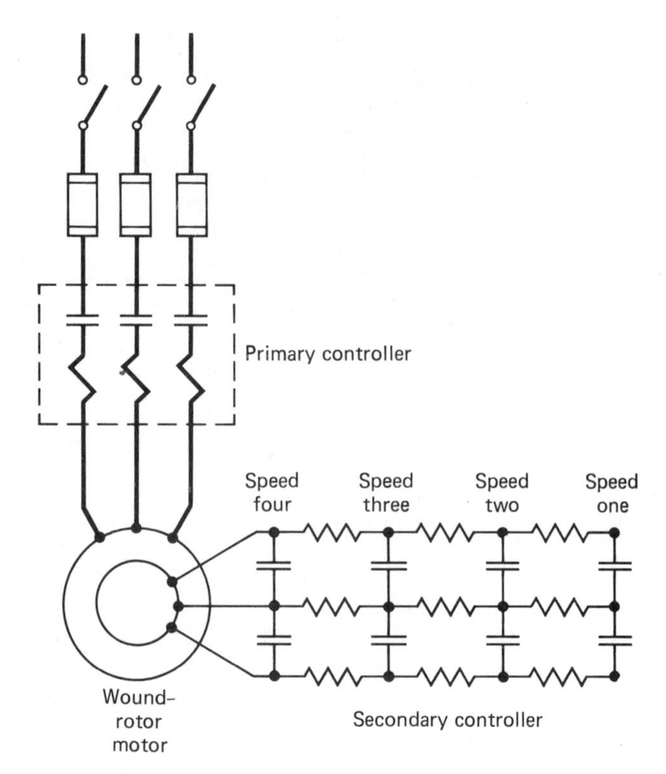

Fig. 3-4 Wound-rotor motor and controller.

3-3 STARTING WOUND-ROTOR MOTORS

The wound-rotor motor is essentially a squirrel-cage motor with definite windings instead of short-circuited bars on the rotor. By the introduction of resistance in series with the rotor windings, using a drum controller or contactors, the speed may be controlled in any number of steps. The most common method of starting wound-rotor motors is by a manual or magnetic full-voltage starter on the primary, interlocked to the secondary controller (Fig. 3-4). The secondary controller may be a manual drum controller, a motor-driven drum controller, a liquid rheostat, or a magnetic contractor designed for secondary control. The secondary controller may be for starting service only and have only two or three steps, or it may be for speed control as well and have any number of steps.

It is necessary that there be an interlock between the primary and secondary controllers that will prevent the motor from starting without all the resistance in the secondary circuit, unless the motor and the machine are designed to start at any speed.

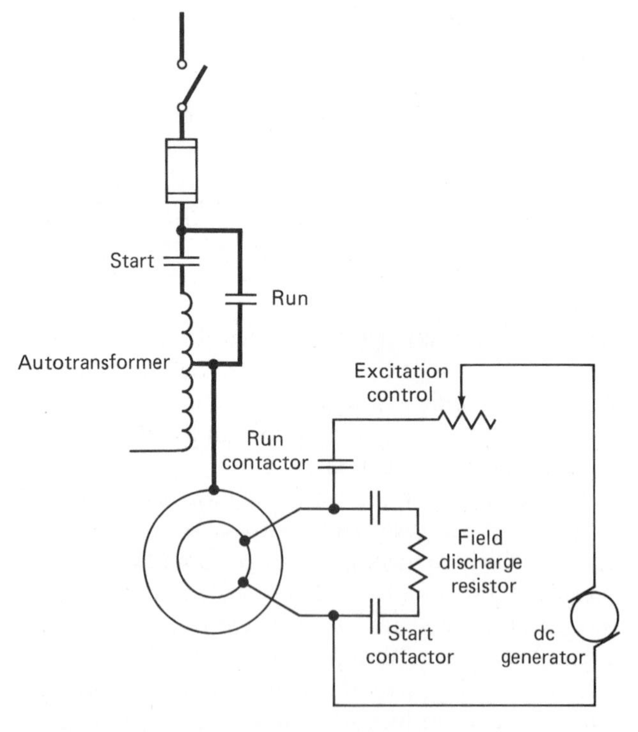

Fig. 3-5 Synchronous motor and controller.

3-4 STARTING SYNCHRONOUS MOTORS

The synchronous motor starts as a squirrel-cage motor with a resistor connected across the field winding to dissipate the power generated in the winding (Fig. 3-5). Usually the stator controller is a reduced-voltage starter combined with a slip-frequency or field-application relay to apply dc voltage to the rotor at about 95 percent of synchronous speed. The slip-frequency relay must also remove field excitation and connect the field resistor if the motor should pull out of step. If the excitation is not removed, the stator winding will be subject to damaging current. The synchronous motor should be provided with an incomplete-sequence relay to protect the starting winding if the starting sequence should not be completed. Provision must be also made to adjust the field excitation.

While the above description of the starting of a synchronous motor may seem to be oversimplified, it is intended to be a general description only of synchronous motors. For a specific application of a definite type of synchronous motor, the manufacturer's literature on the motor should be consulted. Many synchronous motors are designed for specific applications and vary somewhat from this general outline for starting in that they require additional steps or equipment.

3-5 SELECTION OF STARTING CONTROLLERS

There are several points that must be considered when selecting starting controllers. You should ask the following questions when selecting a controller:

1. Is it designed for the type of motor to be used?
2. Does the motor require reduced-voltage starting?
3. Is speed control needed?
4. Does the controller offer all the types of protection that will be needed?
5. Are the line and control voltages and frequency correct?

Analyze the needs of the machine and the motor before selecting any controller, and avoid costly mistakes.

3-6 OVERLOAD PROTECTION

Overloading a motor may be due to mechanical or electrical causes; therefore, overload protection must apply to either. The current a motor draws from the line is directly proportional to the load on the motor, and so if this current is used to activate the overload protection device, the machine as well as the motor will be protected.

Overload protection of multiphase motors is achieved in almost all controllers by placing heating elements in series with the motor leads (Fig. 3-6). These heating elements activate electrical contacts, which open the coil circuit when used on magnetic controllers. When used on manual starters or controllers, the heating elements release a mechanical trip to drop out the line contacts. Older controllers use two overloads, while newer units are required to have three units.

The overload relay is sensitive to the percentage of overload; therefore, a small overload will take some time to trip the relay, whereas a heavy overload will cause an almost instantaneous opening of the circuit. The overload relay does not give short-circuit protection, however. It is quite possible that under short-circuit conditions the relay might hold long enough to allow considerable damage to the motor and other equipment.

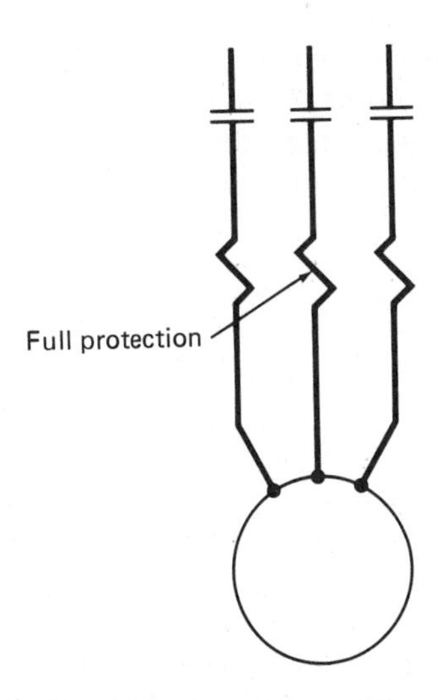

Fig. 3-6 Overload heating elements for multiphase motors.

It would be impossible to overstress the necessity of proper selection of overload protective equipment. The manufacturer's rating of running current for a specific motor should be adhered to in the selection of the heating elements for overload relays. The all-too-frequent practice of increasing the size or rating of the heater element beyond the value called for is probably the greatest single cause of motor failure in industrial plants today. When a motor is tripping its overload units, a careful check of the actual current drawn should be made in order to determine whether the fault lies in the overload-protection device or in the motor itself actually drawing excessive current. Should the motor be found to be drawing excessive current, then it must be determined whether this is caused by mechanical overload or by defective windings within the motor itself. Many times, today's heavy production schedules require that the operator demand more from the machine than its motor is capable of producing. The practice of increasing the allowable current through the overload units will only hasten the time when a shutdown of the equipment becomes necessary in order to rewind or replace the motor.

When one phase of a motor circuit fails, the motor is subject to what is commonly called *single phasing.* This condition causes an excessive current to flow in the remaining motor windings and leads. In most cases, this excessive current will cause the overload units to trip, thus disconnecting the motor from the line and preventing a burnout of its windings. With certain loads, it is possible for the motor to run single phase and burn out its windings, even though there are two overload units in the control device. Modern devices are required to have three overload units.

3-7
SHORT-CIRCUIT PROTECTION

Squirrel-cage motors draw up to 600 percent of full-load current when starting, and some designs draw up to 800 percent or more. The overload relay is designed to pass these large currents for a short period of time. The circuit feeding a motor must have either a fused disconnect or a circuit breaker ahead of the motor to give short-circuit protection since the motor controller is not designed to interrupt short-circuit currents greater than 10 times its rating. Class J fuses (or current limiters with the same characteristics) provide the best limiting of peak short-circuit current.

The dual-element fuse (Fig. 3-7) is so constructed that one of its elements consists of a fuse link. This link will open very rapidly under short-circuited conditions. The second element in this type of fuse consists of a thermal element which allows considerable time lag before breaking the circuit. The net result of using the dual-element fuse on motor circuits is to give short-circuit protection through the fuse link and yet allow a degree of overload protection in the thermal element. This type of fuse is used extensively with small fractional-horsepower motors which have built-in thermal protectors.

The use of thermal-magnetic circuit breakers for short-circuit protection offers a degree of time lag for starting loads in that such breakers have a thermal device which requires some time lag to open on currents less than 20 times rating. This time lag is inversely proportional to the amount of current. The larger the overload, the shorter the time required to open the circuit. Such breakers also have a magnetic trip element for instantaneous trip on currents greater than about 25 times rating.

3-8
LIMIT PROTECTION

Limit protection, as its name implies, limits some function of the machine or its driving motor. The most common type of limit control is used to limit the travel of a cutting tool, table, or other part of a machine tool. When the cutter reaches a predetermined setting, it will activate a limit switch, causing the motor to reverse and the machine to return to the other extreme of travel. There are other types of limit protection, such as limits for over- or underspeed. There are also limit controls

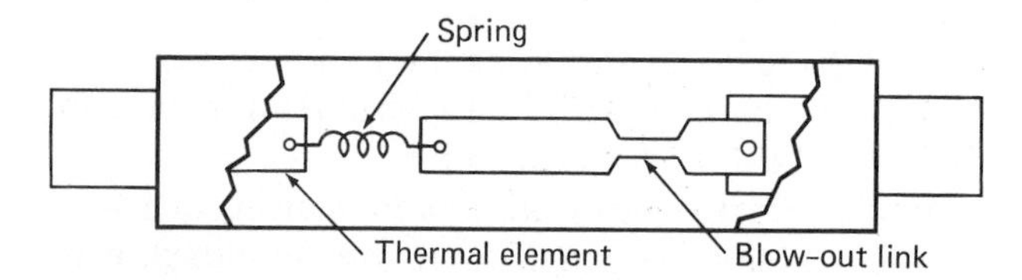

Fig. 3-7 Dual-element cartridge fuse.

which do not reactivate the machinery at all but merely stop the motor until corrections have been made by the operator.

This type of protection is accomplished by the use of limit switches, which will be discussed more fully later. Basically, a limit switch is merely a switch with a mechanical bumper, or arm, that will allow some action of the machine to throw the switch mechanism. Limit switches are among the most frequently used control devices on machines today.

3-9
INTERLOCKING PROTECTION

Interlocking is preventing one motor from running until some other motor or sequence has been activated. A good example of this is found in air conditioning. If the compressor were to run without the cooling-tower pump running, the compressor head pressure would rise dangerously and the compressor would be damaged. To prevent this from occurring, the compressor is interlocked to the pump so that it cannot start until the pump is running.

Interlocking may be done electrically or mechanically or both ways. Reversing starters built with both starters in the same box generally have a mechanical interlock and sometimes have an electrical interlock. When the two units to be interlocked are in separate boxes (Fig. 3-8), electrical interlocking is a necessity.

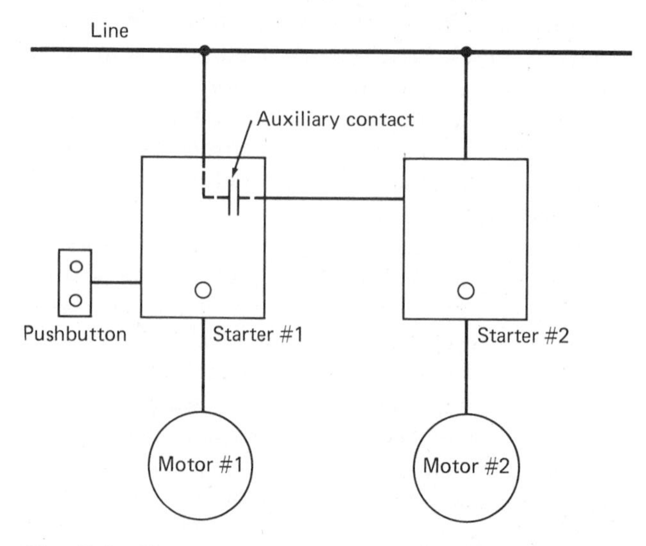

Fig. 3-8 Two motor starters interlocked so that motor 1 must run first and automatically start motor 2.

Electrical interlocking is accomplished by connecting an auxiliary contact on one starter in series with the coil circuit of the second starter.

While we have discussed interlocking in relation only to motors, it is well to realize that interlocking is used in all types of control wiring—whether that involved in starting of motors or in closing valves in a process-control installation. The use of interlocking control ensures that the entire control system operates in the proper sequence.

3-10 SPEED CONTROL

Squirrel-cage motors do not lend themselves to any system of continually variable speed control but rather may be had in only two, three, or four speeds. These limited-speed motors were discussed under acceleration control. The most versatile ac motor, as far as speed control is concerned, is the wound-rotor motor, which is sometimes referred to as the *slip-ring motor*. By the use of secondary control, this type of motor can have as many steps of speed as are desirable. Methods of controlling wound-rotor motors will be discussed in greater detail when we take up their controllers.

Another type of ac motor which gives excellent speed control is the ac commutator-type motor.

There are, in total, four general types of speed control.

CONSTANT-SPEED CONTROL

Many machines require only a reduced speed for starting and then a constant speed for operation. This type of speed control may be accomplished by using a reduced-voltage starter on a squirrel-cage, wound-rotor, or synchronous motor. It must be kept in mind, however, that reduced-voltage starting also invariably gives reduced starting torque.

VARIABLE-SPEED CONTROL

Variable speed is required of motors that must operate at several different speeds. This type of control may best be accomplished by the use of a wound-rotor motor with a secondary controller or a commutator-type ac motor. This type of control requires that a speed change be made under load.

MULTISPEED CONTROL

This type of control differs only slightly from variable-speed control in that it usually does not require speed changes under load. The multispeed squirrel-cage motor is well suited for this type of service.

PREDETERMINED-SPEED CONTROL

With this type of control, the machine is accelerated in steps to a preset operating speed. Both multispeed squirrel-cage and wound-rotor motors are suitable for this type of service.

Any of the above types of speed control may be wired so that the operator may vary the sequence of operation. Quite frequently, however, the control system compels the operator to start at a particular place in the sequence and follow it through without variation. When the control is of this type, it is known as *compelling-sequence control*. This term also applies to any type of control in which the operator has to follow a set sequence of operation.

3-11 UNDERVOLTAGE PROTECTION AND RELEASE

The line voltage supplying motor circuits may drop to dangerously low values or may be shut off at almost any time. When the voltage is too low, severe damage may be done to the motor windings if they are allowed to remain on the line. While some large motors employ a special voltage relay to disconnect the motor under low-voltage conditions, most smaller motors depend on the overload units to open the starter contacts.

If the control circuit allows the motor to restart when the power is restored to its proper value, it provides a form of protection referred to as *undervoltage release*. Using maintained-contact pilot devices on magnetic starters provides this type of protection.

If the type of protection used requires that the motor be restarted manually, then the protection is referred to as *undervoltage protection*. Using momentary-contact pilot devices on magnetic starters gives this type of protection.

Whether to use undervoltage protection or undervoltage release depends upon the requirements of the machine. Ventilating fans, unit heaters, and many other small units may operate more effectively with undervoltage release, which doesn't require restarting these units. In any machine where there is the slightest risk that the machine or operator might be injured by an unexpected start, undervoltage protection should, by all means, be used.

3-12 PHASE-FAILURE PROTECTION

When a three-phase motor has its current interrupted on one phase, the condition is referred to as *single phasing.* Ordinarily, the overload units will trip the starter and remove the motor from the line. For each type of motor, however, some kinds of loading might cause the motor to burn up even though excessive current didn't flow through the overload units. At about 65 percent of load most squirrel-cage motors burn up. For small motors the risk is generally considered too slight to warrant the cost of additional protection. For large motors a voltage relay is placed across each phase, and its contacts are connected in series with the holding coil of the starter. Failure of one phase will drop the starter out at once.

The use of three overload relay units on the starter gives what is generally considered adequate phase-failure protection for most motors of 200 hp or less. Three overload relay units are now required by the National Electrical Code.

3-13 REVERSE-PHASE PROTECTION

Some machines could be severely damaged when the motor runs in reverse, as would occur with a reversal of phasing. While this type of situation is not commonly protected, it can be to prevent costly damage.

Reverse-phase protection can be accomplished by using a phase-sensitive relay with its contacts in series with the holding coil of the starter.

3-14 INCOMPLETE-SEQUENCE PROTECTION

When reduced-voltage starting is used on a motor, there is a danger that the motor windings or the autotransformer, or both, might be damaged through prolonged operation at reduced voltage. To prevent this and to ensure the completion of the starting cycle, a thermal relay is placed across the line during starting. This relay is so designed and connected that prolonged starting will cause the thermal unit on the relay to open its contacts and drop out the starter. This type of protection is also necessary on synchronous motor controllers.

Another method of obtaining incomplete-sequence protection for starting of motors is using a timing relay which will disconnect the motor if it has not completed its starting sequence in the predetermined length of time.

3-15 STOPPING THE MOTOR

There are several factors that must be considered in stopping a motor. On some machines all that is necessary is to break the motor leads and let the motor coast to rest. Not all machines can be allowed to coast, however. For instance, a crane or hoist not only must stop quickly, but also must hold heavy loads. Other machines, such as thread grinders, must stop very abruptly, but need not hold a load.

The method of stopping may be either manual or automatic. Automatic stopping is accomplished by the use of limit switches, float switches, or other automatic pilot devices. Manual stopping is controlled by pushbuttons, switches, or other manually operated pilot devices.

The most common method of stopping is merely to remove the motor from the line by breaking the circuit to the starter coil, if it is a magnetic starter, or by tripping the contacts of a manual starter with the STOP button.

For motors that must be stopped very quickly and accurately but which do not need to hold a load, probably the most widely used method is known as the *plugging stop.* This is accomplished by using an automatic plugging switch or a plugging pushbutton and a reversing starter. With either of these units, the motor starter is dropped out and then momentarily energized in the reverse direction. The momentary reversal plugs the motor to an abrupt stop. This type of stopping will not do for cranes or hoists because it will not hold a load.

When dealing with such equipment as cranes and hoists, we must consider that the load has a tendency to turn the motor. This is known as an *overhauling load.* When ac motors are used, they frequently are of the wound-rotor type and stopping is preceded by a slowing of the motor through one or more steps. This slowing helps to nullify the overhauling load. As the motor is dropped from the line, a mechanical brake (Fig. 3-9) automatically locks the motor shaft connected to the load.

When dc motors are used, the overhauling load is slowed down by the use of dynamic (regenerative) braking; then a friction brake is applied.

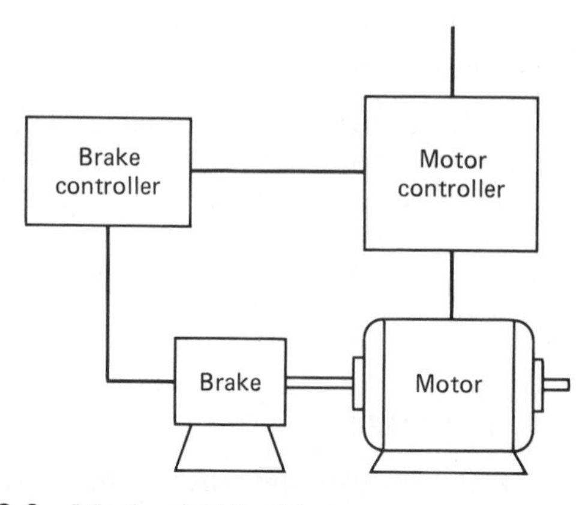

Fig. 3-9 Mechanical braking.

Synchronous motors are sometimes stopped by the use of dynamic braking. This is accomplished by removing the motor from the line and placing a resistance across the motor leads, thus making an ac generator out of the motor. The resistor presents a heavy load to the generator, causing it to come to a rapid stop. Care must be taken to use a resistor unit capable of dissipating the power generated while stopping the motor. It must also be noted that this type of stop cannot be used for frequent stopping because the resistor units must have time to cool between operations.

3-16 STARTING DC MOTORS

Any device used to start a dc motor over about ¼ hp must provide some means of limiting the starting current to approximately 150 percent of the full-load value. An ac motor offers a high impedance to the line, which limits the starting current. Before its armature rotates, the dc motor offers only the low resistance of the armature to limit inrush current.

Once rotating, however, the armature winding begins to cut the flux produced by the field, and a voltage is generated in the armature coil. The voltage generated in the armature coil is opposite in polarity to the applied voltage and is referred to as *counterelectromotive force* or *counter-emf*. The value of the counter-emf increases with speed, until at full speed it is 80 to 95 percent of the applied voltage.

Armature current is calculated by subtracting the counter-emf from the applied voltage and dividing by the armature resistance. When the armature is at rest, the counter-emf is at zero. Therefore, the armature current is equal to the applied voltage divided by the armature resistance, which is very low, generally 1 to 2 ohms (Ω) or less.

If we assume an armature resistance of 0.85 Ω and an applied voltage of 110 V, the initial inrush current would be 129.4 A. At full speed the counter-emf would be about 100 V and would limit the current to 11.8 A. The starting current should be limited to about 150 percent of full-load current by connecting a resistor of 5.35 Ω in series with the armature.

The series resistance must be removed in steps as the acceleration of the motor produces an ever-increasing counter-emf and reduces the resistance required, until at rated speed all resistance is removed. A dc motor develops its greatest power when the counter-emf is at its maximum value.

When series or compound motors are used, the starting resistance is connected in series with the armature and series field. The shunt motor has no series field, and therefore the resistance is connected in series with the armature only (Fig. 3-10).

3-17 SPEED CONTROL OF DC MOTORS

Direct-current motors are used chiefly because of their speed-control characteristics, which make them the best suited for many drive requirements.

When a dc motor has its rated armature voltage and rated field voltage applied, it runs at its base speed. Speeds below base speed (underspeed) are achieved by maintaining the field voltage at rated value and reducing the armature voltage. Speeds above base speed (overspeed) are achieved by maintaining the armature voltage at rated value and reducing the field voltage.

The series motor's speed is controlled by the amount of resistance connected in series with the armature and series field. The resistors used for speed control must be rated for continuous duty rather than starting duty, since they are in the circuit whenever the motor is used at less than base speed.

The most common controller for speed control of series motors is the drum controller with heavy-duty resistance grids.

Shunt and compound motors lead themselves well to applications where speed control is a major consider-

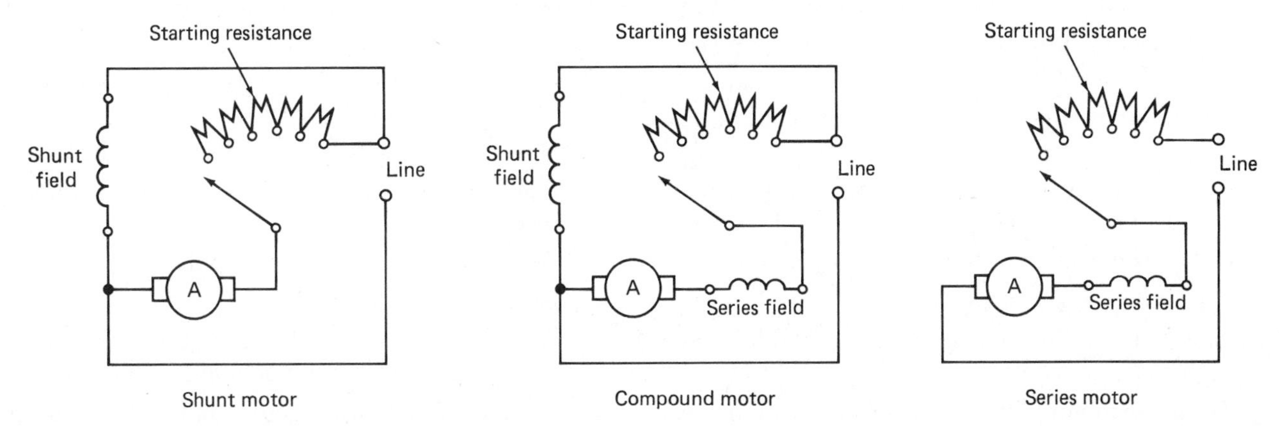

Fig. 3-10 Connections for dc starting resistance.

ation. When the speed desired is over the base speed (overspeed control), resistance is added in series with the armature.

The most popular manual controller for speed control of shunt and compound motors is the combination four-point starter and speed controller (Sec. 4-15).

SUMMARY

Students beginning the study of motor control may feel that they will never learn all the functions that might be performed in the control of a motor or other devices. The advances made in this field are so rapid and far-reaching that new ones are developed almost daily. When analyzed completely, however, most motor controls are merely variations of the basic functions set forth in this chapter. It must be kept in mind that intelligent servicing, development, or installation of control equipment depends upon a thorough understanding of the requirements of the machine and the characteristics of the motor.

REVIEW QUESTIONS

1. What determines whether a motor uses full-voltage starting or reduced-voltage starting?
2. What is the most popular starter for motors up to 600 hp?
3. What is a demand meter used for?
4. What determines whether a motor uses a full-voltage starter or a reduced-voltage starter?
5. What are some of the factors that would affect the selection of a motor for an industrial application?
6. What is the most popular starter for motors up to 600 hp and not over 600 V?
7. For what reasons would reduced-voltage starting be used?
8. How is reduced-voltage starting accomplished?
9. What motor should be used to achieve reduced-voltage starting?
10. What is the function of a secondary-control device?
11. The torque of an induction motor is a function of what?
12. What is meant by increment starting?
13. What do jogging and inching mean?
14. Describe the speed characteristics of a multispeed squirrel-cage motor.
15. What ac motor is most practical in providing adjustable speed?
16. What methods provide automatic acceleration control of motors?
17. What is the chief advantage of definite-time control?
18. What automatic acceleration method would be best suited for belt or gear drives with high-inertia loads?
19. What is the relation of the secondary voltage frequency to the speed of a wound-rotor motor?
20. What is a consequent-pole motor?
21. How are three speeds provided by a squirrel-cage induction motor?
22. What does resistance of various sizes accomplish when connected in series with the rotor windings of a wound-rotor motor?
23. When would the primary control and secondary control for a wound-rotor motor be interlocked?
24. How does the synchronous motor start?
25. What is the purpose of the field-discharge resistor in Fig. 3-5?
26. What is the function of an incomplete-sequence relay in synchronous motor control?
27. How is overload protection of a motor achieved?
28. Why doesn't an overload relay provide short-circuit protection?
29. Explain what single phasing is.
30. Why should a motor starter have three overload relays?
31. How is short-circuit protection of motors provided?
32. What type fuses provide the best current-limiting of peak short-circuit currents?
33. What is the effect of time lag on a thermal magnetic time delay?
34. What is the most common limit protection?
35. What is the purpose of interlocking?
36. How is electrical interlocking accomplished?
37. What is the advantage of a wound-rotor motor over a squirrel-cage motor?
38. How does the wound-rotor motor provide adjustable-speed control?
39. How is variable-speed control accomplished?
40. How does multispeed control differ for variable-speed control?
41. Define compelling-sequence control.
42. Define undervoltage release.
43. What determines whether a motor should have undervoltage release or undervoltage protection?
44. What condition will cause a motor to burn up without excessive current in the overload units?

45. How is phase-failure protection provided and how is it connected to a motor control circuit?

46. How is reverse-phase protection accomplished?

47. Describe incomplete-sequence protection.

48. What is meant by an overhauling load?

49. What is dynamic or regenerative braking used for on a dc motor?

50. How is dynamic braking accomplished on a synchronous motor?

51. What percentage of full-load current should the starting current be for a dc motor?

52. How is the value of the armature current of a dc motor calculated?

53. Define the base speed of a dc motor.

54. How are speeds above base speed achieved for dc motors?

55. How is overspeed and underspeed control accomplished for dc shunt and compound motors?

4
Control Components

As soon as it has been decided what functions of control are needed for a machine, the components or devices to perform these functions must be selected. This selection should be made with care. For instance, if a float switch is needed and its duty cycle is only a few operations per day for a year or so, one of the cheaper competitive units might be satisfactory. If, however, the duty cycle is a few hundred operations per day on a permanent basis, then the best-quality unit available should be used. The small savings gained through the use of inexpensive components are usually soon offset by costly shutdowns due to failure of the components. This chapter will discuss the basic types of control components, how they work, both electrically and mechanically, and at least some of the functions they can perform. Students are urged to obtain manufacturers' catalogs on control components to use with this chapter. The more familiar students become with different manufacturer's control components and their operation, the better prepared they will be to service control equipment on the job.

4-1
SWITCHES AND BREAKERS

The one component common to all but the very smallest motors is a switch or circuit breaker for disconnecting the motor. There are two types of switches in general use on motor circuits. The first of these is the isolation switch, which is rated only for voltage and amperage. This type of switch has no interruption-capacity rating and must not be opened under load. Quite often the switch used for this purpose is a nonfused type. Such switches are permitted by the National Electrical Code only for motors of 2 hp or less and over 100 hp.

The second type of switch is a motor-circuit switch, which is capable of interrupting the motor current under normal overloads. This type of switch is rated for horsepower and, when used according to this rating, is capable of being used as a starting switch. When used for disconnecting and protecting circuits, this switch must be of the fused type.

A circuit breaker offers the same disconnecting features of a switch and the circuit protection of a fuse. The breaker operates on a thermal-released latch so that it may be reset and used again after an overload. Being built all in one unit and offering short-circuit protection, as well as serving as a disconnect, makes this unit more compact than a switch-and-fuse combination. Switches do, however, offer more visual indication that the circuit is open.

Switches and breakers may perform the functions of start, stop, overload protection, and short-circuit protection.

4-2
CONTACTORS

The contactor itself is not generally found alone in a motor-control circuit. It is, however, the basic unit upon which the motor starter is built. Contactors are used to perform the functions of start and stop on many heavy loads, such as electric furnaces, signs, and other types of equipment that do not require running protection.

Perhaps the best way to describe a contactor would be to say that it is a magnetically closed switch. It consists of one set of stationary contacts and one set of movable contacts which are brought together by means of the magnetic force of an electromagnet. The vast majority of contactors use an electromagnet and contact arrangement that falls into one of two general types. The first of these is the clapper type (Fig. 4-1).

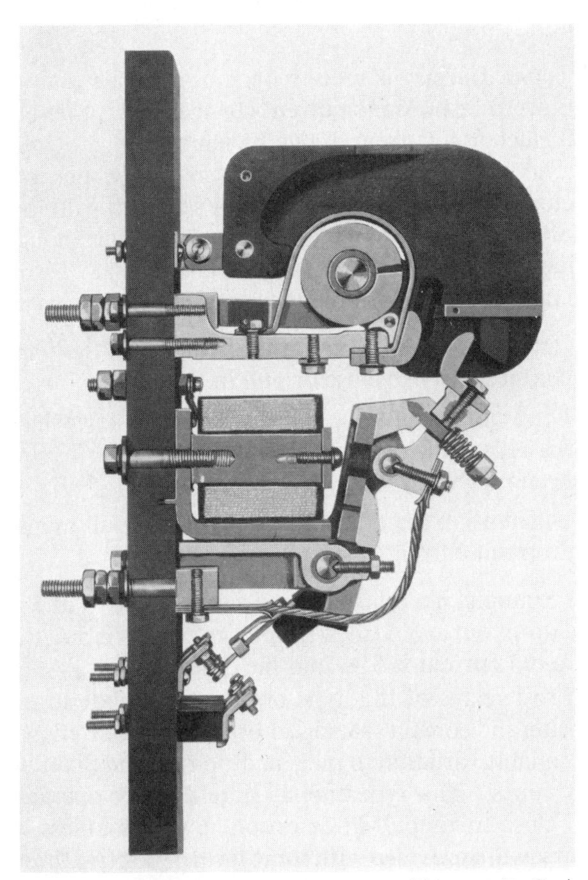

Fig. 4-1 Clapper-type contactor. *(Square D Co.)*

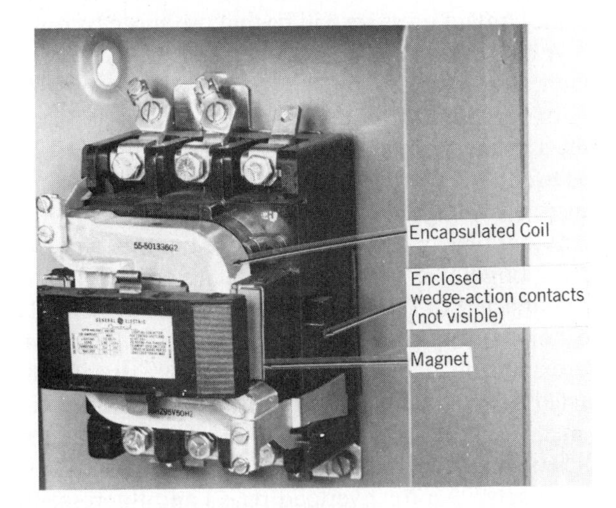

Fig. 4-2 Solenoid-type contactor. *(General Electric Co.)*

The movable contacts are fastened to the pole pieces of the magnet and hinged so that they swing more or less horizontally to meet the stationary contacts.

The second is the solenoid type (Fig. 4-2). On this contactor the movable contacts are coupled to the movable core of a magnet. When the electromagnet is energized, the movable core is pulled to the stationary core, thus closing the contacts.

Regardless of whether the contactor is of the clapper or the solenoid type, the contacts themselves are broken by the pull of gravity or the force of a spring when the electromagnet is de-energized.

All that is necessary electrically to operate the contactor is to provide a voltage of the proper value to the coil of the electromagnet. When the voltage is switched on, the contacts close; when the voltage is switched off, the contacts open.

4-3 RELAYS

Automatic control circuits almost invariably contain one or more relays, primarily because the relay lends flexibility to the control circuits. The relay is by design an electromechanical amplifier.

Let us consider for a moment the meaning of the word *amplify*. It means to enlarge, increase, expand, or extend. When we energize the coil of a relay with 24 V and the contacts are controlling a circuit of 460 V, we are amplifying the voltage through the use of a relay. Relay coils require only a very low current in their operation and are used to control circuits of large currents. So again they amplify the current. The relay is inherently a single-input device in that it requires only a single voltage or current to activate its coil. Through the use of multiple contacts, however, the relay can be a multiple-output device which amplifies the number of operations controlled by the single input.

Suppose we have a relay whose coil operates on 115 V at 1 A, and the contacts of this relay control three separate circuits operating at 460 V and 15 A each. This relay then becomes a power amplifier in that it controls considerably more power in its output circuits than it consumes in its input circuit. It also becomes an amplifier in terms of the number of circuits because its single input controls three separate outputs.

Relays are generally used to accept information from some form of sensing device and convert it into the proper power level, number of varied circuits, or other amplification. The sensing devices are commonly called pilot devices and are designed to sense or detect such things as current, voltage, overload, frequency, temperature, and other quantities. The proper type of relay used in a given circuit is determined by the type of sensing device used to transmit the information to it. For instance, a voltage-sensing device must be connected to a voltage relay, and a current-sensing device must activate a current relay. Each of these will be discussed individually.

VOLTAGE RELAY

This type of relay (Fig. 4-3) is probably the most common because it lends itself to so many applications. The voltage relay is merely a small contactor which opens or closes its contacts, depending on whether they are normally closed or open whenever the proper voltage is applied to its coil. It is available with as many contacts either normally open or normally closed as needed. Voltage relays are used frequently to isolate two or more circuits controlled from one source (Fig. 4-4) or when the control voltage is different from line voltage.

It must be remembered that while a voltage relay is not a primary control device, it does require a pilot device to operate.

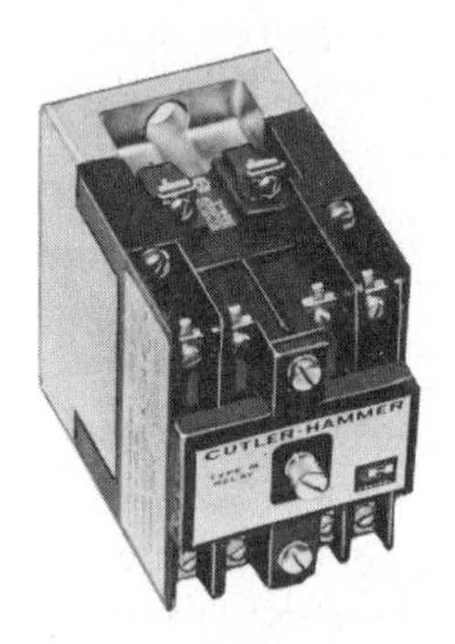

Fig. 4-3 Voltage relay. *(General Electric Co.)*

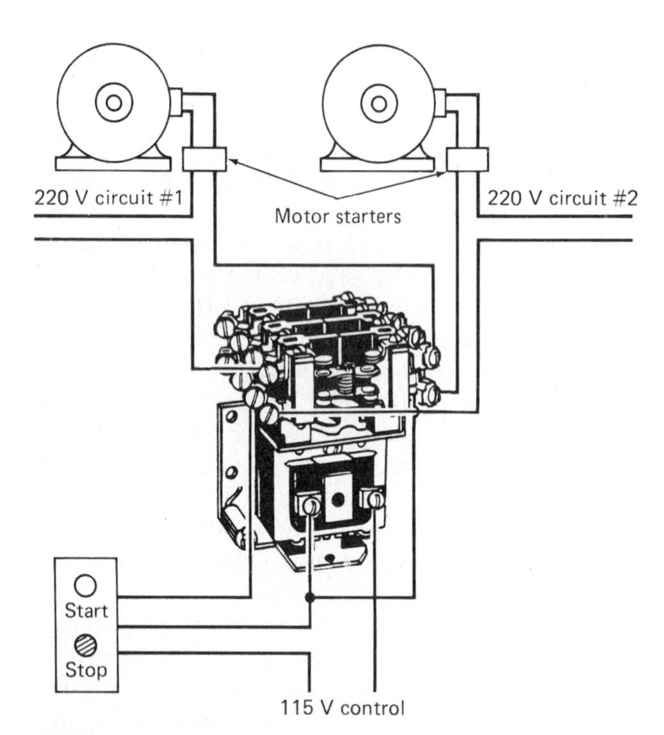

Fig. 4-4 The basic use of voltage relays to control a circuit.

CURRENT RELAY

This type of relay is used to open or close a circuit or circuits in response to current changes in another circuit, such as a current drawn by a motor.

The current relay is designed so that when connected in series with the circuit to be sensed, it will close after the current through its coil reaches a high enough value to produce the necessary magnetic flux. Terms are used in connection with current relays:

Pull-in current: the amount of current in the relay coil necessary to close or pull in the relay

Drop-out current: the value of current below which the relay will no longer remain closed after having been pulled in

Differential: the difference between the pull-in and drop-out currents.

For example, if a relay is energized or pulled in at 5 A and drops out at 3 A, then the pull-in current is 5 A, the drop-out current is 3 A, and the differential is 2 A.

Most relays of this type are provided with spring tension and contact spacing adjustments, which allow a reasonable variation in pull-in, drop-out, and differential values. This type of relay should not be operated too close to its pull-in or drop-out values unless its contacts are provided with some form of positive throw device. This is important because the amount of contact pressure depends upon the difference in actual current and pull-in current. For example, when the above relay is operated with only 5.01 A flowing through the coil, the contact pressure will be that produced by only 0.01 A of current.

Generally, true current relays are used only on circuits of very low current. For larger-current applications, a current transformer is used an its output applied to either a current relay or a voltage relay with the proper coil voltage.

Another type of current relay is the thermal type, in which a bimetallic strip or other device is heated by a coil connected in series with the circuit to be sensed. The bimetallic type depends upon the difference in heat expansion of two dissimilar metals. It is constructed by riveting together two thin strips of dissimilar metals. When the current in the circuit produces sufficient heat, the bimetallic strip expands and releases the contacts. Motor overload relays and fluorescent starters are examples of this type of relay.

FREQUENCY RELAY

The frequency relay is used for field-excitation of synchronous motors and for acceleration control of wound-rotor motors. Most of these units are specially designed for particular applications. One type consists of two balanced coils arranged on a common armature. These coils compare a reference frequency with that of the sensor's circuit. The relay is closed one way when the frequencies are the same or are a predetermined percentage of each other, and is closed the other way when the frequencies differ by a given amount.

TIME-DELAY RELAY

This type of relay is often used for sequence control, low-voltage release, acceleration control, and many other functions.

Essentially, a time-delay relay is a voltage relay with the addition of an air bleed (Fig. 4-5) or a dashpot (Fig. 4-6) which slows down or delays the action of its contacts. This delay in action can be applied when the relay is energized or when it is de-energized.

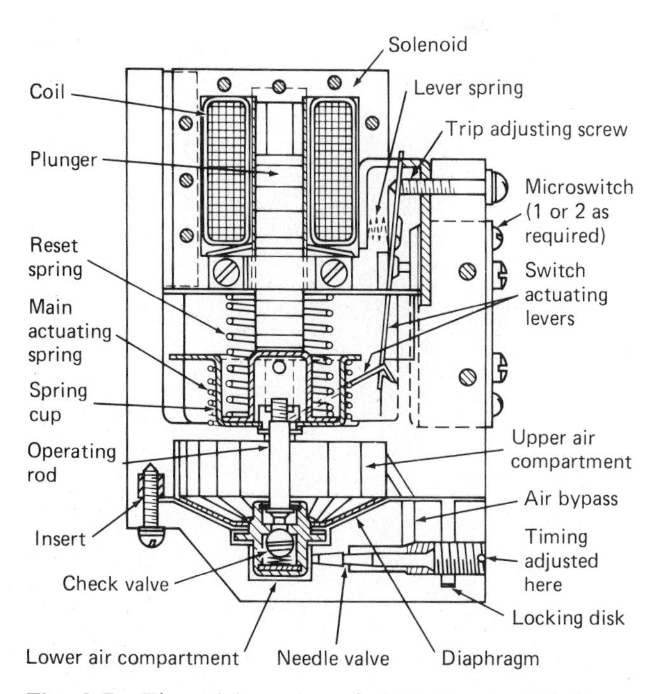

Fig. 4-5 Time-delay relay, air-bleed type. *(Cutler-Hammer Inc.)*

If the delay is to be applied when the relay is energized, it is said to be ON DELAY or *TDOE*. If the delay is to be applied with the relay is de-energized, then it is said to be OFF DELAY or *TDODE*. Both types are provided with an adjuster so that the time delay can be set within the limits of the particular relay. The contacts are always shown in the de-energized position and are either time opening (TO) or time closing (C). These units are built in various sizes depending upon the contact rating needed.

Fig. 4-6 Time-delay relay, dashpot type. *(Square D Co.)*

OVERLOAD RELAY

The overload relay is found on all motor starters in one form or another. In fact the addition of some form of overload protection to an ordinary contactor converts it into a motor starter. This unit performs the functions of overload protection and phase-failure protection in motor circuits. The basic requirement for overload protection is that the motor be allowed to carry its full-rated load and yet prevent any prolonged or serious overload. When a motor is overloaded mechanically, motor current increases, which in turn increases the temperature of the motor and its windings. The same increases in current and temperature are caused by the loss of one phase in polyphase motors or a partial fault in the motor windings. Therefore, to give full overload protection we need only to sense, or measure, the current drawn by the motor and break the circuit if this current exceeds the rated value for the motor.

There are three basic types of overload relays in general use on across-the-line starters. The first is a unit which employs a low-melting-point metal to hold a ratchet (Fig. 4-7*a* on the next page), which, when released, causes the opening of a set of contacts in the coil circuit of the starter. The second type uses a bimetallic strip (Fig. 4-7*b*) to release the trip mechanism and open the coil-circuit contacts.

Regardless of which type of device is used, the overload delay is activated by a heating element placed in series with the motor circuit. The amount of current needed to cause the relay to trip is determined by the size of the heating element used. When used for pro-

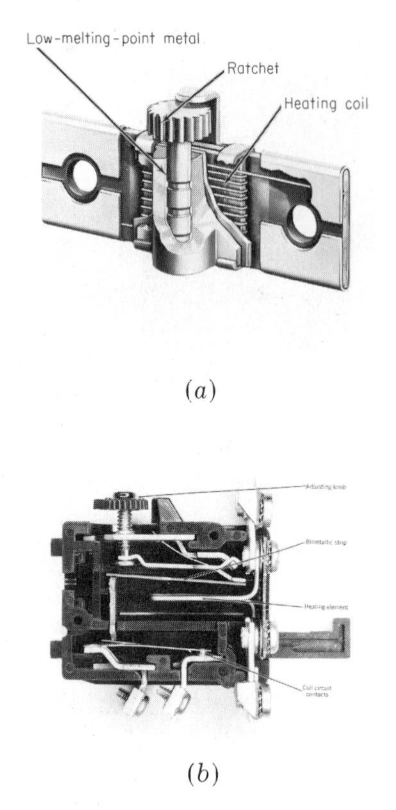

Fig. 4-7 Overload relays. (*a*) Low-melting type. *(Square D Co.)* (*b*) Bimetallic-strip type. *(General Electric Co.)*

tection of small motors drawing low current, a coil of small wire or very thin metal is used as the heating element. On larger motors a heavier coil or strip of metal is used so that the same amount of heat is produced when the rated amount of current flows. Thermal units used in overload relays have an inherent time delay in their action that is inversely proportional to the amount of overload. This should be evident from a study of the curve shown in Fig. 4-8. When the overload is slight, the motor can go on running for some time without tripping the overload unit. If the overload is great, however, the overload relay will trip almost at once, thus removing the power from the motor and preventing damage.

Thermal relays trip on heat and heat alone, and they cannot normally tell whether this heat is from the current to the motor or from the air that surrounds the starter. To offset this, it is sometimes necessary to install oversized heaters in high-temperature locations and undersized heaters in low-temperature locations. Some bimetallic units are designed to compensate for the ambient temperature change. This type of unit is called a *compensated overload relay.*

The third type of overload relay is magnetic (Fig. 4-9). This unit has a magnetic coil so connected that it senses the motor current either by the use of current

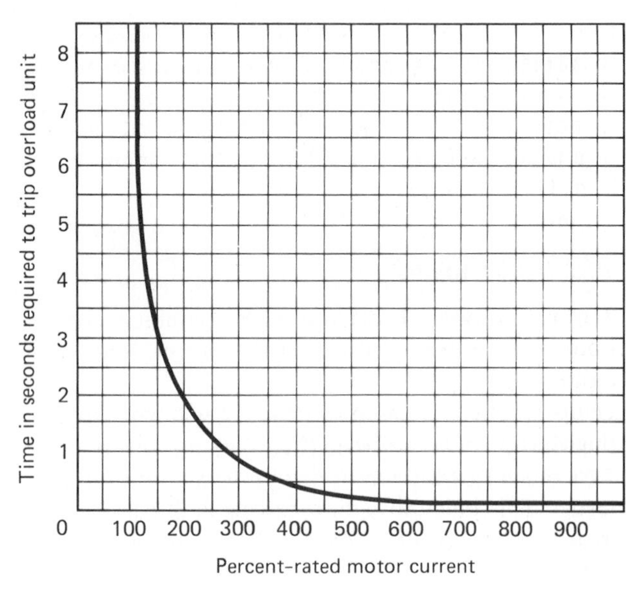

Fig. 4-8 Overload-relay current curve.

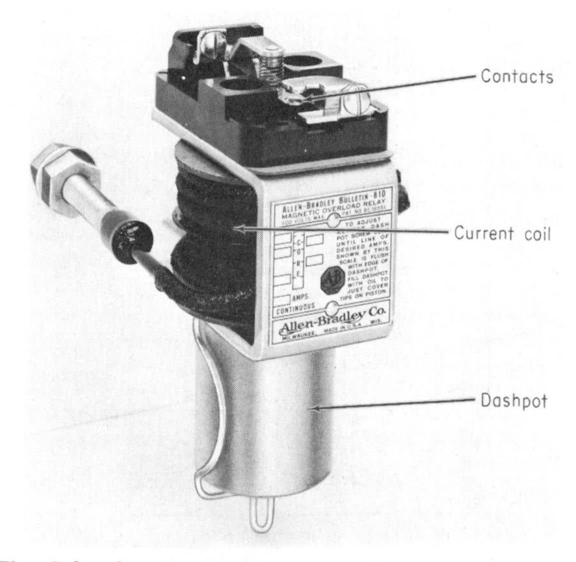

Fig. 4-9 Overload relay, magnetic type. *(Allen Bradley Company)*

transformers or by direct connection. When the current exceeds the rating of the motor, the overload coil lifts the plunger that forms its core and opens the contacts in the control circuit. Magnetic overload relays are generally found only on large motor starters.

Overload relays must be reset after each tripping, either automatically or manually. The automatic-reset type should be used only on equipment for which it is expressly designed. It must never be used where automatic restarting presents a danger to life or equipment. After the overload relay has been tripped, it requires a little time to cool, so that there is some delay before resetting can be accomplished.

4-4 MOTOR STARTERS

A motor starter in its simplest form consists of some means of connecting and disconnecting the motor leads from the line leads, plus overload protection for the motor. Many other refinements are added to this basic unit to achieve the desired degree of control and protection. There are many types and classifications of motor starters. Each of these types draws its name from the method or classification of operation of the motor that it starts. Some of these classifications are manual or automatic, full-voltage or reduced-voltage, single-phase or three-phase, and dc or ac. To describe a particular motor starter, it is necessary to combine several of these terms or classifications. For instance, a particular motor might require a reduced-voltage, automatic, three-phase, ac motor starter. Even this does not completely describe the unit, because it must be of a definite size and be rated for the proper voltage. Then we must know whether it is to be remotely controlled or to have a pushbutton in the cover and many other features. In this chapter, we shall discuss many of these classifications, and the student should keep in mind that any particular starter might well be a combination of several of the types we discuss here.

Keep in mind also that there is a difference between a motor starter and a motor controller. While it is difficult to draw a fine line between them, for the purpose of the discussion in this book a starter consists of a means of connecting the motor to the line and of providing needed protection. By contrast, a controller contains not only the motor starter but at least a major part of the sensing devices and relays necessary for more complete control.

Motor starters are built to NEMA standards. These standards include such things as sizes, so that a purchaser may expect the equipment to be built to handle its rated load. For instance, the size 0 starter in 460-V, three-phase service is rated at 5 hp. At 660 V a size 1 starter is rated at 10 hp; size 2, at 25 hp; size 3, at 50 hp; and size 4, at 100 hp. For lower voltage service, each size starter has a smaller horsepower rating because of the increased current demand by the motor running at a lower voltage.

Also included in the NEMA standards are types of enclosures for starters to satisfy code requirements for atmospheric conditions existing in the place of installation. The NEMA type 1 enclosures are for general-purpose indoor use wherever atmospheric conditions are normal. They are intended primarily to prevent accidental contact with the control apparatus and energized circuits. The NEMA type 3 enclosures are weather-resistant and protect against dust, rain, and sleet in outdoor applications. The NEMA type 4 enclosures are watertight and suitable for outdoor applications on piers and in dairies and breweries. They may be washed down with a hose. The NEMA type 7 enclosures are intended for use around hazardous gases such as are found in oil fields and must satisfy the code requirements for class 1, group D locations. The NEMA type 8 enclosures are also intended for such locations and thus have oil-immersed contacts. The NEMA type 9 enclosures are built for use around dust such as is found in flour mills and are classified as class 2, groups E, F, or G. The NEMA type 11 enclosures are corrosion-resistant enclosures that protect the enclosed equipment against the corrosive effects of gases and fumes by immersing the equipment in oil. The NEMA type 12 enclosures are dust-tight industrial enclosures and are designed for use where the enclosure is required to provide protection against dirt and oil.

If you were requested to order the starter for a 5-hp motor connected to a 230-V, three-phase line in ordinary service, you would need to specify several things. You would order a size 1, three-phase, 230-V across-the-line starter in a NEMA type 1 enclosure. Additional information would be needed as to whether the starter should be manual or automatic, depending upon the type of control to be utilized in the installation.

4-5 MANUAL MOTOR STARTERS

To be classified as manual, a motor starter must depend upon the operator closing the line contacts by pressing a button or moving a lever which is physically linked to the contacts in some manner. To illustrate, suppose we take the size 0 starter available in either manual or automatic type, having the pushbutton in the cover. The manual type (Fig. 4-10) is so con-

Fig. 4-10 Manual motor starter. *(Square D Co.)*

structed that when the START button is pressed, a mechanical linkage forces the contacts to close. Once closed, the linkage is latched in this position. When the STOP button is pressed or the overload units opened, the linkage is tripped and the contacts open.

By contrast, when the START button is pressed on the magnetic starter, it merely energizes the starter coil, which in turn magnetically closes the line contacts. The STOP button or the overload relays break the circuit to the coil, thus allowing the line contacts to open.

The chief disadvantage of the manual starter is the utter lack of control flexibility. It must be operated from the starter location, and it is definitely limited even as to protection-control possibilities. When the degree of control it offers is satisfactory for the installation, it does have the advantage of being less expensive. The vast majority of manual starters found in service will fall into one of three classes, namely, the thermal switch for use on very small single-phase motors, the size 0 and size 1 manual across-the-line starters in single- and three-phase motors, and the manual reduced-voltage compensator for large motors.

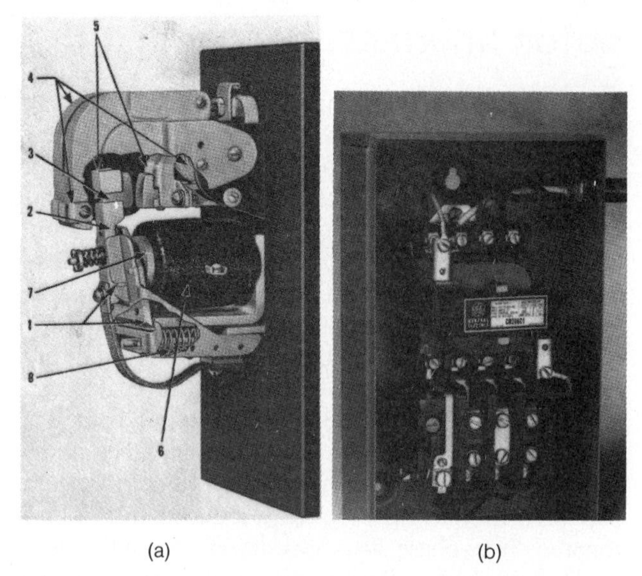

Fig. 4-11 Types of contactors. (*a*) Clapper-type construction: (1) hinge; (2) contact arm; (3) contact; (4) arc shield; (5) contact-holders; (6) coil; (7) pole face; (8) tension spring. *(Square D Co.)* (*b*) Solenoid-type construction. *(General Electric Co.)*

4-6 AUTOMATIC MOTOR STARTERS

The automatic starter, also known as a *magnetic starter,* consists of a contactor with the addition of protective control. This starter depends upon the force of an electromagnet to close and hold the line and auxiliary contacts and offers unlimited control flexibility. With a reasonable amount of maintenance, it is dependable and has a long life expectancy. There are many mechanical arrangements used on this type of starter. They fall into two general classifications, however, depending on the movement of the magnetic coil core.

The first of these is the clapper type (Fig. 4-11*a*), which has the movable contacts attached to the hinge along with the magnetic core or a section of the core. The hinge is so arranged that the pull of the magnetic circuit swings the pole piece and the contacts in a more or less horizontal direction, and the stationary line contacts are mounted on the vertical backboard of the starter.

The second is the *solenoid* type (Fig. 4-11*b*). With this type, the contacts are mounted to the movable portion of the core so that when the core closes, the movable contacts move to meet the stationary contacts.

The all-important magnetic circuit consists generally of an adaptation of one of the three basic magnetic shapes (Fig. 4-12). The E or C type is used on most clapper-type starters, and the modified E, or solenoid type, is used on vertical-action starters.

When ac coil circuits are used, the pole faces of the magnet are equipped with a shading coil (Fig. 4-12*d*).

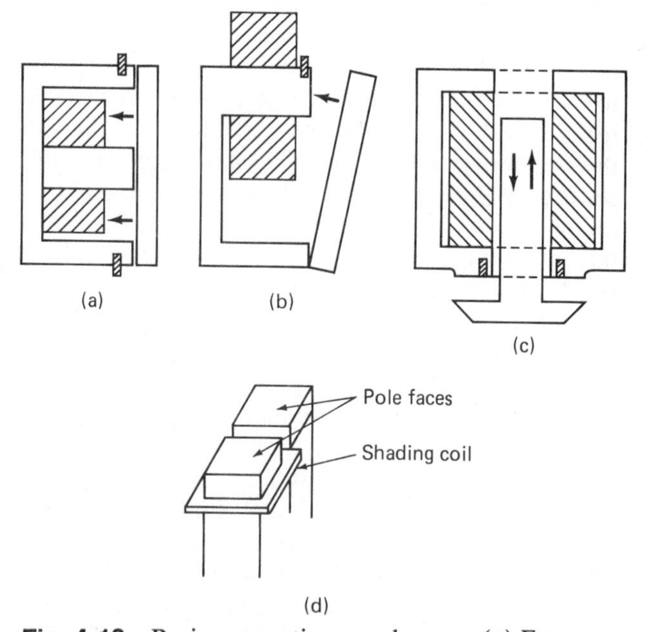

Fig. 4-12 Basic magnetic core shapes. (*a*) E type; (*b*) C type; (*c*) solenoid type; (*d*) magnetic pole piece showing shading coil.

This gives an out-of-phase flux to hold the magnet closed during the zero points of current, thus preventing chatter of the contacts. While this method of preventing chatter is effective, many starters for large motors employ a dc coil circuit because of its constant magnetic pull and freedom from any tendency to chatter.

There are also two basic types of contacts in general use. Most small starters employ a bridge-type contact (Fig. 4-13). The bridge-type contact offers good con-

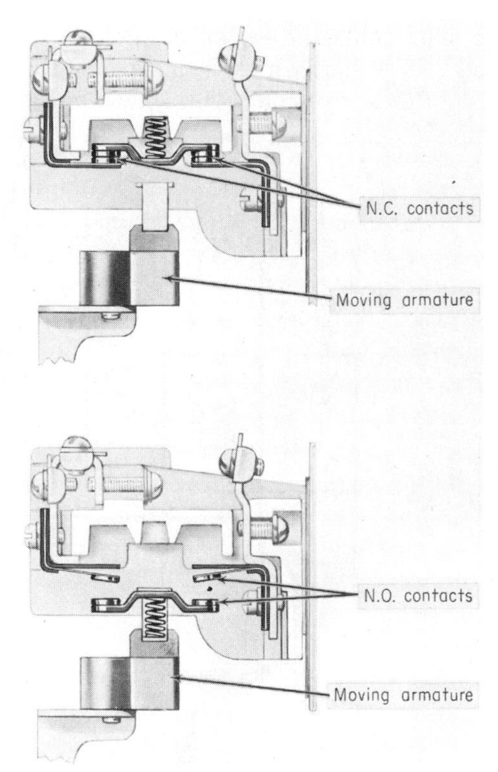

Fig. 4-13 Bridge-type contacts. *(Square D Co.)*

tact alignment and a natural wiping action, which helps to prolong contact life. Most large starters employ the required number of spring-mounted movable contacts that meet a corresponding number of rigid stationary contacts (Fig. 4-11*a*). The necessary wiping action is obtained by making the contacts into a curved shape that allows them to slide into alignment as they close and open. Good contact alignment is necessary in order to prevent excessive arcing and contact pitting.

The greatest single contribution to modern-day production machinery is the magnetic starter. The flexibility of control offered by the magnetic starter allows automation and automatic control accuracy never dreamed of, nor possible, with manual operation. This flexibility stems from the fact that all that is necessary to start a magnetic-controlled motor is to provide electric energy to the coil of the starter. The source of energy may be independent of the motor circuit and can be turned on and off from any point and by any of many means.

4-7 FULL-VOLTAGE STARTERS

Full-voltage, or across-the-line, starters (Fig. 4-11*b*) are the most widely used type. They are used on almost all three-phase, squirrel-cage, and single-phase motors. This type of starter is also used extensively as the primary control on wound-rotor motors that employ manual secondary control. Rated for use on motors up to 600 hp and up to 600 V, they can give full protection to the motor, the machine, and the operator. The only limitation on the use of this starter for squirrel-cage motors is the strain imposed on the wiring system and the machine by the starting current and torque of the motor. Across-the-line starters are available in a variety of enclosures to meet the needs of starter location conditions. These enclosures conform to the standards published by NEMA to suit every condition of the location. Availability in either manual or magnetic types to suit the customer's needs adds to the flexibility of these units.

Any starter which connects the motor leads directly to the line voltage without any means of reducing an applied voltage or limiting the starting current is classed as a full-voltage starter.

4-8 REDUCED-VOLTAGE STARTERS

As the name implies, the reduced-voltage starter contains some means of reducing the line voltage as it is applied to the motor during the starting period. This is done in order to limit the inrush of current during the starting cycle. The requirements for using reduced-voltage starting depend upon several factors. These units are built in either manual or automatic types, and, as with full-voltage starters, the manual type is cheaper but less flexible.

Manual reduced-voltage starters for squirrel-cage motors, more commonly referred to as *compensators* (Fig. 4-14), consist of a double-throw switch and an

Fig. 4-14 Manual reduced-voltage starter, or compensator. *(General Electric Co.)*

autotransformer. The START position of the switch applies power to the motor through an autotransformer. The operating handle is held in this position until the motor is running at its highest speed; full voltage is then applied by throwing the handle to the RUN position. The switch mechanism is held in the RUN position by a latch which can be released by either the undervoltage release, the overload units, or the hand trip. Generally, these units are in a self-contained metal enclosure that is designed to be wall-mounted.

Automatic reduced-voltage starters (Fig. 4-15) take many forms and are generally designed for a particular

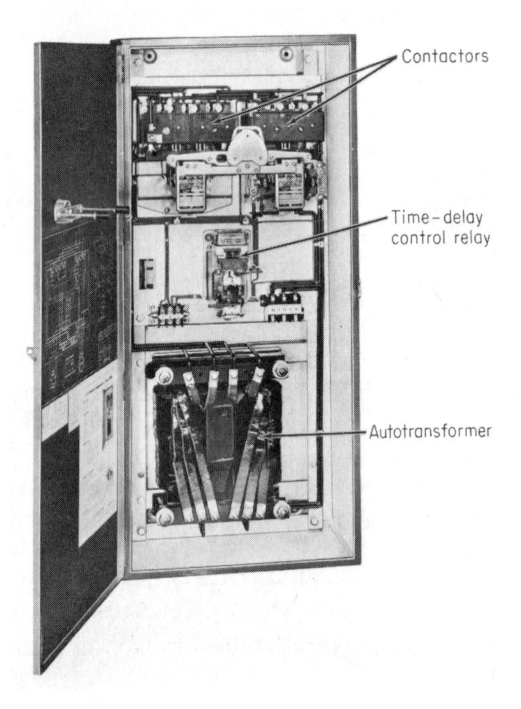

Fig. 4-15 Automatic reduced-voltage starter. *(Square D Co.)*

type of motor and for a particular application. The essential requirements are that some means be provided for connecting the motor to a source of reduced voltage and then automatically connecting it to the full-line voltage after it has had time to accelerate.

The start contactor on a primary-resistance starter need be only a three-pole unit. This contactor connects the motor to the line in series with resistor units designed to limit the motor-starting current. The run contactor in this starter is also a three-pole unit which shorts out the resistors in order to apply full voltage to the motor.

The reactor-type reduced-voltage starter employs exactly the same contact arrangement as the primary-resistance starter. The only difference between a starter for primary resistance and a reactor-type reduced-voltage starter is that the latter uses reactors in place of resistors.

The start contactor for an autotransformer-type reduced-voltage starter must be a five-pole contactor. The use of these five contacts is shown in Fig. 4-16.

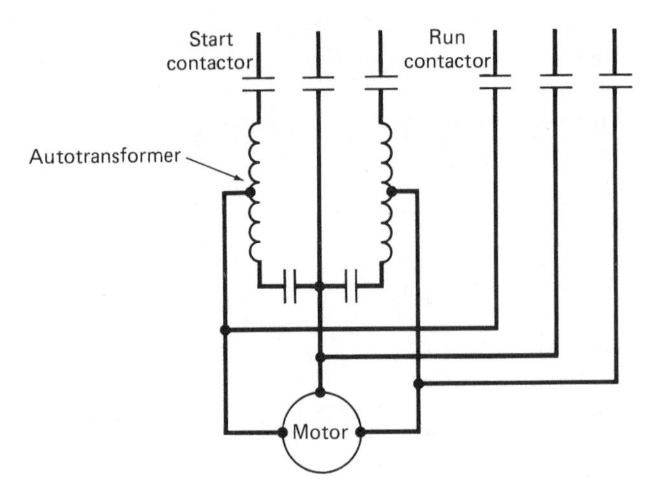

Fig. 4-16 Connections for starter autotransformer.

These contacts connect the motor to the line through the autotransformer connected open-delta. The run contactor is this starter consists of three contacts which apply full line voltage to the motor.

Regardless of its type, an automatic reduced-voltage starter must contain some means of automatically changing from START to RUN at the proper time. Generally, this is accompanied by the use of a time-delay relay. In the case of primary-resistance or reactance starters, this relay is required only to energize the coil of the run contactor. When used on an autotransformer starter, this relay must break the circuit to the start contactor and then close the circuit to the run contactor. The use of a time-delay relay in this service gives definite-time control. Another method which is sometimes used employs a form of current relay which will open the start contactor and close the run contactor when the motor current drops to a predetermined level. This gives current-limit control.

When primary-resistance or reactor-type reduced-voltage starting is employed, there is no interruption of current to the motor. With the autotransformer-type reduced-voltage starting, however, the current may or may not be interrupted momentarily before the motor is placed across the line. When the current is not interrupted in the transition from reduced-voltage to full-voltage operation, it is referred to as a *closed-transition start.* When the motor is disconnected momentarily from the line and the current interrupted, it is referred to as an *open-transition start.* When the transition is of the open type, it is quite possible to have an inrush of current as high as twice the full-voltage starting current at the instant of voltage application. This surge of current is referred to as *transition current.*

Any of these controllers may contain almost any or all of the protection control functions, such as undervoltage release, phase failure, or incomplete-sequence protection.

One thing that you must keep in mind when choosing between manual and automatic controllers of this type is that when a manual unit is installed, it must be located so that the operators cannot only see but also hear the motor, in order that they may properly judge when to apply full voltage. This limited selection of location can be overcome to some extent by installing a remote indicating tachometer. This enables operators to determine the degree of acceleration of the motor from the control location.

Another method of obtaining the effect of reduced-voltage starting is to use a wound-rotor motor with secondary control. This arrangement gives a higher starting torque with reduced starting current than does the squirrel-cage motor in primary reduced-voltage starting. The starter for such an arrangement would consist of an across-the-line starter connected to the primary winding of the motor. The secondary, or rotor, winding of the motor is connected to resistance grids by means of either a manual or automatic drum controller (Fig. 4-17). A second advantage to this arrangement is that it can offer speed control as well as limited starting current.

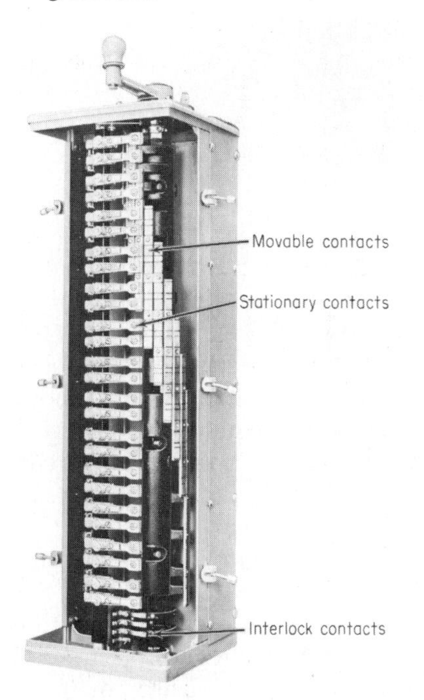

Fig. 4-17 Manual drum controller. *(Cutler-Hammer Inc.)*

4-9 SPEED-CONTROL STARTERS

Besides the secondary speed control offered by wound-rotor motors, there are methods of controlling the speed of squirrel-cage motors, provided these motors are designed for multispeed operation.

One type of two-speed motor is wound with two separate stator windings and requires a starter with two sets of interlocked line contacts. This interlock may be either mechanical, or electrical, or both. With electrical interlock two separate across-the-line starters may be used when a special unit is not available. A reversing starter makes an excellent unit, provided the reverse phasing is eliminated. The two sets of contacts or starters are wired so that each set connects one speed winding to the line. As in other types of starters, they may be either manual or automatic.

Another type of two-speed motor is the consequent-pole motor, which has only one stator winding but gives two speeds by means of regrouping the stator coils to provide for a different number of poles. While the dual-stator-winding motor might have almost any ratio of high to low speed, the consequent-pole motor gives a 2 to 1 speed ratio. Three speeds may be obtained by using two stator windings. One of these windings gives one speed, and the other is regrouped for a second and third speed. To obtain four speeds, both windings must be regrouped. To obtain many speeds and variable torque, the high-speed windings are connected parallel-star and the low-speed windings are connected series-star (wye) (Fig. 3-3).

For constant horsepower and many speeds, the connection should be series-delta for high speed and parallel-star for low speed.

Because of the variety of possible connections, the starter for this motor must be designed for the particular type of motor to be used. One of the most popular manual arrangements requires the use of a drum-type controller to make the necessary changes in connections. This controller, however, must be preceded by an across-the-line motor starter which is interlocked through a set of contacts on the drum controller. The interlocks must disconnect the motor whenever a speed change is made by rotating the drum controller. The use of the across-the-line starter also provides the necessary protection for the motor which is not available on the drum controller.

Magnetic starters for multispeed motors must have a contactor for each speed (Fig. 4-18 on the next page). For the motor to be used, the contacts on each contactor must be so arranged that they will make the proper connections to the stator windings. These starters can be built to give any one of three types of control. The first and simplest of these is selective speed control. With selective speed control, operators may start the motor at any speed desirable and increase the speed merely by selecting any higher speed. To reduce speed, however, they must press the STOP button first and allow the machine to lose speed before the lower speed control is energized. This is done to prevent undue stress and strain on the motor and the machine.

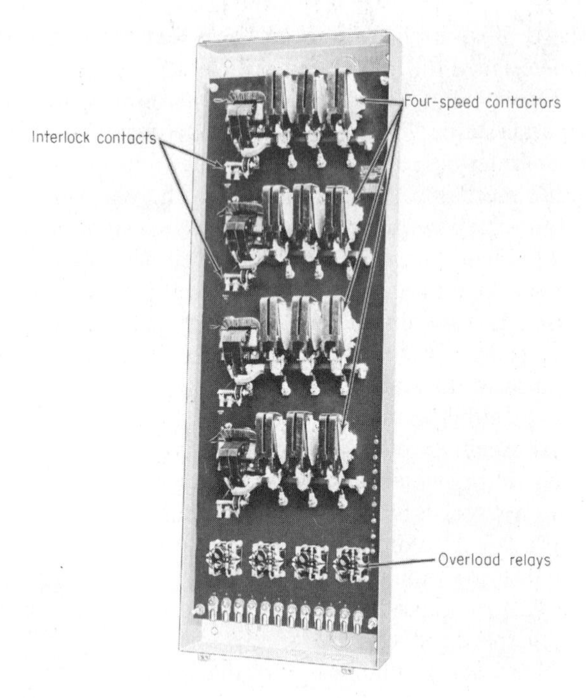

Fig. 4-18 Speed-control starter. *(Cutler-Hammer Inc.)*

The second type is sequence-speed control, which requires that the machine be started in its lowest speed and brought up to the desired speed through a set sequence. The acceleration to the desired speed requires that the operator press the button for each speed in proper sequence until the desired speed is reached. To reduce speed, the motor must be stopped and the sequence started over at the lowest speed.

The third type is automatic speed control, in which the operation is like sequence-speed control, except that the operator needs to press the button only for the desired speed. The controller will automatically start in the lowest speed and accelerate through each successive speed to the one selected. To reduce speed, the STOP button must be pressed first. Then the button for the new speed should be pressed, which will recycle the controller to the new speed through each successive lower speed.

The choice of controller depends upon the type of load and the required operating conditions of the motor. Keep in mind that the basic difference among controllers is that selective speed control allows starts in any speed while the other two require a start at the lowest speed. It is not possible to describe the physical structure of this type of unit in general terms, because of the many possible variations. In any case, however, a magnetic contactor with the required number of contacts will be required for each speed to make the connections for that speed and to give the desired protection control.

4-10 COMBINATION STARTERS

The National Electrical Code requires a disconnect switch or breaker within sight of each motor. The combination starter includes this switch or breaker in the same enclosure with the starter itself. Combination starters are available as across-the-line or reduced-voltage and single-phase or three-phase types. In fact, almost any type of starter can be had in a combination form. The most common form of combination starter, however, includes a breaker or switch and an across-the-line starter (Fig. 4-19). A combination starter offers several advantages in that its compact size lends itself very well to a neat mechanical installation and reduces the wiring required at the job site. Electrically, the combination starter protects operators or service personnel in that it generally includes a mechanical interlock which requires that the switch or breaker be in the OFF position before the door can be opened. This ensures that the circuit is dead whenever the door to the starter is open.

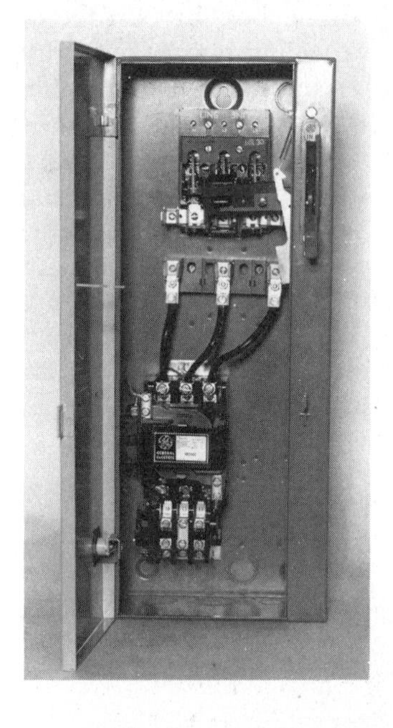

Fig. 4-19 Combination starter. *(General Electric Co.)*

The switch used in this type of unit may be unfused or fused. If the unfused switch is used, then the motor circuit must be protected by another fused switch or breaker to give short-circuit protection. The use of a fused switch or circuit breaker in the combination starter adds short-circuit protection to the other control functions offered by the starter itself.

4-11 REVERSING STARTERS

The basic requirement of a reversing starter for three-phase motors is that it be capable of connecting the motor to the line in one phase rotation for forward and in the opposite phase rotation for reverse. A magnetic reversing starter (Fig. 4-20) incorporates two magnetic starters in one enclosure. The line sides of these starters are so connected (Fig. 4-21) that line 1 on starter 1 is connected to line 3 of starter 2 and line 3 of starter 1 is connected to line 1 of starter 2. Line 2, or center phase, of both starters is connected.

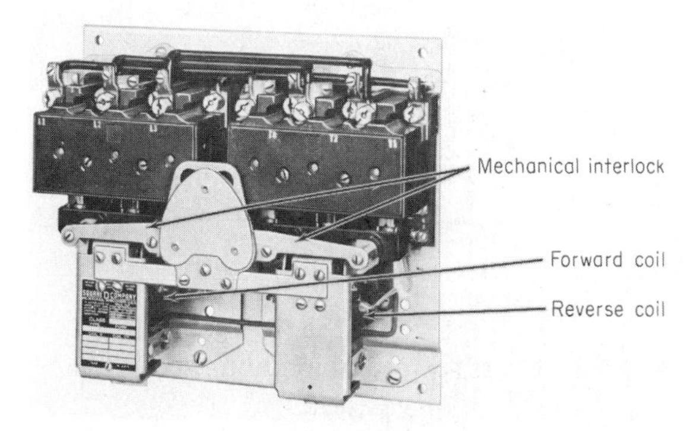

Fig. 4-20 Magnetic reversing starter. *(Square D Co.)*

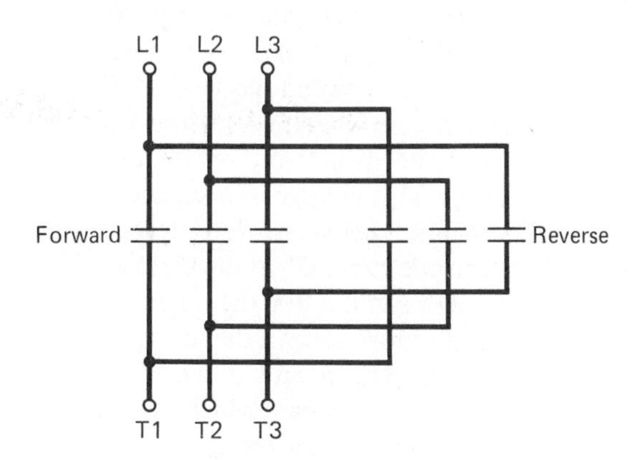

Fig. 4-21 Connection for reversing starter.

Thus, when starter 1 is energized, L1 and T1 are connected and L3 and T3 are connected. When starter 2 is energized, however, L1 and T3 are connected and L3 and T1 are connected, thus accomplishing a reversal of phase rotation of the motor itself. These units are provided with a mechanical interlock consisting of a lever or arm which prevents either starter from closing when the other is energized. Many of these units also incorporate an electrical interlock to achieve the same purpose.

Remote control of a magnetic reversing starter requires only that the pushbutton energize the coil of the starter, which gives a desired rotation of the motor. The STOP button must be wired so as to de-energize whichever coil is in use at the time. The normal wiring of a FORWARD, REVERSE, and STOP pushbutton station requires that the STOP button be pushed first when going from one direction to the other. This allows the motor to be disconnected from the line before being reversed and prevents plugging of the motor. Plugging of a motor is the sudden reversal of rotation without first removing it from the line.

If plugging is desirable, then the FORWARD and REVERSE buttons must be of the double-pole type, with one set of normally open and one set of normally closed contacts activated by one button. The normally closed contacts are so wired that the stop circuit is broken before the start circuit is closed, regardless of which button is pushed. Caution should be exercised in employing plugging on any machine because not all machines can withstand the severe strain imposed by the sudden reversal of the motor. Plugging can damage machinery. It may damage the motor and at times can endanger the personnel operating the machine. Plugging is used extensively in industry on presses, grinders, and many other pieces of machinery, but these should be so designed that they cannot be damaged by the stresses and strains encountered in this type of operation.

Manually reversing three-phase motors of the squirrel-cage type is generally accomplished by use of a drum controller or switch between the line starter and the motor. This type of reversal requires that the motor be disconnected from the line before the drum switch is moved from forward to reverse or from reverse to forward. This prevents severe arcing of the drum-switch contacts. Proper wiring of the drum switch requires that it have a set of contacts interlocked with the line starter so that any time the handle of the drum switch is rotated, it will disconnect the motor from the line. This system of wiring also prevents plugging of the motor.

Reversal of single-phase fractional-horsepower motors may also be accomplished by using a drum switch or even a toggle switch. Reversing this type of motor generally requires only that the starting windings be reversed in relation to the running winding. Suitable precautions must be taken to ensure that the motor comes to rest before attempting to reverse. Automatic or magnetic control for reversal of single-phase fractional-horsepower motors are accomplished by the use of relays or starters. The possible connections for drum switches, toggle switches, or relays for reversing this type of motor are shown in Fig. 4-22 on the next page.

The reversing starter as such will offer the same control functions as any other starter of the manual or magnetic type and, in addition, provide the control function of reversing the motor. Should a reversing starter not be available, two across-the-line starters properly connected may be used. When two starters are used in this service, electrical interlock must be used with them to prevent both starters from closing at one time. The electrical connection between the starters should be the same as that used in a reversing starter unit. Factory-built reversing starters generally are so wired that they require only one set of overload relays.

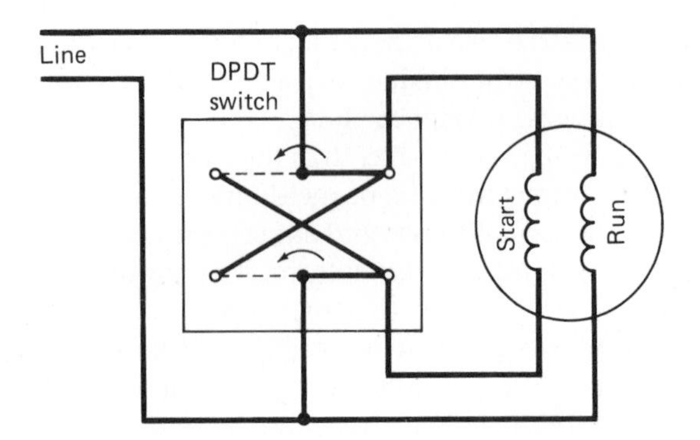

Fig. 4-22 Basic circuit for reversal of small motors.

When reversal is required on multispeed squirrel-cage motors or wound-rotor motors, it is generally accomplished by the use of a drum switch between the speed-control line starter and the motor itself.

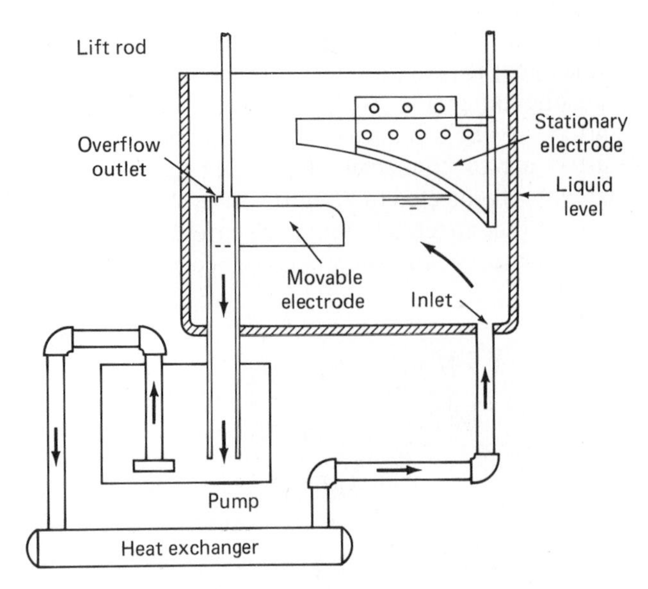

Fig. 4-23 Cross section of a liquid rheostat.

4-12 WOUND-ROTOR MOTOR STARTERS

The starter for a wound-rotor motor consists of a full-voltage across-the-line starter used to energize the field, or primary, winding of the motor and some form of secondary control. The use of the primary starter provides overload and undervoltage protection of the motor. The primary starter may be either manual or automatic and may be of the reversing type if primary reversal of the motor is desired. The primary starter should be interlocked with the secondary controller in such a manner that the motor cannot be started unless all the secondary resistance is in the circuit.

There are several possible types of secondary control and controllers for wound-rotor motors. The most common component used for secondary control is the drum controller (Fig. 4-17). This unit is merely a set of rotating contacts operated by a handle attached to a shaft with movable contacts. The movable contacts engage a set of stationary contacts which short-out resistance in the secondary circuit as needed for speed control. The drum controller may be of the reversing type or the nonreversing type. It may have several steps for speed control or may have only two or three steps for reduced-voltage starting. It is also possible to use a magnetic contactor similar to a speed-control contactor as a secondary controller.

Another type of secondary controller which gives the smoothest speed regulation, over the widest range of speed, for a wound-rotor motor is the liquid rheostat (Fig. 4-23). This type of secondary speed control is generally limited to very large equipment such as fans, blowers, and pumps where a constant torque or load is involved. The liquid rheostat consists basically of three tanks of water or another electrolyte. These cells or tanks must be made of insulating material. Inside each tank is a stationary contact or electrode and one movable electrode. The resistance is varied by increasing the distance between electrodes. The maximum resistance occurs when the electrodes are the furthest apart, and the minimum resistance when the electrodes mesh. This change in resistance is affected both by the distance between the electrodes and the area of the electrode exposed to the electrolyte. One critical feature of this unit is that the water or electrolyte level must be maintained above the movable electrode even though it moves up and down. This level is maintained by the use of a gate which is fastened so as to move up and down with the electrode. The water or electrolyte must be kept in constant circulation and provided with some means of cooling. Generally, a heat exchanger is used for this purpose.

The various types of speed control, such as definite-time control, sequence control, current-limit control, and frequency control, as discussed under speed-control starters, also apply to wound-rotor motor starters.

4-13 SYNCHRONOUS MOTOR STARTERS

The synchronous motor in its most commonly found form starts as a squirrel-cage motor. During the starting period, the stator is energized by alternating current and may be placed either directly across the line or through a reduced-voltage starter. It is necessary during the starting time to short-out the dc field winding through a field-discharge resistor. This resistor protects the field from high induced voltages and also serves to increase the starting torque by serving as a secondary resistance. When the motor reaches synchronizing speed, which is usually between 93 and 98 percent of its synchronous speed, the starting resistance must be disconnected and the dc voltage applied to the field winding of the motor. The application of the dc field excitation will cause the motor to pull into step and synchronize.

The field application relay, or slip-frequency relay, as it is sometimes referred to, is probably the most critical component in a synchronous motor starter. Its function is to apply the dc field excitation at exactly the proper time. The sensing of the proper time to apply field excitation is accompanied in the field application relay by sensing the induced alternating current that flows in the field winding during the starting period. This current is at a maximum when the motor first starts and diminishes in strength and frequency as the motor approaches synchronous speed. At approximately 95 percent of synchronous speed, the current induced in the field winding has reached a weak enough value to allow the field application relay to pull in and apply excitation to the motor. The use of a field application relay prevents application of excitation current when the motor is out of step more than approximately 75 to 80 degrees. This unit also applies the field discharge resistor after removing the excitation whenever an overload or other trouble causes the motor to pull out of synchronism.

The starter for a synchronous motor of standard design consists of a starter, either an across-the-line or reduced-voltage type, similar to that required for a squirrel-cage motor, and the necessary field control equipment (Fig. 4-24).

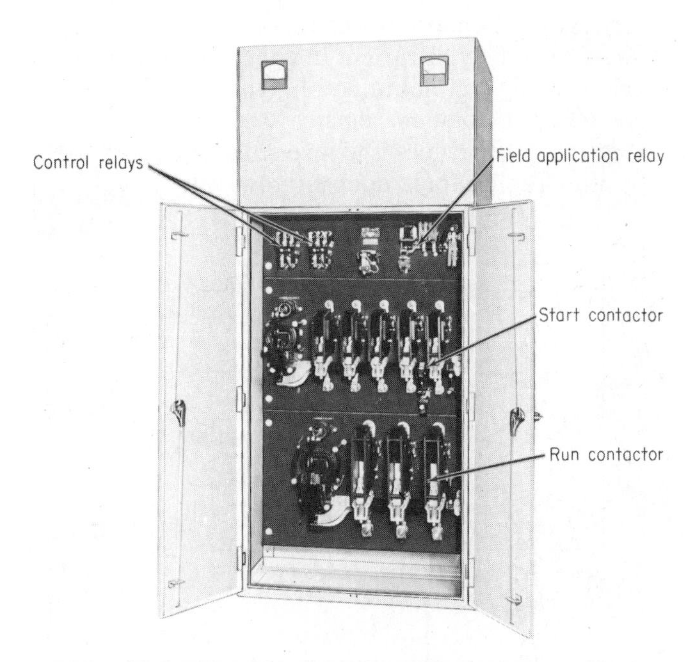

Fig. 4-24 Synchronous motor starter. *(Cutler-Hammer Inc.)*

A manual starter consists of a manual compensator with overload and undervoltage protection, a field application relay and its contactor, a field rheostat to control excitation, and a field-discharge resistor.

An automatic starter consists of an automatic reduced-voltage starter with definite-time control, a field application relay and contactor, a field-discharge resistor, and overload and undervoltage protection.

There are several factors involved in the starting of synchronous motors where extraheavy loads are to be handled, such as in rubber mills and cement plants. Caution should be used in determining the starting requirements of this type of motor in its particular installation. It is beyond the scope of this book to go into the various possibilities involved in the starting and control of synchronous motors in special applications.

4-14 SERIES MOTOR STARTERS

The manual starter for a series motor consists of a tapped resistor and a contact arm arranged so as to short out progressive steps of the resistance as the handle is moved from step to step. When all of the resistance has been cut out of the circuit, the motor is connected across the line. The handle must be held by the holding coil in the RUN position against the tension of a spring.

When the holding coil is connected in series with the motor (Fig. 4-25), it has a few turns of heavy wire and carries full motor current. When the load is removed

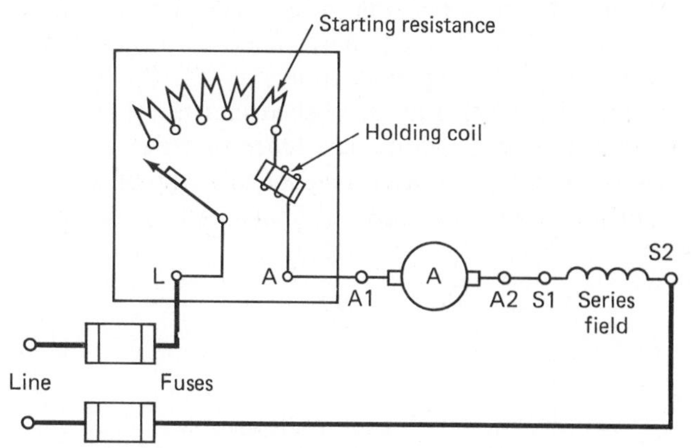

Fig. 4-25 Series motor starter with no-load protection.

or greatly reduced, the current drops to a low value, which allows the spring to return the control arm to the OFF position. This arrangement provides no-load protection.

When the holding coil is connected across the line, it has many turns of fine wire and draws its own current from the line. Any serious drop or failure of the supply voltage will drop out the holding coil and allow the control handle to return to the OFF position. This arrangement provides no-voltage protection (Fig. 4-26).

Direct-current drum controllers used with resistance grids are generally employed for the large series motors used on cranes, elevators, and other heavy-load devices which require speed and reversing control.

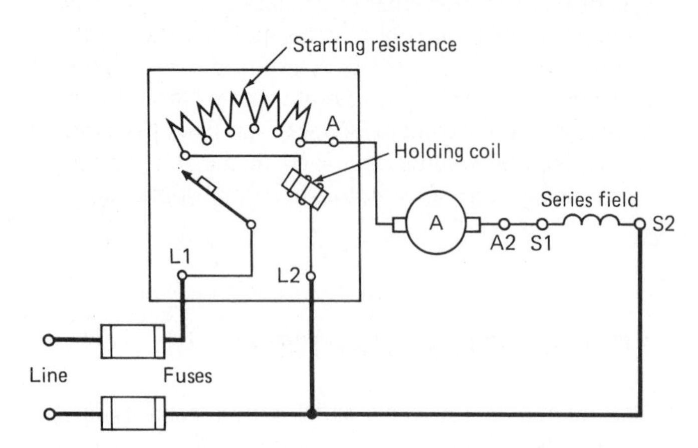

Fig. 4-26 Series motor starter with no-voltage protection.

4-15 DIRECT-CURRENT SHUNT AND COMPOUND MOTOR STARTERS

When speed control is not desired, the basic starters for these motors are the three-point starter (Fig. 4-27) and the four-point starter (Fig. 4-28). The action of these starters is basically the same as that of the series starter providing no-voltage protection.

Speed control is accomplished in a shunt or compound motor by adding resistance in series with the shunt field. The starter used for this type of service, which is to provide above-normal speeds, is the four-

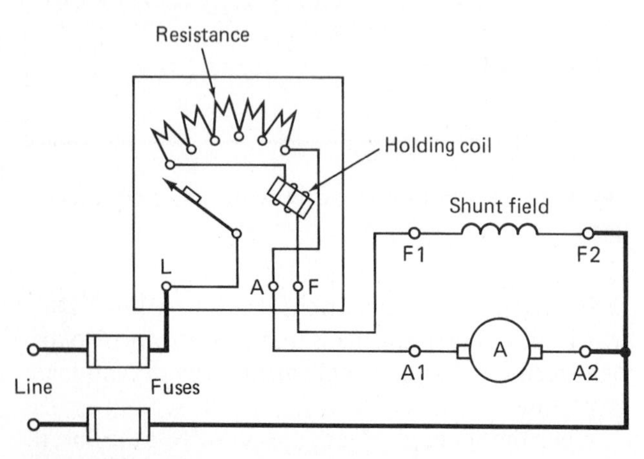

Fig. 4-27 Three-point starter.

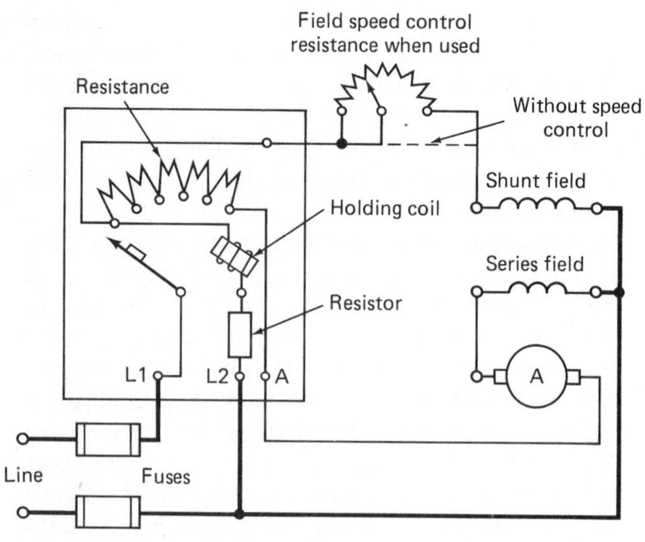

Fig. 4-28 Four-point starter.

point starter (Fig. 4-28). The motor is started in the same manner as with the three-point starter. The field rheostat is used to adjust above normal speeds.

Manual speed controllers are available which will provide both above- and below-base-speed control. These starters use mechanical linkage to position two tapped resistances, one in the armature circuit for low speeds and one in the field for high speeds. The armature resistance must be rated for continuous duty.

4-16 DIRECT-CURRENT REVERSING STARTERS

Reversal of a dc motor is accomplished by reversing the direction of flow of current in either the armature or field winding, but not in both. When both windings are reversed, rotation remains the same as before. Generally, the practice is to reverse the armature leads by using a double-pole, double-throw switch (Fig. 4-29)

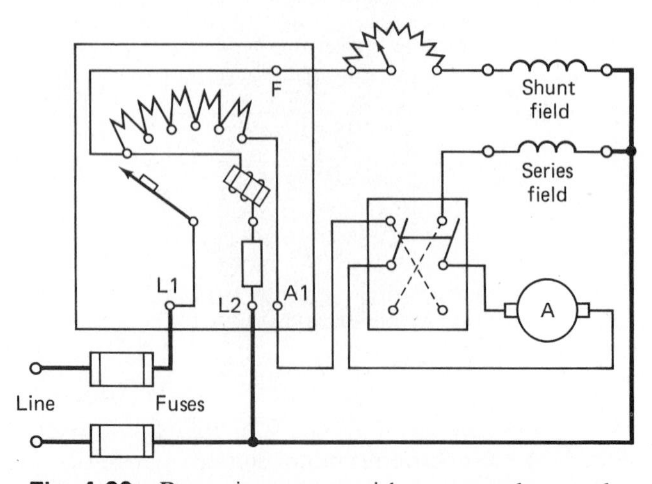

Fig. 4-29 Reversing starter with overspeed control.

and a four-point starter. When a drum controller is used as a reversing speed controller, it has the ability to rotate in either direction from stop and thus reverses the armature connections as needed for forward or reverse.

4-17 AUTOMATIC DC CONTROLLERS

The requirement of an automatic controller for dc motors is the same as that for manual controllers, except that contacts are used to replace the control arm of the manual unit.

For starting, we have learned that a tapped resistance must be progressively shorted out to bring the motor up to speed. When a relay contact is connected across each section of the starting resistance, we have another means of shorting out other than using resistance.

When the coils of the starting relays are connected to the line and the relays are time-delay relays, we have time-limit acceleration control. When the coils of the starting relays are connected across the armature, they will be subject to armature voltage changes. When these relays are adjusted to close on progressive steps of voltage, we have counter-emf acceleration control.

There are many other arrangements possible for automatic starting and speed control of dc motors, but they are beyond the scope of this book.

SUMMARY

This chapter on control components is intended to give students an insight into the variety of devices used to control the functions of motors. Always remember that a motor can perform only as well as the components in its control circuit. The fact that a particular component is a quality product does not necessarily mean that it will perform well in a particular circuit. The component must be selected to fit the motor to be controlled and the function to be performed by that motor.

REVIEW QUESTIONS

1. How important is duty cycle in the selection of control components?
2. What is the difference between switches intended for isolation purposes and those intended for use as disconnecting means for motors?
3. How are isolation switches rated?
4. The National Electrical Code permits isolation switches for motors of what ratings?
5. How is a motor-circuit switch rated?
6. What are the requirements for a motor-circuit switch if it is used for disconnecting and motor-circuit protection?
7. According to the National Electrical Code, switches and breakers perform what type of control functions?
8. What is the function of a contactor in control work?
9. Why is a relay considered a multiple-output device?
10. What does a pilot device do and what are its functions in control work?
11. Describe the operation of a voltage relay.
12. What are some applications of voltage relays?
13. Describe a current relay and its functions in control circuits.
14. Define pull-in current, drop-out current, and differential as applied to current relays.
15. When would a current transformer be used with a current relay?
16. What type of relay uses a bimetallic strip?
17. What are time-delay relays used for?
18. What is the function of a dashpot?
19. Define the meaning of letters *TDOE* and *TDODE.*
20. What is the function of an overload relay?
21. What are the three basic types of overload relays in general use?
22. Explain the inverse-time characteristics of overload relays.
23. What is the purpose of a compensated-overload relay?
24. What are the classifications of motor starters?
25. Why do motor starters have lower horsepower ratings at low voltage as compared with high voltage?
26. What environments are NEMA type 1, type 3, type 4, type 7, type 8, type 9, type 11, and type 12 motor starter enclosures used in?
27. What three classes of manual motor starters are most often used today?
28. Compared to manual starters, what is the advantage of automatic motor starters?
29. What is the purpose of a shading coil on a pole face of an ac motor starter?
30. Describe a compensator and what it is used for.
31. Describe a primary-resistance starter.
32. How does a reactor-type reduced-voltage starter differ from a primary-resistance starter?

33. What are the requirements for the starting contactor for an autotransformer-type reduced-voltage starter?
34. What is the difference between closed-transition and open-transition starting?
35. How are two speeds obtained from a three-phase ac motor?
36. Explain how a reversing starter can be used as a two-speed motor starter.
37. How are the windings of a consequent-pole motor connected to provide multispeed and variable torque?
38. How are the windings connected in a consequent-pole motor to provide constant horsepower and two speeds?
39. What is the function of interlocks in a motor controller?
40. What determines the type of speed control used for a motor?
41. What are the National Electrical Code requirements concerning disconnect switches and breakers?
42. Describe a combination starter.
43. What is the basic requirement for a reversing starter for a three-phase motor?
44. How is plugging control accomplished?
45. How is manual reversing of a single-phase motor accomplished?
46. If a reversing starter is not available for a three-phase motor, how can two across-the-line starters be used to provide reversing control?
47. What is the most commonly used secondary control for a wound-rotor motor?
48. What type secondary control would be used for a wound-rotor motor if smooth-speed regulation over a wide range of speeds is required?
49. When is the dc voltage applied to the field of synchronous motor?
50. What is the function of a slip-frequency relay?
51. What type starter can be used for synchronous motors?
52. What control does a rheostat in series with the shunt field of a dc shunt or compound motor provide?
53. What is the rule for reversing dc motors?
54. Describe how time-limit acceleration control is provided by automatic dc controllers.
55. How is counter-emf acceleration control provided by dc automatic controllers?

5
Pilot Devices

All components used in motor control circuits may be classed as either primary-control devices or pilot-control devices. A primary-control device is one which connects the load to the line, such as a motor starter, whether it is manual or automatic. Pilot-control devices are those which control or modulate the primary-control devices. Pilot devices are such things as pushbuttons, float switches, pressure switches, and thermostats.

5-1 DESCRIPTION OF PILOT DEVICES

An example of a primary pilot-control device is a magnetic across-the-line starter controlled by a single toggle switch. This device is used to energize and de-energize the coil of the starter. When the switch is closed, the starter is energized and the motor starts. When the switch is open, then the starter is de-energized and the motor stops. The starter, in that it connects the motor or load to the line, would be classed as a primary-control device. The switch does not connect the load to the line but is used to energize and de-energize the coil of the starter. Therefore, it would be classed as a pilot-control device.

For any given motor, two primary-control devices are generally used. These are the disconnecting means or switch and the motor starter. There may be many pilot devices used in parallel and series combinations to control the functions of start and stop performed by the primary-control device. The overload relays, for instance, which are included in the motor starter, are actually pilot devices used to control the primary device whenever the motor is overloaded.

The requirements of pilot devices vary greatly with their function and their proposed use. For instance, a float switch must open and close its contacts on the rise and fall of a liquid in some form of container. A pressure switch must open and close its contacts when the pressure in some vessel, pipe, or other container varies through the limits built into the pressure switch. Perhaps the best picture that can be drawn to show the difference between primary devices and pilot devices would be the comparison of a contactor and a voltage relay. The contactor is built to carry relatively large currents; therefore, it has heavy contacts capable of interrupting these currents. The relay, designed for pilot duty, has relatively small contacts because the current it is expected to interrupt is very small. In general, pilot devices might better be termed *sensing devices* because they are generally used to sense such quantities as pressure, temperature, liquid level, or the pressure applied to a pushbutton. The function of these pilot devices is to convert the information that they sense into control of the primary-control device with which they are connected.

5-2 FLOAT SWITCHES

Float switches have many different mechanical structures. Basically, they consist of one or more sets of contacts either normally open or normally closed, operated by a mechanical linkage. Many float-switch units, and other pilot devices, employ a mercury switch in place of metallic contacts. The simplest mechanical arrangement for a float switch (Fig. 5-1 on the next page) would be a pivoted arm having the contacts fastened to one end and a float suspended from the other end. As the water level rises, it would lift the float, thus moving the contact end of the lever downward and either making or breaking the contact, depending on whether the stationary contact were mounted above or below the arm. If a single-pole double-throw action of the contacts were desirable, then one stationary contact could be mounted above, and one below, the center of the arm. If the float were all the way up, it would join the lower set of contacts; and if the float were all the way down, it would join the upper set of contacts.

Float switches require some means of adjusting their range of operation, that is, the amount of float travel

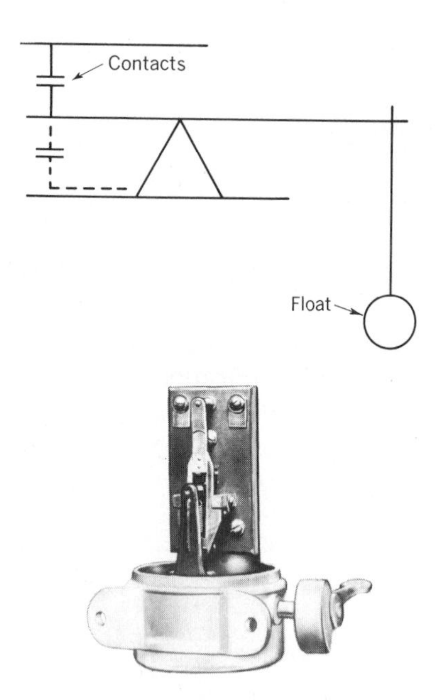

Fig. 5-1 Float switch, pivoted arm type. *(Cutler-Hammer Inc.)*

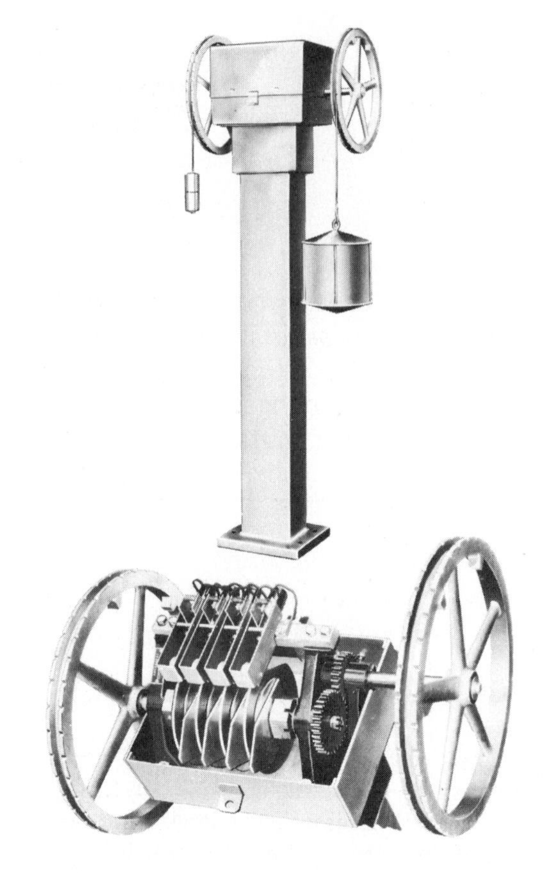

Fig. 5-2 Float switch, drum-switch type. *(Cutler-Hammer Inc.)*

between the joining and breaking of the contacts. In the simple float switch, this is usually accomplished by suspending the float on a rod which passes through a hole in the arm of the switch itself. Then its stops are placed above and below the arms on the float rod, the amount that the float travels before it operates the switch being adjusted by moving the stops further apart or closer together.

Another system used in float-switch construction to give an even greater range of adjustment is to have the float suspended on a chain or cable which winds up on a reel. The action of the float is then transformed into a rotary motion which actuates a drum-type switch (Fig. 5-2). As may be noted from the figure, these are only two of the many possible ways that a float may be made to actuate a set or sets of contacts. Any arrangement that will accomplish this may be properly classified as a float switch and used for pilot duty.

It should be noted here, however, that float switches are also made with heavy contacts and are suitable for primary control of small fractional-horsepower motors. When used for primary control, they are inserted in the line leads ahead of the motor and merely make and break the motor circuit in response to the action of the float.

It is highly desirable when studying pilot devices, if it is at all possible, to obtain several units made by different manufacturers and study the mechanical devices employed in their operation. The student will find that they vary greatly in actual mechanical design but fall into the same basic type of operation as described herein.

5-3 PRESSURE SWITCHES

Pressure switches, like float switches, are generally considered to be pilot devices. In the heavy-duty models, however, they are sometimes built for primary control of fractional-horsepower motors. Again, as with all pilot devices, they vary considerably in their mechanical design. Basically, they fall into three general classes according to their means of operation. The first of these is a bellows, which is expanded and contracted in response to an increase and decrease in pressure. The contacts are mounted on the end of a lever, which is acted upon by the bellows (Fig. 5-3). The bellows expand, moving the lever, and either join or break the contacts, depending on whether they are normally open or normally closed.

The second general type uses a diaphragm in place of the bellows (Fig. 5-4). Otherwise, the action of the switch is identical whether it contains a bellows or a diaphragm. The advantage of one type over the other depends greatly upon the type of installation and the pressures involved.

It should be noted that pressure switches have a definite range of pressure where they are designed to oper-

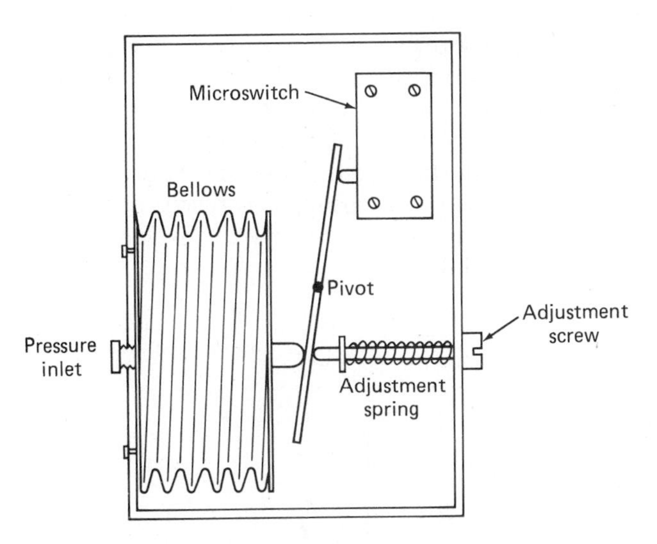

Fig. 5-3 Pressure switch, bellows type.

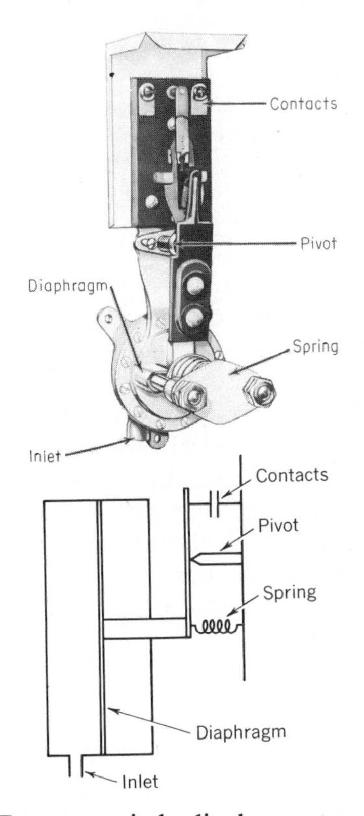

Fig. 5-4 Pressure switch, diaphragm type. *(Cutler-Hammer Inc.)*

ate. For instance, a pressure switch made to operate from a vacuum up to possibly several pounds of pressure would not be suitable for use on a line which normally would carry from 100 to 200 lb of pressure.

A third general type of pressure switch, the bourdon tube, employs a hollow tube in a semicircular shape so designed that an increase in pressure tends to straighten the tube. This action is transformed into a rotary motion by a linkage which trips a mercury switch mounted within the enclosure.

5-4 LIMIT SWITCHES

Limit switches are designed so that an arm, lever, or roller protruding from the switch is bumped or pushed by some piece of moving equipment. The movement of this arm is transformed through a linkage to a set of contacts. Movement of the arm causes the contacts to open or close, depending upon whether they are normally closed or normally opened (Fig. 5-5).

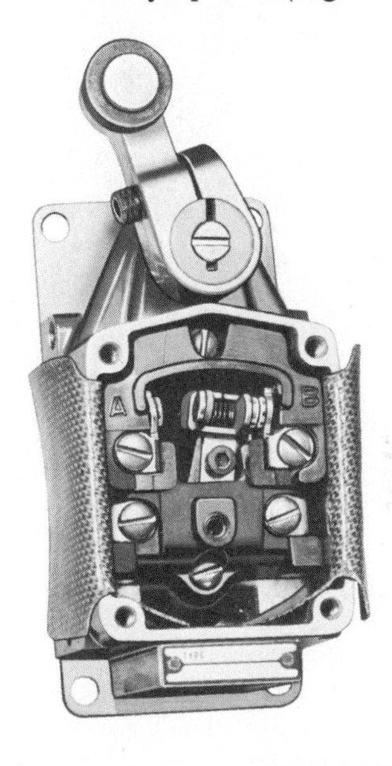

Fig. 5-5 Limit switch with cover removed. *(Square D Co.)*

There is a great variation in the internal design and action of these units, but, again, they fall into two general mechanical design classifications. The units intended for rugged use but not for precision control generally have metallic contacts operated directly from the lever action of the switch. Most manufacturers build a more accurate, or precision, unit which employs a microswitch to allow operation on very minute movements of the external lever of the switch. As with float switches, there are limit switches which are built so that a cable or chain is wound up on a reel and forms part of the limit switch. This movement of the chain or cable is transformed to a rotary motion which actuates a drum-type switch. This type of limit switch is used where a great deal of travel must be allowed before the switch is activated.

Another type of limit switch (Fig. 5-6 on the next page) employs a drum-type switch and is designed for direct shaft mounting where the rotation of the machine directly causes a rotation of the switch. The contacts in this type of limit switch must be designed to resemble a cam so that they can close and open with a continuous rotation in the same direction. Limit switches of this type are coupled through a reduction gear so that many revolutions of the machine are required to produce one revolution of the limit switch, thus extending the range of control offered by the limit switch.

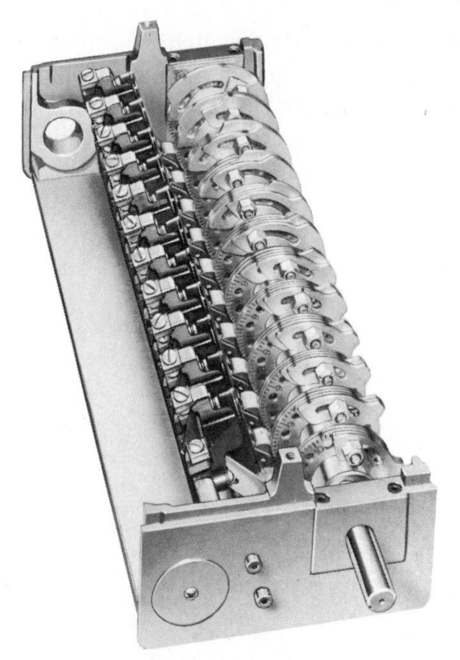

Fig. 5-6 Drum-type limit switch. *(Allen Bradley Co.)*

5-5 FLOW SWITCHES

The purpose of a flow switch is to sense the flow of a liquid, air, or gas through a pipe or duct and to transform this flow into the opening or closing of a set of contacts. One type of flow switch (Fig. 5-7) utilizes a pivoted arm having contacts on one end and a paddle or flag on the other end. The end with the paddle or flag is inserted into the pipe so that the flow of liquid or gas over this valve causes a lever to move and open or close the contacts.

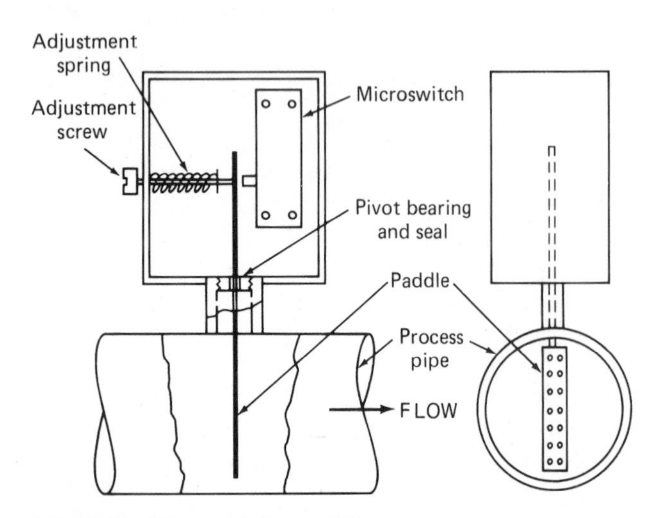

Fig. 5-7 Flow switch, paddle type.

Another type of flow switch uses a difference in pressure across an orifice flange which is installed in the pipe. A pipe is run from each side of the orifice to a pressure switch. The corresponding difference in pressure actuates the pressure switch in one direction or the other, opening or closing its contacts, depending upon their arrangement. Such a flow switch is illustrated in Fig. 5-8.

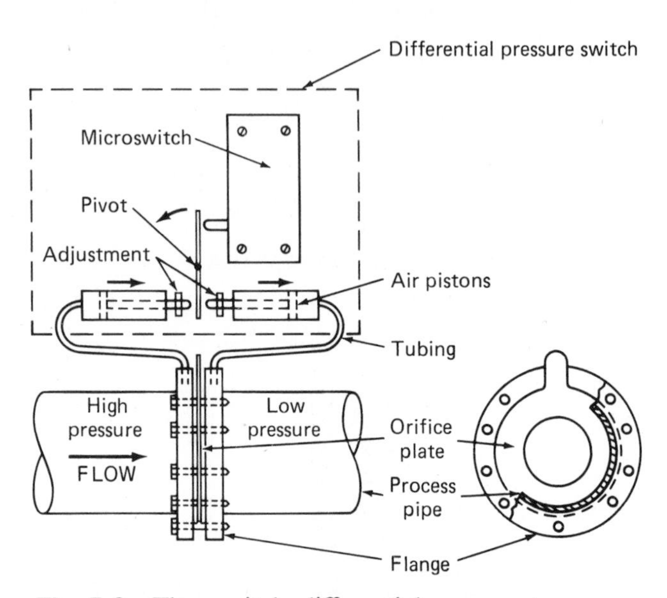

Fig. 5-8 Flow switch, differential-pressure type.

As with other types of pilot control, there are many other possible mechanical arrangements of flow switches. The student should consult manufacturers' catalogs and study the diagrams and illustrations in them to acquire a broader knowledge of the design and application of flow switches.

5-6 THERMOSTATS

The thermostat is probably the pilot device which is built in the greatest variety of mechanical arrangements. Some are made to employ the action of the bellows to move the contacts. Some employ bimetallic strips to sense temperature and actuate the contacts. Many other arrangements are possible with this type of unit. A study of Fig. 5-9 will help the student to visualize a few of the arrangements found in common circuits using thermostats. Thermostats for use in motor-control circuits merely open or close a set of contacts in response to temperature changes, regardless of their mechanical construction and action.

The modulating thermostat moves a contact across a resistance in proportion to the change in temperature, thus varying the relative resistance of the circuit.

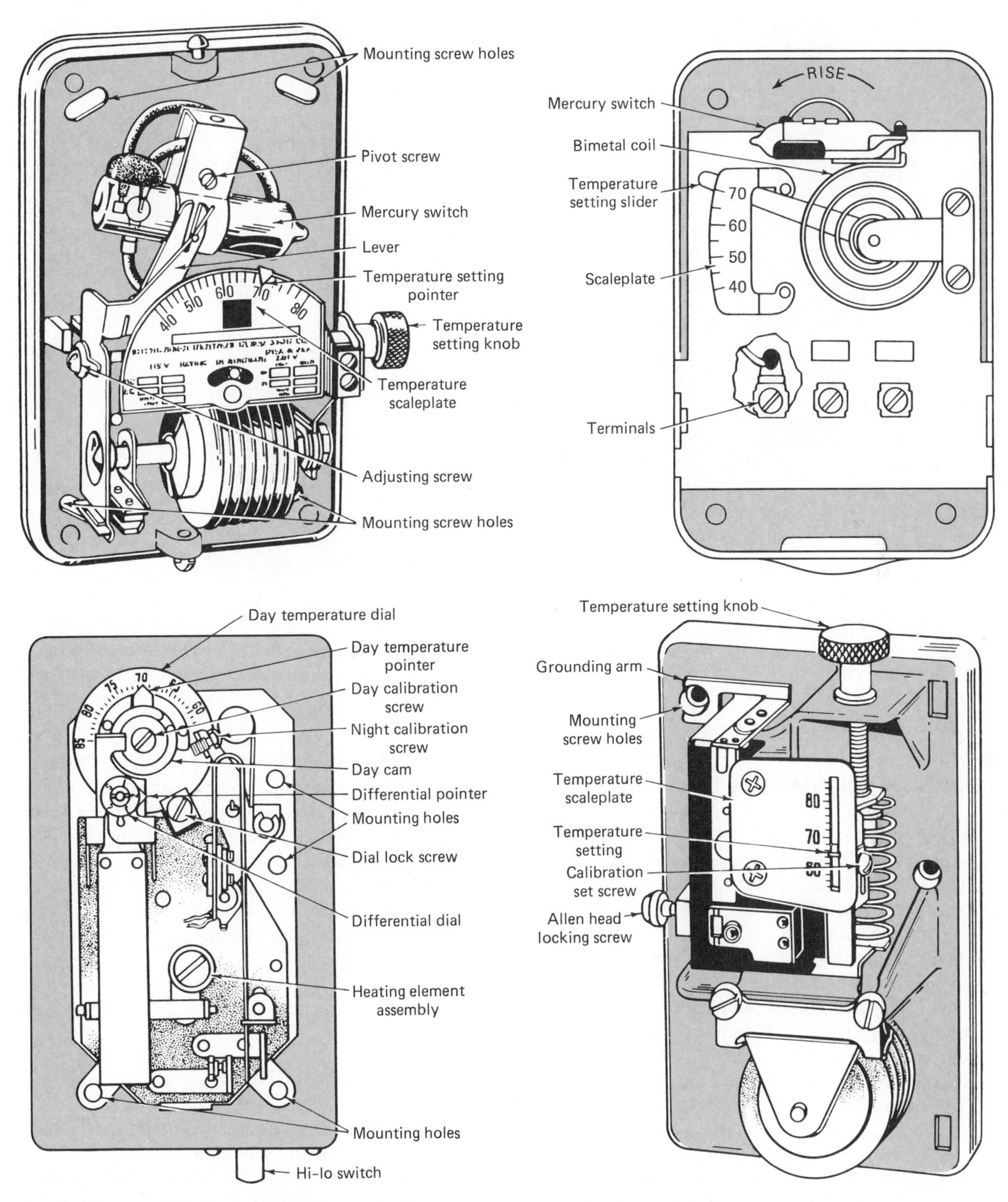

Fig. 5-9 Thermostat arrangements. *(Minneapolis-Honeywell Regulator Co.)*

When connected to a modulating motor (Fig. 5-10 on the next page), it controls motor position in direct response to changes in temperature. The movement of an arm on the motor shaft is directly proportional to the amount of change in temperature. When connected to a damper, the modulating motor can control the amount of air flowing through a duct. When connected to a valve, it can control the flow of water or other liquids or gases through a pipe. While this type of thermostat is seldom, if ever, used for the direct control of a motor, it can initiate control through contacts mounted on the shaft of the modulating motor.

Fig. 5-10 Modulating motor. *(Minneapolis-Honeywell Regulator Co.)*

Fig. 5-11 (*a*) Assortment of pushbutton stations; (*b*) cutaway view of a single pushbutton. *(Cutler-Hammer Inc.)*

5-7 PUSHBUTTON STATIONS

The pushbutton station (Fig. 5-11), probably the simplest of all pilot devices, is the most commonly used in motor-control circuits. Pushbutton stations are of two general types: the maintained-contact type and the momentary-contact type. When the START button is pushed on the maintained-contact type, the contacts close and remain closed until the STOP button is pushed. This action is accomplished through a mechanical linkage from the button to the set of contacts, which are so arranged that they will remain in either position until moved to the opposite position.

The normally open momentary-contact pushbutton, such as used for START buttons, merely closes its contacts for whatever period of time the button is held down. The normally closed momentary-contact pushbutton opens its contacts for whatever period of time the button is held down. Pushbuttons also are available in the double-pole style. This style of pushbutton has one set of contacts that are normally closed and one set that are normally open.

Pushbutton stations are made up of individual pushbuttons which may be normally opened, normally closed, or double-pole units to give whatever combination of contacts that are needed. The most common pushbutton station is the start-stop station. Pushbutton stations are available, however, in most standard labelings to cover normal control operations and are available with special labels to fit special needs. Also found on pushbutton stations are pilot lights to indicate when the motor is running, or possibly when it is not running, and selector switches. A selector switch may be used for hands-off automatic control, or it may be simply an on off switch.

5-8 PLUGGING SWITCHES

The plugging switch, sometimes referred to as the *zero-speed switch* (Fig. 5-12), is a special control device which is operated by the shaft of the motor or a shaft or pulley turned by some part of the machine. The rotation of the shaft causes a set of contacts to close; and when the power is removed, these contacts cause a momentary reversal of the motor. When the motor is running in reverse, the forward set of contacts is closed. When the power is removed, this set of contacts momentarily energizes the motor in the forward direction. This sudden, momentary reversal of direction of rotation of a motor is known as a *plugging stop.*

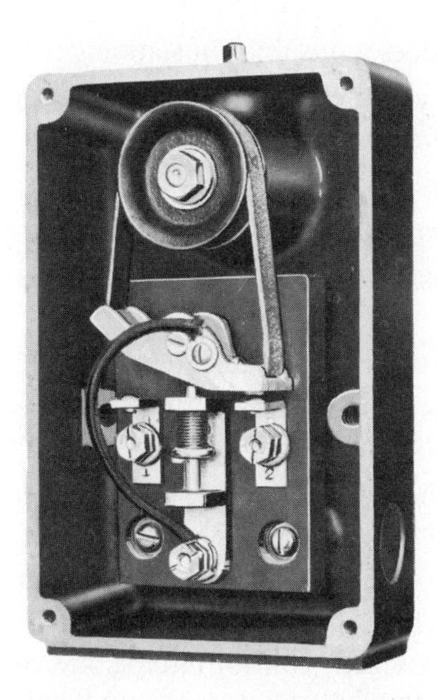

Fig. 5-12 Plugging switch with cover removed. *(Cutler-Hammer Inc.)*

Plugging is used on many precision machines, such as presses, grinders, and other machine tools. The purpose of plugging a motor is to bring it to an abrupt stop, and so the plugging switch must not hold the starter closed for any appreciable length of time. Before a plugging switch is installed on a motor, it should be determined that the machine and motor are built for this rugged operation and that plugging will not endanger the operator.

5-9 TIME CLOCKS

The time clock used for motor-control circuits consists of an electric clock to drive adjustable cams which open and close contacts at any preset time. There are many versions of the time clock in general use. The simplest clock has a set or sets of contacts which can be adjusted to turn the external circuit on once and off once each day. There are several more elaborate versions—from the simple time switch to the elaborate program clock. A program clock may be used to open and close several circuits independently at any desired time. This clock may also be set to skip undesired days so that a program for a period of time, generally consisting of a week, may be set up on the tape, and the clock will make and break the circuit by opening and closing its contacts at each predetermined time.

SUMMARY

The discussion in this chapter is not by any means a complete list of all pilot devices used in motor controls. We have covered the most common and frequently encountered types. Most special pilot devices, such as aquastats, stack switches, relative-humidity controllers, airstats, high-pressure cutouts, and suction-pressure controls, are merely adaptations of the basic types that we have discussed. Control technicians must above all else be able to look at an unfamiliar control device and analyze its mechanical and electrical functions so that they will understand its operation in a control circuit. A thorough understanding of the basic types of control components will enable you to handle most, if not all, new components that you will find on the job.

The best method of becoming familiar with all types of control components is to make a study of the literature which manufacturers are happy to supply, free of charge. This literature generally contains illustrations showing the mechanical and electrical operations of the various devices and is usually accompanied by written descriptions of these operations, their ranges, and their possible uses.

Quality should never be sacrificed in control components, particularly in pilot devices, where the temptation to use cheap competitive units is the greatest. While a thermostat, float switch, or limit switch seems to be an insignificant part of an overall control system, the failure of one of these pilot devices can shut down the operation of a whole industrial plant. This is especially so if it is located on a critical machine.

Many times a technician is overwhelmed by the complexity and size of control systems for such industrial application as central refrigeration plants, central heating plants, automatic manufacturing assembly lines, and other multiple-machine environments. This fear of complexity is, in fact, groundless if you understand the basic functions of control and the principles of operation of the basic types of control components. The overall system of complex control is made up of a series of individual control circuits involving these basic functions and basic components which are relatively easy to master.

REVIEW QUESTIONS

1. How may all components used in motor-control circuits be classified?
2. What is a primary-control device?
3. What is a pilot device?
4. What are the primary-control devices used with any given motor?
5. What determines the type of pilot device to be used?
6. Give an example that shows the differences between a primary device and a pilot device.

7. Why is a pilot device sometimes referred to as a sensing device?
8. What type of pilot device may employ a mercury switch?
9. What does the range adjustment on a float switch accomplish?
10. When would a float switch be used in primary control?
11. What are the three general classes of pressure switches?
12. What is the function of a microswitch used in a limit switch?
13. What is the function of a drum-type limit switch?
14. What is the purpose of a flow switch?
15. What is the function of thermostats in motor-control circuits?
16. What is the function of a modulating motor?
17. What are the two general types of pushbuttons?
18. What type of momentary-contact pushbutton is used for motor start and stop switches?
19. What does a pushbutton station consist of?
20. What type of switch is used for hand-off automatic control?
21. What is another name for a plugging switch?
22. What is the purpose of plugging control?
23. What is the function of a time clock in motor control?
24. What is a program clock used for?
25. What is the best method of becoming familiar with control components?
26. Why should quality never be sacrificed in control components?
27. What is the best method of overcoming fears of complex motor-control systems?
28. Most complex control circuits are merely interlocked in what way to perform what function to several machines?

6
Control-Circuit Diagrams

If you were to find yourself in a foreign land and unable to read or speak the language, you would still recognize such familiar objects as buildings, automobiles, newspapers, and people, but you would not be able to fully understand and communicate what was going on around you.

If you could understand and speak the oral language but could not read the printed words, you would still be dependent upon someone else for a full understanding. The same thing applies to control work. If you have mastered the first five chapters of this book, you will now be able to understand and speak the oral language of controls. Until you master the written language of control diagrams, however, you will be dependent upon someone else to understand them.

This chapter deals with the written language of control and control circuits. Study it until you can read and understand control prints readily and with reasonable speed. There are only a few basic symbols that are used to express the meaning and functions of control circuits. The chief difficulty is that while these are standard symbols, there is no real standard for their use and sometimes it is necessary to do a little guessing as to what these symbols mean in individual control diagrams.

The symbols used in this chapter are those in most common use.

6-1 SYMBOLS

In Fig. 6-1 on the next page, symbol 1 shows a normally open contact that is automatically operated. It might represent a line contact on a starter, the contact of a limit switch, the contact of a relay, or any control device that does not require manual operation. Symbol 2 a normally closed automatic contact, and all that applies to symbol 1 applies to symbol 2 also, except its on and off positions are opposite those of symbol 1.

Symbol 3 represents a manually operated, normally open contact of the pushbutton type. Symbol 4 represents the same type of contact except that it is normally closed. Symbol 3, the normally open pushbutton, should be drawn so that there is space between the dots and the cross bar, but it is not always drawn with such care. If the cross bar is above the dots, the symbol is for a normally open contact, even though the bar may be touching the dots. Symbol 4 should be drawn so that the cross bar just touches the bottom of the dots, but again this is not always done. If the cross bar is below the dots, then it is normally closed even though the bar does not touch the dots. A good way to remember this is to picture the symbol as being a drawing of a pushbutton, which it is. If you push on the button part, represented by the vertical line, the cross bar will move downward. When it is above the dots or contact points, the pressure will close them. When it is below the dots, it will move them apart and open the circuit.

Symbols 5 and 6 represent manual contacts of the toggle-switch type, 5 being normally open and 6 being normally closed.

Symbol 7 is a toggle switch of the single-pole, double-throw (SPDT) type, where one contact is normally open and the other normally closed.

When more than one set of contacts are operated by moving one handle or pushbutton, they are generally connected by dotted lines, as in symbol 8 and 9. The dotted lines represent different forms of mechanical linkage that will make the two contacts operate together. One other method that is used frequently to show pushbuttons that have two sets of contacts is shown in symbols 10 and 11. Symbol 10 has two normally open contacts, and symbol 11 has one normally open and one normally closed contact.

Symbol 12 is a pilot light, which is identified chiefly by the short lines radiating out from the center circle.

Symbol 13 represents a coil. It might be a relay coil or a solenoid coil or the closing coil on a starter. Later we shall discuss how to tell which it is.

Symbol 14 also is used to represent an operating coil.

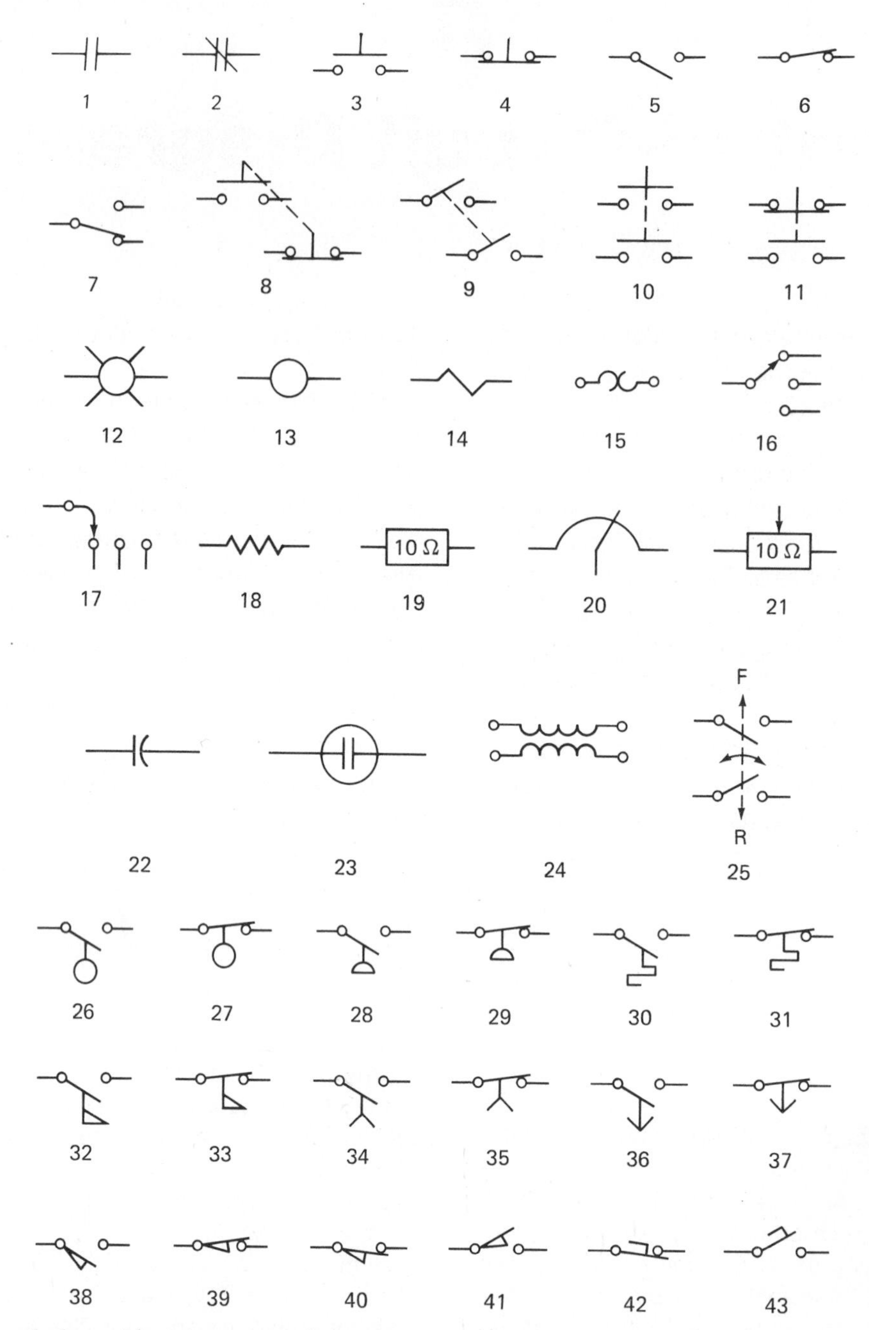

Fig. 6-1 Basic symbols used in motor control circuits.

Symbol 15 represents the heating element of an overload relay.

Symbol 16 is a rotary selector switch. The same type of switch is shown by symbol 17.

Symbols 18 and 19 show two ways that resistors are drawn. Symbols 20 and 21 are variable resistors. Capacitors are shown in symbols 22 and 23. Symbol 24 is used to represent a transformer. Symbol 25 shows a plugging switch, which stops plugging action after the drive has practically come to rest.

There is a need in some diagrams for clarification of specific devices and their operation; the additional symbols 26 to 43 are NEMA approved. Symbols 26 and 27 represent, respectively, normally open and normally closed liquid-level switches. Symbols 28 and 29 represent, respectively, normally open and normally closed vacuum or pressure switches. Symbols 30 and 31 represent temperature-activated switches. Symbols 32 and 33 represent flow switches.

Symbols 34 and 35 represent timer contacts which have delay on energizing (TDOE). Symbols 36 and 37 represent timer contacts which have delay on de-energizing (TDODE).

Symbols 38 and 39 represent direct-actuated limit switches. Symbol 40 represents a normally open limit switch which is held closed. Symbol 41 represents a normally closed limit switch which is held open.

Symbols 42 and 43 represent foot switches.

With English, or any other language, the meaning of a word depends to some degree upon how it is used, and so it is with the language of control symbols. As we progress in our study of control circuits, we shall develop these few basic symbols into words and sentences that will tell the story of what functions are to perform the control components represented in diagrams by these symbols.

6-2 DIAGRAMS

The control diagram is the written language of control circuits, and it takes several different forms to fit the particular needs for which it is used. As with all languages, the same form will not suit all needs. Some things are better expressed in poetry, while others are best written in prose. There are three general types of control diagrams in use.

The first of these is the wiring, or schematic, diagram (Fig. 6-2*a*), which is best suited for making the initial connections when a control system is first wired or for tracing the actual wiring when troubleshooting.

The second type is a line, or ladder, diagram (Fig. 6-2*b*), which is by far the easiest to use in trying to understand the control circuit electrically. Most circuit diagrams are first developed by drawing a ladder diagram.

The third type is the wireless-connection diagram (Fig. 6-3 on the next page), which has very few claims to usefulness except that it is compact and eliminates confusing lines when many wires must be shown. Its chief advantage is for installing already-formed wiring harnesses for factory assembly lines.

6-3 WIRING DIAGRAMS

Wiring diagrams are developed by drawing the symbol for each component in its two-dimensional relation to the other components and then drawing the wires connecting the components. In other words, it is a drawing of the equipment and wires more or less as they will be run on the job. Its chief advantage is that it helps to identify components and wires as they are found on the equipment.

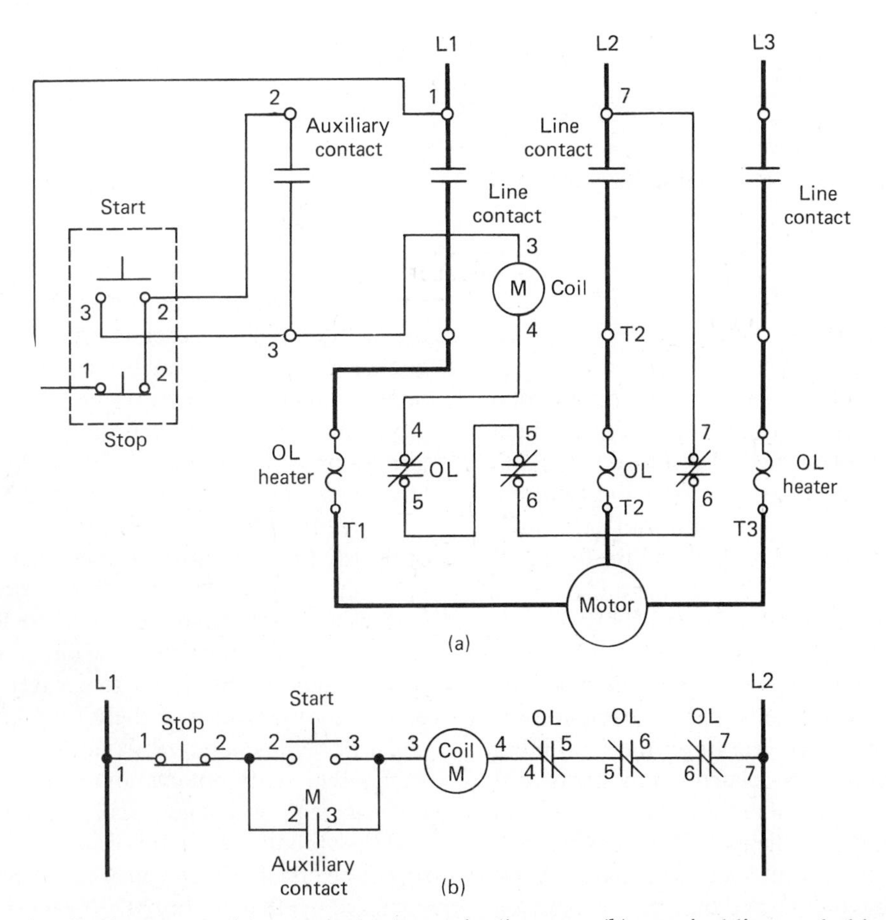

Fig. 6-2 Motor circuit diagrams. *(a)* A typical schematic diagram; *(b)* a typical line, or ladder, diagram.

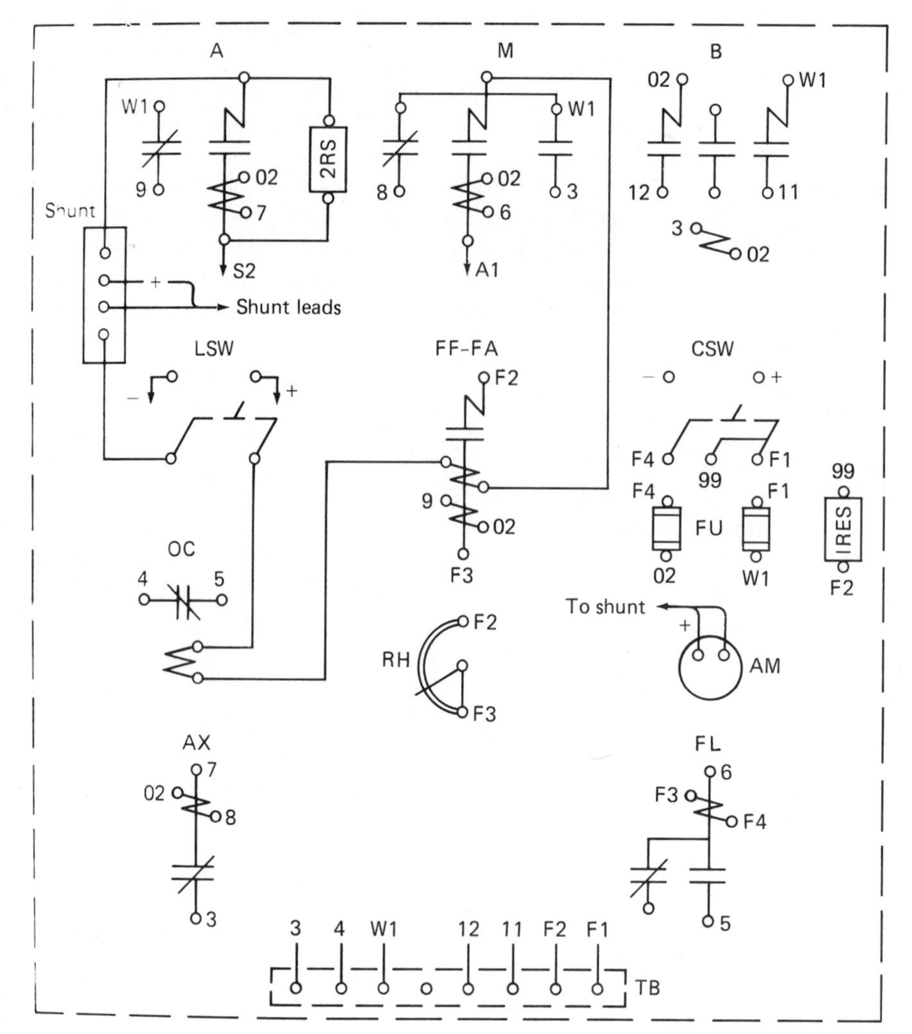

Wire		Connect to
+		LSW, CSW
−		LSW, CSW
F1	TB	CSW, FU
F2	TB	IRES, RH, FF-FA
F3		FF-FA, RH, FL
F4		FL, CSW, FU
02		FU, M, A, AX, FF-FA, B
W1	TB	FU, M, A, B
3	TB	M, AX, B
4	TB	OC
5		OC, FL
6		FL, M
7		A, AX
8		M, AX
9		A, FF-FA
11	TB	B
12	TB	B
99		IRES, CSW

Wire table

Fig. 6-3 A typical wireless-connection diagram.

6-4 LADDER DIAGRAMS

The ladder diagram (Fig. 6-2*b*) is a representation of a control circuit in a form that shows the order in which its components are connected electrically. Assume that you have wired part of a control circuit beginning at line 1 and you continue to wire each contact, switch, and coil until you reach line 2. With all the contacts, switches, and coils free of their mountings and the wire out in the open, you then take the ends of each connecting wire and stretch the wire tight. What you would see would be a straight wire, broken in places by the contacts, switches, and coils. This is what you see in a ladder diagram. Each line from line 1 to line 2 represents a wire and its associated components as it would appear if stretched out in the above manner. A careful study of this type of diagram in this chapter will show you that the more complex circuits have several interconnecting wires or lines and that each of them is a small circuit within itself.

The chief advantage of this type of diagram lies in the fact that it shows each component in electrical sequence in the circuit. Each component is shown in the order it falls in the electric circuit without regard to its physical location. There is no better diagram for obtaining an understanding of a control circuit or for locating trouble in a control circuit.

To read a ladder diagram, start at the left-hand side of the top line and proceed to the right. If a contact is open, the current will not go through; if it is closed, the current will go through. In order to energize a coil, for example, or other device in the circuit, you need every contact and switch closed to form an uninterrupted path. In other words, if there is an open contact, the coil will be de-energized; if not, it will be energized. Remember that contacts and switches are shown in their normal, or de-energized, position.

The symbols used in this type of diagram must have some means of telling you what or who operates the components they represent. Since we have put them into their electrical instead of their physical position in the circuit, the several contacts of a relay, for example, might be scattered from one end of the diagram to the other. In order to identify the relay coil and its several contacts, we put a letter or letters in the circle that represents the coil (Fig. 6-4). Each of the contacts that

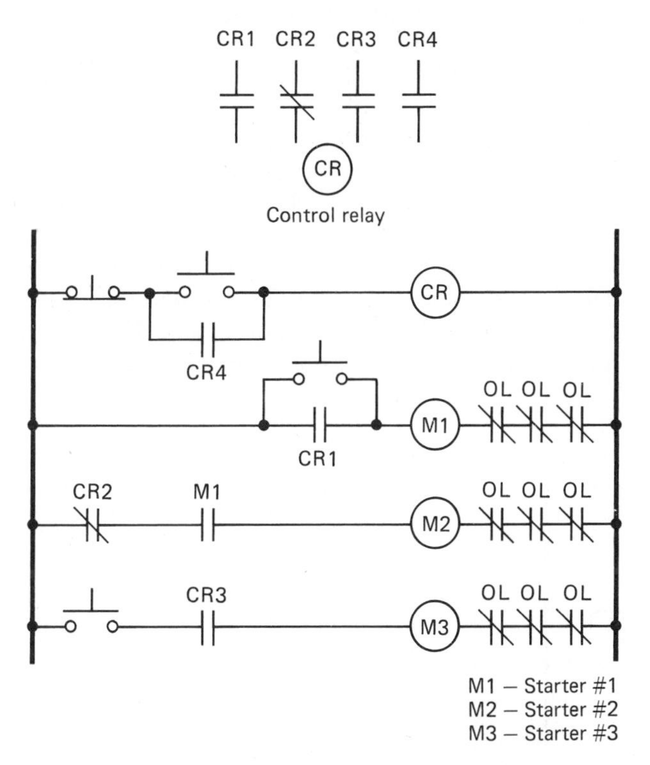

Fig. 6-4 Identification of contacts and coils.

are operated by this coil will have the coil letter or letters written next to the symbol for the contact. Sometimes, when there are several contacts operated by one coil, a number is added to the letter to indicate the contact number, generally counted from left to right across the relay.

While there are standard meanings of these letters, most diagrams provide a key list to show what the letters mean—generally they are taken from the name of the device. For instance, the letters *CR* generally are used to indicate the coil of the control relay. The letters *FS* are used frequently to show a float switch. The letters *LS* are used to show a limit switch. Quite frequently, when several motor starter coils are shown on one control diagram, such as a circuit for sequence operation of several motors, the starter coil may be shown with the letters *M*1, *M*2, *M*3, etc.

6-5 DEVELOPING A LADDER DIAGRAM

In order to see the relationship between the schematic and the ladder diagram, suppose we develop a ladder diagram from a schematic diagram (Fig. 6-2). This method of development is highly recommended for use in the field when a control diagram is needed but not available. The first step is to number the wires of the control circuit. Start where the control wire leaves L1, and number each end of each wire. Change numbers each time you start another wire until you reach L2, or the end of the control circuit. In Fig 6-2*a* we numbered the wire from L1 to STOP button 1 and placed the numeral 1 at each end. The wire going from the other side of the STOP button to the START button and to terminal 2 of the starter auxiliary contact is numbered at each of its three ends with the numeral 2. The wire that connects the other side of the START button to terminal 3 of the starter auxiliary contact and the starter coil is numbered 3. A wire from the starter coil to the first overload contact is numbered 4. The wire between the two overload contacts is numbered 5. The wire from the second overload contact to L2 is numbered 6.

To draw the control diagram of this circuit, the first step is to draw two vertical lines, one on each side of the paper (Fig. 6-2*b*). Now draw a short horizontal line to the right from L1 and number each end with numeral 1. This represents wire 1 of the schematic diagram, which ends at the STOP button. Draw the symbol for STOP button at the end of this line. Now draw wire 2 from the STOP to the START button and down to the auxiliary contact. Note that this is shown as an automatic contact operated by coil M; therefore, it is labeled M to show what operates it. Continue the circuit horizontally across the paper, following the numbers on the schematic diagram, until you reach L2. Be sure to show each contact in its normal, or de-energized, position. When a schematic diagram of the control circuit is needed and only a ladder diagram is available, the reverse of the above method should be used.

To read the control diagram in Fig. 6-2*b*, start at L1, which is a hot line, and follow the circuit across the page. First we come to the STOP button. It is normally closed so that the current can flow through, and we can proceed to the START button and auxiliary contact M. Both of these are normally open, so the current cannot go any farther. Contact M closes when coil M is energized, so we cannot complete the circuit that way. The START button can be pushed, which will close its contacts and allow current to flow to the coil M and on through the two normally closed overload contacts, marked OL, to L2. This completes the circuit to coil M, and it closes the starter and contact M. When we release the START button, thus opening its contacts, the coil does not drop out, because contact M is now held closed by coil M. The motor is now running and will remain so until the control circuit from L1 to L2 is broken.

To stop the motor manually, all that is needed is to push the STOP button, which interrupts the circuit at this point, causing an interruption of current to coil M and dropping out the contacts of the starter. Contact M being operated by coil M is now open, so that when we release the STOP button, the coil is not energized again. Note that when the motor draws too much current, one or both overload contacts will be opened, thus interrupting the circuit between coil M and L2. The result of opening the circuit at this point is the same as that of pressing the STOP button.

While this is a simple circuit and fairly easy to follow on either diagram, the same system for developing the schematic diagram and analyzing the control circuit will work regardless of the complexity of the circuit.

Suppose now that we add one more start-stop station to the control circuit of Fig. 6-2*a*. The new schematic diagram is shown in Fig. 6-5*a*. If you look at the numbering of the wires in the diagram, you will see that it goes from 1 to 8. This increase in total numbers is caused by the insertion of the extra STOP button in the circuit. If we follow the same technique used on Fig. 6-2*b*, we will develop the ladder diagram of Fig. 6-5*b*, which starts at L1 and proceeds horizontally through the first STOP button, the second STOP button, and the first START button, which is paralleled by the second START button and auxiliary contact. From there it proceeds to the coil of the starter marked M and thence to the first overload contact and the second overload contact to L2.

As in the preceding circuit, all STOP buttons will be closed in their normal position so that current can flow from L1 to as far as the parallel group of START buttons and the auxiliary contact. Current can flow from L2 through the normally closed overload contacts to the coil M, so that now all that is needed to start the motor is to close one of the START buttons. Since the START buttons are in parallel, either of them will complete the

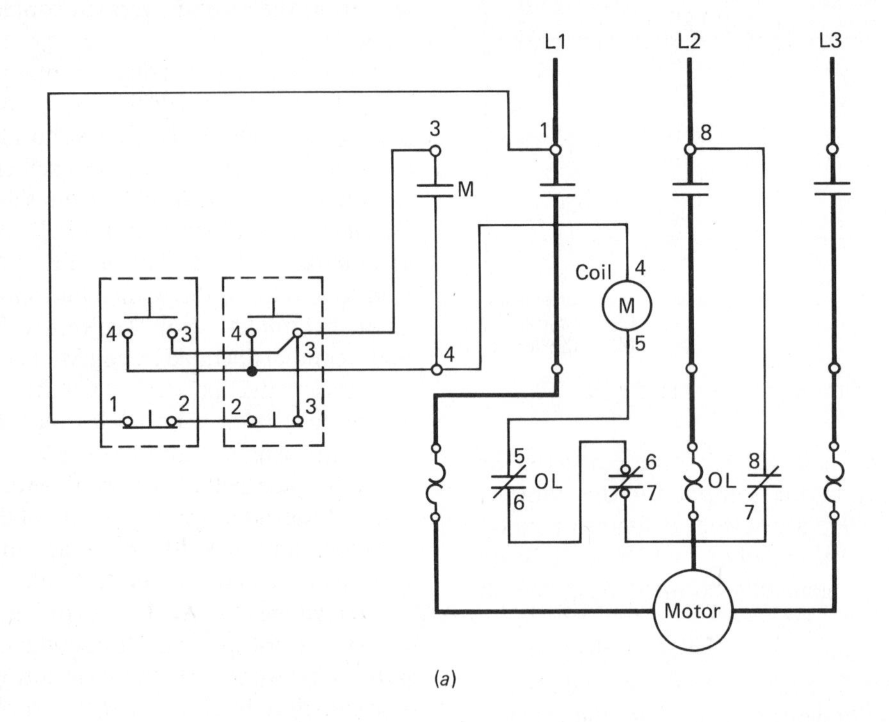

(*a*)

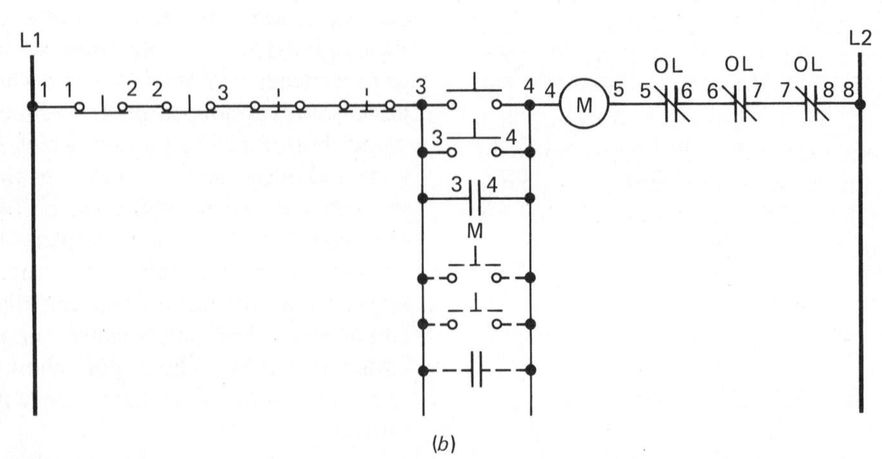

(*b*)

Fig. 6-5 The basic T-formation.

circuit from L1 to coil M, so that it makes no difference which one is pushed in order to energize the coil.

It should be noted that there are two additional STOP buttons, two additional START buttons, and an additional contact shown dotted on this diagram. They indicate additional controls as they would be added to this circuit. Careful note should be taken of this diagram, since it is the basic T formation used for any circuit with multiple control components used to control a single coil. Note that all the STOP buttons are connected in series from one side of the line or the other. The start components, in this case consisting of two or more START buttons and one or more contacts, are in parallel. The value of this T formation lies in understanding that if any control component, regardless of its type, is to be used to stop the motor, it will be placed in series with the STOP button; if it is to be used to start the motor, it will be placed in parallel with the START button. You can develop more complex circuits by placing additional components in series or parallel with existing circuits. To perform the function of STOP, add additional stop components in a series with the original STOP circuit. To perform the function of START add additional start components in parallel with the existing START circuits.

6-6 ADDING CIRCUIT ELEMENTS

Suppose that you are instructed to add a limit switch, float switch, or pushbutton to an existing circuit. Then, if this component is to be used to stop the motor, all that is necessary is that you locate the wire connecting L1 to the STOP button or other components and break it at some point with the new component.

Suppose instead that you are required to install a control component such as a limit switch, float switch, or pushbutton to perform the function of start for the motor. Then all that is required is that you parallel the new component with the existing start component. These additional components are represented in Fig. 6-5*b* by dotted lines.

It should be noted also that components used to perform the function of stop are normally closed components. That is to say, their contacts are in the closed position whenever the components are deactivated. Those components which are to perform the function of start are normally open components. In other words, their contacts are open in their deactivated state. There is no limit to the number of components that can be added in series with the STOP button of the simple circuit shown in Fig. 6-2*b* to perform the function of stop, nor is there any limit to the number of components that can be added in parallel with the START button to perform the function of the start for the coil M.

Consider the circuits shown in Fig. 6-6. The top circuit is the same as that shown in Fig. 6-5*b*. The

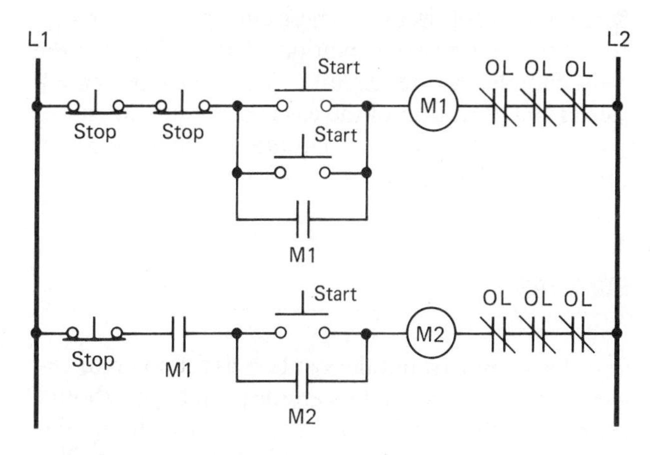

Fig. 6-6 Interlocking.

current can flow through both of the normally closed STOP buttons to as far as the START buttons. All that is required to energize the coil is to press the START button, thus closing its contacts and energizing coil M1. Coil M1, in turn, closes contact M1 in parallel with the START buttons, thus maintaining the circuit to coil M1.

Now look at the bottom circuit, and you will see that the current can flow from L1 through the normally closed contact only as far as the normally open contact M1. This contact must be closed in order to energize coil M2 through the START button. This contact has an identification M1, which indicates that it would be closed whenever coil M1 is energized. This means, then, that the motor which is energized by coil M1 must be running before we can start the motor which is energized by coil M2. If we start motor M1 by pressing the START button, coil M1 is energized, thus closing both contacts labeled M1. The contact in parallel with the START buttons is used to maintain the circuit to coil M1. The contact in the lower circuit, labeled M1, will be closed and will allow current to flow as far as the START button. When this START button is pushed, current can reach coil M2. Energizing this coil closes contact M2, maintaining the coil circuit and permitting the second motor to run.

Consider what happens if we push the STOP button of coil M1. This will break the circuit and de-energize coil M1, thus dropping out all its contacts. This will open the maintaining contact in parallel with its START buttons and the contact in series with the STOP button of coil M2. Opening of the contact in series with the STOP button will de-energize coil M2, which will drop out its contact M2, and both motors will be stopped, even though the button we pressed was in the circuit for motor M1. Circuits of this type are frequently used for multiple conveyor-belt operation, where the first conveyor must not run unless the following conveyor is running, thus preventing material from piling up where the two conveyors converge.

Actually, we have considered only three fairly simple basic control circuits. These circuits, however, represent the majority of conditions that are found in the most complex control circuits. The same type of analysis of the operation of the electric circuit will enable you to understand many circuits which now might puzzle you considerably.

SUMMARY

In order to understand the symbols as found on drawings or diagrams made by various people, you should study manufacturers' booklets and control-circuit diagrams. This study will enable you to become familiar with the many types of symbols used to represent a single component just as your understanding of the spoken or written word in English depends upon the size of your vocabulary. The knowledge of words and phrases in English makes it easier for us to understand the spoken and written word as it is presented by various people, and so it is with control circuits. The greater your knowledge of the symbols and the components that will be used to perform the functions of control, the better will be your understanding of the various diagrams and circuits as drawn by the many people engaged in this kind of work.

In this section we have discussed in some detail the control, or ladder, diagram. Emphasis has been put on the ladder diagram because it transmits concisely to the reader understandable electrical information about control circuits. The same procedure for reading and understanding the control functions in a circuit from the ladder diagram also applies to the schematic, or wiring, diagram. When it is necessary to use a wiring diagram to analyze or understand a control circuit, it is necessary that you trace each wire, beginning at the source of power and noting each component or contact that is in the circuit and what its function might be. It is highly recommended that on the more complicated circuits, if a control diagram is not available, you develop such a diagram using the methods put forth in this chapter. Understanding the circuit will be much easier when this procedure is followed.

To read a wireless-connection diagram, the same principles apply except that you must find the proper component by comparing the number of the wire indicated as it leaves a terminal of each piece of equipment. Caution is needed to be sure that you have found all of the places where a particular wire is connected by finding all the points labeled with the same numeral. Again it is suggested that you develop a control diagram from the wireless-connection diagram before attempting an analysis of the control circuit if it contains more than a very few components.

Should it be desired to have a wiring diagram when only a control diagram is available, one can be developed by applying the reverse of the procedure outlined for the development of a control diagram from a wiring diagram. Draw each component to be used in the circuit in its proper physical relationship to the others. Now number each wire on the control diagram as we have been doing. Then number each terminal of each component on the wiring diagram as it is numbered on the control diagram. All that is left is to connect corresponding numbers by wires or lines on the drawing, and you will have a wiring diagram which includes the control circuits shown on the ladder diagram.

REVIEW QUESTIONS

1. What chief difficulty is encountered when standard symbols are not used in control circuits?
2. Identify symbols 1 to 43 in Fig. 6-1.
3. Identify the following abbreviations:

   T.D.
   TDOE
   SPDT
   TDODE
   T.C.
   DPST

4. What are the three different general types of electrical diagrams?
5. What is the main use of the wiring diagram?
6. What is another name for the line diagram?
7. What is the main use of the wireless-connection diagram?
8. How is a wiring diagram developed?
9. What is the chief advantage of the wiring diagram?
10. What is the chief advantage of the ladder diagram?
11. What is the procedure to follow in reading a ladder diagram?
12. How are the relays and contacts identified on a ladder diagram?
13. When a relay has several contacts, how are they identified?
14. Do the letters within the coils of a ladder diagram have any special meaning?
15. What are the steps in developing a control diagram?
16. How is the ladder diagram read?

17. What is the rule for adding starts and stops to a control diagram?

18. What is the rule for adding limit switches to a control diagram?

19. What is the purpose of interlocking?

20. What is the procedure for tracing through ladder and wiring diagrams?

21. How is a ladder diagram developed from a schematic diagram?

22. How is a wiring diagram developed from a ladder diagram?

PROBLEMS

P6-1. For each of the circuits in Fig. 6-7*a* through *e*, redraw the control circuit in the form of a ladder diagram.

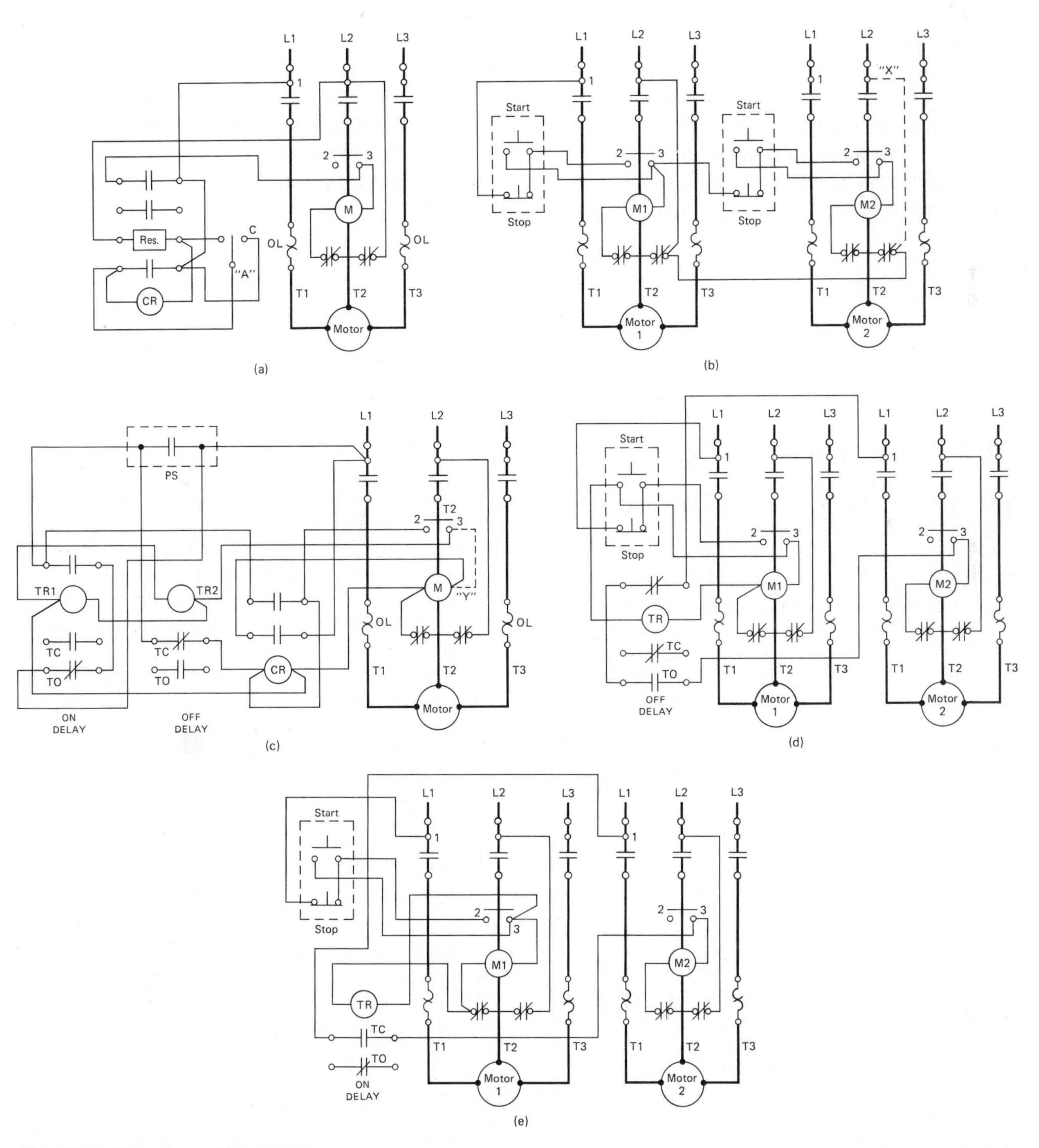

Fig. 6-7 Circuits problem P6-1.

7

Development of Control Circuits

A control circuit is very seldom drawn or designed as a complete unit. Rather, it is developed one step at a time to provide for each control function that it will be expected to perform. It is very much like writing a letter. When you have a general subject in mind, you proceed, sentence by sentence, to put this idea on paper. The same procedure should be followed in developing a control circuit. You must have all of the control functions in mind when you start, so as to provide for each function in its proper relationship to all other functions as the circuit is developed.

Schematic diagrams are used to develop all types of control circuits. Control circuits are also developed into wiring diagrams that are used for the electrical construction of circuits.

In this chapter we will see how the various types of diagrams are developed and expanded.

7-1 TYPES OF CONTROL CIRCUITS

There are two basic types of control circuits: three-wire circuits and two-wire circuits. These designations stem from the fact that for three-wire circuit control only three wires are required for the ordinary across-the-line motor starter to the control components, and for two-wire circuit control only two wires are required (Fig. 7-1).

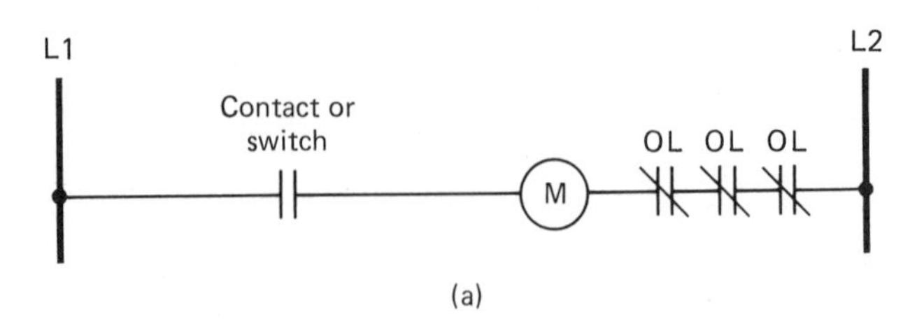

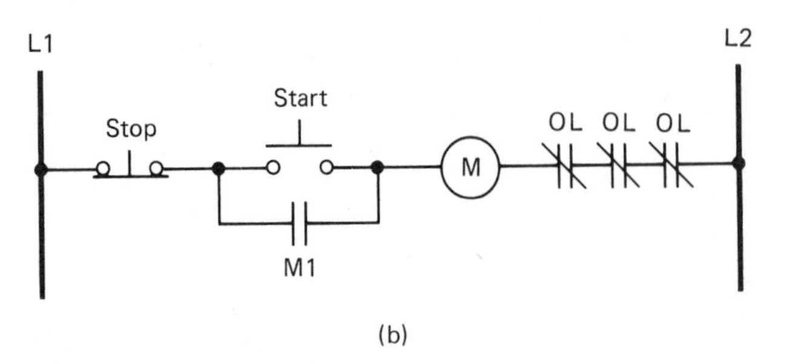

Fig. 7-1 Basic control circuits. *(a)* Two-wire; *(b)* Three-wire.

The three-wire control circuit requires that the primary pilot-control components be of the momentary-contact type, such as are momentary-contact pushbuttons. Maintained-contact devices, such as limit switches and float switches, may be used in various parts of the circuit to supplement the primary start and stop control devices. This type of control is characterized by the use of an auxiliary contact on the starter to maintain the coil circuit during the time that the motor is running.

The two-wire control circuit uses a maintained-contact primary pilot-control component, which may be a simple single-pole switch, a maintained-contact pushbutton station, or any type of control component which will close a set of contacts and maintain them in that position for as long as the motor is to be running. The opening of this contact or contacts stops the motor by dropping out the coil of the starter.

All control circuits, regardless of how complex they may be, are merely variations and extensions of these two basic types. In this chapter each of these basic circuits is developed by the addition of contacts or pushbuttons operated by various control components. The ladder diagram is used for the development of all control circuits.

The simplest method for the development of a control circuit is to start with the coil and the overloads. Add the primary start and stop control device, which, generally, in a three-wire circuit consists of a STOP and START pushbutton used in conjunction with the auxiliary contact of the starter. To this circuit all contacts or pushbuttons that are to be used to perform the additional control functions are added one at a time until the final circuit has been developed.

Keep in mind, when considering a three-wire control circuit, that all devices intended to perform the function of stop must be normally closed devices and will be located in series with the original STOP button. All devices which are to perform the function of start must be normally open control devices and will be connected in parallel with the original START button.

Sometimes a circuit requires that two or more normally open contacts or pushbuttons be closed before the function of start can be performed. These contacts or pushbuttons would be connected in series, and then the series connection paralleled with the original start components. If it is desired that several contacts or pushbuttons be opened before the function of stop is performed, then these normally closed contacts or pushbuttons should be wired in parallel and then connected in series with the line to perform the function of stop.

When there is a definite sequence to the action of the various control components, you should add them one at a time to your control circuit in the same order as their operating sequence. Be sure to check the circuit for proper electrical operation after each contact or pushbutton has been added to be sure that you have not interfered with the proper function of any control component which has already been placed in the circuit.

The preceding chapters of this book provided the basic knowledge of control functions, control components, and circuit diagrams. Using these building blocks, you are now ready to develop specific control circuits. Experience in developing your own circuits to perform desired functions is a necessary prerequisite to interpreting and analyzing the operation and functions of someone else's circuit.

7-2 EXPANDING AN EXISTING CIRCUIT

In order that the step-by-step method of circuit development can be made more clear to you, we shall consider our first circuit as a series of jobs done at different times to improve the operation of the original circuit.

The existing control circuit is to control a pump which pumps water from a storage tank into a pressure tank. The physical arrangement of the pump and the two tanks, along with the final control components, is illustrated in Fig. 7-2*a*. As the original circuit stands, it is a manual operation requiring that the START button be pushed whenever the water is too low in the pressure tank (Fig. 7-2*b*). The pump is allowed to run until the tank is observed to be full. The operator then pushes the STOP button, securing the pump and stopping the flow of water into the pressure tank.

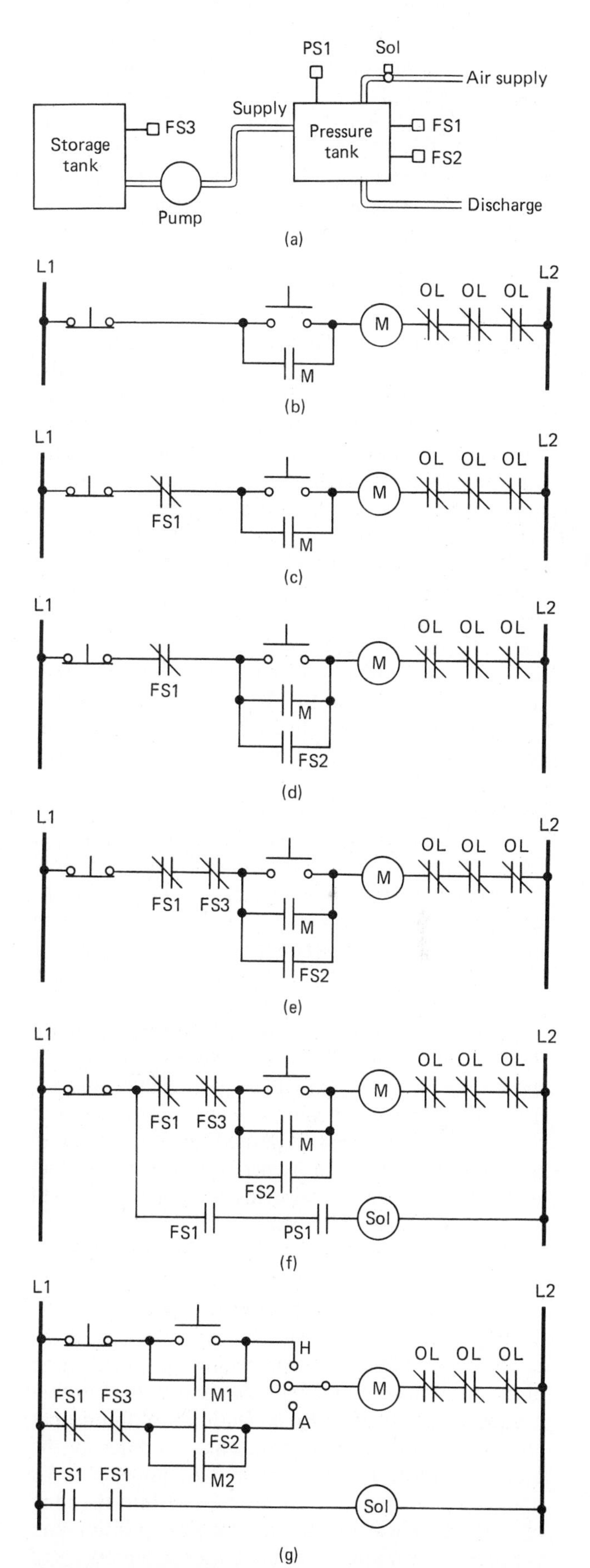

Fig. 7-2 Expanding an existing circuit. Automatic control for a water pump.

The owner now desires that a float switch be installed in the pressure tank near the top so that the operator need only press the START button, thus energizing the pump and starting water to flow into the tank. When the level of the water has reached float switch 1, its contacts will be opened, thus stopping the pump and the flow of water. The function to be performed by float switch 1 is that of stop. Therefore, it must be a normally closed contact and must be connected in series with the original STOP buttons, as shown in Fig. 7-2*c*.

After operating with this control for some time, the owner decides that it will be more convenient if the pump is started automatically as well as stopped automatically. He requests that another float switch be installed to maintain the lower level of the tank. This version of the control circuit requires that the pump be started whenever the water reaches a predetermined low level. The control function desired is that of start, so the float switch must have a set of normally open contacts that will be closed whenever the water drops to the lowest desired level. These contacts must be connected in parallel with the original START button so as to perform the function of start for the motor. This connection is shown in Fig. 7-2*d*.

After some time operating with the new control circuit, it is discovered that occasionally the storage tank drops so low in water level that the pump cannot pick up water. The owner requires a control to prevent the pump from starting whenever the storage tank is low in water. Even though this control does not necessarily stop the pump while it is running, it must prevent its starting whenever the water is low. It must also stop the pump if it is running and the water reaches this low level in the storage tank. Thus, the new control performs the function of stop for the pump.

This function can be obtained by the installation of a float switch to sense the extreme low level of water in the storage tank. Float switch 3 was installed and adjusted to open a set of contacts whenever the water in the storage tank reached the desired low level. Because the control function to be performed is STOP, float switch 3 must have normally closed contacts which will be opened whenever the water level drops to the set level of the float switch. It is wired in series with the other stop components. (See Fig. 7-2e.)

Later it is decided that the pressure placed on the line by the pressure tank when it is full is insufficient for the needs of the plant. The owner requests the installation of the necessary components and controls to maintain a pressure on the tank by the addition of the proper amount of air to the top of the tank. In order for the proper balance of water level and air pressure to be maintained at all times, air must be let into the tank only when the water level is at its highest position and the pressure is under the desired discharge pressure of the tank.

In order to achieve this, suppose that we install a solenoid valve in the air supply line which will allow air to flow into the tank only when the coil of the solenoid valve is energized. Now we can install a pressure switch in the top of the tank which will sense the pressure in the tank at all times. This pressure switch will perform the function of start for the solenoid valve. When the pressure is lower than the set point of the pressure switch, its contacts must close and complete the circuit through it to the solenoid. If, however, the water is below its top level when the pressure drops, we do not want the solenoid valve to open. Therefore, we require the function of stop in regard to water level, to prevent air being put into the tank when it is not desired.

If float switch FS1 is of the double-pole variety, having one normally open and one normally closed set of contacts, we can wire it into the circuit as shown in Fig. 7-2*f*. The circuit for the solenoid valve is a two-wire control circuit requiring that both FS1 and pressure switch PS1 be closed in order that air will be placed in the tank by the energizing of the solenoid valve. When the water level reaches its highest point, FS1 will be activated. The normally closed contact in the pump circuit will open and the normally open contact in the solenoid circuit will close. If the air pressure is low, the contacts of PS1 will be closed and air will flow into the tank until either the water level drops and opens FS1 or the pressure increases to normal and opens PS1, thus satisfying the requirements of the control circuit as they have been specified by the owner of the plant.

While the circuit of Fig. 7-2*f* gives a degree of hand operation because the pushbuttons were left in the circuit, it would be preferable to have either a definite hand operation or a definite automatic operation, as desired by the operator. The necessary changes required to give hand, off, and automatic operation to the circuit are shown in Fig. 7-2*g*.

If you had been charged with the responsibility of developing the final circuit of Fig. 7-2*g*, you would have had certain specifications or requirements as to the proper function or operation of the completed circuit. The first of these probably would have been that it have hand, off, and automatic control selection; the second, that the pump be controlled so as to maintain the water level in the pressure tank between high and low points; third, that the pump be prevented from running whenever the water levels in the storage tank were below a given point; and fourth, that the pressure on the pressure tank be maintained by adding air to the tank whenever necessary. To develop this circuit properly from this set of specifications, the procedure would be the same as that we have followed, assuming that the circuit was built up a little at time by going back and adding control components to the original manual circuit.

7-3 SEQUENCE-CONTROL CIRCUITS

Our second circuit will be for the control of three conveyors so arranged that conveyor 1 dumps material onto conveyor 2, which in turn dumps its material onto conveyor 3, which is used to load trucks or other vehicles at a shipping dock or a warehouse. The specifications for the operation of this circuit are:

1. One pushbutton is to start all conveyor motors in sequence from 3 to 1 or, in other words, from the last to the first.
2. An overload on any conveyor will stop all conveyors.
3. One STOP button will stop all conveyors in sequence from 1 through 3, or, in other words, from first to last.

An additional requirement is that there be a 2-min delay between the stopping of each conveyor in the sequence so that the material on the following conveyor clears each conveyor before it is stopped.

If we are to develop this circuit step by step, then our first step is to meet the requirements of specification 1, that a single pushbutton start all conveyors in sequence starting with conveyor 3. The circuit for this will be found in Fig. 7-3*a*. Here you will find a control relay which is started and stopped by a three-wire pushbutton control. It is maintained during the run operation by a set of contacts on the control relay, identified on the drawing by CR1. Since conveyor 3 is required to be the first conveyor to start, the contacts identified on the drawing by CR2, which are closed by the control relay, are connected between the starter coil and the line, giving two-wire control for conveyor 3. This conveyor will start when the control relay is energized and stop when the control relay is de-energized.

In order that conveyor 2 be prevented from starting until conveyor 3 is running, we can use the auxiliary contacts on starter M3 for conveyor 3 to energize the coil of conveyor 2. These contacts are identified as M3 to indicate that they are closed by energizing coil M3. The use of this contact satisfies the condition that conveyor 2 start in sequence, following conveyor 3.

By the same token, if we use the auxiliary contact of the starter for conveyor 2 to energize the starter of conveyor 1, it must follow in sequence behind conveyor 2. The contacts identified on the drawing as M2 are connected in series with the coil M1 for conveyor 1, thus satisfying the condition that conveyor 1 start after conveyor 2 in its proper sequence. We have now satisfied the conditions of specification 1 and are concerned with specification 2, which requires that an overload on any one conveyor will stop all conveyors.

The conditions of specification 2 can be obtained by series connection of all overload contacts between the

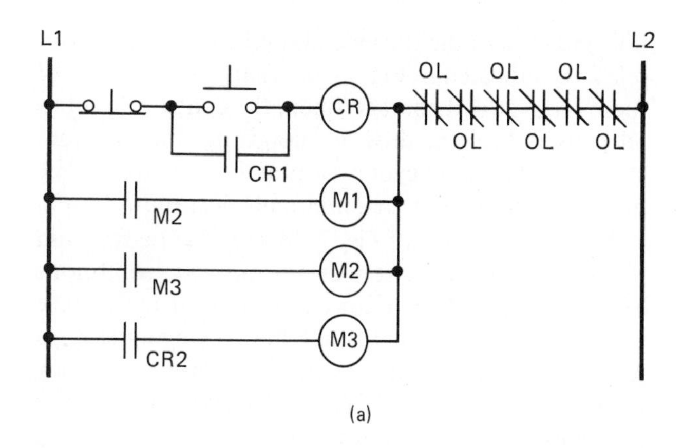

(a)

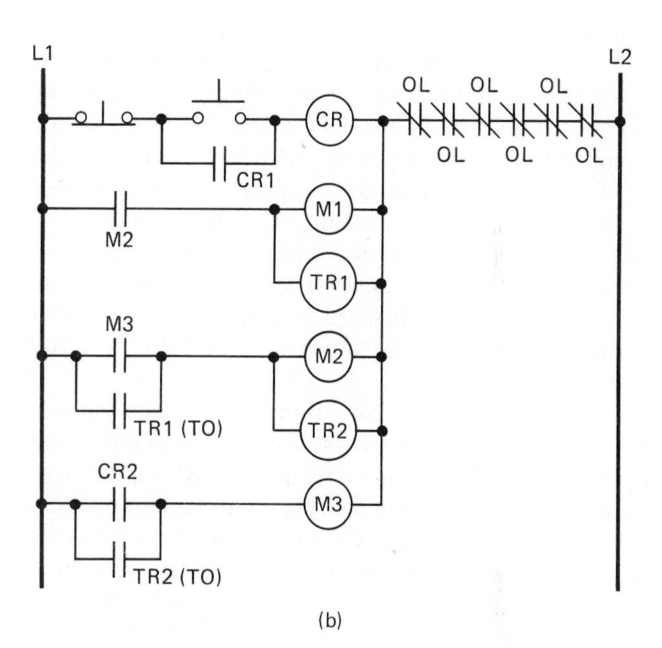

(b)

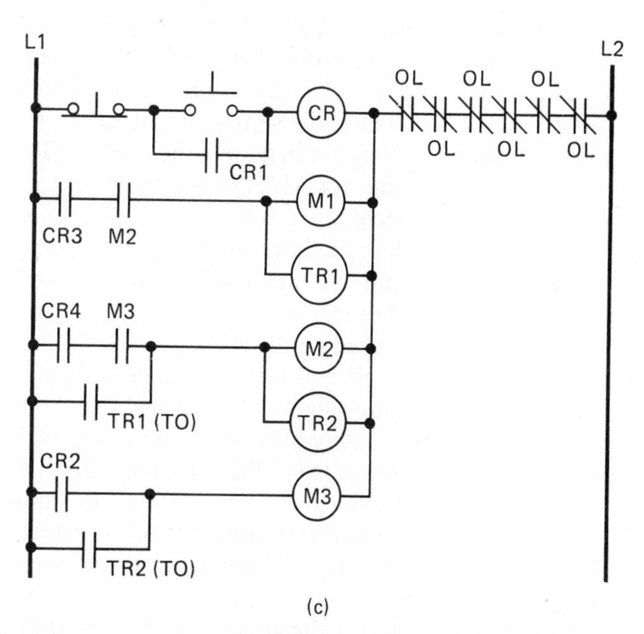

(c)

Fig. 7-3 Development of a sequence control circuit.

line and each of the starter coils and coil of the control relay, as indicated in Fig. 7-3*a*. If any one or more of these six overload contacts open, the control circuit to all coils is broken, thus de-energizing the coils and stopping all the conveyor motors at the same time. We have now fulfilled the requirements of specification 2.

While the circuit of Fig. 7-3*a* satisfies the first and second specifications, it does not meet the conditions of specification 3, that conveyors stop in the reverse order. The requirement that the conveyors stop in reverse order and that they have a 2-min time delay between the stopping of each conveyor in sequence indicates the use of timing relays with time opening (TO) contacts. The first inclination is to connect them as indicated in Fig. 7-3*b*. A careful study of this circuit, however, should reveal to you that when the STOP button is pushed, the control relay will drop out, opening contacts CR1 and CR2, which will result only in the control relay being de-energized, because contact M2 is still closed, maintaining the circuit to coil M1, contact M3 is still closed, maintaining the circuit to coil M2, and contact TR2 is closed, maintaining the circuit to coil M3. Thus, all conveyor motors continue to run. A modification of this circuit must be made in order that the conveyors be stopped by pushing the STOP button.

In order to satisfy condition 3 of the specification, the circuit in Fig. 7-3*b* must be modified to the circuit in Fig. 7-3*c*. In this circuit we have added two contacts, normally open, operated by the control relay and identified on the drawing as CR3 and CR4. Now when the STOP button is pressed, the control relay drops out, opening all its contacts and isolating each conveyor-motor starter from the line except for the time-delay relay contacts, which are held closed by time-delay relay 1 and time-delay relay 2. The opening of contact CR3 breaks the circuit to coil M1, thus stopping conveyor 1. The contact identified as TR1 is time opening; therefore, the circuit to coil M2 is maintained for a period of 2 min, which is the setting of time-delay relay 1. At the end of this 2-min period, the contact TR1 will open and drop out coil M2, thus stopping the second conveyor in accordance with the specifications. This starts the timing action of time-delay relay 2, and after a period of 2 min its contact (TR2) will open and drop out coil M3, thus stopping conveyor 3.

We have now satisfied all the specifications for this circuit. The conveyors will start in sequence, beginning with 3 and progressing to 1 by the pushing of the single START button. Any overload on any conveyor will drop out all starter coils, thus stopping all conveyors. When the STOP button is pressed, the conveyors will stop in the reverse order to that in which they started, with the delay of 2 min between the stopping of each conveyor.

This circuit performs the functions of start, stop, sequence control, overload protection, and time-delay action. The use of the control relay with its three-wire control circuit provides low-voltage protection not possible with the two-wire control circuit to each of the conveyor starters.

7-4 FORWARD AND REVERSE MOTOR CONTROL CIRCUIT

The specifications for this circuit state that it must have three-wire control to give low-voltage protection. It must have electrical interlock, and the STOP button must be pushed in order to change direction of rotation of the motor.

The first step in the development of this circuit is to provide start and stop in the forward direction. The circuit for this is shown in Fig. 7-4*a*. You will notice that this is the ordinary three-wire pushbutton control circuit and satisfies the requirement that the motor start and stop in the forward direction with three-wire control.

The second provision of the circuit is that it start and stop in the reverse direction, which is accomplished by the addition of a START button and an auxiliary contact as shown in Fig. 7-4*b*. The START button is wired behind the STOP button so that only one STOP button will be required to stop the motor, regardless of the direction in which it is running.

The requirement that electrical interlock be used is satisfied by the addition of the contacts shown in Fig. 7-4*c* and identified by the letters R2 and F2, which are auxiliary contacts on the forward and reverse starters. The normally closed contact R2 will be opened whenever coil R is energized, thus preventing coil F from being energized at the same time. Contact F2 will be opened whenever coil F is energized, thus preventing the reverse starter from being energized at the same time. This circuit satisfies fully the specifications that the motor be able to start and stop in either the forward or reverse direction and that the STOP pushbutton be pressed in order to change from forward to reverse or from reverse to forward. Electrical interlock has been provided so that both starters cannot be energized at the same time.

Suppose now that plugging reversal is required on this machine. The circuit would have to be modified as shown in Fig. 7-4*d*. The START pushbuttons for forward and reverse would need to be of the double-pole type, having one set of normally open and one set of normally closed contacts. When we push the forward START button, it closes the circuit for the forward starter and at the same time breaks the circuit for the reverse starter. When the reverse START pushbutton is pushed, it not only will complete the circuit to the reverse starter but also will break the circuit to the forward starter, thus giving plugging action.

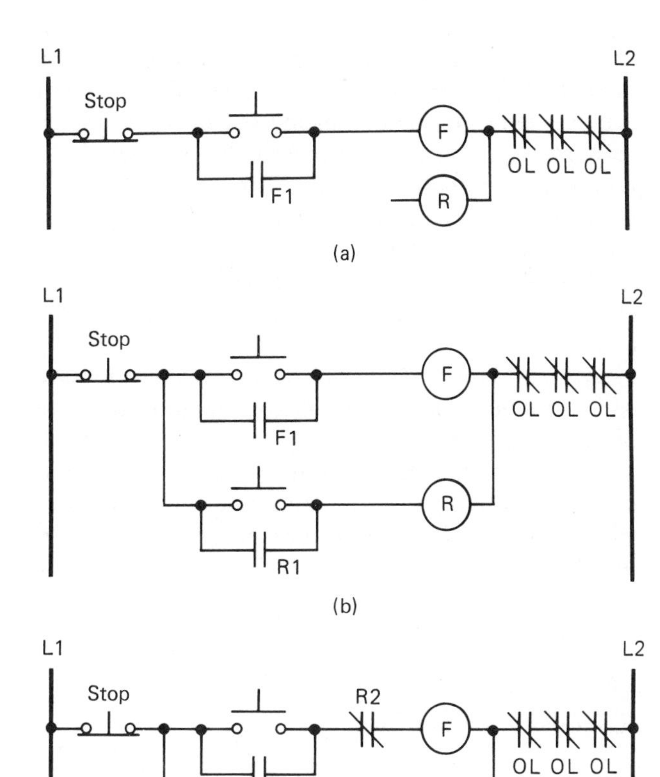

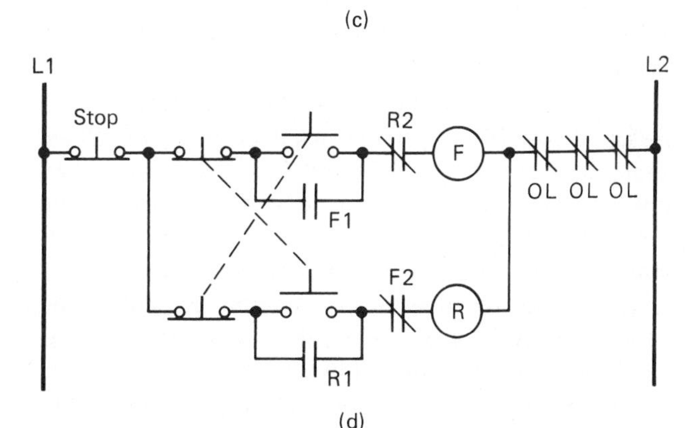

Fig. 7-4 Forward and reverse control of a motor.

This circuit performs the control functions of forward-reverse control, start, stop, interlock, overload protection, plugging, and low-voltage protection.

7-5 MOTOR CONTROL CIRCUIT WITH LIMIT SWITCHES

The specifications for this circuit are as follows. It must give limit-switch control for forward and reverse running of the motor by the use of momentary-contact limit switches. It must also provide low-voltage protection. The initial start and stop for the control system will be by momentary-contact START and STOP pushbuttons.

The requirement that START and STOP pushbuttons be used to initiate a control of the circuit by limit switches would indicate the use of a control relay. The wiring for this is in Fig. 7-5*a* on the next page. Contact CR1 is used to maintain the circuit to the control relay during the running operation of the circuit. Contact CR2 is used to make and break the line circuit to the forward and reverse control circuit, thus satisfying the provision that the START and STOP buttons initiate and terminate the automatic control of the motor by limit switches. Using the control relay and its START and STOP buttons also provides low-voltage protection.

The specifications call for the use of momentary-contact limit switches, which would require a three-wire control circuit for forward and reverse. These limit switches necessarily have two sets of contacts, one normally open and the other normally closed. When wired as shown in Fig. 7-5*a*, the normally closed contact of limit switch 2 would act as the stop for the forward controller, and the normally open contact of limit switch 1 would act as the start contact for the forward controller. The auxiliary contact of the forward starter must be connected in parallel with the normally open contact of limit switch 1 in order to maintain the circuit during the running of the motor in the forward direction.

Figure 7-5*b* shows the additional wiring required for the reverse starter. The normally closed contact of limit switch 1 is wired as a stop contact for the reverse starter, and the normally open contact of limit switch 2 is wired as a start contact for the reverse starter. The auxiliary contact on the reverse starter is wired in parallel with the normally open contacts of limit switch 2 to maintain the circuit while the motor is running in reverse.

This circuit satisfies all the requirements of the specifications with the exception of electrical interlock, which is shown in Fig. 7-5*c*. This electrical interlock is accomplished by the addition of a normally closed contact in series with each starter and operated by the starter for the opposite direction of rotation of the motor.

Plugging reversal is provided in this circuit by the action of the limit switches themselves. When limit switch 1 is moved from its normal position, the normally open contact closes energizing coil F and the normally closed contact opens and drops out coil R. The reverse action is performed by limit switch 2, thus providing plugging in either direction.

The circuit of Fig. 7-5*c* would work perfectly and satisfy all the conditions of operation if it were always to stop in a position that would leave either normally

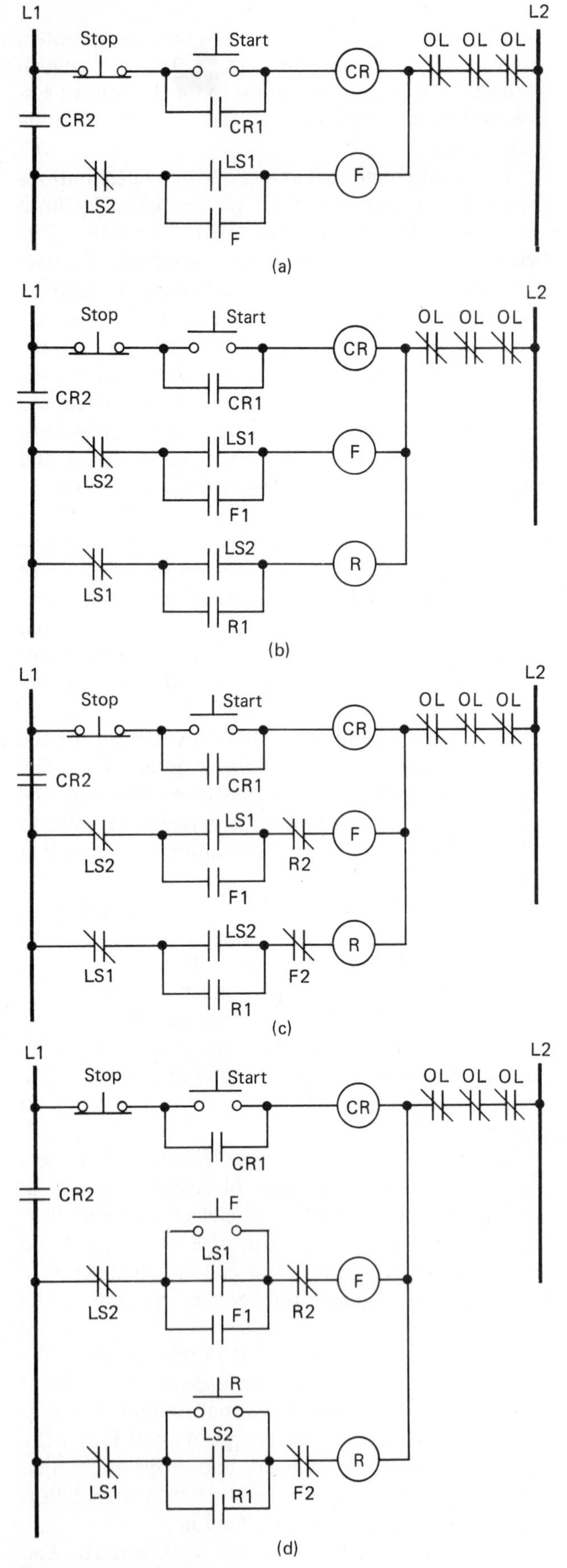

Fig. 7-5 Forward and reverse control using limit switches.

open limit switch contact closed. This is not very likely to be the case, however, and therefore we must provide some means of starting the motor in either forward or reverse in order that the limit switches can take over automatic control. The circuit additions necessary to accomplish this are shown in Fig. 7-5*d*. Here we have added a pushbutton in parallel with the other start component in the forward and reverse circuits. The function of these pushbuttons is to start the action of the motor in the desired direction so that it can run until the first limit switch is actuated and then will continue to operate automatically until the STOP button is pushed.

Circuits similar to this are used frequently for the control of milling machines and other machine tools which require a repeated forward and reverse action in their operation.

7-6 JOGGING CONTROL

The requirements of this circuit are to add jogging control in both forward and reverse to the circuit which is shown in Fig. 7-4*d*. In order to jog a motor, the pushbutton must connect the line to the starter coil while the pushbotton is held down, causing the motor to run. It must also prevent the auxiliary contact of the starter from maintaining the circuit when the JOG button is released.

The circuit of Fig. 7-6 shows two JOG buttons identified as JF and JR. If you follow this circuit, you will see that the JOG-FORWARD button has a normally open contact which is connected from the STOP button to the coil side of the auxiliary contact for the forward starter. The normally closed contact of this pushbutton is connected between the STOP button and all the other control devices. When the JOG-FORWARD button is pressed, the circuit is made from line 1 through its normally open contacts to the coil of the forward starter through the normally closed electrical interlock. At the same time, the normally closed contact of this JOG button breaks the circuit between the STOP button and all the other pushbuttons and contacts in the circuit, thus preventing the motor starter from sealing in when the JOG button is released.

The installation and wiring of the JOG-REVERSE button are identical with those of the JOG-FORWARD button, except that it is connected to the reverse starter; its action electrically and mechanically is the same as that for the JOG-FORWARD button.

This circuit incorporates many of the functions of control. It has start and stop in both forward and reverse, manual plugging duty, jog service in forward and reverse, electrical interlock, low-voltage protection, and overload protection.

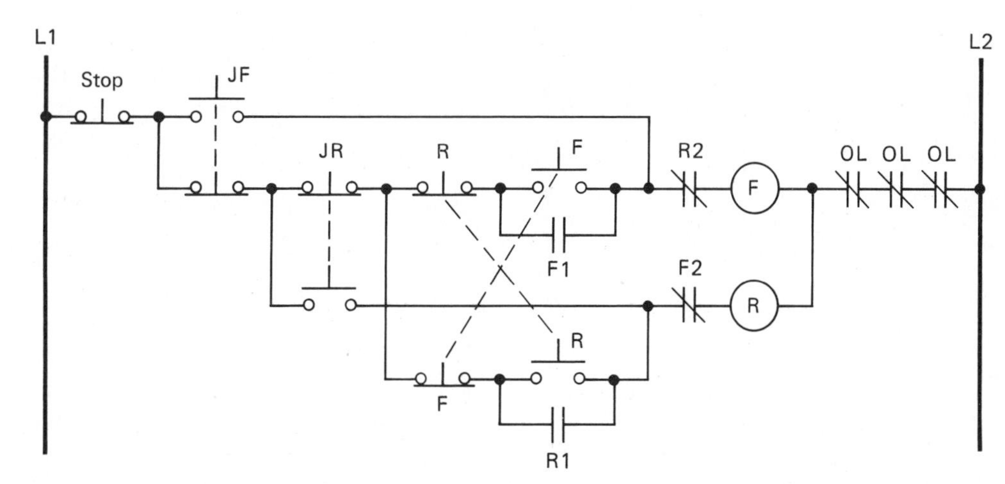

Fig. 7-6 Forward and reverse control with jogging.

7-7 AUTOMATIC PLUGGING CONTROL

The assignment here is to add automatic plugging reversal to the circuit of Fig. 7-4*c*. The easiest method of obtaining automatic plugging reversal is by the use of a plugging switch. The connection for this switch is shown in Fig. 7-7. The action of this switch is such that

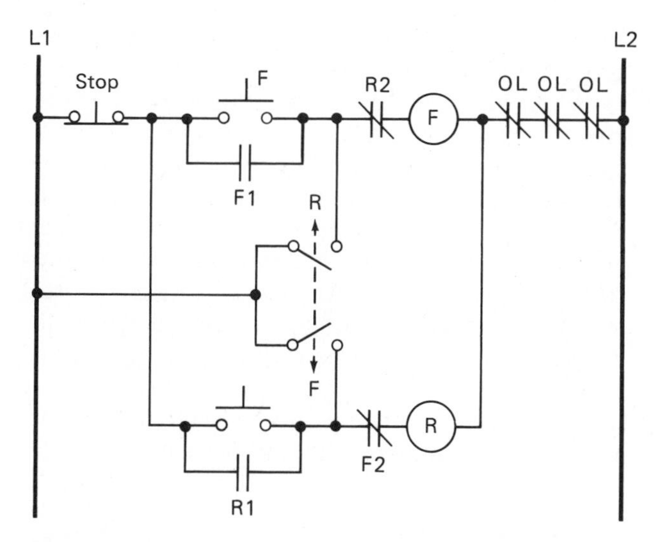

Fig. 7-7 Forward and reverse control with automatic plugging.

when the motor is running in the forward direction, the movable arm of the switch is held in the direction shown by the arrow marked with the letter F. When the STOP button is pressed, the circuit is broken to the forward starter, allowing it to drop its contacts and thus closing the interlock contact marked F2. At this instant, the circuit is made from line 1 through the plugging switch through the normally closed interlock contact F2 to the coil R on the reverse starter. This will plug the motor in the reverse direction. The closing of the maintaining contact R1 will energize coil R, which ensures the running of the motor in the reverse direction. Should the STOP button be held down, the flow of current through R1 to coil R would not be possible and the result is plugging stop for the motor. The action of the plugging switch when the motor is running in reverse is such that its arm is in the position marked R. When the STOP button is pressed, the circuit functions in exactly the same way that it does when the motor is running in forward, except that now the motor is plugged by the energizing of the forward starter.

7-8 THREE-SPEED MOTOR CONTROL

This circuit is to control a three-speed motor, and the requirement is that it provide selective speed control. To satisfy the requirement that the circuit give selective speed control would indicate the use of three simple start circuits, one circuit for each speed, so that the operator can start the motor in any desired speed. To increase speed, the operator need only to press the button for the desired higher speed. Such a circuit is shown in Fig. 7-8*a*.

This circuit, however, ignores any form of interlock which will prevent two speeds from being energized at the same time unless such interlock is provided mechanically in the starter. The necessary electrical interlock has been added in Fig. 7-8*b*. A careful study of this circuit will reveal that it is possible to increase speed by merely pushing the button for the next speed. For instance, to increase the motor's speed from first to its second, the normally closed interlock contact identified as M3 will be closed and the coil M2 will be energized. This will break the normally closed contact identified as M2, thus dropping out coil M1 and deenergizing the contactor for speed 1.

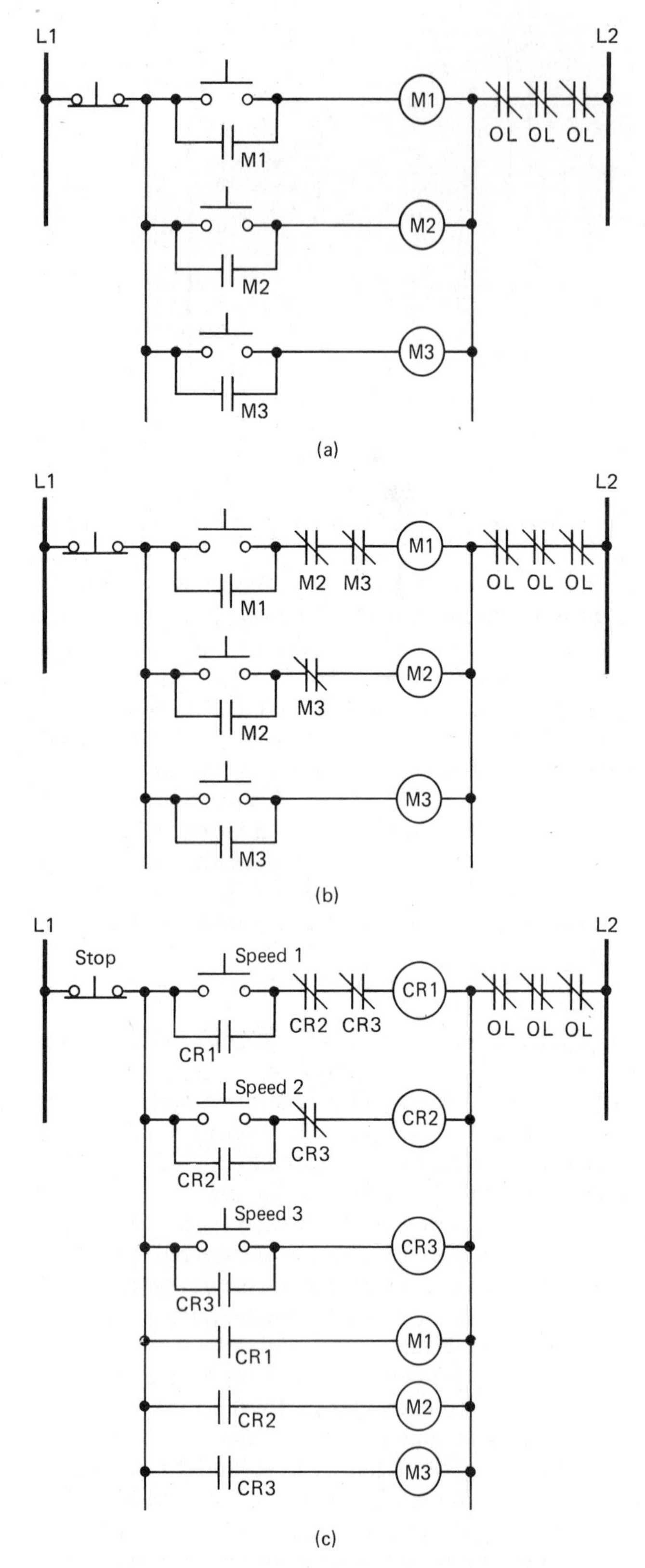

Fig. 7-8 Selective speed control for a three-speed motor.

If this circuit is to function properly, then contact M2 must be built so that it will break before the line contacts of contactor M2 are made. If this is not done, then the starter will energize two speeds at one time, causing damage to the motor and the wiring. The action of the circuit for speed 3 is similar, in that by energizing coil M3 the normally closed contact M3 is broken before the line contacts for speed 3 are made, thus dropping out either speed 1 or speed 2, whichever is energized at the time.

In order to reduce speed, the STOP button must be pressed first. Analyzing the circuit, we see that if we try to go from speed 3 to speed 2, the pressing of the START button for this speed will result in current flowing only as far as contact M3. The same is true if we try to reduce from speed 3 to speed 1. If we try to reduce speed from speed 2 to speed 1, current can flow only as far as contact M2. The pressing of the STOP button drops out any coil that is energized, thus returning all contacts to their normally closed position and allowing the circuit to be energized in any desired speed.

Chiefly because it is difficult to obtain contact arrangements on starters which will provide the break-before-make action necessary in this circuit for interlocking, starters of this type generally employ control relays for each speed, which in turn energize the proper coils. The circuit for use of control relays is shown in Fig. 7-8*c*. It should be noted that the circuit for the three control relays is identical with the circuit of Fig. 7-8*b*, with only the addition of a contact on the relay for each contactor coil. Thus, it gives a three-wire control to the control relays and essentially a two-wire control to the contactor coils.

While this circuit has been developed for speed control of a single motor, it is equally applicable to sequence control of three motors. If coils M1, M2, and M3 were coils of individual starters for individual motors, they would start in selective sequence. This means that the operator could start any motor desired and could progress upward in the sequence of motors at will. To go backward in the sequence, however, the operator must first stop whatever motor is running and then select the motor desired. This would be selective sequence control of three motors and could be expanded to any number of motors desired.

7-9 SEQUENCE-SPEED CONTROL

To give sequence-speed control the circuit discussed in Sec. 7-8 will be modified. The requirement of sequence-speed control is that the motor be accelerated by pressing the START button for each successive speed in order until the desired speed is reached. Figure 7-9 is a circuit to accomplish sequence-speed control of a three-speed motor using control relays. The contact arrangement on these relays in this type of service is

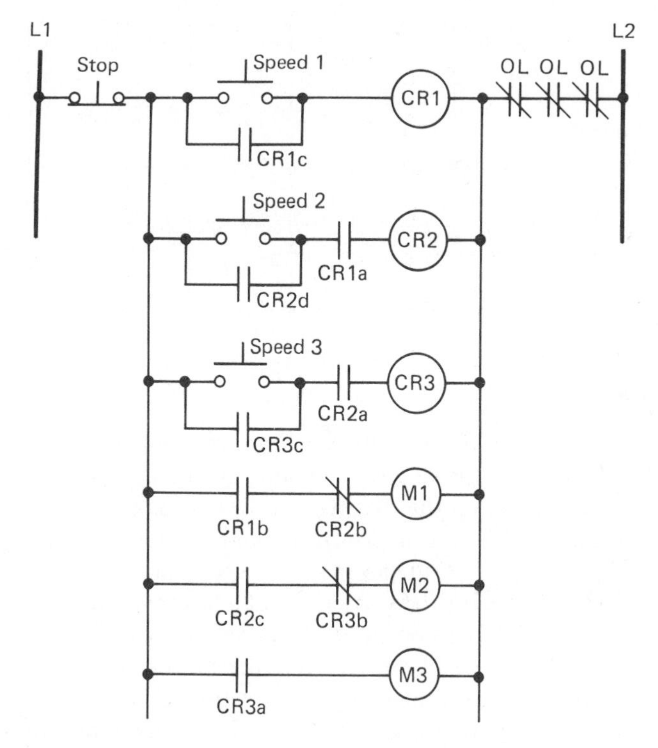

Fig. 7-9 Sequence-speed control using control relays.

critical, and it must be pointed out that normally closed contact CR2b must break before normally open contact CR2 closes. Also, contact CR3b must break before contact CR3a closes.

This circuit will not be developed step by step, because a similar circuit for speed control was developed in Sec. 7-8. Rather, we shall analyze the operation of this revised circuit. Suppose that the operator wishes to run the machine in its third speed. Then the button for speed 1 must be first pressed, which will energize coil CR1. Energizing the coil causes contacts CR1a and CR1b to close. The closing of contact CR1b energizes the contactor for speed 1, and the motor starts and accelerates to this speed. The closing of contact CR1a sets up the start circuit for speed 2, and when this button is pressed, the circuit is complete to coil CR2, thus energizing this coil and closing contacts CR2a and CR2c. Also, it opens contact CR2b. The opening of contact CR2b drops out the contactor for speed 1.

Immediately thereafter, contact CR2c closes, energizing the contactor for speed 2 and allowing the motor to accelerate and run at the second speed. The closing of contact CR2a sets up the start circuit for the third speed. When the START button for the third speed is pressed, the circuit is complete to coil CR3, which in turn first opens contact CR3b, which drops out the contactor for speed 2. Immediately thereafter, contact CR3a closes, thus energizing the contactor for speed 3, which will allow the motor to accelerate and run in speed 3.

It should be noted that the control relays remain energized until the STOP button is pressed and that the only way to reduce speed is to press the STOP button. To progressively accelerate the motor, start with speed 1 and increase speed as desired. While this circuit is designed for only three speeds, it could be extended to include as many speeds as desired.

This circuit is not presented as the only or most desirable method of providing sequence-speed control. There are many factors involved in the design of control circuits for a given motor and controller. The control designer will find many variations of circuits to accomplish the same purpose and should try to develop an overall understanding of the operation of components and circuits to accomplish an end.

7-10 ACCELERATION CONTROL

The circuit for acceleration control will be a magnetic control for a wound-rotor motor. The customer desires four steps of automatic definite-time acceleration when the RUN button is pressed. He also desires the option of running the motor at any one of the four reduced speeds by pressing a pushbutton for that speed. He also wishes to be able to change speeds at will, either up or down, after the motor has accelerated to run speed.

In order to visualize what contacts and control relays will be needed, an elementary drawing of the secondary circuit for the motor should be considered (Fig. 7-10*a* on the next page). This circuit provides the essentials of a four-step acceleration or four independent speeds, provided the contacts are properly controlled. To provide definite-time control, these contacts would be time closing (TC).

The first step in development of this circuit would be to provide the four-step definite-time acceleration, using a RUN button to initiate the control process. Figure 7-10*b* shows our circuit as it is developed. The provision of definite-time acceleration requires the use of time-delay relays for each speed. A control relay with three-wire control seems to be required for the run condition.

When the RUN button is pressed, the circuit is complete to the coil of CR, thus closing CR1 to maintain the circuit. Contact CR2 closes and energizes the primary contactor PC, thus energizing the primary of the motor. Some form of interlock should be provided to prevent the motor from starting unless all the resistance is in the secondary circuit, speed 1; however, this can best be provided after the balance of the speed control is developed.

A third contact on the control relay (CR) (Fig. 7-10*c*), CR3, can be used to energize a time-delay-on-

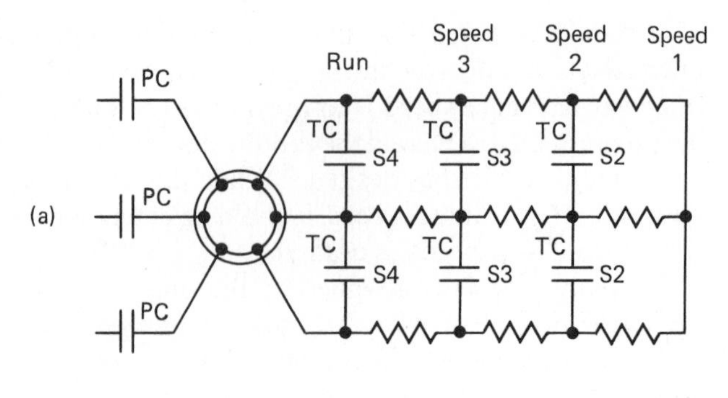

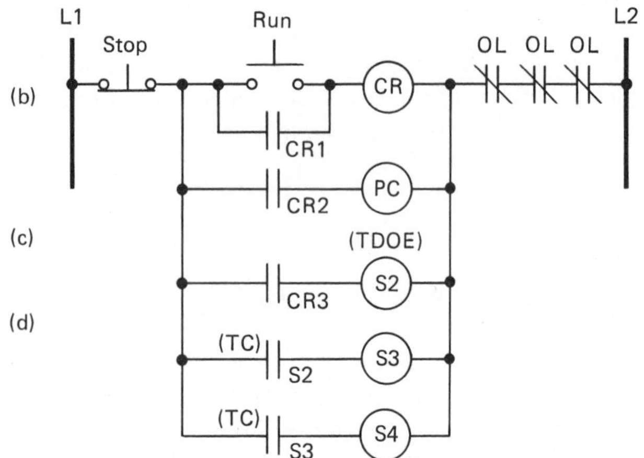

Fig. 7-10 Definite-time acceleration control.

energizing (TDOE) relay S2, which has two TC contacts connected to short out the first section of the resistance grid (Fig. 7-10*a*), thus providing acceleration to the second step when S2 times out.

To provide the third step of acceleration (Fig. 7-10*d*), it is obvious that a second TC contact on S2 will be required to energize a second TDOE relay S3, which has two TC contacts connected to short out the second section of the resistance grid (Fig. 7-10*a*). The fourth step of acceleration is provided by a similar circuit.

The circuit of Fig. 7-10*d* provides a satisfactory degree of interlock in that the contacts of S2, S3, and S4 open whenever the STOP button is pressed and restore all resistance to the secondary circuit of the motor. This circuit satisfies the first specification for the development of the circuit.

The second specification for independent speed control after acceleration will require some modification of the circuit of Fig. 7-10*a*. To be able to select any speed at will, the operator must be able to open any closed contacts and close any open contacts in the secondary circuit as required for that speed.

Adding a normally closed contact in series with each TC contact in the circuit of Fig. 7-10*a* would allow the equivalent of opening and closed contacts in the secondary circuit.

Adding a normally open contact in parallel with each group of two series contacts would provide the effect of closing any open contacts in the secondary circuit. These contacts are shown in Fig. 7-11.

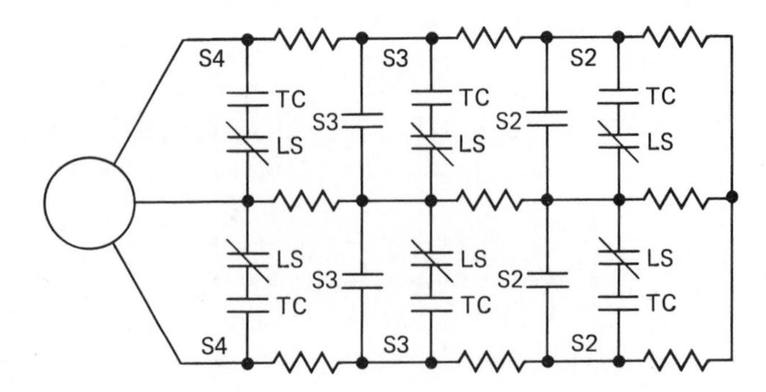

Fig. 7-11 Wound-rotor-motor secondary circuit.

Figure 7-12 provides pushbutton control of the added contacts of Fig. 7-11 and satisfies the second specification for the circuit. A time closing contact on S4 should be used in a stop function for the pushbuttons to ensure that the motor has accelerated to run speed before it can be operated at a lower speed.

The low-speed pushbutton energizes a relay LS in a three-wire control circuit. Six normally closed contacts on LS are connected in series with the TC contacts in the circuit of Fig. 7-11. When the low-speed button is pressed, relay LS is energized and opens all six normally closed contacts, which restores all resistance in the secondary circuit and results in low-speed operation.

To increase speed from low, or first, speed to second speed, relay 2S must be added in three-wire circuit with a normally open pusbutton. One normally open contact 2S is used to seal in LS when moving up in speed and to energize LS when moving down from run to second speed. Two normally open contacts on 2S are used to short out part of the resistance grids (Fig. 7-11).

The third speed is added by duplicating the second-speed components using contacts indicated by S3 on the circuits of Fig. 7-11 and 7-12. There is no need for interlock when going from second speed to third speed because the contacts S2 in Fig. 7-11 do not affect the operation even though they remain closed.

The fourth, or run, speed can be restored by merely dropping out any and all instantaneous contacts and relays associated with the lower speeds, since all TC contacts in Fig. 7-11 are now closed. This can be done by adding a normally closed pushbutton as indicated in Fig. 7-12.

The circuit as developed up to this point would work as long as the operator always wanted to go from a low speed to a higher speed. To reduce the speed, it is necessary to open the contacts S2 and S3 of Fig. 7-11. Interlocking by the use of contacts could involve more relays in the circuit, but normally closed contacts as part of the speed-control pushbuttons, as indicated in Fig. 7-12, provide positive drop-out of all higher speeds whenever a lower speed is desired.

This circuit may seem slightly impractical in its requirements, but is used only to indicate that any

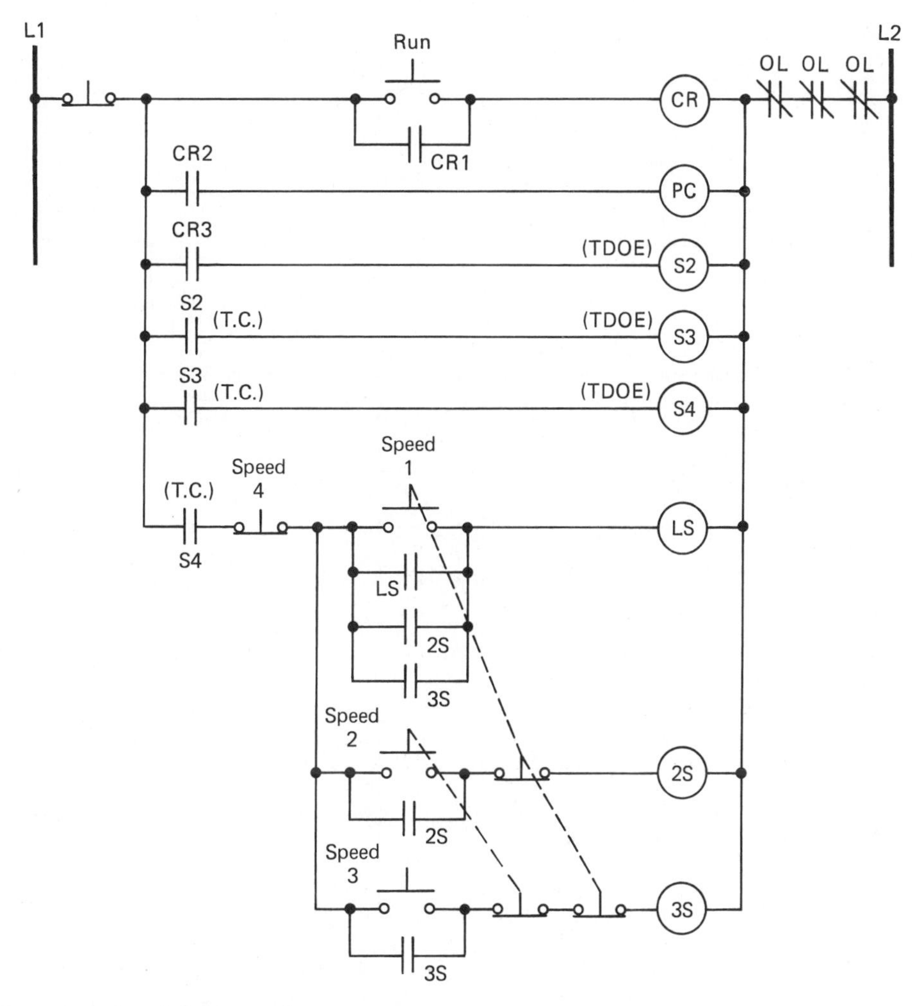

Fig. 7-12 Final control circuit for wound-rotor motor.

requirements of the system can be met by systematic development.

7-11 PRESELECTED SPEED CONTROL

The requirements for this circuit are to develop a control system which will provide definite-time speed control for a dc shunt motor. The owner wishes to be able to start the motor and have it accelerate to any one of five preselected speeds. Speeds 1 and 2 are below base speed (underspeed control). Speed 3 is base speed, and speeds 4 and 5 are above-base speed (overspeed control).

The requirement of two speeds below base speed indicates three resistance grids in series with the armature (Fig. 7-13*a* on the next page). One of the resistances would be short-circuited for speed 1; the second, for speed 2; and the third, for base speed by contacts S1a, S2a, and S3a, respectively.

The requirement of two speeds above base speed indicates two resistance grids in series with the shunt field and shorted out by normally closed contacts S4a and S5a until base speed has been reached and overspeed is required.

The addition of an overload heating element and line contacts completes the motor circuit. This leaves only the development of the control circuit to activate the contacts properly.

Step 1 in the development of the control circuit would be to provide for speed 1 by connecting a pushbutton to energize the first TDOE relay S1. This circuit must remain energized; therefore, it requires a three-wire control.

When a dc motor is started, considerable resistance is required in the armature circuit until the motor begins to accelerate. This requires a time delay before R1 is shorted out. If contact S1a is a time closing contact, it will provide time for the motor to reach the desired speed.

The line contacts must be closed before any current can flow through the motor circuit. If the line contact is an instantaneous contact on S1, this requirement will be met. An overload contact and STOP button must be provided to protect and stop the motor. Contact S1c, used to maintain the circuit to coil S1, must be an instantaneous closing contact on S1.

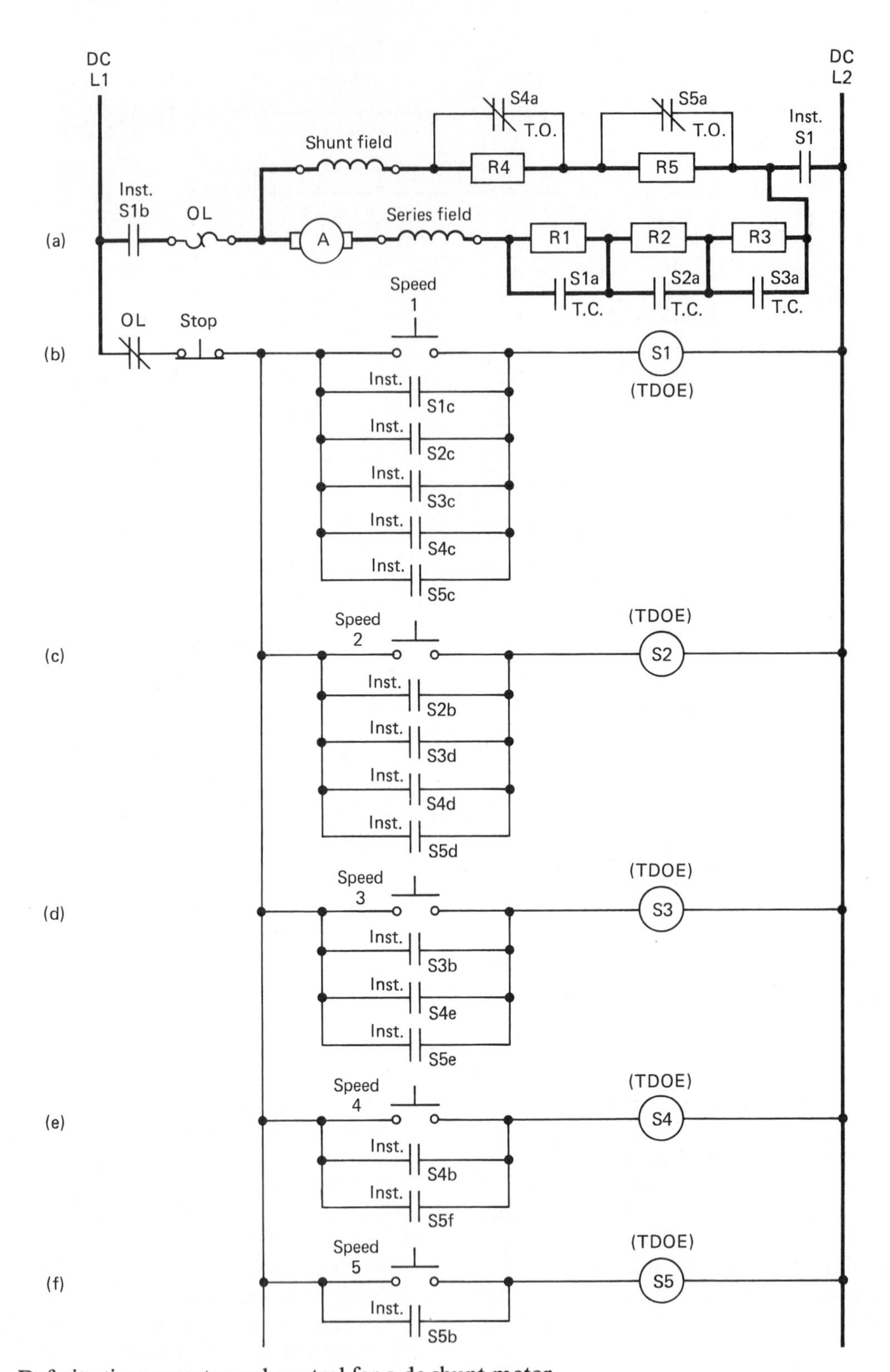

Fig. 7-13 Definite-time-preset speed control for a dc shunt motor.

At this point it is obvious that S1 must be energized regardless of the speed selected; therefore, we can add a maintaining contact for each speed in parallel with S1c. All these contacts must be of the instantaneous closing type.

The circuit of Fig. 7-13*b* will start the motor and allow it to accelerate to the desired speed, at which time S1a times out and removes R1 from the circuit. The motor will run at this speed unless other buttons are pressed.

The requirement of a pushbutton for each speed indicates a three-wire circuit and a TDOE relay for each speed. Figure 7-13*c* is a three-wire circuit to energize and maintain S2. The instantaneous contact S2c will energize S1 and thereby close the line contact S1b. The relay S2 must provide a time delay twice as long as that of S1 so that S1 may time out and allow acceleration time for speed 2. Instantaneous contacts must be provided parallel to S2b in order to bring in speed 2 whenever a higher speed is selected.

The other speeds are controlled by circuits similar to the one for speed 2. Each higher speed must have a longer setting on the time-delay relay and provide contacts to energize all lower speeds.

There are other circuits which could provide this control; this was chosen to show the degree to which a circuit can be dependent upon proper interlocking.

7-12 CHILLED-WATER AIR-CONDITIONING SYSTEM CONTROL

This circuit will be one for a chilled-water air-conditioning system. The compressor is a 500-hp squirrel-cage motor driving a centrifugal compressor. Head pressure is controlled by condenser water furnished by a condenser-water pump. The chilled water is pumped to the heat exchanger by a chilled-water pump. Protective control must be provided to shut down the compressor whenever flow ceases in either the chilled water or condenser water.

The starter for the compressor is a reduced-voltage autotransformer type using air circuit breakers requiring that a coil be energized to close the breaker and a second coil be energized to trip the breaker. Definite-time control of reduced-voltage starting is provided by a TDOE relay in the starter.

The first step in the system is to establish chilled-water flow and to sense by means of a flow switch that there actually is flow in the pipe. The flow switch (CWF) is usually located in the return water line to ensure that flow occurs throughout the system. Figure 7-14*a* provides the START and STOP buttons for the system and energizes the chilled-water pump.

When chilled-water flow is sensed by the flow switch, the oil pump on the compressor must be ener-

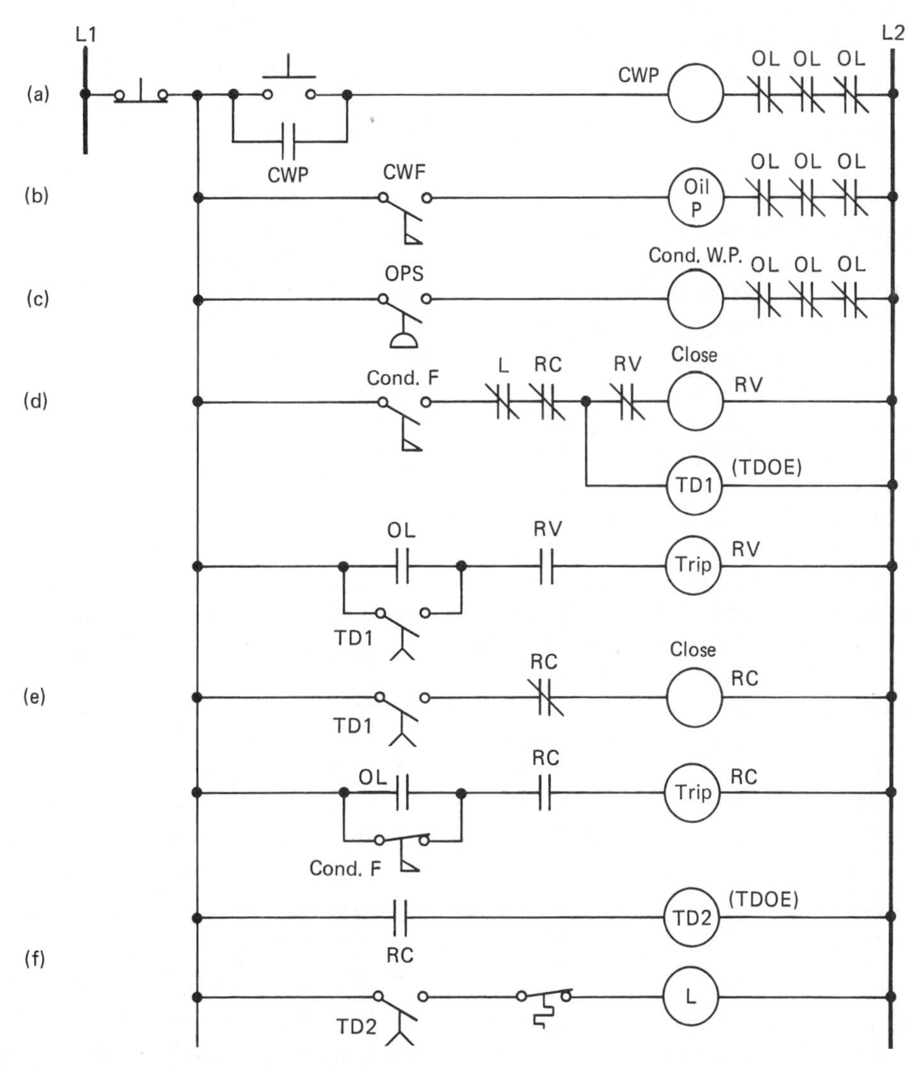

Fig. 7-14 Chilled-water air-conditioning system.

gized and correct pressure established (Fig. 7-14*b*). Once the oil-pressure switch senses sufficient pressure, it must energize the condenser-water pump (Fig. 7-14*c*) to provide cooling for the compressor and keep the head pressure from rising to a dangerous value.

The compressor has chilled-water flow, oil pressure, and condenser-water flow and is ready to be started under reduced voltage if the unloader ensures that the load is less than 10 percent. There are many unloaders and vane positioners used. We shall indicate the unloaders as a contact L which is normally closed when the load is 10 percent or less. The unloader contact L should not allow the compressor to start if it is open. Figure 7-14*d* is the circuit for the reduced-voltage contactor closing coil (RV). Remember that this contactor is latched closed; therefore the coil must be energized only briefly.

The trip coil on the breaker must be energized if an overload occurs during reduced-voltage start and when the run contactor is energized by the time-delay relay. The trip coil must be energized only momentarily; therefore, a contact on RV must remove the coil from the line. The trip-coil circuit must be held open when the run contactor is closed; therefore, normally open (N.O.) contact RV is connected in series with this circuit.

When the reduced-voltage start timer TD1 times out, it closes its contacts to trip the reduced-voltage contactor and briefly energize the run-contactor closing coil (RC). The run contactor is not latched closed (Fig. 7-14*e*).

Protective control for the compressor while running must be paralleled to normally open contacts in the trip-coil circuit. The chilled-water flow contact and the oil pressure switch (OPS) contact will cause a shutdown of the condenser-water pump. The condenser-water flow contacts (Cond. F) in the start circuit will not shut down the compressor when they open; therefore, a set of contacts on this switch needs to be connected in the trip circuit of the run contactor for protection. These contacts would be normally closed when there is no flow, so that they will be open when flow is established. Normally, open contact RC on the run contactor prevents the energizing of the trip coil unless the run contactor is closed.

All the circuit lacks at this point is the energizing of the loader device L after a short time delay to ensure that the compressor is completely up to speed. The thermostat can then control the load of the compressor (Fig. 7-14*f*).

SUMMARY

This chapter has presented a method of developing control circuits. While the few circuits developed here do not in any way approach the limitless number of possible circuits that the student of control will find in actual practice, they should provide the basic principles that are necessary for the development of any control circuit.

Develop the circuit one function at a time, adding only the components necessary to perform that function. Analyze the circuit after each addition to see that it has not interfered with any previous operation and that it actually does perform the function intended before proceeding with any further additions to the circuit. If you follow these simple rules, you should have no trouble in developing a circuit to perform any desired function. The greatest danger in developing control circuits is to try to draw a complete circuit at one time. Therefore, it cannot be overstressed that step-by-step development will lead to fewer hours spent in trying to find out why a circuit did not work after it was wired.

It is highly recommended that the student practice developing circuits of various types and checking them to see that they actually operate when wired. If a setup of control components is available to the student, he or she should at this stage of study develop circuits of various types and then actually wire them and test them to see that they work. Should a circuit not operate as expected, a student should then troubleshoot this circuit and determine why it did not operate. If the student can continue to develop circuits in this manner, he or she will have a great advantage when applying these principles on the job.

The techniques used in the development of control circuits have been discussed here, and all that is required to perfect these techniques to a satisfactory degree is practice. Your own desires as to your degree of proficiency will dictate how much additional time you spend practicing the development of control circuit.

REVIEW QUESTIONS

1. What are the requirements of two-wire and three-wire control circuits?
2. Where should you start when developing a control circuit?
3. How must devices intended for the function of stop be connected?
4. How must devices intended for the function of start be connected?
5. When would contacts or pushbuttons be connected in series-parallel or parallel-series?
6. Describe a solenoid valve and its application in control circuits.
7. Why are overload relay contacts for several motors sometimes connected in series?
8. Why is low-voltage protection not possible with two-wire control circuits?

9. What is the function of an electrical interlock?
10. Describe what is meant by plugging.
11. What is the purpose of limit switches in control circuits?
12. Define jogging control.
13. Why are relays used to provide each speed in Fig. 7-8?
14. What is meant by definite-time acceleration?
15. How does the circuit of Fig. 7-13 provide speeds below and above base speed?
16. Draw a diagram showing a motor controlled by a three-wire start-stop station.
17. Add to the above circuit a second pushbutton for starting the motor from a different location.
18. Add to the above circuit a limit switch to stop the motor.
19. Draw a diagram showing a motor controlled by a three-wire pushbutton station. When this motor stops, it starts a second motor which runs until stopped by pressing a STOP button.
20. Revise the circuit in question 19 so that the second motor runs for only 2 min and then stops automatically.
21. Draw a control diagram showing three motors connected so that they are all started by one START button and interlocked so that if any one of them should fail to start or should drop out, all will stop. The STOP button stops all motors.
22. Draw a control diagram showing two pumps started and stopped one at a time by a pressure switch. Provide a manual switch to run the pumps alternately.
23. Add to the circuit in question 22 a second pressure switch to start the idle pump if the pressure continues to drop.
24. Replace the manual switch in the circuit of question 23 with a stepping relay to alternate the pumps automatically each time they start.
25. Replace the stepping relay in the circuit of question 24 with a time clock to alternate the pumps every 24 h.
26. Draw a control diagram showing four motors started in compelling sequence. Provide a 20-s time delay between the starting of each motor.
27. Draw a control diagram showing four motors started in selective sequence.
28. Draw a control diagram for a three-speed motor with selective-sequence starting. Provide control so that the speed may be reduced without pressing the STOP button. (*Hint:* This is similar to plugging without reversal of the motor.)
29. Draw a diagram satisfying the following conditions: There are four exhaust-fan motors in a building. Each fan is also equipped with a thermostat known as a firestat. Should any one of the firestats, which have normally closed contacts, open from high heat, it will stop all fans.

PROBLEMS

P7-1. On a separate sheet of paper redraw the schematic diagram of Fig. 7-15*b*. To this add a schematic diagram of the control circuit shown in the ladder diagram of Fig. 7-15*a*. Label the color of each control circuit wire using the color code shown in Fig. 7-15*a*.

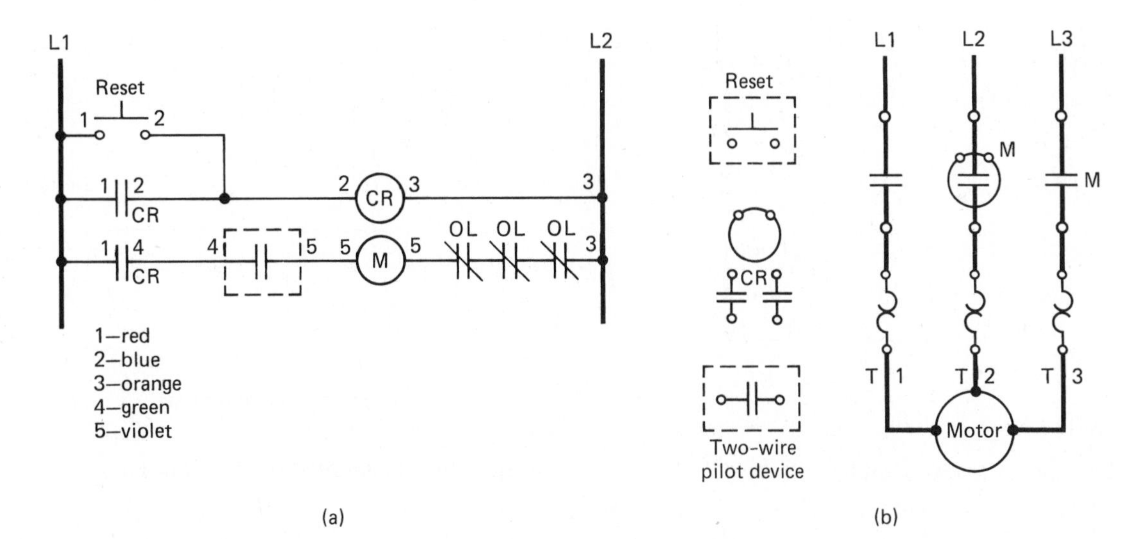

Fig. 7-15 Diagrams for Problem P7-1.

P7-2. On a separate sheet of paper redraw the schematic diagram of Fig. 7-16*b*. To this add a schematic diagram of the control circuit shown in the ladder diagram of Fig. 7-16*a*. Note that coil M1 operates the main contacts in the lines feeding motor 1, and coil M2 operates the contacts in the line feeding motor 2.

P7-3. On a separate sheet of paper redraw Fig. 7-17*a*, *b*, and *c*. Show the connections of the control diagram in part a and the schematic diagram in part b. In part c indicate the number of wires in the wire runs indicated. Assume that only the wires shown in part a are in the wire runs.

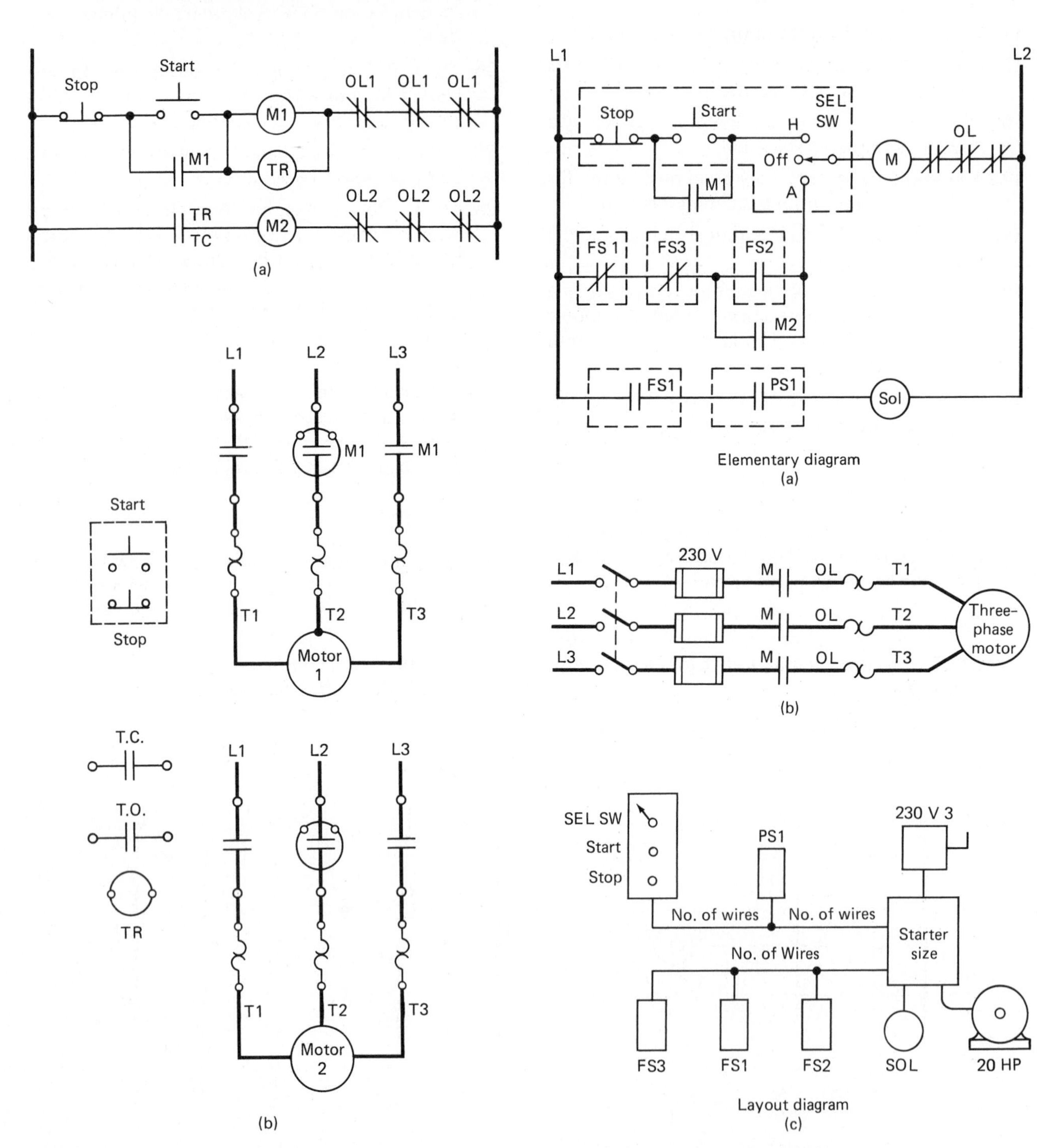

Fig. 7-16 Diagrams for Problem P7-2.

Fig. 7-17 Diagrams for Problem P7-3.

P7-4. On a separate sheet of paper redraw the ladder diagram of Fig. 7-18*a* and add it to the schematic diagrams of Fig. 7-18*b*.

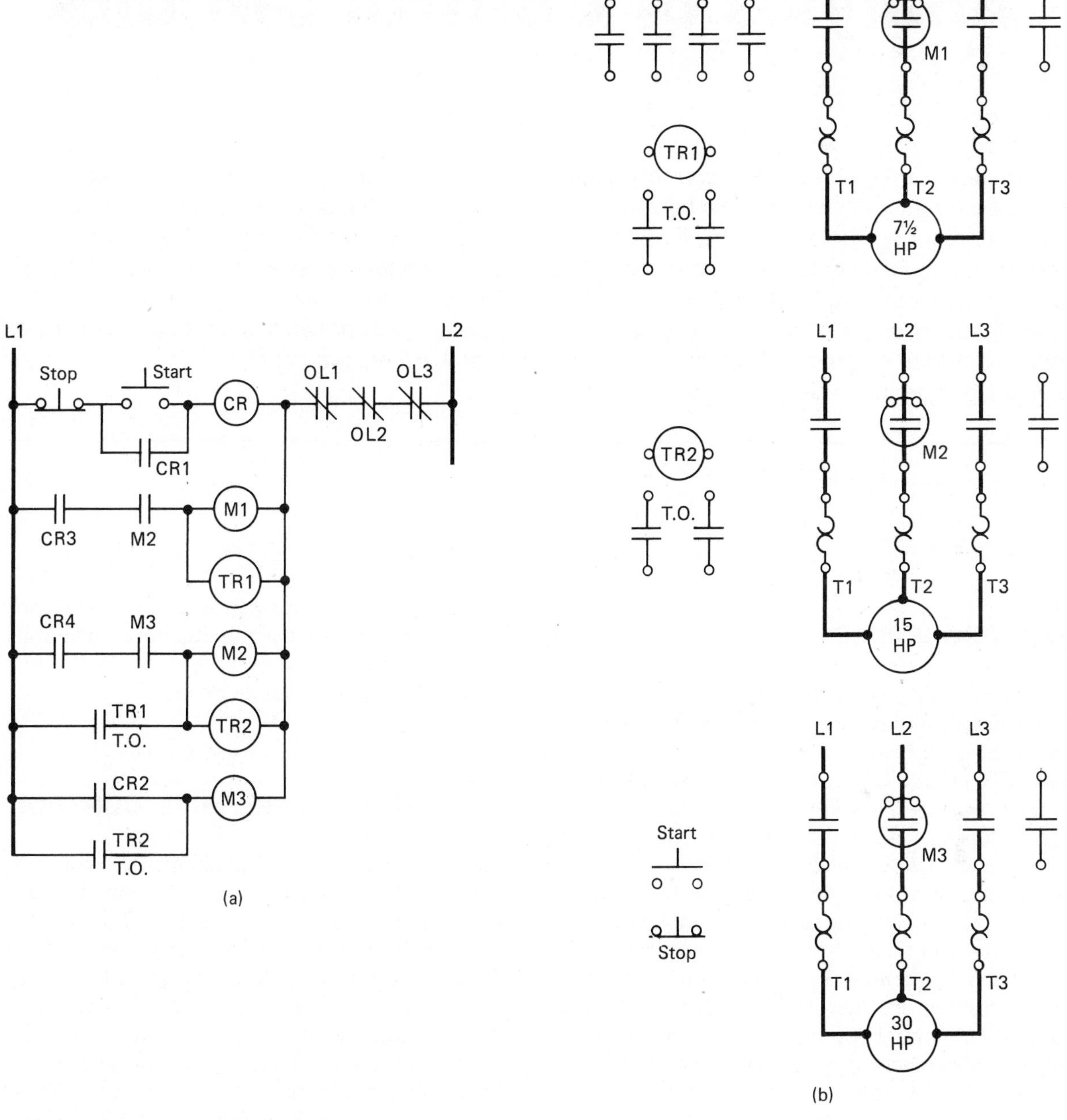

Fig. 7-18 Diagrams for Problem P7-4.

8 Analysis of Control Circuits

The first step in analyzing a motor control circuit is to determine as much as possible how the machine or other equipment which the motor drives operates, so that the functions of the circuit can be more readily understood. To analyze any given circuit, it should be depicted in a control, or ladder, diagram. As stated earlier, if the drawing is properly made, the sequence of control operations should appear along a horizontal rule that proceeds more or less from the upper left of the drawing, with each line of the circuit appearing as a vertical rule at the point it intersects the horizontal rule. Not all diagrams, however, are drawn this way, so do not expect that it will always apply.

8-1 BASIC PROCEDURE

The basic procedure for analyzing a motor control circuit is relatively straightforward. That is, consider one component of the circuit at a time and decide what happens if a pushbutton is pressed or if a contact closes or opens, realizing that you must have a closed circuit from one line through the coil to the other line in order to energize any relay, contactor, or starter. If the circuit is open at any point, the coil will be de-energized and its contacts, wherever they may be found in the circuit, will be in their normal, or de-energized, position. When the circuit is closed to the coil, contactors, relays, or starters are energized, and their position is opposite their normal position. Thus, if they are normally closed contacts, they are now open; if they are normally open contacts, they are now closed.

If a time-delay relay is used in the circuit, you must keep in mind whether its contacts are time-opening or time-closing to determine their normal position and their function in the circuit. When relays are used in the circuit, be sure that you consider every contact which is closed or opened by the relay whenever the coil is energized. Failure to consider one contact of a relay may lead to a misunderstanding of the whole circuit. When analyzing a circuit, be sure that you consider every component in its normal and energized positions so that you understand the whole operation of the complete circuit. Do not jump to conclusions when halfway through the analysis.

In the following section we shall analyze several circuits, using a step-by-step procedure which should give you the basic fundamentals of this operation so that you can apply it in actual job situations. The ability to analyze a circuit is a prerequisite to the efficient troubleshooting of motor control circuits.

8-2 FORWARD AND REVERSE CONTROL

Figure 8-1 shows a control circuit for a forward and reverse starter. To analyze its operation, we shall start with the upper left-hand side at L1. The first component is a STOP button which is normally closed. Therefore, the current may flow through as far as the normally open START button marked FWD (for forward). Also, the current may flow downward in the wire to the right of the STOP button to a single-pole switch, which is shown in the idle, or OFF, position, and also to the normally open pushbutton marked REV (for reverse).

If now we press the FORWARD button, the current can flow through that button, through the normally closed contact of the REVERSE button, and through the normally closed contact identified with the letter R2 to the coil F, which is the contactor coil for the forward direction. From there it will proceed through the normally closed overload contacts identified as OL to L2. The circuit is therefore complete from line 1 through the forward-starter coil to L2 and coil F is now energized. The energizing of this coil will open the normally closed contact F2 and close the normally open contact F1. The opening of the normally closed contact has no immediate effect on the circuit, because the normally open pushbutton for reverse has the circuit

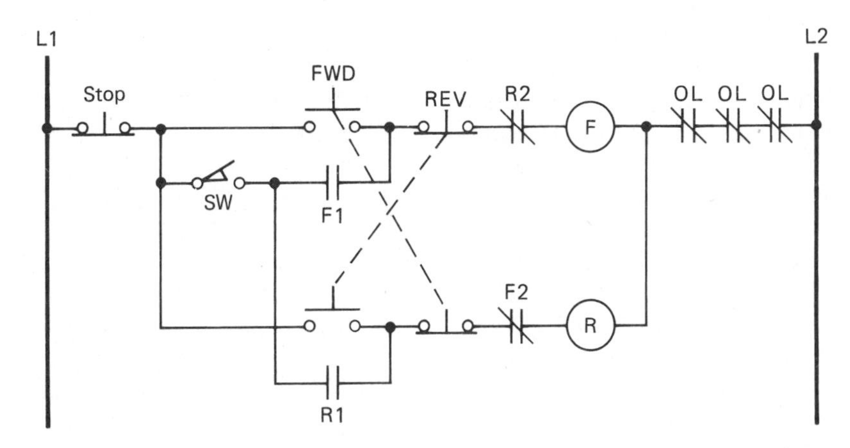

Fig. 8-1 Forward and reverse control for a single motor.

broken ahead of this contact. The closing of the normally open contact accomplished no immediate results, because the switch on the line side of this contact is open and there is no voltage present.

When the FORWARD button is released, the circuit from line 1 to coil F is broken at this point and, because there is no maintaining contact around this break in the circuit, the coil will drop out. Suppose now that we close the switch so that it connects line 1 to one side of the normally open contact F1 and again press the FORWARD button. The action of the circuit is the same as previously discussed, except that now when the normally open contact F1 is closed, it completes the circuit from line 1 around the normally open pushbutton contact. When this pushbutton is released, the circuit is maintained through contact F1 and the motor will continue to run in the forward direction.

Suppose that we now press the REVERSE button. It will open its normally closed contact and close its normally open contact. The result of this action will break the circuit to coil F and complete the circuit through the normally open contacts of the REVERSE button, through the normally closed contact F, through coil R to line 2, thus plugging the motor from forward to reverse. The running of the motor in reverse is maintained through the normally open contact R1, which is now closed. The forward starter is prevented from running by the opening of the normally closed contact R2. If the switch is in the closed position and the REVERSE button is pressed, we have exactly the same operation as when we pressed the FORWARD button and the switch, except now the reverse starter is maintained in an energized state when the REVERSE button is released.

Now that we have analyzed the operation of each component of this circuit, we can summarize by saying that this circuit provides forward and reverse run. It also provides plugging in either direction and positioning the switch will also provide jogging in either direction. The normally closed contacts R2 and F2 provide electrical interlock between the forward and reverse starters. The switch shown in this diagram is known as a jog-run switch because in one position it allows the motors to be jogged in either direction and in the other position permits the motor to run in either direction.

8-3 START, STOP, AND JOG OPERATION

In Fig. 8-2, the single contactor, or starter, coil indicates that this is a circuit for the control of a single motor running in only one direction. Again applying our

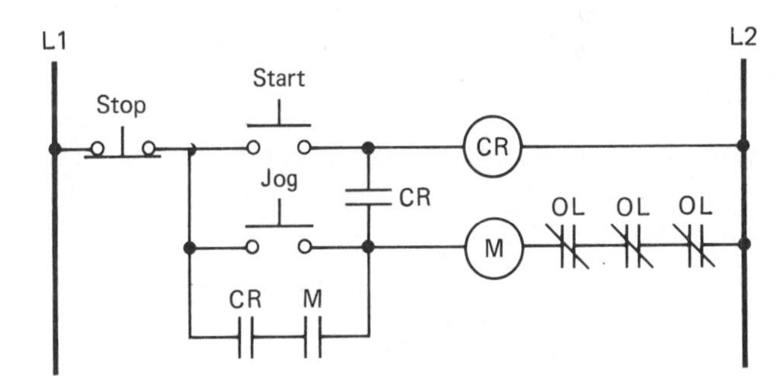

Fig. 8-2 Start, stop, and jog operation for a single motor using a jogging relay.

principles of analysis to the circuit to determine its operation, we see that the STOP button is normally closed so current can flow through it to either of two normally open pushbuttons (START and JOG) and a normally open contact CR.

When we press the START button, the circuit through coil CR is completed to line 2. The designation CR indicates a control relay which in this case has two normally open contacts. These normally open contacts close when coil CR is energized. The one energized from L1 to the STOP button allows current to flow only as far as the normally open contact labeled M.

The other control relay contact, which connects the wire following the START button down to the second horizontal line of the diagram, permits current to flow through this normally open pushbutton, through the now-closed relay contacts to coil M, and through the normally closed overload contacts to L2, thus energizing coil M and the motor. The energizing of coil M causes the normally open contact M to close, which will allow current to flow from line 1 through the normally closed STOP button, through the now-closed relay contact CR, through the now closed contact M, and through coil M, maintaining the circuit to this coil and keeping the motor running even though we release the normally open START button. The motor can be stopped only by pressing the STOP button, which breaks the circuit from L1, allowing both the control relay and the starter coil to drop out.

Suppose that we now press the normally open JOG pushbutton. Current will flow directly from line 1 through the normally closed STOP button, through the JOG button, to coil M, then through the overloads to line 2, and the motor will be energized. Energizing coil M again closes its normally open contact; but this will not maintain the circuit when the pushbutton is released, because the normally open contact CR is open and has the circuit broken from line 1. When we release this pushbutton, the motor is disconnected from the line.

This circuit provides jogging with the additional safety protection of a relay which definitely prevents the starter from locking in during jogging service. When the START button is pressed, both the control relay and the starter are energized, and the starter is locked in through the relay contacts. When the JOG button is pressed, only the starter is energized, and it is definitely prevented from locking in by the normally open relay contacts.

8-4 DYNAMIC BRAKING

Figure 8-3 shows the motor feeders containing the line contacts of the starter identified by the letter M, the overload heater elements, and the three motor terminals identified as T1, T2, and T3. The next circuit contains the contacts DB and then the primary of a transformer identified as PT. The secondary of this transformer is connected to a full-wave bridge rectifier with the dc terminals marked with a plus and a minus sign. The output of this rectifier is applied to terminals T1 and T3 of the motor through two other DB contacts. The part of the circuit so far considered is part of the internal wiring of the controller; the remaining section of the circuit contains the external start-stop control.

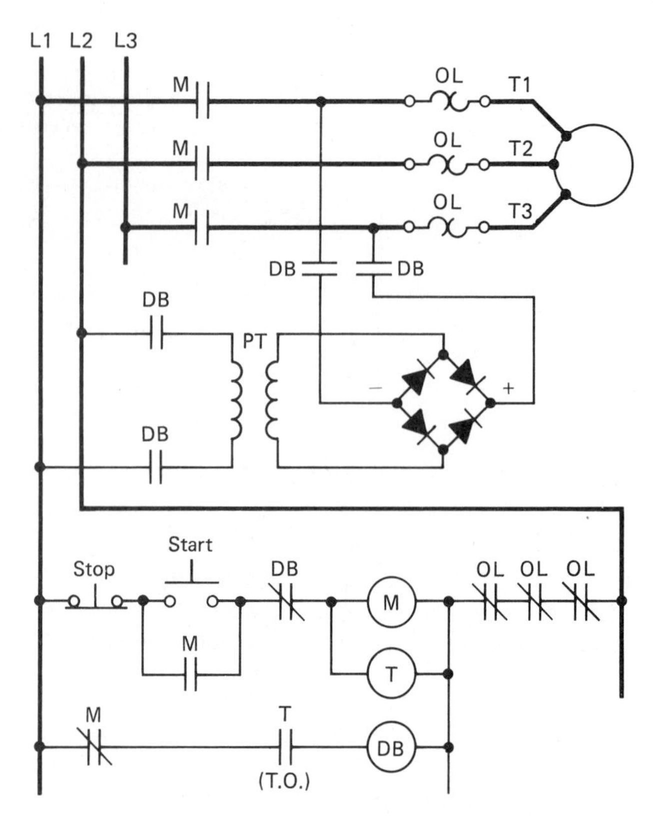

Fig. 8-3 Dynamic braking of a squirrel-cage motor. *(Courtesy of Cutler-Hammer Inc.)*

For the balance of this circuit, pressing the START button will energize coil M because all the other components in this circuit are normally closed. Energizing coil M will close all its contacts, and the motor will be energized through the closing of the three line contacts M. The auxiliary contact in parallel with the START button will close, thus sealing in the circuit and maintaining the motor in the running position. The opening of the normally closed contact M, located on the bottom line of the drawing, will prevent coil DB from being energized.

Simultaneously with the energizing of coil M, coil T is energized. This is a time-delay relay because its contact T is indicated to be time opening. If we now press the STOP button, coil M is dropped out along with all its contacts, which will return to their normal position. The opening of the line contacts M breaks the circuit to the motor and stops the flow of current.

The auxiliary contact in parallel with the START button opens, which has no effect on the circuit at this time. The returning of the normally closed contact M to its closed position, however, will energize coil DB because contact T is still closed. We know that this contact is closed because it is designated to be time opening, and even though its coil is now de-energized, the timer would maintain this contact in a closed position.

With coil DB energized, all contacts indicated by the letters DB will now be in their operating position. The normally closed contact in series with coil M will be open, thus preventing a re-energizing of this coil until the time-delay relay has opened contact T. The closing of the four normally open DB contacts associated with the transformer and rectifier will, in effect, apply dc voltage to T1 and T3 and hold this voltage on the motor until the time-delay relay has timed out, thus opening contact T, which returns the circuit to its normal, unenergized condition.

What is the purpose of applying dc voltage to a motor when you press the STOP button? The application of dc voltage to a rotating squirrel-cage motor has the effect of producing magnetic poles in the stator. The turning rotor acts as a generator armature, which causes high currents to flow in the shorted rotor bars. The effect of the current is to oppose the direction of rotation of the rotor. This produces a smooth but positive braking action and will bring the motor to a rapid but very smooth stop. The time-delay relay is necessary in this circuit because if the dc voltage were not disconnected from the motor at almost zero speed, the low dc resistance of the motor winding would allow excessive current to flow, thus overheating and possibly damaging the motor windings. This time-delay relay should be adjusted so that it will apply the dc voltage to the motor windings practically down to zero speed and remove it so that the motor stops immediately.

This circuit seems to provide a normal across-the-line start for a squirrel-cage motor, but in addition provides a rapid, smooth-braking effect on its stop. It can be applied to any piece of equipment whenever a smooth, fast stop is required or where it is desired to have the motor shaft free for manual rotation when the power is disconnected. It also provides a stop without any tendency to reverse, such as is encountered with a plugging stop. This type of braking is also an advantage where the braking effect must be applied frequently. It requires less maintenance than a mechanical brake, thus reducing maintenance cost. It also provides less shock to the drive system than a mechanical brake and less heating than with a plugging stop. This type of braking is known as *dynamic braking.*

8-5 WYE-DELTA CONTROLLER

In Fig. 8-4 on the next page, we see a double set of line contacts identified as 1M and 2M, which connect lines 1, 2, and 3 to the motor terminals. Also in this part of the circuit we have contacts identified as S, which seem to connect some of the motor windings. In the lower control section of the diagram, we have a START button, a STOP button, and a coil S, which seems to be some sort of auxiliary contactor. Also, we have coil 1M, which apparently is a line contactor for the motor. Coil TR appears to be a time-delay relay. Coil 2M appears to be a second-line contactor for the motor.

In analyzing the circuit, suppose we press the START button, which will energize coil S, since all the contacts and pushbuttons in this circuit are closed. The energizing of this coil will operate all its contacts, which will energize coil 1M and also will prevent the energizing of coil 2M by the opening of its normally closed contact. The two normally open contacts S, which connect the three motor terminals, will now be closed, forming a wye connection for the motor coils. The energizing of coil 1M closes all its contacts, three of which are line contacts for the motor, thus energizing the motor and starting it running. One such contact is in parallel with the START button and acts to maintain the control circuit. Another contact is in series with coil 2M, but at this time it has no effect on the circuit, because the normally closed contact S is now open.

At this point, we have a wye-connected, squirrel-cage motor running across the line. At the time coil 1M was energized, coil TR was energized and the timing action of its normally closed contact TR was started. When this contact times out and opens, it breaks the circuit to coil S and returns all its contacts to their normal position. An opening of the two contacts connecting the motor windings breaks the wye connection for the motor windings. The opening of the contact in series with coil 1M has no effect on the circuit, because this circuit is completed through contact 1M in parallel with it. The closing of the normally closed contact in series with coil 2M now completes the circuit to this coil and causes its contacts to close, connecting the motor terminals directly to the line and forming a delta connection for the motor.

If you have any trouble visualizing these motor connections, you should draw them separately on a sheet of paper to see that the first connection was a wye and the second a delta connection of the three motor windings. Of course, pressing the STOP button de-energizes all the coils and returns the circuit to its normal, at-rest condition. This circuit shows three resistors and three contacts connecting them, along with a coil and other associated contacts which would be necessary to establish a closed transition for the starting of this motor.

Our analysis of this circuit shows that it is a wye-delta-type motor controller used for the purpose of giving a reduced-voltage effect to the starting of this motor, as discussed in Chap. 4. In applications where a closed transition is necessary or desirable, the additional connections are shown for adding resistance to bridge the motor connections during the transfer from wye to delta. This is a rather common circuit and deserves some concentrated study as to its principle of operation. Again, however, a warning is in order: don't memorize this circuit as being the only possible

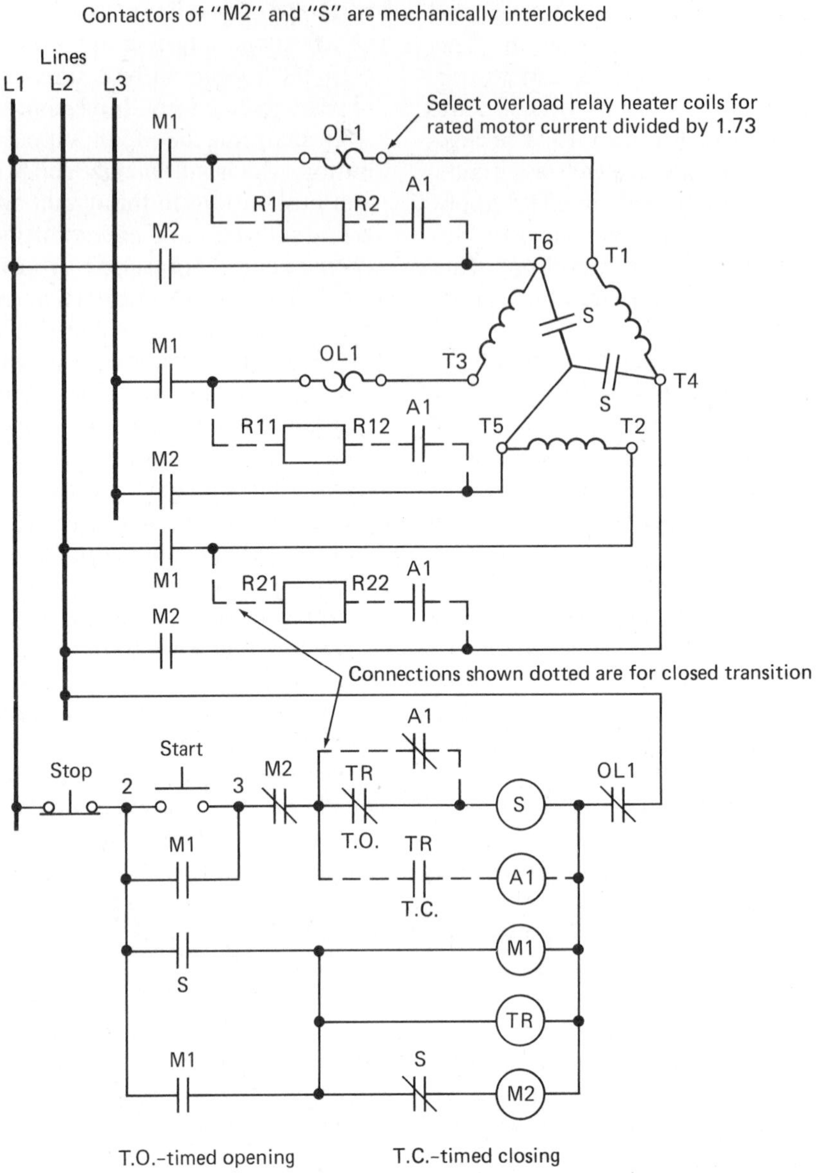

Fig. 8-4 A wye-delta controller for a squirrel-cage motor. *(Courtesy of Cutler-Hammer Inc.)*

way to give wye-delta starting to squirrel-cage motors. The use of the time-delay relay with its time-opening contacts gives us a definite-time type of acceleration for this control circuit. This controller, as you may have noticed, involves a two-pole and two three-pole magnetic contactors along with the necessary mechanical interlock to ensure a sequence of operation and to prevent two connections at the same time, which would cause a short circuit.

8-6 PRIMARY-RESISTANCE, REDUCED-VOLTAGE STARTER

Considering the circuit of Fig. 8-5, we find that the resistance in series with the motor leads would seem to indicate that this is a primary-resistance, reduced-voltage starter. Looking at the control section of the diagram, we have what seems to be an ordinary three-wire control circuit to energize coils 1CR and TR. If we press the START button, current may flow through the normally closed STOP button, the START button, and contact R2, and coils 1CR and TR will be energized. The energizing of coil 1CR will cause all its contacts to close. Contact 1CR1 is in parallel with the START button and will perform the function of maintaining the circuit to the coil. Contact 1CR2 will close and energize coil S. The energizing of this coil will cause the line contacts S to close and will energize the motor through the series resistances. The presence of resistance in series with the motor leads will cause a reduced voltage to be applied to the motor, thus reducing the

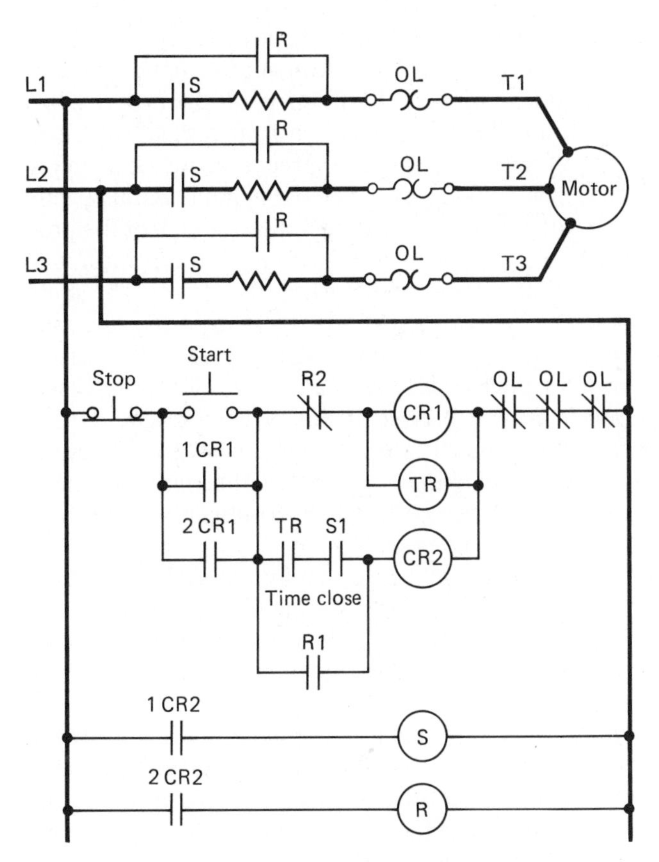

Fig. 8-5 Primary-resistance, reduced-voltage starter.

inrush current to the motor and providing reduced-voltage starting.

The motor is now accelerating under reduced voltage, and the time-delay relay TR is timing out. When relay TR times out, it will close contact TR. When this contact closes, it will energize coil 2CR because contact S1 is closed by coil S. The energizing of coil 2CR will cause contact 2CR1 to close. This contact is in parallel with the START button and forms an additional maintaining circuit for the coil. The closing of contact 2CR2 causes coil R to be energized, closing the line contacts identified as R. These contacts are in parallel with the resistors and effectively short them out of the circuit, thus applying full-line voltage to the motor, which will enable it to accelerate to its full speed and run across the line.

The energizing of coil R also closes contact R1, which is in parallel with contacts TR and S1. The opening of contact R2 will cause the dropping out of coils 1CR and TR. The contacts associated with these two coils will now return to their normal position, and we shall have the motors running through the control circuit, which consists of the STOP button, contact 2CR1, contact R1, and coil 2CR. This circuit maintains the circuit to the run coil through contact 2CR2. Should the STOP button now be pressed, all contacts would return to their normal position and all coils would be de-energized, thus opening the line contacts to the motor, and the motor would come to a stop.

This circuit obviously is one for a primary-resistance, reduced-voltage starter. Again, it must be pointed out that this is only one of the many arrangements of coils and contacts which could be used to achieve the same results. Different manufacturers will use variations of a similar circuit in the control of their starters, but the basic principle of operation is the same if a definite-time sequence of starting is employed.

This circuit could be expanded to give several more stages of acceleration by the addition of more units of resistance in series with the motor, with a control relay and a time-delay relay for each step, or stage, of acceleration. This is a two-stage, or two-step, starter, since it provides two steps of acceleration, one under reduced voltage and one under full voltage.

The only critical adjustment in this circuit will be found in the time-delay relay TR, which must not be allowed to maintain the motor under reduced voltage longer than the time it takes to accelerate to its maximum speed under reduced-voltage conditions. Prolonged operation of the motor under reduced voltage may quite possibly overheat and damage the motor windings and cause the resistance elements to be seriously damaged or burn out.

This controller consists of a start contactor S, which must be three-pole, and a run contactor R, which also must be three-pole. In addition to the two contactors, there are two control relays and one time-delay relay. This equipment would be found generally mounted in one enclosure with the START-STOP station either mounted on the door of this enclosure or separately wired to any convenient location in the building.

At this stage in your study of controls and analysis of control circuits, you should consider a circuit from the standpoint of what would happen if a particular coil burned out or a particular contact failed to open or close. For instance, suppose that the time-delay relay TR were to have a burned-out coil. What would be the effect on this circuit? The circuit would function up through the closing of the start contactor S, and the motor would be energized under reduced-voltage conditions. If contact TR does not close, however, then the second control relay cannot be energized and the run contactor cannot be energized. Thus, the motor would continue to run under reduced-voltage conditions. The current under this condition is such that it will open the overload relay contacts and will drop out coil 1CR, thus stopping the motor and returning it to its normal position. These overload units should be manual-reset units so that the operator will have to reset them and determine the cause of trouble before restarting the motor. This control circuit will give overload protection and incomplete-sequence protection (Sec. 3-15).

8-7 LOCKOUT CIRCUIT

The circuit of Fig. 8-6 is a partial circuit used to illustrate a lockout circuit, which is extensively used where malfunction of some part of the equipment must require attention by the operator before the equipment is restarted. Looking at the circuit of Fig. 8-6, you will see dotted a contact which represents the normal control components, such as START and STOP buttons, limit switches, or other devices which normally start and stop the machine. This circuit concerns itself only with the lockout components. The normally closed contacts represented as PS1, 2, and 3 are pressure switches which will open only when the required pressure is not maintained in their particular part of the machine or process. Coils A, B, and C are relay coils paralleled by pilot lights.

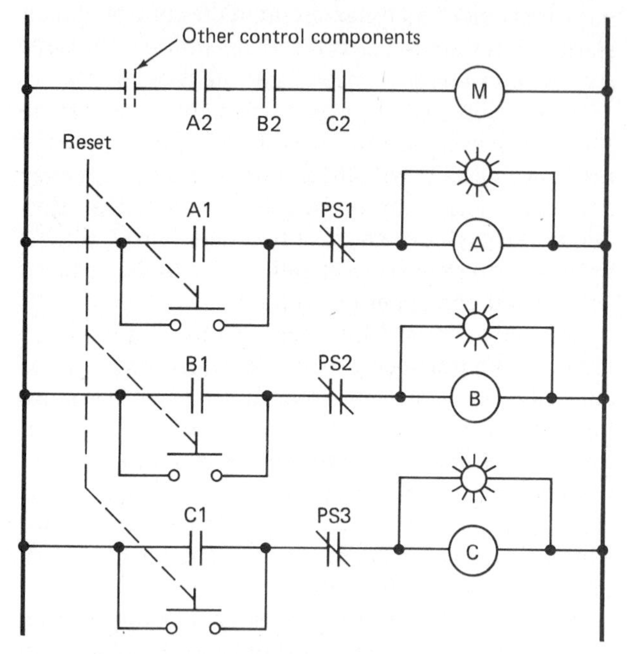

Fig. 8-6 Lockout circuit.

To start the operation of this equipment, it is necessary to press the RESET button, which will close the three associated and mechanically interlocked switches. The three relays will be energized, thus closing their normally open contacts. Contacts A1, B1, and C1 are used to maintain the coil circuit. Contacts A2, B2, and C2 are maintained in their closed position as long as the operation of the pressure switches is normal, thus enabling the normal control components to energize coil M at will. If the pressure drops or rises, as the case may be, from its normal value at any one of the three places where pressure switches are located, it will open one of the normally closed contacts. For instance, if PS1 opens, coil A will be de-energized, which will open contact A2 and drop out the motor. At the same time, contact A1 will open. If the proper pressure returns to pressure switch 1, its contact will close but coil A will not be re-energized, because the circuit is broken at contact A1. The pilot light in parallel with this coil will be out and will indicate which of the protective circuits is not functioning. The operator will know that pressure switch 1 was the cause of the shutdown of the equipment. In order to restore the machine to operation, the pressure sensed by pressure switch 1 must be restored to normal. Then the RESET button must be pressed in order to energize coil A, thus closing its contacts and allowing the normal operation of the control circuit. Of course, the same procedure will apply to the second and third pressure switches and associated contacts and coils.

This type of circuit is generally applied to fully automatic equipment, where the machine or process is allowed to start and stop by itself under the control of pilot devices which sense the condition of the process or material as the machine does its work. When machinery operates under these conditions, it is generally desirable to have some means to stop the process whenever a malfunction occurs and to prevent it from restarting until it has received the attention of an operator.

8-8 DIRECT-CURRENT CONTROL CIRCUITS

As defined by NEMA, an electric motor controller is a device or group of devices which serves to govern, in some predetermined manner, the electric power delivered to a motor. Thus, the purpose of a controller is to connect (or disconnect) electric power to a motor. The circuits in this chapter illustrate this basic function as applied to the control of dc motors.

LINE DISCONNECT

Figure 8-7 illustrates a simple circuit whose function is to disconnect the motor from the power source. A

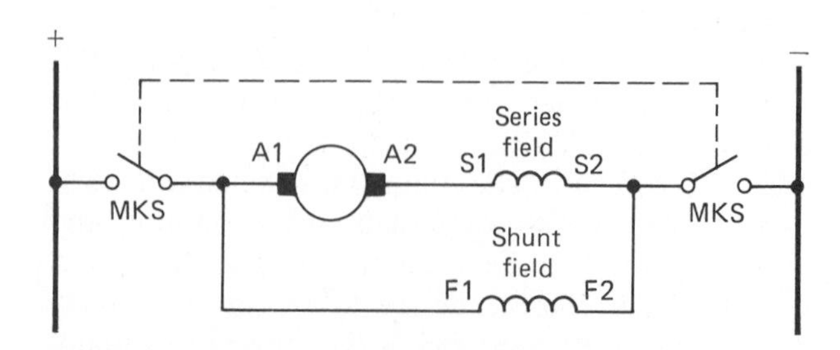

Fig. 8-7 Direct-current motor started with a manual switch.

two-pole main knife switch, or MKS, or some other disconnect device, such as a circuit breaker, a safety switch, or even some type of snap switch, could be used.

Manual starters which contain a suitable interrupting mechanism are approved by NEMA for use with dc motors having power ratings of 2 hp or less.

LINE CONTACTOR

Because even the most basic dc motor controller uses magnetically operated devices, a main line contactor, M, is the next logical addition to a starter circuit, as shown in Fig. 8-8. Since a coil operates the contactor, a control circuit and its supply must be added.

To complete the circuit to the contactor coil, a pilot-type device such as a pushbutton or master switch is required. For simplicity, a two-unit START-STOP pushbutton is shown in Fig. 8-8.

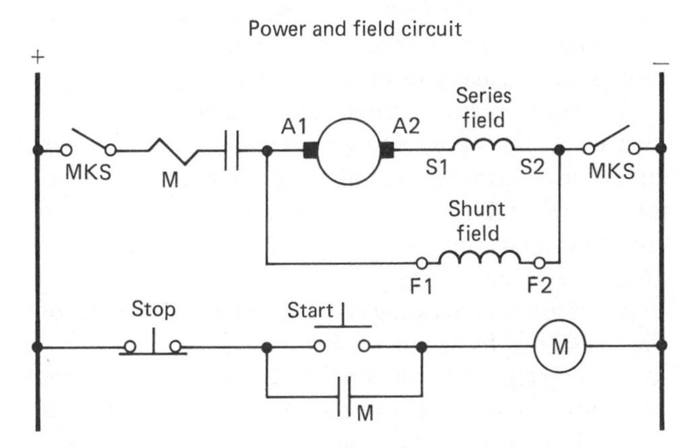

Fig. 8-8 Direct-current motor with a magnetic starter.

Pressing the START button energizes the contactor coil and thereby closes the main power contacts of the contactor to complete the power circuit to the motor armature.

To prevent the contactor from dropping out (that is, opening) as soon as the START button is released, an auxiliary contact, or control-circuit contact, is provided on the contactor to bypass the START pushbutton.

To stop the motor, all that is necessary is to momentarily press the STOP pushbutton. This de-energizes the operating coil of the contactor, opens the contactor, and removes power from the motor armature circuit.

FIELD-DISCHARGE MEANS

If MKS in Fig. 8-8 is opened, the shunt-field circuit is also opened. However, under these conditions, the highly inductive shunt-field winding can induce such a high voltage that the winding insulation is damaged. Therefore a path must be provided so that the induced voltage can produce a current flow. This path prevents the discharge voltage from rising too high. Figure 8-9 shows the addition of a field-discharge resistor.

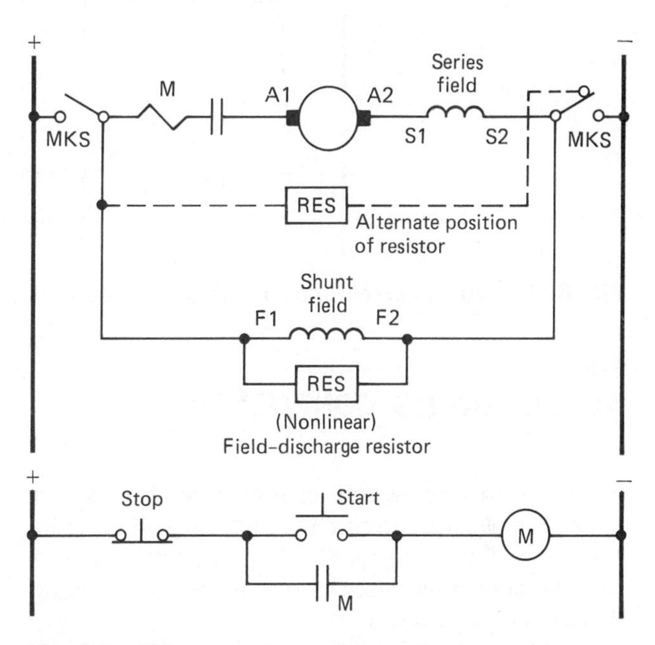

Fig. 8-9 Direct-current compound motor with a field-discharge resistor.

ACCELERATING RESISTOR

To prevent a dangerously high current inrush when the motor armature is connected to the power lines, a resistor is placed in the armature circuit. The necessity for adding a resistor can be illustrated with a typical 50-hp motor which has a full-load current rating of 180 A and an armature resistance of 0.13 Ω. If the armature of this motor is directly connected to a 230-V power source, the inrush current would be 230/0.13, or 1770 A. This is almost 10 times the rated full-load current of the motor and would surely cause "flashing" of the commutator and damage to the armature. In addition, the torque produced by the surge of current might also cause mechanical damage either to the motor or the connected load.

A resistor in the armature circuit (Fig. 8-10 on the next page) can reduce this abnormally high inrush current to a reasonable level. Usual practice is to select a resistor which limits inrush current to 150 to 200 percent of the full-load current rating of the motor.

For the same, typical motor, limiting current inrush to 150 percent of 180 A would permit a current of 270 A to flow. Then, the total resistance that must be added to the circuit is 230/270, or 0.85 Ω. This resistance value includes the internal resistance of the armature, but for simplicity armature resistance will be neglected here.

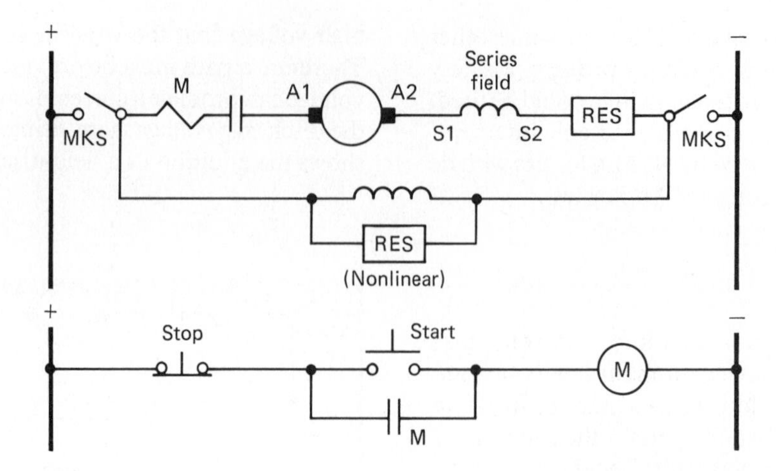

Fig. 8-10 Direct-current motor with a senes current-limiting resistor and shunt field discharge resistor.

8-9 ACCELERATING CONTACTORS

If the current-limiting resistor is allowed to remain in the motor armature circuit, the motor would never be able to reach its rated speed. This can best be shown with the same motor used as an example in discussing the accelerating resistor.

From one to six accelerating contactors may be used, but usually two or three are sufficient for most applications. The first and second accelerating contactors, 1A and 2A, are shown in the accompanying circuit of Fig. 8-11. Because these "accelerators" are magnetically operated, they have operating coils which must be added to the control circuit.

For satisfactory operation, contactor 1A must close at a certain interval after M closes, and 2A in turn must close at a suitable interval after 1A closes. The intervals may be determined by such methods as definite time, motor speed, motor counter-emf, or decrease in current flowing through the motor armature. Conventional practice is to base the interval on a definite time period, usually in the order of 1 s per interval.

Although the time periods can be determined in various ways, the accelerating contactors shown in the circuit have a time delay built into them. Hence, separate relays or sensing devices are not necessary. The control-circuit contacts on M and 1A ensure the proper order or sequence of operation.

After both accelerating contactors have closed, all of the resistor is bypassed or shorted out so that, effectively, no resistance remains in series with the motor armature. Therefore, the motor will run at its rated speed, having been "accelerated" by the successive closing of contactors M, 1A, and 2A.

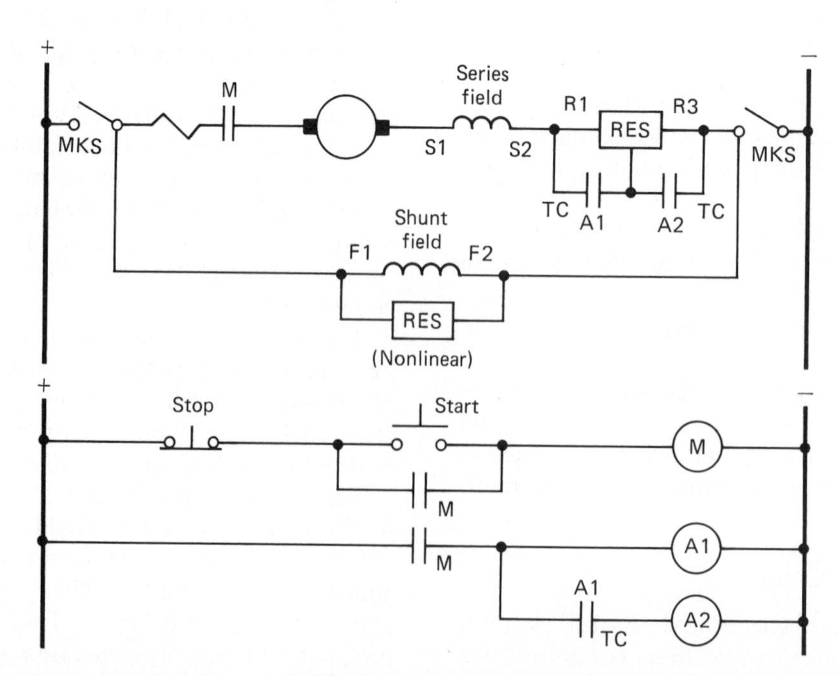

Fig. 8-11 Direct-current compound motor with motor starter providing two steps of acceleration.

OVERLOAD RELAY

A greater-than-normal load thrown on the connected machinery, either by operator fault or mechanical fault in the equipment, represents one type of undesirable condition. This overload would increase the current through the motor as the motor attempted to support the load. Such a condition might result in damage to either the driven machinery, the motor, or both, especially if allowed to continue for an indefinite period of time.

Detecting an overload condition and acting to remove the motor from the power line requires a sensing device that has its operating coil connected in the armature circuit. It has a normally closed contact which is connected in series with the operating coil of contactor M. This is illustrated in Fig. 8-11 also.

FIELD-LOSS RELAY

Another fault condition that must be sensed and acted upon is the loss of shunt-field excitation. With the loss of shunt field excitation the motor will tend to run away.

As illustrated in Fig. 8-12, a field-loss relay, FL, protects against this fault. The relay operating coil is connected in series with the shunt-field circuit. A normally open contact is connected in series with the operating coil of contactor M in the control circuit. Therefore, if the shunt-field circuit is broken or opened, no current flows through the coil of the FL relay, its contact opens, and contactor M is de-energized and removes the motor from the source of power.

FIELD RHEOSTAT

When a motor is shunt-wound or lightly compound-wound, its speed can be increased above rated base speed by decreasing the strength of its shunt field. Accordingly, shunt-field current can be varied, primarily from rated value to reduced values, with a field rheostat as shown in Fig. 8-13 on the next page.

Essentially, a field rheostat is an easily operated adjustable resistor. It may be driven by a pilot-type motor, in which case it is known as a motor-operated rheostat. The conventional field rheostat is manually adjusted by the operator to increase or decrease the speed of the dc motor.

FIELD-ACCELERATING RELAY

If the motor speed range is more than 2 to 1, the presence of a field rheostat creates the following two demands on the controller:

1. To provide rated or required torque when accelerating from standstill to base speed, the shunt field must be at full strength.
2. If the motor is running at base speed because the rheostat is in its all-resistance-out position and the rheostat is then rapidly moved toward or to its all-resistance-in position, an excessive current will be drawn by the motor from the power lines. To avoid this high armature current condition, it is desirable that the field strength be weakened slowly, in a multiplicity of steps, or by some manner which accomplishes the same result. Figure 8-14 illustrates the field-accelerating relay that protects against excessive armature current.

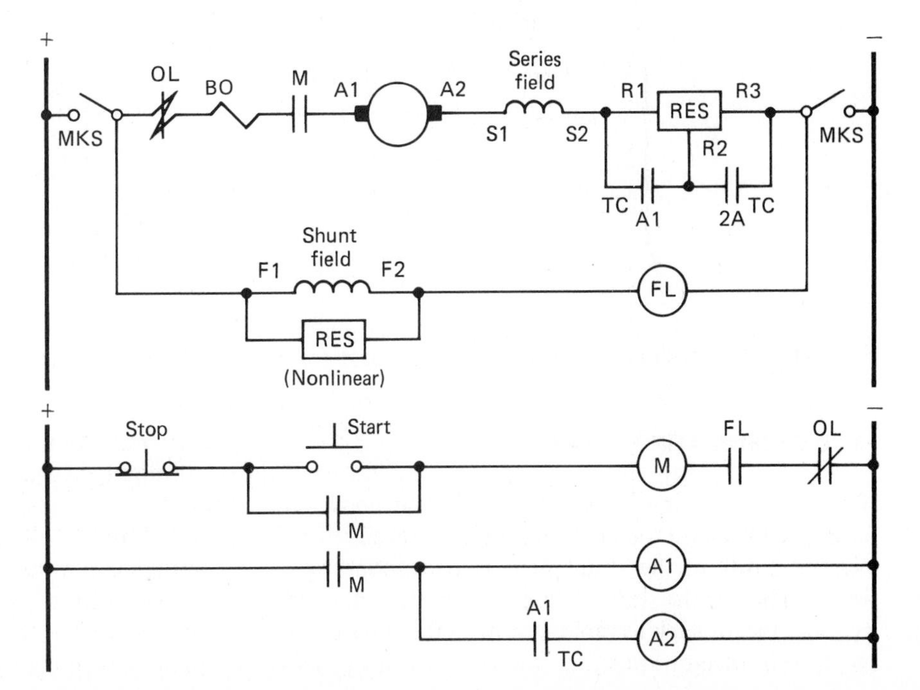

Fig. 8-12 Direct-current compound motor with overload relay and field-loss protection. Coil BO is used to minimize arcing at the contacts. All large dc motors require these coils (called *blow-out coils*) in series with contactors in power circuits.

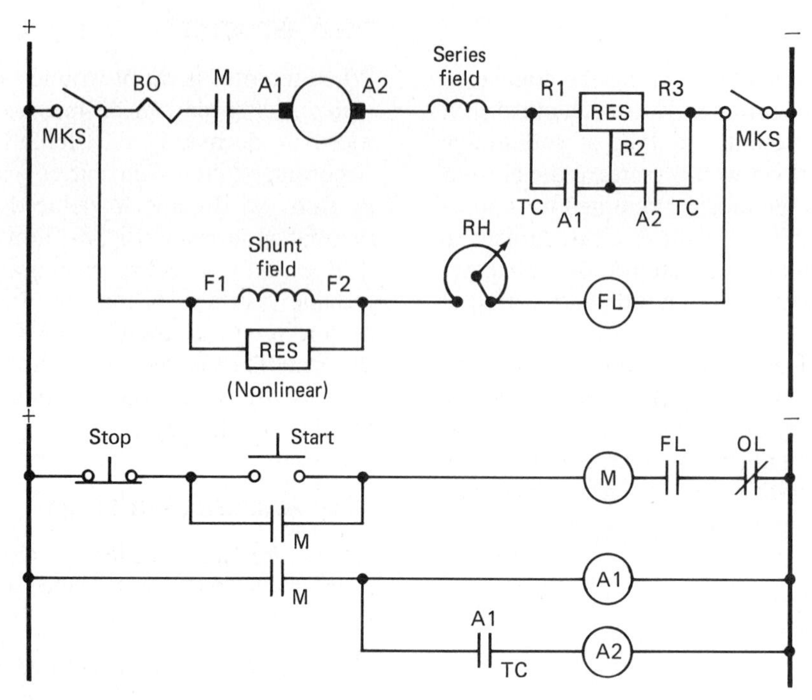

Fig. 8-13 Direct-current compound motor and controller with a field rheostat for speed control.

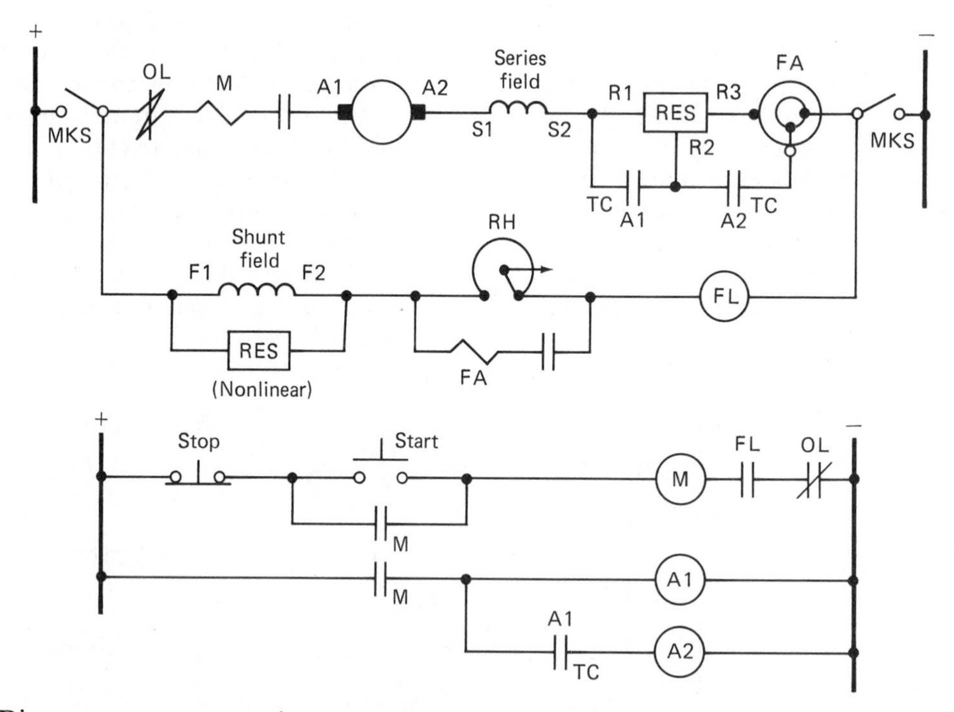

Fig. 8-14 Direct-current compound motor and controller with a field-acceleration relay.

FIELD-DECELERATING RELAY

If a motor is running at a speed considerably above its base speed because the field rheostat is in its all-resistance-in position, the shunt field is correspondingly weak. Then, if the shunt field is rapidly strengthened by some means such as rapidly moving the rheostat to its all-resistance-out position, the counter-emf generated by the motor will also rise rapidly. This counter-emf might exceed the voltage of the power supply and result in a reversal of current through the motor armature. Usually, if the value of the reversal current is not too high and if the speed range is less than perhaps 2 to 1, the condition is not harmful. However, if the speed range is greater, the magnitude of the reversal current might be high enough to cause flashing at the commutator, a reaction on the driven machine, or possibly a disturbance of the power supply.

Reversal current can be held to an acceptable value with a field-decelerating relay FD and an associated resistor. The type of relay used in this circuit has a series coil, which is connected in the armature circuit, and a shunt coil, which is connected across the control-circuit supply. The relay also has a normally closed contact which is connected across the resistor in the shunt-field circuit. This is illustrated in Fig. 8-15.

FIELD-PROTECTIVE RELAY

Some motors are so designed that the shunt field is not suitable for continuous duty at full field excitation when the motor is at standstill. When the motor is running, however, the movement of air is sufficient to prevent the shunt field from overheating. The controller must therefore provide protection for the shunt field when the motor is at rest.

A field-protective relay FP and its associated resistor, usually called the *field-economizing resistor,* can provide shunt-field protection. The relay is a conventional type and has a normally open contact. The operating coil is located in the control circuit, while the resistor and relay contact are connected in the shunt-field circuit. When the motor is stopped, the relay is de-energized and the resistor is inserted in the circuit. When the motor is started and during running, the relay coil is energized, thereby shorting out the economizing resistor. This is illustrated in Fig. 8-16 on the next page.

8-10 DYNAMIC-BRAKING CONTACTOR

A common method of retarding and stopping a motor is dynamic braking. This feature is provided by dynamic-braking contactor DB and a resistor. The combination of resistor and dynamic-braking contact are connected across the armature terminals, as shown in Fig. 8-17.

The conventional dynamic-braking contactor is a spring-closed, magnetically opened device. Therefore, the main contact of the contactor is shown in Fig. 8-17 as normally closed. The DB operating coil is added to the control circuit to give the following operation.

When the motor is running, contactor coil DB is energized. Therefore, the contacts of the contactor are open and the dynamic-braking resistor is not in the circuit. When the controller functions to remove power from the motor, coil DB is de-energized. Contact DB then closes and connects the dynamic-braking resistor across the armature of the motor. Thus, the dynamic-braking loop is established, and current produced by motor counter-emf circulates in the armature, thus retarding motor rotation.

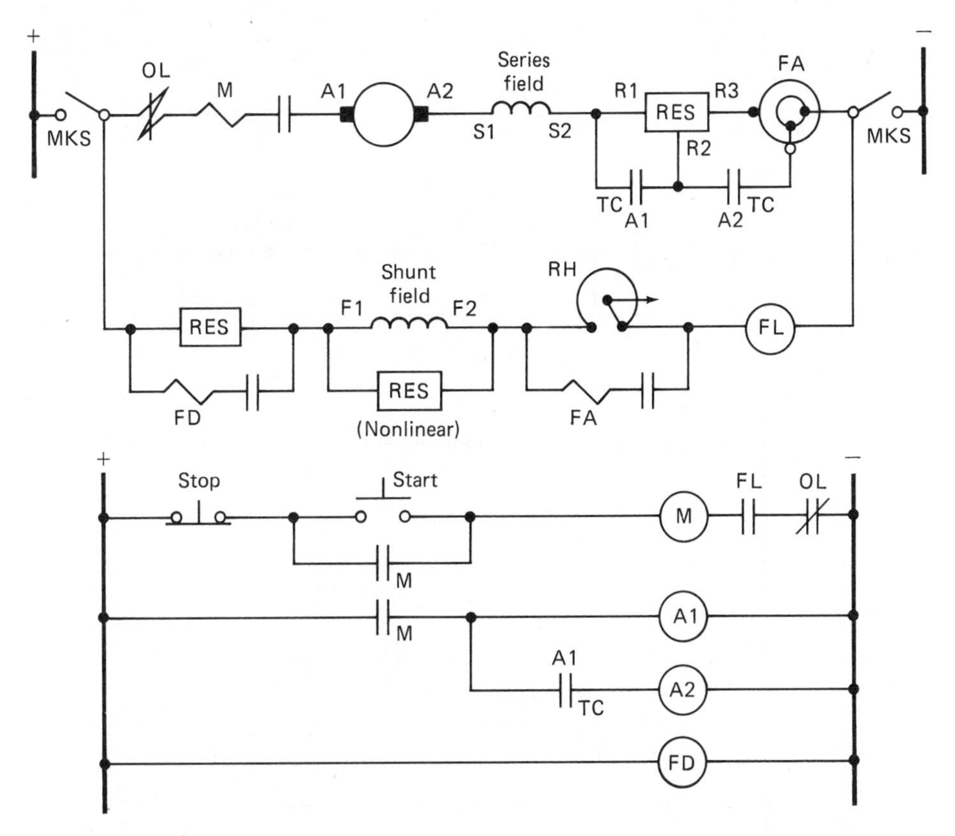

Fig. 8-15 Direct-current compound motor and controller with a field-deceleration circuit.

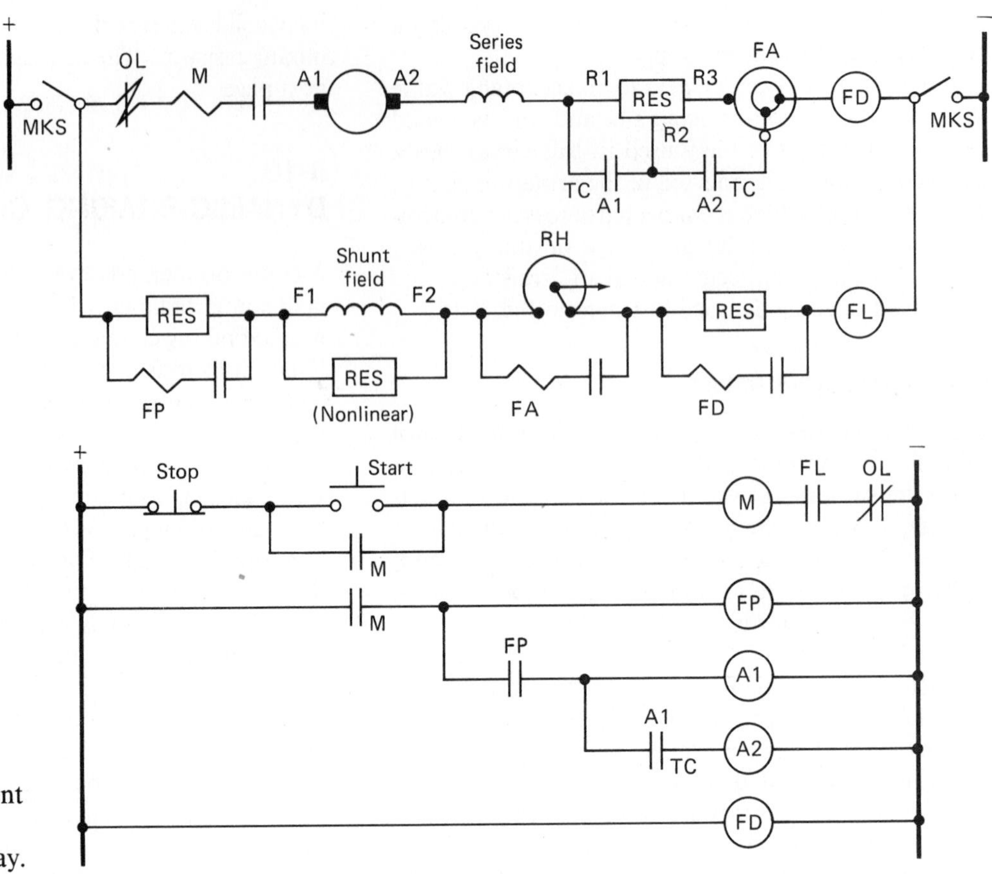

Fig. 8-16 Direct-current compound controller with field-protective relay.

Fig. 8-17 Direct-current compound motor and controller with all protective relays and dynamic braking.

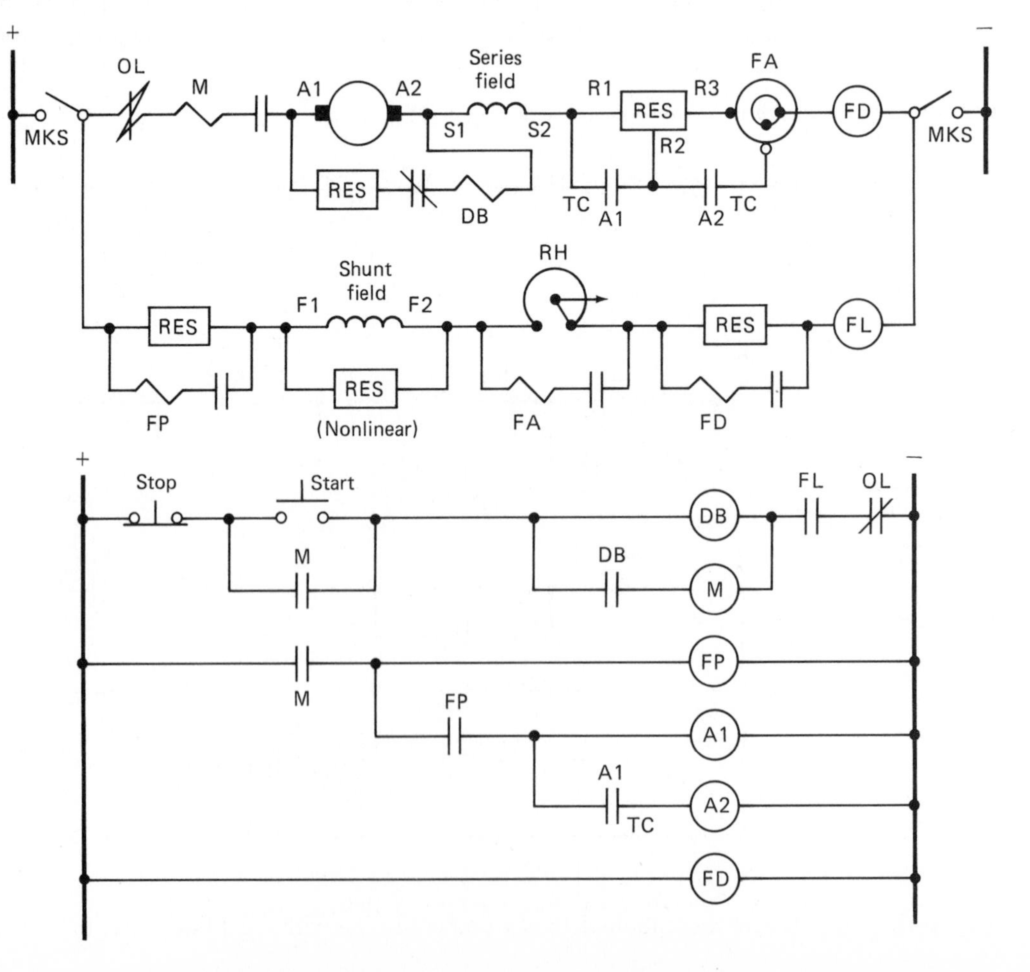

8-11 REVERSING CONTACTORS

In many controller applications the motor must be operable in both forward and reverse directions. This is normally accomplished by replacing contactor M with a set of contactors whose function is to reverse armature polarity with respect to power-line polarity. The reversing-contactor system shown in Fig. 8-18 includes forward contactor F and reverse contactor R. Several previously introduced control devices are omitted in this circuit only for clarity.

Two contacts are necessary for the forward function and two for the reverse function. Here, each set of two contacts is located on one two-pole contactor. This arrangement is often used on smaller-size contactors. However, for larger contactors, the usual practice is to use two separate single-pole contactors for the forward function and two separate single-pole contactors for the reverse function. It is conventional to interlock directional contactors electrically and mechanically.

The use of a master switch as shown in Fig. 8-18 is common practice, especially for reversing controllers. Therefore, to illustrate the basic application of such a device, a two-point switch is incorporated as the pilot or control device. Depending on the user's preference and the number of accelerating contactors, the type of master switch could vary from a one-point up to a five- or six-point switch.

An undervoltage or loss-of-voltage relay UV has also been added. It serves as an interlocking relay which requires the operator, after an overload trip, to return the master switch to the OFF position before the motor can restart. This operating and safety feature is conventionally used with master-switch-operated controllers. The UV relay also drops out on a power failure. The master switch must be returned to the OFF position for the UV relay to energize.

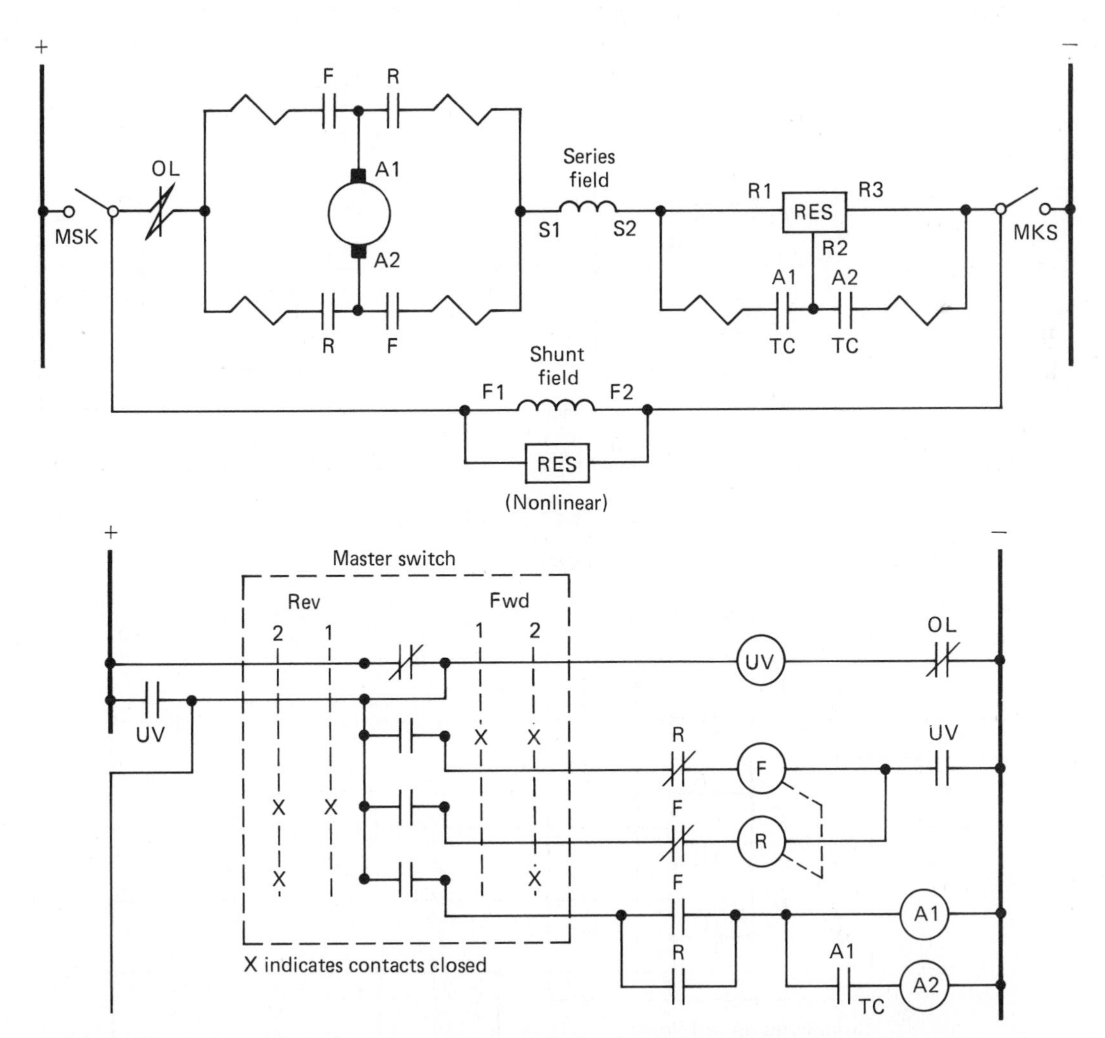

Fig. 8-18 Direct-current compound motor with reversing controller. Controller provides two-step acceleration forward or reverse. Blow out coils, though not shown, are in series with contactors F, R, 1A, and 2A in power circuits.

8-12 ANTIPLUGGING RELAY

If motor rotation is suddenly reversed with a master-switch type of reversing controller, the counter-emf produced by the motor would add to the line voltage, and an objectionably high current would flow through the armature and accelerating resistor. A common method for avoiding such a condition uses some means to prevent closing of the opposite-direction contactors until the motor has come approximately to a standstill either by coasting or by some type of braking.

Nonplugging or antiplugging operation is provided by the antiplugging relay AP, as shown in Fig. 8-19. Because it is a voltage relay, this device has its coil connected across the armature terminals so that it measures and responds to the counter-emf of the motor. Its normally closed contact is connected into the operating coil of the forward and reverse contactors in the control circuit.

If armature counter-emf is above a desired value, the relay contact will be opened, thus preventing the opposite-direction contactor from having its operating coil energized. For normal motor operation after the permissible start or reversal, relay contact AP is bypassed by normally open control-circuit contacts on reversing contactors F and R.

This type of reversal control is useful and finds wide application, especially where rapid reversal is not necessary or perhaps not desirable.

8-13 PLUGGING RELAY

Instead of adding an antiplugging relay to a basic reversing controller with a master switch, another method is available for preventing an abnormally high current when a motor is reversed. Suppose the opposite-direction contactors are allowed to close so that the counter-emf of the motor is added to the line voltage. If the resistance of the accelerating resistor were increased an appropriate amount at this time, the current flowing through the motor could still be maintained

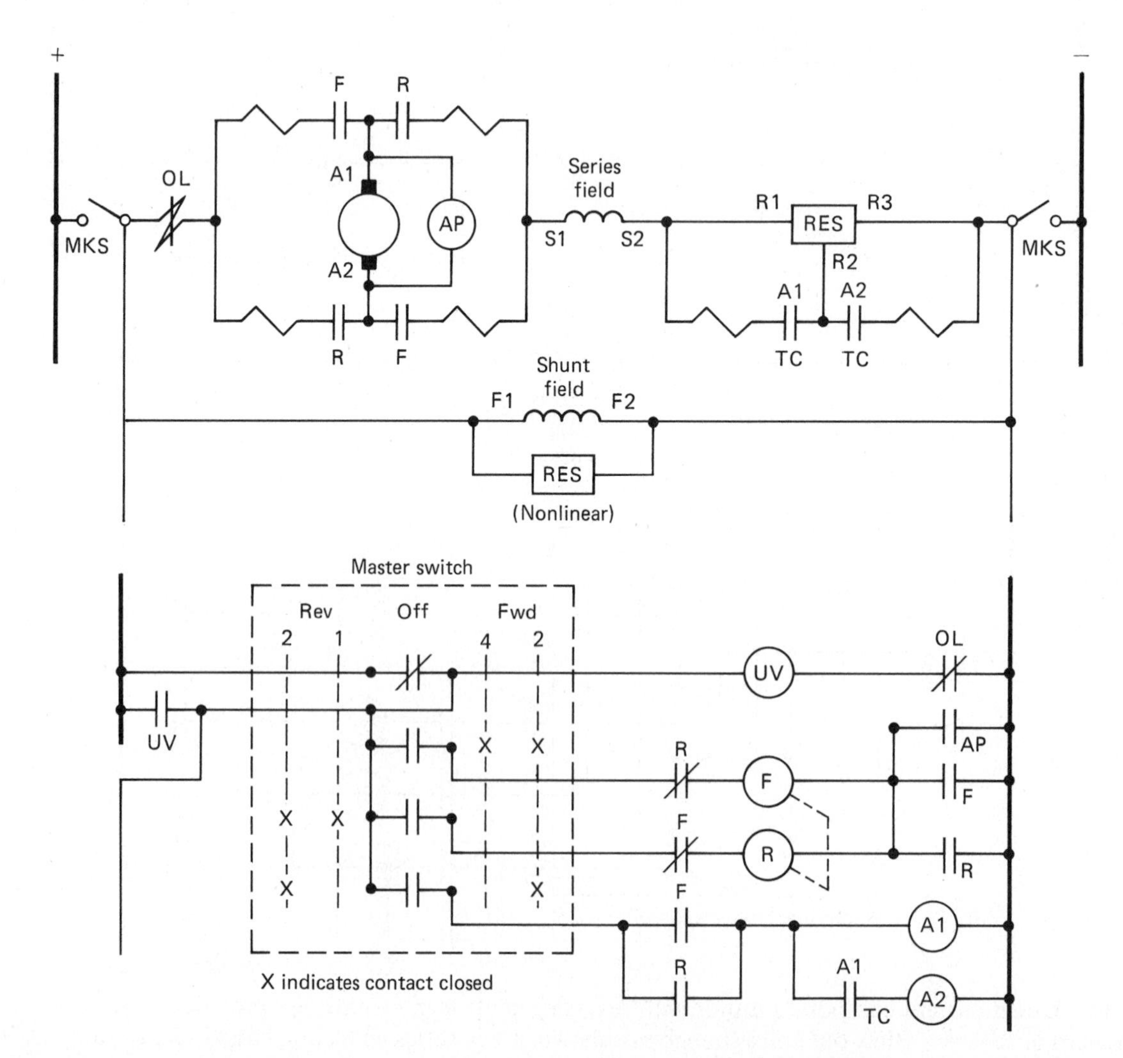

Fig. 8-19 Direct-current compound motor and controller with antiplugging control.

within reasonable limits even though the voltage across the resistor increases.

A plugging controller, or perhaps more descriptively, a controlled-plugging type of controller, which functions in this manner is shown in Fig. 8-20. The equipment for providing this reversing action consists of an additional section of resistor, an additional contactor P, and a plugging relay PR. Relay PR has an open contact which is connected in the control circuit in series with the operating coil of contactor P. Thus, the contactor cannot close unless the relay contact is closed.

The relay design used in the circuit has two coils and is built on a two-path magnetic structure. The resulting characteristic is such that, when counter-emf is high and in the same direction as line voltage, the relay does not operate and its contact remains open. However, when counter-emf falls to a comparatively low value, the relay operates. The relay remains in the operated condition as the counter-emf builds up in the opposite direction, opposing line-voltage polarity.

Relay PR senses that condition at which time the motor has approximately slowed to a standstill because of the reverse current flowing through the armature. In this condition, contactor P is allowed to close and the normal accelerating cycle becomes effective.

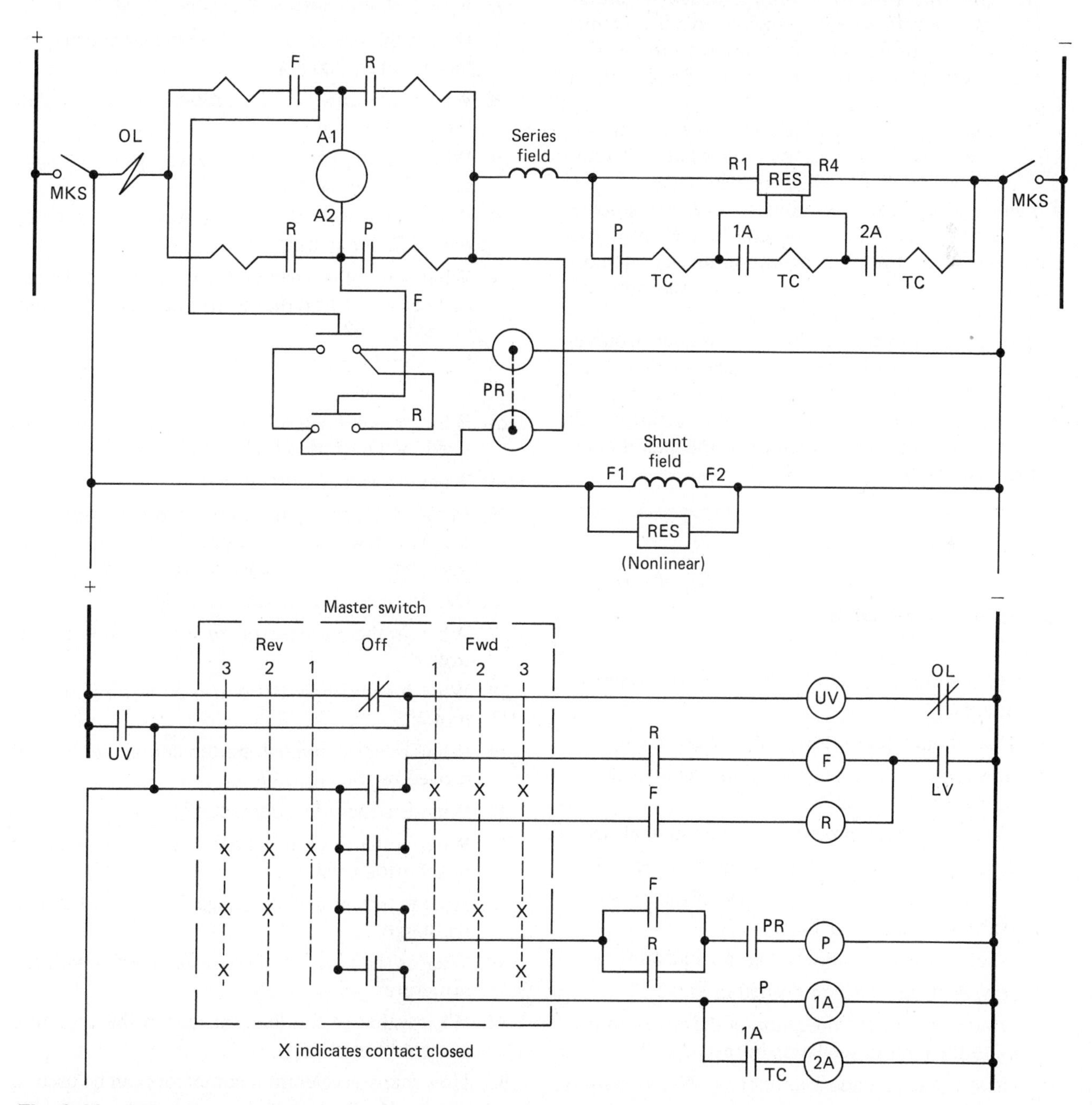

Fig. 8-20 Direct-current compound motor with controller. Controller provides three steps of acceleration as well as plugging control.

SUMMARY

The circuits covered in this chapter and their analysis should form a basis upon which you can build the skills for analyzing circuits used everyday in industry. While these circuits do not in any way represent all or even a major part of the possibilities in motor control, the procedure and method given here of analyzing their operation, if properly understood, may be applied to any and all control circuits, giving you a complete understanding of the operation of the equipment and control components associated with it.

Students who wish to become proficient in motor control can apply these basic principles to other circuits until they are satisfied that they can, with reasonable speed, interpret and analyze control circuits of all types.

The danger in circuit analysis lies in the tendency to jump to conclusions, that is, to decide what the circuit does and how it operates when you have analyzed only a fraction of all of its possibilities. Learn to study a circuit contact by contact and coil by coil until you have completely traced its operation through its normal sequence from beginning to end, and you will avoid many headaches in the future.

The mark of distinction between a good troubleshooter and a poor troubleshooter generally lies in the electrician's, technician's, or maintenancer's ability to analyze a control circuit and determine quickly which of the many components could cause the malfunction of the machine being repaired.

REVIEW QUESTIONS

1. What is the first step in the analysis of control circuits?
2. What should be done to analyze a given circuit?
3. How is the sequence of operation of control developed from electrical drawings?
4. What is the basic procedure for circuit analysis?
5. What types of interlocking are used in Fig. 8-1?
6. In Fig. 8-1, what is the purpose of the switch in series with contacts F1 and R1?
7. What do the dash lines in Fig. 8-1 indicate?
8. How is plugging accomplished in Fig. 8-1?
9. What would be the operation of the circuit in Fig. 8-2 if the control relay was burned open?
10. How would you add another START-STOP station to Fig. 8-2?
11. What would happen in the circuit in Fig. 8-3 if the rectifier burned open?
12. If coil T was open in the circuit in Fig. 8-3, what would this do to the operation of the circuit?
13. What would happen to the operation of the circuit in Fig. 8-3 if the relay T was replaced with an ordinary relay and its contact was made a normally closed contact?
14. What is the sequence of operation when the START button is pressed in Fig. 8-4 and the motor is wired for open-transition operation? What is the sequence if it is wired for closed-transition?
15. How is the motor in Fig. 8-4 connected when starting? How is it connected when it is running?
16. What is the advantage of wye-delta motor control?
17. How is the size of the overload heater selected for the circuit in Fig. 8-4?
18. What is the sequence of operation for the circuit in Fig. 8-5?
19. When would a primary-resistance reduced-voltage starter be used with a motor?
20. What determines the time setting for time delay relay TR in Fig. 8-5?
21. What would the effect on the operation of the circuit in Fig. 8-5 be if the coil for the time delay relay burned open?
22. Why should the overload relays in Fig. 8-5 be of the manual reset type?
23. What protection besides overload protection is offered by the overload relays in Fig. 8-5?
24. What is a lockout circuit used for?
25. In Fig. 8-6 assume the circuit is in operation with coil M energized. What would be the sequence of operation if PS2 normally closed contact opened?
26. Define an electric-motor controller.
27. What is the purpose of an electric-motor controller?
28. What is the simplest circuit used to disconnect a motor?
29. What must a manual starter contain to be approved by the NEMA?
30. What is a main-line contactor?
31. What is the auxiliary contact paralleled with the pushbutton used for?
32. How is the inductive voltage from a motor field discharged?
33. What is the purpose of resistors in series with the armature?
34. What value of full-load current is the armature current limited to?
35. How many acceleration contactors can be used to accelerate a dc motor?
36. What is the typical time delay between the steps of the acceleration contacts?

37. What is the function of an overload relay?
38. Where is the contact of an overload relay located in a motor-control circuit?
39. What is used to protect a dc motor from an open-field circuit?
40. What effect on motor speed does adding resistance in series with the shunt field have?
41. What are motor-operated rheostats used for in motor control?
42. What is the purpose of a field-decelerating relay?
43. What is the purpose of a field-protective relay?
44. What is the name of the resistor that is associated with the field-protective relay?
45. Describe dynamic braking in a dc motor.
46. Describe a dynamic braking contactor.
47. What is the function of the reversing contactors in reversing the direction of rotation in a dc motor?
48. How are the forward and reverse contactors on a dc controller interlocked?
49. What determines the number of points on a master switch for a dc controller?
50. What is the purpose of an undervoltage (UV) relay?
51. What is the purpose of an antiplugging relay in a dc controller?
52. What is the function of a plugging relay in a dc controller connected to a dc motor?

9 Maintaining Control Equipment

There are probably more electricians and technicians involved with the maintenance of industrial motor control systems than there are with the installation of control systems. While a sound knowledge of motor theory and control fundamentals is essential as a foundation for further study in industrial electricity, an early understanding of the practices and procedures of maintenance is essential. In fact, an understanding of good maintenance procedures, and the ability to follow them, is often one of the prerequisites for employment in the electrical field.

If there is a single rule which applies to all maintenance procedures in all plants and under all conditions, it is be careful. Carelessness and failure to observe safety precautions are two things that the maintenance technician cannot afford.

9-1 GENERAL PROCEDURE

The first procedure to follow in any organized maintenance program should be periodic inspection of equipment and systems to prevent serious trouble from arising. This inspection should include not only the electrical equipment but the associated machinery as well. It should reveal points of wear and tear on the electrical equipment, and should provide a basis on which replacement of parts and correction of danger spots can be made before they cause serious trouble.

One of the greatest causes of failure of control systems is the presence of dust, grease, oil, and dirt, which must be removed periodically in order that the equipment may function properly. The removal of dust and dirt may be accomplished by wiping the equipment with rags, but this is not always effective with oil and grease. These substances generally should be removed by using a grease solvent. Care should be exercised whenever these solvents are used, because the inhaling of any appreciable quantity of their fumes could present a serious health hazard. Adequate ventilation should always be provided when solvents are being handled.

Periodic inspection should always include a check for overheating of electrical equipment and mechanical parts, because excessive heat is often an indication of trouble to come. The value of checking for excessive heat depends upon your knowledge of the proper operating temperature of bearings, coils, contacts, transformers, and the many other pieces of equipment associated with machinery, motors, and controllers.

Bearings of motors and mechanical equipment should be checked for proper lubrication. It is very seldom, however, that bearings of electrical equipment such as starters and switches are oiled. They are designed to operate dry, and, generally speaking, oiling the bearings will eventually cause a gum to form, causing the equipment to malfunction.

Another frequent cause of failure of control equipment is loose bolts and electrical connections. Each connection should be periodically checked for tightness, and the inspection should include the checking of possible loose bolts and nuts on the equipment.

Short circuits and grounds in the electrical wiring may be prevented by inspecting insulation and by using an insulation tester such as a Megger on motor and cables.

If you are to maintain the same equipment over a period of time, you must become familiar with your equipment. Know your equipment mechanically and electrically so that you will sense trouble before it develops.

Be observant. Whenever you pass a piece of equipment for which you are responsible, listen and look. Odor and touch also play an important part in uncovering possible problems. Quite often it is necessary to use all your senses to tell you that trouble is on its way. A good maintenance procedure can be summed up in a very few words: Keep it tight, keep it clean, keep it lubricated, and inspect it frequently and thoroughly.

9-2 MAINTENANCE OF MOTOR STARTERS

The most frequent trouble encountered with motor starters is associated with contacts. Contacts should be inspected for excessive burning or pitting and for proper alignment. If they are pitted, copper contacts may be filed, but care must be exercised not to remove too much contact surface or to change their shape appreciably. Copper contacts are subject to heat and oxidation with the closing and opening of the circuit, and copper oxide may be formed on contact surfaces. This oxide is an insulator which must be removed if it covers a large part of the contact surface. Most contacts made of copper are arranged to be of the wiping type, which allows the mechanical closing of the contacts to remove this oxide as it forms. If the contacts are silver-plated, the silver oxide is a good conductor and need not be removed; in fact, silver contacts should never be filed.

The contacts should be inspected not only for pitting but for improper alignment and for improper contact pressure. Improper alignment or lack of contact pressure will cause excessive arcing and pitting of the contacts.

9-3 CAUSES OF TROUBLE

One of the most frequent causes of failure of automatic equipment is the improper adjustment of contacts and time-delay circuits. Generally, the manufacturer of controllers for automatic equipment will supply the proper contact-clearance distances and other information necessary for the proper timing of the circuit. This information should be readily available to maintenance personnel so that they can periodically correct these adjustments. A check of these adjustments should be part of the regular inspection of this type of equipment.

The second most prevalent cause of trouble in motor starters and contactors is coil burnout. Coils on modern starters are well built and well insulated, which has eliminated a considerable amount of trouble due to vibration and moisture. Coils are still subject to burnout, however, because of either of two things. The most frequent cause of coil burnout is the failure of the contactor magnetic circuit to close, creating a gap in this circuit which increases the normal current through the coil to dangerous levels. The normal current needed to start the movement of a magnetic pole piece may be as high as 40 or 45 A, but as the magnetic circuit is closed, this current usually drops to a very low value of 1 to 1½ A, which is all that is required to maintain the magnetic circuit. If this circuit does not close, the coil will maintain a current somewhere in between these two values, which can very easily cause it to overheat and burn out its windings.

The second most frequent cause of coil burnout is improper voltage. If the voltage applied to the coil is exceedingly high, the current through the coil can reach dangerous levels and cause it to burn out. If the voltage applied to the coil drops so low that the magnetic circuit cannot be completed, the gap condition described above will create exceedingly high currents, again resulting in coil burnout. In view of the above-mentioned causes, the proper procedure to use when it is found that a coil has burned out on a starter is to check the mechanical linkage to see that the contactor can close completely and to check the voltage applied to the coil under load to see that it is sufficient but not excessive. Check for spring tension to see that the springs themselves are not causing the magnetic circuit to remain partially open.

Should the contactor be equipped with flexible leads, they should be checked for fraying and broken strands and should be replaced if these conditions exist. Should the starter be equipped with arc shields, they should be inspected for proper alignment around the contacts. They should be checked for accumulation of dust and dirt, and if carbon deposits have built up on the inside of these shields, these deposits should be carefully removed, since carbon reduces the arc path and can be the cause of serious arc-overs, particularly under high-voltage conditions.

Having sufficient spring tension for proper contact pressure is very important in a starter, and this tension should be checked against manufacturer's standards if they are available. The starter should at least be checked to see that each contact has approximately the same spring tension so that the contact pressure will be the same for all contacts. Improper or unequal spring tension is one of the most common causes of contact chatter and starter hum. Be sure that when these conditions exist, the spring tension on every contact is checked to determine if it is sufficient and that it is the same on all contacts.

9-4 MAINTENANCE OF RELAYS

Generally speaking, the maintenance of voltage relays is the same as that for motor starters and contactors with only the additional precaution that, in general, relays operate on lower currents and thus be provided with less power. This lower power demands a smoother, easier operating mechanical linkage and mechanism and thus requires more careful attention to matters concerning them.

Current relays must be checked to see that they are receiving the proper amount of current for closing their contacts and that the spring tension and contact spacing are correct to give the proper pull-in and drop-out currents. Wear of contact surface and change in spring tension can cause a great deal of variance in these values of pull-in, drop-out, and differential currents, which may make the circuit operate in a manner harmful to the equipment.

Overload relays are devices which normally do not operate for long periods of time; therefore, they are subject to corrosion and accumulations of dust and dirt, which must be removed after periodic checking to see that the overload relays can operate when needed. If proper equipment is available, overload relays should be tripped by current periodically to see that they do function. Excessive tripping of overload relays is generally not an indication of relay failure so much as it is of overload on the circuit. The maintenance technician should first determine the current value at which the overload unit actually trips and compare this with the allowable current to determine whether the fault lies with the overload unit or with the circuit itself.

Time-delay relays, whether of the pneumatic or the dashpot type, require periodic adjustment to compensate for normal changes in their operating characteristics. The dashpot relay should be checked for dust and other foreign matter in the oil reservoir, since any impurities in the oil will affect the accuracy of the timing of the dashpot.

Quite frequently, relay contacts may be of the make-before-break or break-before-make type, and here again spring tension and contact spacing become very important and require a check to determine that they are what they were intended to be.

9-5 MAINTENANCE OF PILOT DEVICES

Generally speaking, pilot devices require very little maintenance other than a check of their mechanical operation and their contact condition. Where the pilot device is a form of pressure switch or vacuum switch, the range of its operation should be checked occasionally to see that the contacts open and close at the pressure they were set up to operate on. Contact surfaces should be examined to see that they have not accumulated a coating of copper oxide, dust, or oil. They should be operated through their pressure range several times to check for consistency of operation.

Float switches are subject to troubles because of bent float rods or leaks in the float. A check for the proper operation of the float, the float rod, and the mechanical linkage to the float switch itself will determine the amount of wear and can generally indicate a replacement of parts before any serious trouble can develop. Of course, the condition of the contacts on these and other pilot devices must always be checked.

When limit switches are an integral part of a control system, they are a very likely source of trouble because they perform many thousands of operations per day on an active piece of equipment. They are prone to mechanical failure because of worn bearings and cam surfaces as well as altered contact surfaces and improper spring tension. The only solution to prevent limit-switch failure is frequent and accurate inspection to determine the mechanical and electrical condition of limit switches. When the mechanical condition of a limit switch becomes questionable, it should be replaced or repaired before it causes serious trouble with the other equipment.

9-6 MAINTENANCE OF BRAKES AND CLUTCHES

The chief cause of brake failure is, of course, worn brake linings or brake disks. Either is an inexcusable cause of failure, however, when periodic inspection is performed. Never allow brake lining to become worn.

The second most prevalent cause of brake failure is excessive wear of the shoe or disk and thus improper linkage between the electric solenoid or other operating device to the brake shoe or brake disk. These brake components must be maintained in their proper mechanical alignment and condition. Improper linkage is a frequent cause of coil burnout on brake solenoids, since it may prevent the proper closing of the magnetic circuit, which in turn causes excessive current to flow into the solenoid coil.

Solenoid-operated clutches are subject to the same types of trouble as solenoid-operated brakes. Therefore, the inspection and maintenance procedure for these units should be the same as that for brakes.

SUMMARY

While this chapter has attempted to cover some of the basic principles of good maintenance, the actual maintenance of a specific piece of equipment must be determined by its operating cycle, its complexity and the amount of time available to maintain it. The chief difficulty in most maintenance situations is a lack of understanding of the word "maintenance." The purpose of maintenance is to prolong the safe and efficient operation of equipment, not to repair equipment after it has broken down. Again, inspect it, keep it clean, and keep it tight, and you will be performing maintenance, not repair.

REVIEW QUESTIONS

1. What single rule applies to all maintenance procedures in industrial plants, mills, and factories?
2. What is the first procedure to follow in any organized maintenance of industrial electrical equipment?
3. What should be included in a maintenance inspection?
4. What is one of the greatest causes of control-system failure?
5. When using solvents for cleaning, what safety procedure should be observed?
6. What possible danger conditions does excessive heat indicate?
7. When should the moving parts of a controller be oiled?
8. How important is the lubrication of motors?
9. What mechanical problems can cause electrical troubles?
10. How are short circuits and grounds prevented?
11. What is the first rule to follow in maintaining equipment?
12. What is the second rule to follow in maintaining equipment?
13. What is the most frequent trouble encountered with motor starters?
14. What contact problems can be found by a careful inspection of motor starters?
15. When should contacts be filed?
16. What problem can copper oxide on contact surfaces give?
17. What problem does silver oxide cause?
18. When should silver contacts be filed?
19. What problem can improper alignment of contacts cause?
20. What are some of the frequent causes of trouble in automatic control equipment?
21. What are the two major causes of motor starter coil burnout?
22. What is the relation between closing current and holding current in the coil of a motor starter?
23. How can low voltage cause a motor starter coil to burn out?
24. How important is spring tension in the operation of motor starters?
25. What conditions would affect contact pressure in a motor starter?
26. What should be checked in the maintenance of current relays?
27. What is checked when doing maintenance work on overload relays?
28. How is the current setting of an overload relay checked?
29. What conditions are pilot devices checked for when doing preventive maintenance work?
30. What conditions can cause trouble with float switches?
31. Why do limit switches give so much trouble in control circuits?
32. What mechanical problems are limit switches subjected to?
33. What are some of the problems that occur with electrical brakes?
34. What is the second most prevalent cause of brake failure?
35. What can cause brake coil burnout?
36. What types of trouble are solenoid-operated clutches subjected to?

10
Troubleshooting Control Circuits

In high-production industries, downtime can often cost hundreds to thousands of dollars a minute. This chapter will help you develop the skills necessary to perform a successful troubleshooting procedure.

Many workers capable of doing a beautiful job of wiring a new control circuit from a circuit diagram are lost if the circuit fails to function as expected. The chief asset of a troubleshooter is an analytical mind familiar with the functions of control components and circuits. The secret to efficient and accurate troubleshooting lies in determining the section of the control circuit that contains the defective component and then in selecting the proper component to be checked. This can be accomplished only by efficient and accurate circuit analysis, not by trial and error, long, extended wire tracing, or random checking of components.

10-1 GENERAL PROCEDURE

First let us consider a new circuit which has just been wired but fails to function as expected. Here there is a possibility that the wiring was misconnected or even that the circuit was not properly designed. If we were to check all the connections in all the wiring, however, it would become a trial-and-error process and generally involve a considerable waste of time.

The procedure to follow should be to first analyze the circuit (using the circuit diagram) to determine that it has been properly designed and works as expected. The next step is to follow the operation of the equipment through the expected sequence until you find the section of the circuit which is not operating properly. When you have located the section of the circuit that is causing trouble, check the wiring and operation of the components in this section and clear up whatever the trouble might be.

In this process you are required to make use of your knowledge of control-circuit analysis and your knowledge of components and their proper functions. A lack of knowledge of the functions of control components or circuits or of analysis skills will cause undue delay and wasted time. When you have located the trouble in one section of the control circuit, the operating sequence should be started again and run through until either the circuit operates successfully or another section of the control circuit has been determined to be malfunctioning also.

When troubleshooting an existing circuit, you can generally eliminate the possibility of improper connections. If the circuit has been improperly wired, it would not operate at all. It is surprising, however, how many technicians will begin their troubleshooting procedure by checking the wiring, connection by connection, to determine if it has been properly done. This procedure is unfair to the plant owner and the operator of the machine, who are interested in speedy and efficient repair rather than time-consuming experimentation. However, this does not eliminate the necessity of checking the mechanical or electrical connections themselves. Over a period of time connections do deteriorate because of corrosion, vibration, chemical reactions between the connecting metals, and other factors.

The first step in troubleshooting an existing circuit which has developed trouble is to understand that circuit and to understand the operation of the machine it controls. With a complex circuit time constraints generally do not allow the servicer or troubleshooter to study the complete circuit. With the help of the operator, however, you can determine how much of that circuit is operating properly. Follow the machine through its cycles until it reaches that point. You can then analyze the circuit, starting with the first section that does not operate. A careful check of this circuit and the components involved in this section of the circuit will generally lead you to the source of the trouble. The malfunction of some control components must be the cause of the control circuit failure.

In cases where insulation breakdown is the cause of the trouble, a visual inspection of the components and the wiring would indicate so. Quite frequently, however, the grounding of a wire in the control circuit may escape detection in a visual inspection. If it is suspected that a ground is the cause of the trouble, careful checks should be made with the power off. With an ohmmeter determine the resistance to ground of the wires in this particular section of the control circuit.

Let us assume that you have now located the section of the control circuit which seems to be causing the trouble. The first step is to locate the components involved in this part of the circuit. There must be a coil of a relay, a contactor, or some other device which is energized by this section of the control, and the machine should be run through its sequence to determine if such a coil does receive energy.

If the contactor or relay does not close as it should, the circuit should be disconnected and the wires removed from the coil of the relay or contactor so that a voltage check can be taken. Apply a voltmeter across the wires which were connected to the coil and again energize the circuit operating the control sequence up to this point. If the voltmeter indicates that the proper voltage is applied, then the trouble most likely is in the windings of the coil itself. Do not attempt to check the voltage or resistance of the coil while it is connected in the circuit, since false readings are likely to result from feedback and parallel paths in the control circuit.

If it is suspected that the coil is at fault, disconnect the power from the circuit and with an ohmmeter check the resistance of the coil, which should have a low dc value. If the coil is burned out, you will receive a high-resistance reading or a reading of infinity on the ohmmeter, indicating that the coil needs to be replaced. Do not depend on the coil smelling burnt or showing any visible evidence of being burned out, since this is not always the case.

Suppose that our voltage check showed that the voltage did not reach the coil when it should have in the sequence of operation of the control circuit. This indicates that some contact is not closing when it should, thus de-energizing the circuit to the coil. A careful study of this section of the control circuit following the principles outlined in Chap. 8 should easily show what contacts should close in order to energize this coil. You must now locate the components which contain these contacts and again operate the machine through its sequence, observing the operation of the relay, limit switch, float switch, pressure switch, or other device that contains these contacts, to determine whether it operates mechanically as it should. If this component does operate, there are two possibilities for its failure. The first and most likely is that the contacts involved are not properly closing or are coated with copper oxide or other insulating material which prevents their passing current to the coil as they should. The other possibility is an open circuit due to a broken or burnt wire. Generally this is the less likely cause of trouble. Having checked the contacts and eliminated the trouble, which probably will be found there, again operate the control circuit with all coils connected, and if it does complete its sequence, then proceed to follow the above procedure to the next section of the control which does not function.

The above procedure is based on the fact that magnetic and electromechanical control circuits are made up of only two basic components: contacts, which make and break the circuit, and coils, which operate these contacts. If the contacts close and open as they should, then the proper voltages should appear in the coils. If this is true, then the malfunction must lie in the coil itself. If the contacts do not operate properly, however, then the trouble must be in the contacts or in the wire associated with them which carries current from the contact to the coil.

The most important rule in troubleshooting is to change only one circuit item at a time. If you find a set of contacts that you suspect are not properly functioning, correct this trouble and try the circuit again before changing anything else. If you find a coil you suspect to be burnt or otherwise defective, repair or replace it and try the circuit again before attempting any other changes. One of the most common mistakes of troubleshooters is to change or correct several apparently defective components at one time before testing the circuit for proper operation. Quite frequently several changes made at one time may introduce more trouble than was in the circuit originally. This should be a cardinal rule in troubleshooting. It is unusual for several parts of a machine to wear out at the same time. Therefore, even though the overall condition of the control components may be poor, it still remains probable that only one component has failed completely.

If the machine that you are troubleshooting is not thoroughly familiar to you, do not underestimate the value of the operator in your process of determining the cause of trouble. The operator's knowledge of the normal operation of this piece of equipment can be put to work to avoid wasting time in determining how the machine should operate. Depend upon operators to help you locate components which may be hidden by parts of the machine, since the operators probably know where these components may be found. In short, make use of every available source of information to shorten the time necessary for you to find the source of trouble.

All failures of electric control circuits are not necessarily caused by electrical troubles. Quite frequently, a mechanical malfunction of some component may be the sole source of trouble, and so remember to examine suspected components not only for electrical but also for mechanical trouble.

It must also be pointed out that a technician who is attempting control troubleshooting who is not equipped with a voltmeter, an ammeter, and an ohmmeter is wasting valuable time and money. Troubleshooters must also be trained and competent in the proper use of these instruments and the proper interpretation of their readings.

10-2 TROUBLESHOOTING CONTROL COMPONENTS

In this chapter, the trouble spots recommended to be checked when performing maintenance, troubleshooting, and the repair of electric control components are the same as those outlined in Chapter 9. Trouble spots are detected during the process of troubleshooting circuits that fail to perform as they should. Knowing how to locate trouble spots will make the repair of faulty circuits faster and more efficient.

Again, the best way to efficiently and properly troubleshoot individual motor control components is to acquire a complete knowledge of their proper operation and a familiarity with as many versions of each component as possible. Much of this knowledge will, of necessity, have to be gained through experience. Students may obtain a sizable portion of this required knowledge by a study of manufacturers' literature and by making a sustained effort to familiarize themselves with the various components they come in contact with in their daily work.

10-3 STEP-BY-STEP PROCEDURE

In order to make the procedure outlined in Sec. 10-1 clearer, we shall now consider a specific circuit and determine the probable causes of some troubles which we shall assume have occurred in this circuit.

The circuit of Fig. 10-1 is that of a chilled-water air-conditioning compressor. The coil CR is a control relay. The coil M1 is the starter for the chilled-water pump. The coil M2 is the starter for the condenser-water pump. The coil M3 is the starter for the oil pump on the compressor itself. The coil M4 is the compressor-motor starter. The contact identified as T is a thermostat which senses the temperature of the chilled-water return. Its function is to start the condenser-water pump when this temperature reaches a predetermined high level. The contact identified as PS1 is an oil-pressure switch whose function it is to stop the compressor should the oil pump fail and also to

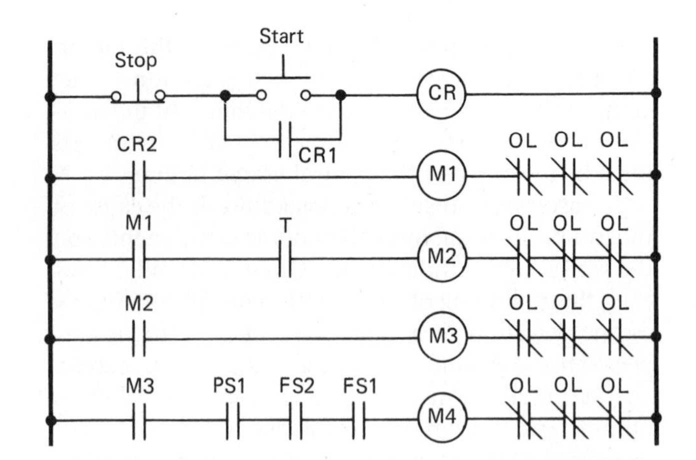

Fig. 10-1 Circuit for chilled-water air-conditioning compressor control.

prevent the compressor from starting before the proper oil pressure has been obtained. The contact identified as FS1 is a flow switch in the chilled-water piping system. Its function is to prevent the compressor from running unless there is a sufficient flow of chilled water. The contact identified as FS2 is a flow switch in the condenser-water piping system.

Suppose now that you are called in to troubleshoot this circuit. The first step should be to determine from the owner or operator what trouble he or she is having with the circuit. Suppose that you are told that the condenser-water pump does not start as it should. Then from a study of a diagram we can assume that the section of the circuit for the control relay is functioning properly, that contact CR2 closes, and that the chilled-water pump runs as it should. Something must be wrong in the third line of our control diagram.

The first step is probably to check the overload relays and determine that they were not tripped. Having done this, we next check the thermostat to see that its contacts are closed as they should be. Here, it must be pointed out that determining the setting of this thermostat and the actual water temperature will indicate whether the contacts should be open or closed. We are assuming that, because of the shutdown of the machine, the water temperature has increased to a point where these contacts must be closed. Let us assume that the thermostat contacts are closed; then inspection of the starter for the chilled-water pump is indicated to determine if contact M1 is closing when these contacts are energized.

If our inspection of this starter shows that these contacts seem to be closing properly, then the next procedure is to disconnect the wire from coil M2 and apply an ohmmeter to the coil to determine whether it is open or not. From the preceding analysis, it is almost certain that this coil will be found open, and for the sake of this illustration we will assume that it is. Before you replace this coil, the starter should be examined for proper mechanical operation. Determine that the contact arm, which raises and lowers or swings to move the contacts, is free to move and that the spring tension is not excessive. Also, examine the faces of the magnetic pole pieces to see that they have not been abused and possibly damaged by someone forcing them. The normal operations of opening and closing can also lead to contactor failure. When all mechanical problems have been eliminated, install a new coil in the starter.

It would be good practice to check the voltage at the ends of the wires which feed this coil before putting the coil back into service. This can be done by connecting a voltmeter between the ends of these wires and operating the control circuit up to this point. If the voltage is excessively low or excessively high, then the cause of this trouble must be determined and eliminated. Otherwise, the new coil will also burn out.

Suppose that this circuit did not malfunction in this way. It was reported instead that everything seemed to work except the compressor itself. Then our procedure would be to energize the circuit and watch its sequence of operations to determine for ourselves where it failed. We would see that the control relay operates, the chilled-water pump starts, the condenser-water pump starts, and that the oil pump on the compressor starts.

Here, we shall assume that our sequence stopped and the compressor did not come on the line as it should. Again, examining our circuit, we find that we have a contact on the oil-pump starter which could cause trouble. We have a pressure switch and two flow switches which also might be the source of trouble. So again we must determine which of these components is not properly functioning. If these components are readily accessible, a physical examination of each of them may immediately disclose the trouble. If they are inaccessible, however, a good procedure to follow is to disconnect the wires from the starter coil and operate the control circuit to determine if voltage is reaching the coil, thus eliminating the possibility of trouble being in the coil itself.

Let us assume that the contact M3 is properly functioning and we have checked it. The two flow switches have been determined to be properly functioning and their contacts closed. Then the examination of the pressure switch is the only remaining possibility. It may even be necessary to recalibrate pressure switches with known pressure to see that they are operating at the settings which show on their indicating dials.

SUMMARY

While the procedures in this chapter may seem oversimplified, they are the basis upon which good troubleshooting practice is based. No matter how complex the control circuit is, it can be separated into simple branches, as we have illustrated in this and other chapters of this book. The efficient troubleshooter will narrow the problem down to a simple branch even in a very complex circuit, so that the actual process of locating the troublesome component will be as simple as outlined here.

REVIEW QUESTIONS

1. What is the chief asset of a good troubleshooter?
2. How vital is circuit analysis in troubleshooting?
3. What is the first step in troubleshooting a newly wired circuit?
4. Why is the knowledge of control functions and control components vital in troubleshooting?
5. What is the procedure for troubleshooting a circuit that has been in existence for a long time?
6. How is insulation breakdown determined?
7. How would you troubleshoot a contactor or relay?
8. If you intend to use an ohmmeter to check components, what should you check first before using it?
9. How would you check a coil to determine whether it is burned out?
10. How important in the troubleshooting process is determining a machine's sequence of operation?
11. How would you determine if contacts are coated with copper oxide?
12. Control circuits are made up of two basic components. What are they?
13. When contacts do not open or close properly, where would you expect the trouble to be?
14. What is the most vital rule in troubleshooting?
15. What is one of the most common mistakes made by troubleshooters?
16. Why is it important to seek the aid of the machine operator in determining a machine's trouble?
17. Are troubles only in electrical control circuits?
18. What part does appropriate test equipment and knowledge of their use play in the efficiency of troubleshooting?
19. What meters are recommended for efficient and proper troubleshooting?
20. What is a good source of information on control components?
21. What is the first step you would use in troubleshooting Fig. 10-1?
22. What method would you use to check the coils in Fig. 10-1?
23. How would you check the circuit of Fig. 10-1 to determine if the circuit is providing the correct sequence of operation?
24. When is it necessary to recalibrate pressure switches?

11
Motor Control Centers

In this chapter we will see how motor starters are combined into motor control centers. These centers control multimotor machines or groups of machines.

For many years, motor controls were always located alongside the machines they controlled. This approach is still used today where the operations or processes performed by the machines are relatively simple or limited in scope. But where complex manufacturing processes which require large in-plant electrical installations are involved, the trend has been toward increasing centralization of motor starters and associated equipment. Motor control centers are now universally used by industry to house the motor starters and motor control equipment.

11-1 HISTORY OF MOTOR CONTROL CENTERS

During the late 1930s, Westinghouse had the idea to mount groups of starters (usually of the combination type) into a common enclosure. Also included in the enclosure was a bus system to which the supply power feeder was connected and to which the line side of each starter was cable-connected, eliminating the costly conduiting used in the past to direct ac power to each combination starter. This general arrangement became known as a *control center.*

Over the years, further improvements were added to this arrangement, resulting in modular assembly. Other features appeared such as drawout construction, wiring troughs, complete isolation of one starter from another, and standardized structures into which the starters were mounted. These elements—all bolted together and serviced by a main, three-phase bus—make up today's motor control center (Figs. 11-1 and 11-2).

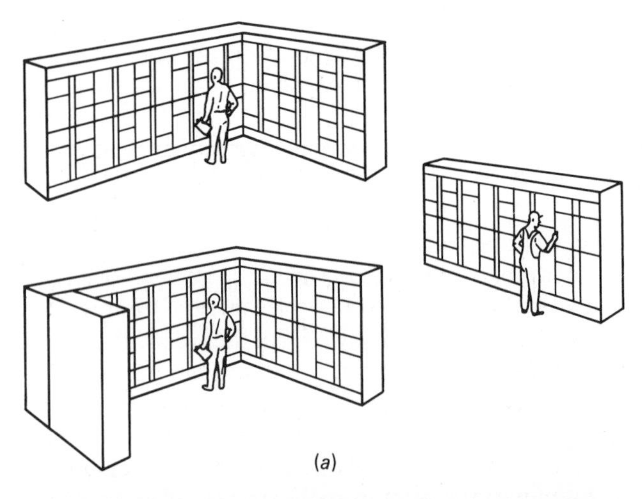

(*a*)

(*b*)

Fig. 11-1 Motor control centers. (*a*) Typical configurations; (*b*) portion of the front of a motor control center. (*Courtesy of Westinghouse*)

11-2 MOTOR CONTROL CENTER STARTERS

To conform with NEC standards, each starter in a control center must be a combination-type using either a fusible disconnect switch or a circuit breaker as the branch-circuit protective device. Very seldom is there a mixture of the different type starters in the same control center.

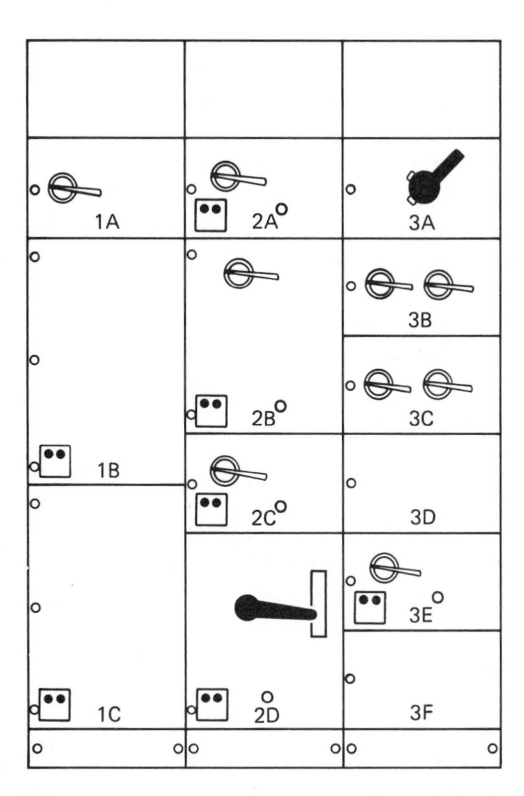

Fig. 11-2 Front view of a motor control center showing the layout of the individual starter compartments. (*Courtesy of Westinghouse*)

The kind of motor control furnished, of course, depends on the type of ac motor to be started and stopped, and the type of motor starting equipment needed for each application. Most control centers are used to centralize control of squirrel-cage motors using the following standard starting methods:

Full voltage
- Nonreversing
- Reversing
- Multispeed
- Multimotor

Reduced voltage
- Primary resistor
- Autotransformer
- Part winding
- Star (wye)-delta

When control for synchronous or wound-rotor motors is also required in the control center, these combination starters may be constructed so that they match the control-center construction standards.

11-3 POWER DISTRIBUTION

Control centers generally have a main, three-phase bus to distribute incoming ac power to each structure (Fig. 11-3). In turn, some means of distributing the incoming ac power throughout each structure is provided,

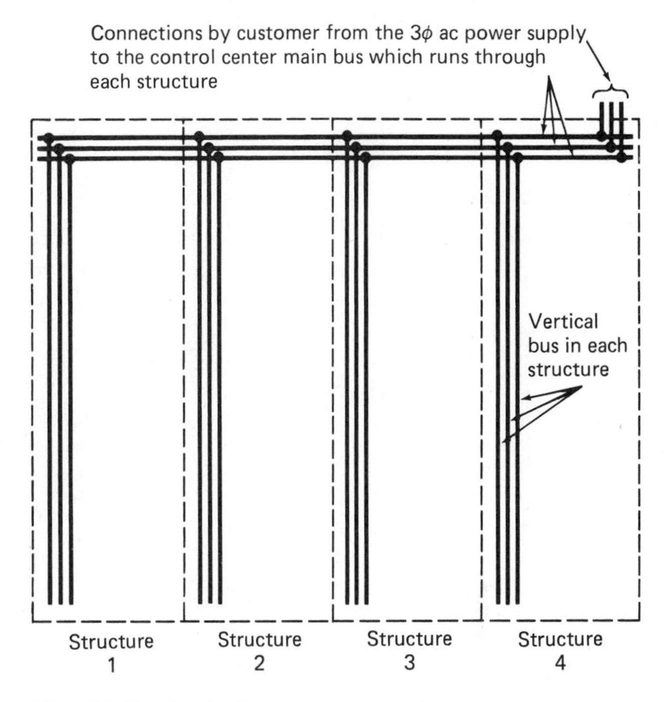

Fig. 11-3 Typical arrangement of main bus.

that is, a vertical bus or some other means of connection to the main bus by various cabling techniques. A typical horizontal and vertical bus arrangement is shown in Fig. 11-4.

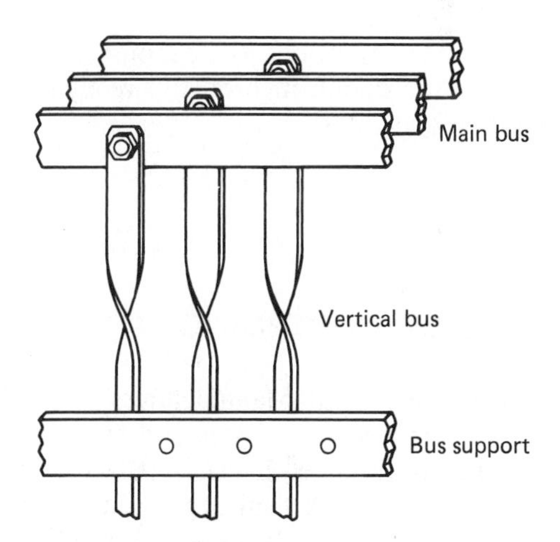

Fig. 11-4 Typical arrangement of main and vertical bus.

Ways of further distributing incoming power to the line side of each starter in a structure containing vertical bus include:

- A fixed or bolted-type connection

- A disconnecting-type connection, such as a stab arrangement (Fig. 11-5)

The present trend is toward the disconnecting-type connection arrangement because it lends itself mechanically to the use of the unit-designed starters that can be easily inserted or withdrawn from the structure while the control center is energized and operating.

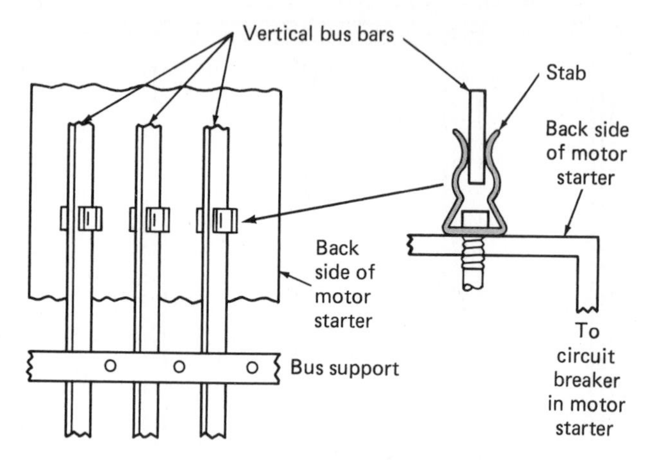

Fig. 11-5 Power stab and vertical bus.

Usually, the stab itself is mounted on the combination starter unit and so arranged as to contact the vertical bus positively when the starter unit is locked into its proper position in the structure. In the starter unit itself the stabs are cabled to the line side of the circuit-protective device.

The external connections to the motor and to a remote pushbutton station are brought directly to terminals standard on starters.

11-4 CLASSES OF CONTROL CENTERS

The National Electrical Manufacturers Association (NEMA) has defined two classes and three types of control centers within each class. Class I control centers are essentially mechanical groupings of combination motor controls, feeder taps, and/or other components arranged in a convenient assembly. They include connections from the common horizontal bus to the starting units. They do not include interwiring or interlocking of units or of remotely mounted devices and units, nor do they include control-system engineering. Diagrams of the individual units only and sketches showing overall dimensions of the control center are supplied in this chapter. When master terminal blocks are specified, a sketch showing the general location of terminals is also provided here.

Class II control centers are basically the same as Class I, but are designed to form a complete control system. They include the necessary electrical interlocking and interwiring of units and of remotely mounted devices and units. A connection diagram of the complete control assembly showing remote control devices and sketches showing overall dimensions are provided. When master terminal blocks are specified, terminal arrangements and all wiring connections are shown on diagrams.

The types of control centers in each class are defined in terms of the following NEMA wiring specifications:

Type A. Includes no terminal blocks. Combination-line starters are factory-wired and -assembled in the structure in the most efficient arrangement. Auxiliary devices can be supplied, but no wiring external to the unit is furnished. All feeder circuit-breaker or fusible-disconnect units are in this classification.

Type B. Essentially duplicates type A, except that all control wires terminate at blocks near the bottom of each unit. Load terminals are all conveniently located adjacent to the control terminal blocks. Plug-in-type terminal blocks for all control wiring and load wiring through size 2; control wiring above size 2 can be supplied when specified.

Type C. Utilizes type-B units. Factory wiring of all control wiring and load wiring through size 3 is extended from the unit terminals to master terminal blocks located at the top or bottom of each vertical compartment.

11-5 IMPORTANT TERMS

Some common and important terms used with motor control centers are *no-voltage release* or *low-voltage release two-wire control* and *no-voltage protection* or *low-voltage protection three-wire control.*

Two-wire control means that the starter will drop out when there is a voltage failure and will pick up again as soon as voltage returns. Reference to Fig. 11-6 will show how this occurs. The pilot device is unaffected by the loss of voltage and its contact remains closed, ready to carry current as soon as line voltage returns to normal.

Three-wire control means that the starter will drop out when there is a voltage failure but will not pick up automatically when voltage returns. The control circuit (Fig. 11-7) is completed through the STOP button and also through a holding contact (2-3) on the starter.

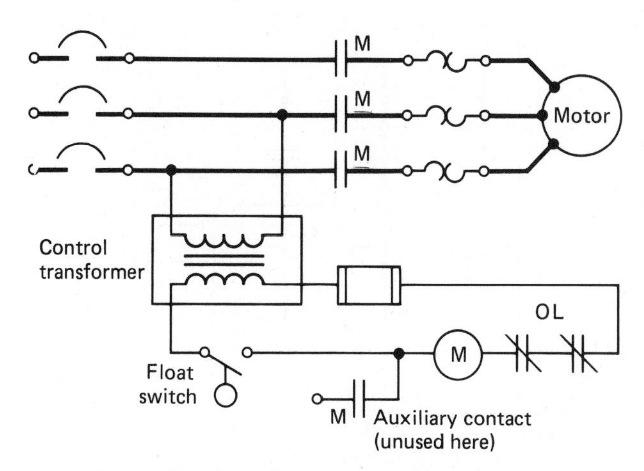

Fig. 11-6 Typical motor starter with two-wire control (low-voltage release) used in motor control centers.

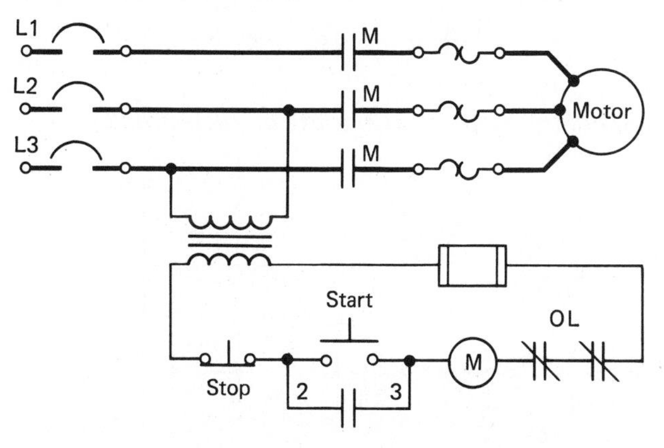

Fig. 11-7 Typical motor starter with three-wire control (undervoltage protection) used in motor control centers.

When the starter drops out, this contact opens, breaking the control circuit until the START button is pressed once again.

The main distinction between the above two types of control is that with no-voltage release the coil circuit is maintained through the pilot switch contact and with no-voltage protection it is maintained by a stop contact on the pushbutton station and also by an auxiliary contact on the starter. The designations *two-wire* and *three-wire* are used only because they describe the simplest applications of the two types. Actually, in other systems, there might be more wires leading from a pilot device to the starter, but the principle of two-wire or three-wire control would still be present.

11-6 CONTROL DIAGRAMS

Most of the diagrams in this book are shown in two ways. There is a schematic diagram and adjacent to it a ladder, or control, diagram. Control diagrams are included because their use is becoming more widespread and we believe it is advantageous to learn to use them.

A special type of schematic diagram called a *wiring diagram* includes all of the devices in the system and shows their physical relation to each other. All poles, terminals, coils, etc., are shown in their proper place on each device. These diagrams are helpful in wiring a system, because connections can be made exactly as they are shown on the diagram. In following the electrical sequence of any circuit, however, the wiring diagram does not show the connections in a manner that can be easily followed. For this reason a rearrangement of the control circuit elements to form a control diagram is desirable.

The control diagram is a representation of the control system, showing everything in the simplest way. No attempt is made to show the various devices in their actual relative positions. All control devices are shown between vertical lines which represent the source of control power, and circuits are shown connected as directly as possible from one of these lines to the other. All connections are made in such a way that the functioning of the various devices can be easily traced.

A wiring diagram gives the necessary information for actually wiring a group of control devices or for physically tracing wires when troubleshooting is necessary. A control diagram gives the necessary information for easily following the operation of the various devices in the circuit. It is a great aid in troubleshooting because it shows, in a simple way, the effect that opening or closing various contacts has on other devices in the circuit.

A wiring diagram of the Allen-Bradley Bulletin 709 starter is shown in Fig. 11-8 on the next page. The principle parts are labeled so that the wiring diagram can be compared with the actual starter. This should aid in visualizing the starter when studying a wiring diagram and will help in making connections when it is actually wired. Note that the wiring diagram shows as many parts as possible in their proper relative positions. It is not necessary to show the armature and crossbar or the overload-reset mechanism in the wiring diagram since these parts need not be considered from the wiring standpoint.

Figure 11-9 shows control diagrams for larger Allen-Bradley Bulletin 709 starters. Notice that control circuits of these diagrams do not attempt to show the physical location of any of the components found in the motor starter, nor do they usually show the motor itself. Circuits for motor control centers can be shown in wiring diagrams, elementary schematic diagrams, or control diagrams, as illustrated in Figs. 11-8 and 11-9. A typical motor starter showing local pushbuttons is shown in Fig. 11-10 on page 127.

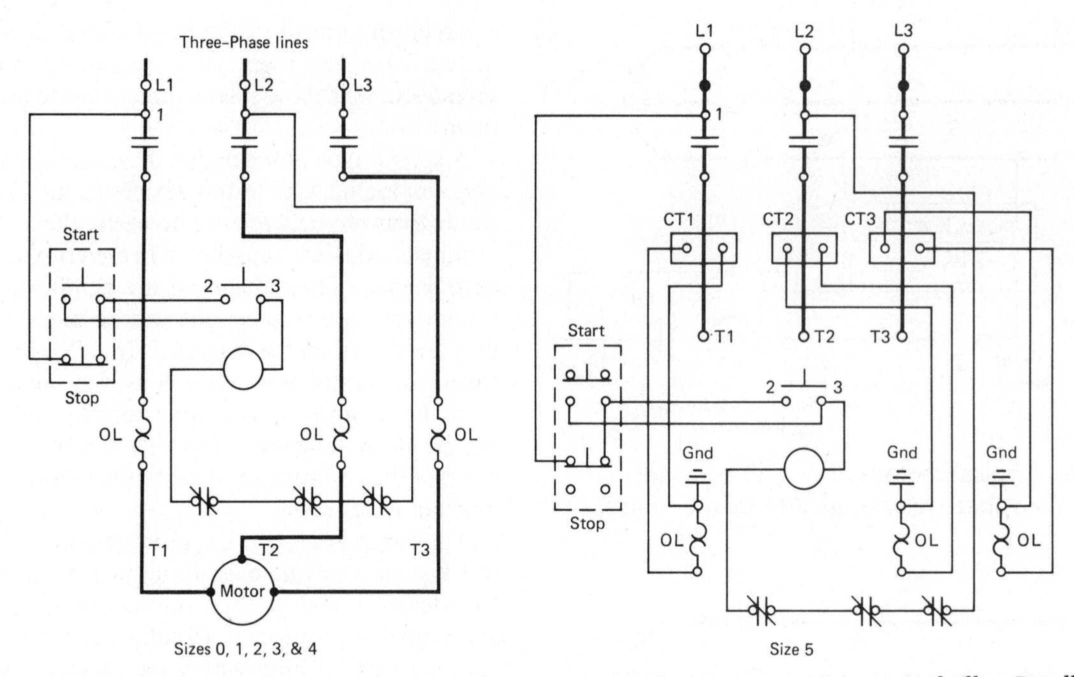

Fig. 11-8 Typical wiring diagrams for starters used in motor control centers. (*Courtesy of Allen-Bradley*)

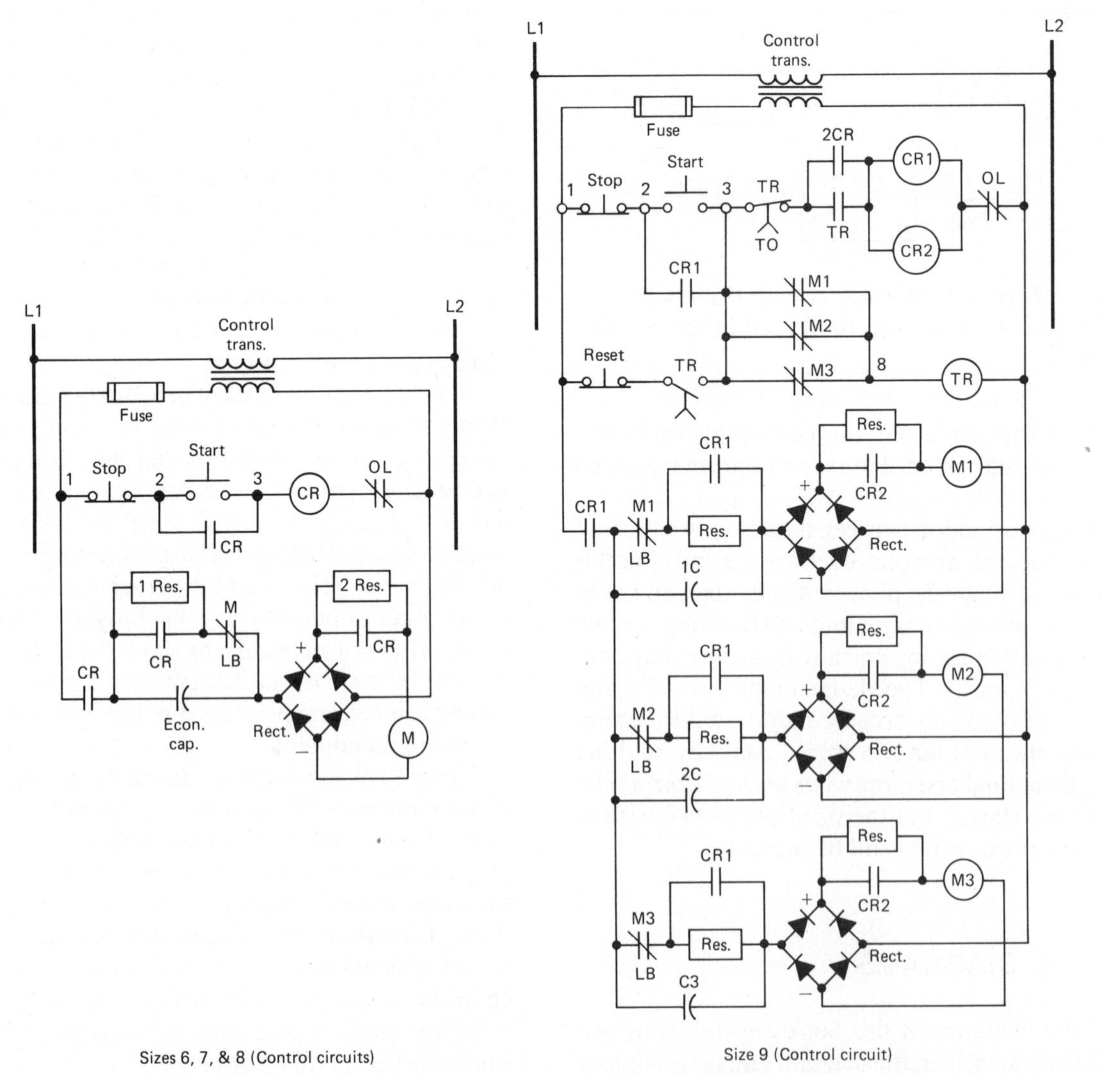

Fig. 11-9 Typical control diagram used for motor starters in motor control centers. (*Courtesy of Allen-Bradley*)

Fig. 11-10 A motor starter installed in a motor control center. *(Courtesy of Westinghouse)*

SUMMARY

In the past motor controllers and starters were always located close to the machines they controlled. This approach is still used in small industries where there are relatively few motor-driven machines. As the manufacturing processes in industry became more complex, the number of motor-driven machines increased dramatically. Mounting the motor controllers and starters next to the machines they controlled was impractical and centralization of the motor starters became necessary. Grouping the starters in a common enclosure with a bus system to supply power to the starters eliminated costly conduit and construction costs. The motor control center was created out of necessity and made motor control more convenient, safer, and less costly.

Over the years improvements in the design of motor control centers brought about modular construction, drawout construction, wiring troughs, and complete isolation of one starter from another. Modern control center design has made it safer and easier to install and maintain motor control starters.

Motor control centers furnish all the standard starting methods used for full-voltage and reduced-voltage starting. They can also be used for special motors such as synchronous and wound-rotor motors.

Motor control centers are designated as Class I and Class II. The two classes differ in mechanical and electrical construction. Class II includes provisions for electrical interlocking and interwiring of remotely mounted devices and units.

Motor control centers are also defined by type A, type B, and type C. Type A includes no terminal blocks; starters are factory-wired. In the type B motor control center, all wiring terminates at terminal blocks at the bottom of each unit. In the type C motor control center, the terminal blocks are at the top or bottom of each vertical compartment.

Motor control center starters can be wired to provide no-voltage release or no-voltage protection, whichever is specified in the design of the control system. Interlocking, time delay, sequencing, dynamic braking, and other control features can be wired into the control of the starters found in modern motor control centers.

REVIEW QUESTIONS

1. When were motor control centers first developed?
2. What is the advantage of a motor control center over individually mounted starters?
3. What is meant by a modular assembly?
4. What are the NEC standards for control centers?
5. What are the different starting methods that can be supplied by motor control centers?
6. Why are disconnecting-type starters preferred over the mechanically bolted type for motor control centers?
7. What is a stab?
8. Name the two classes of control centers.
9. What are the three types of motor control centers, and how do they differ from one another?
10. Define no-voltage release.
11. Define no-voltage protection.
12. What is the main distinction between no-voltage release and no-voltage protection?
13. What type of control do two-wire and three-wire control systems offer?
14. When is a wiring diagram preferred over a control diagram?
15. When is a control diagram preferred over a wiring diagram?

12
Solid-State Motor Control

In this chapter we will cover the portion of the motor control system block diagram that is concerned with information decisions and control. Since solid-state devices are used in all of the sections of the control of motors, the theory and application of the many different types of solid-state devices will be studied. We will see the dramatic changes in electric motor control that have taken place since the introduction of solid-state devices.

12-1 INTRODUCTION TO SOLID-STATE MOTOR CONTROL

In recent years solid-state devices have been finding more and more applications in industrial motor control. Motor control is becoming a very important application of semiconductors. Many industrial machines and mills require adjustable-speed drive motors. The dc motor is the best motor on the market for giving adjustable speeds. It can also supply many other properties required of industrial drives. Besides giving a wide range of speeds, the dc motor can provide for dynamic braking, soft starts and stops, plugging, jogging, inching, and reversing. The dc motor has one disadvantage in that it requires a dc power supply. Alternating current and voltage are of course the main source of power obtained from generating plants and utilities in this country.

Having only ac power available and the need for dc power for dc-drive industrial machines has not been a problem for years. The industry took the approach of using large motor generators to convert ac power to dc power. Figure 12-1*a* illustrates how power conversion used to be accomplished using a motor-generator (M-G) set. Figure 12-2 shows how it is done in modern industry with solid-state control.

Notice how inefficient the M-G method was. In this method, alternating current was converted to mechanical power by an ac motor. The mechanical power was then converted to dc power by the dc generator. The dc power was then connected to a bus and circuit breakers for power distribution. The circuit breakers contained short-circuit protection. From the circuit breaker the dc power was connected to the dc motor by a suitable controller-disconnect. The speed of the dc

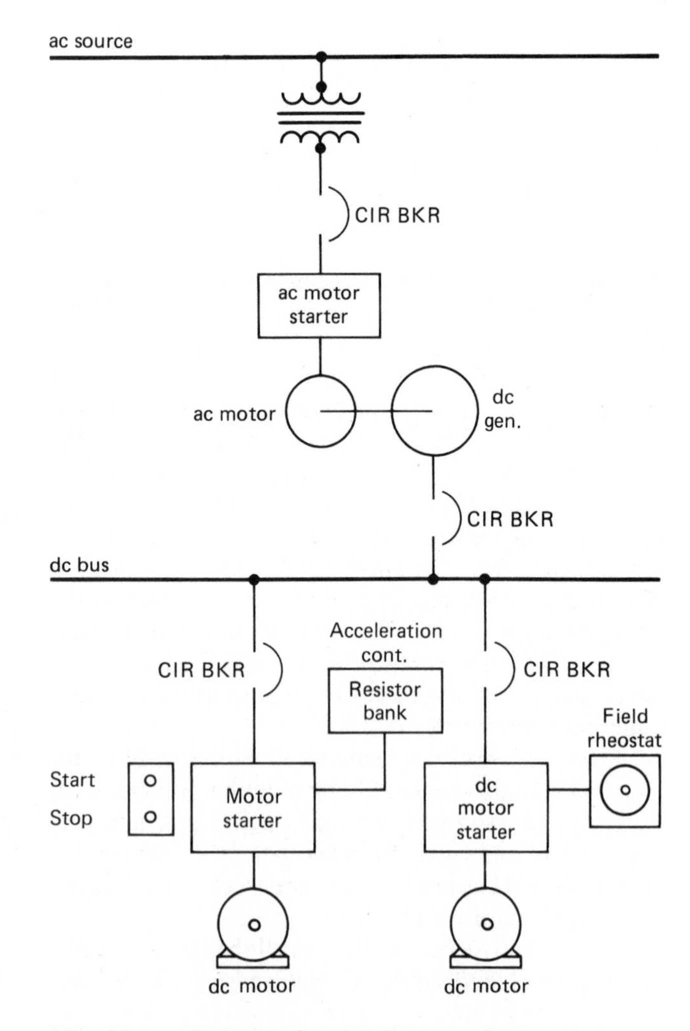

Fig. 12-1 The use of an M-G set to change ac to dc. Speed control in dc motors is accomplished by adjusting resistance in the motor's armature and field circuits.

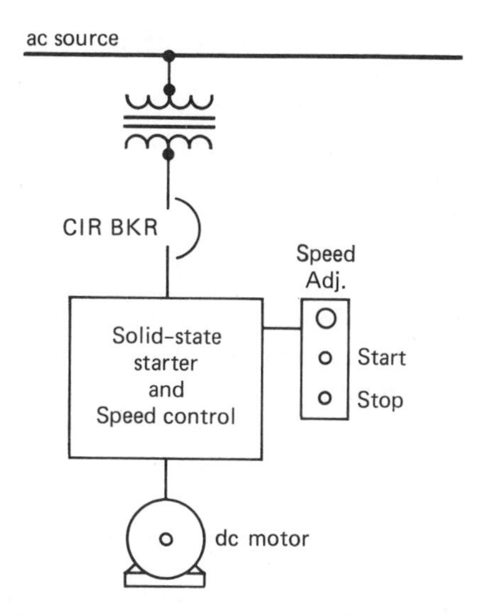

Fig. 12-2 The modern method of ac/dc conversion using a solid-state drive.

motor could then be controlled by conventional means in the controller. The speed of the motor could be controlled by series resistors in the armature or in the field circuit of the motor or by both methods of speed control. When an M-G set for ac-dc power conversion is used, considerable power is lost as heat when it travels from the ac source to the power output of the dc motor. This method of obtaining dc power was very costly as well as inefficient.

In the modern, solid-state drive (Fig. 12-3), the ac power enters the motor controller. The ac power is converted to a dc voltage by the rectifying action of diodes and SCRs (silicon-controlled rectifiers). The speed of the motor is controlled by the gating time of the SCRs. Direction control is determined by gate signals received by the SCRs. It is obvious that the solid-state drive system is considerably more efficient than the older M-G power conversion system. All that is needed for an adjustable-speed drive today is an ac power source, a solid-state motor control, and a dc motor. The solid-state drive can be installed on any machine or mill in any location where an ac power source is available.

In mines that have explosion hazards due to gases, the solid-state motor starter is preferred because of its safety features. Solid-state starters have no contacts to arc and spark. Solid-state starters help minimize explosion hazards when starting and stopping the conveyor belt drives of mining machines.

Large steel processing mills may use 5000-hp motors controlled by solid-state reversing drives. Thus solid-state devices are not limited to small dc motors. In the control room for the 5000-hp motors, there may be no noise other than the ventilation fan. The motor can go from full-speed forward to full-speed reverse with no noise at all coming from the solid-state controllers. With older systems, which used contactors, relays, and M-G sets, considerable noise was created by the opening and closing of these devices. There was also considerable arcing and flashing when a motor with such control was started, accelerated, and then stopped and reversed.

Fig. 12-3 Solid-state motor starters installed in a motor control center.

12-2 BASIC SEMICONDUCTOR THEORY

Copper is a good conductor because it is made up of atoms that contain many free electrons. When a voltage source is applied to copper, its electrons can freely move from atom to atom and an electric current results.

Electrical insulators are made up of atoms which have few free electrons. A voltage source applied to an insulator results in almost no current flow through the insulator.

Certain types of materials with few free electrons or a moderate number of free electrons can be either conductors or insulators. They are called *semiconductors.* They conduct some current when a voltage is applied and so they are not insulators, but they also do not conduct enough current to be good conductors. It would appear that there are no electrical uses for semiconductors. But this is not true; semiconductors play a key role in modern solid-state electronics.

If a material has a surplus of electrons; it is said to have a *negative charge.* If it has a surplus of holes (that is, a deficiency of electrons), it has a *positive charge.* A semiconductor that has free electrons is called a *N-type* material and is said to contain a negative charge. If a semiconductor has a deficiency of electrons (or a surplus of positive charges), it is called a *P-type* material.

Semiconductor devices are usually constructed from such metals as germanium and silicon. Materials added to the semiconductor metals to change their level of conductivity are called *impurities*. Materials such as arsenic or indium are impurities. Adding impurities to semiconductors is called *doping*.

12-3 DIODES AND RECTIFIERS

A *diode* or *rectifier* consists of a layer of N-type material and a layer of P-type material meeting at a junction area. A drawing that shows this construction is called a *junction schematic*. Figure 12-4 shows a junction schematic and the standard symbol for a rectifier or diode.

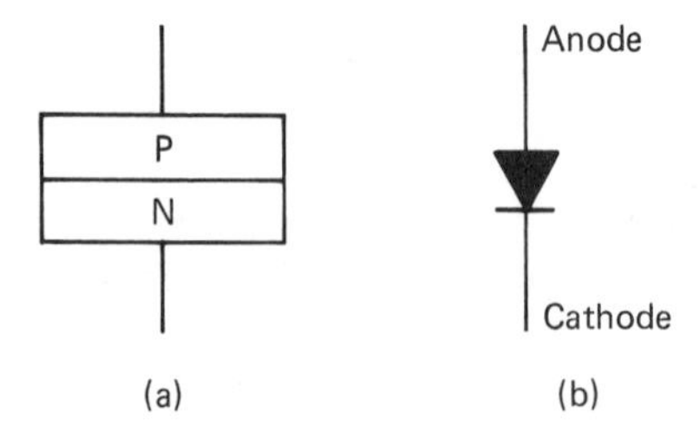

Fig. 12-4 Rectifier or diode. (*a*) Junction schematic diagram; (*b*) symbol.

The junction schematic is very useful in explaining the operation of a diode. Figure 12-5*a* shows a junction diode with no voltage (called *bias*) connected to the cathode and anode. Without a voltage source no conduction through the diode can occur. If it is desired to know the condition of the applied voltage that would cause conduction in a diode, the junction schematic can show this effectively. Figure 12-5*b* and *c* illustrate this. The P-type material has a surplus of holes, or a positive charge. The N-type material has a surplus of free electrons, or a negative charge. At the junction the charges are not sufficient to cause the combination of the N and P charges. When a voltage source is applied such as is illustrated in Figure 12-5*b*, their is no current flow across the junction because of a condition called *reverse bias*. When the negative side of the power source is connected to the P-type material and the positive side of the power source is connected to the N-type material, the charges are attracted away from the junction and no current can flow. When the power source is connected to a diode with the negative source to the cathode and the positive source to the anode, free electrons are repelled to the junction; the holes are also repelled to the junction. The holes and electrons combine at the junction and current flows through the diode. The diode is said to be *forward-biased* when the applied voltage causes current to flow through it. Thus, a diode can be made to behave like a switch.

From the previous discussion it is obvious that when the junction of the diode is reverse-biased, the diode acts like an open switch. Current flows when the diode is forward-biased (cathode negative and anode positive) and therefore the diode acts like a closed switch.

The direction of current flow is a matter of convention. The flow of current from the positive terminal of a voltage source to the negative terminal (+ to −) is called the *conventional flow of current*. The flow of current from the negative terminal of a voltage source to the positive terminal (− to +) is called the *electron flow of current*. The actual amount of current is the same in both cases, but the indication of current flow through particular devices will be in opposite directions. In this book we will describe current direction in terms of electron flow. Therefore, current flow will be indicated from negative to positive. Electrons, or negative charges, will flow to a positive potential. Positive charges, or holes, will flow toward a negative potential.

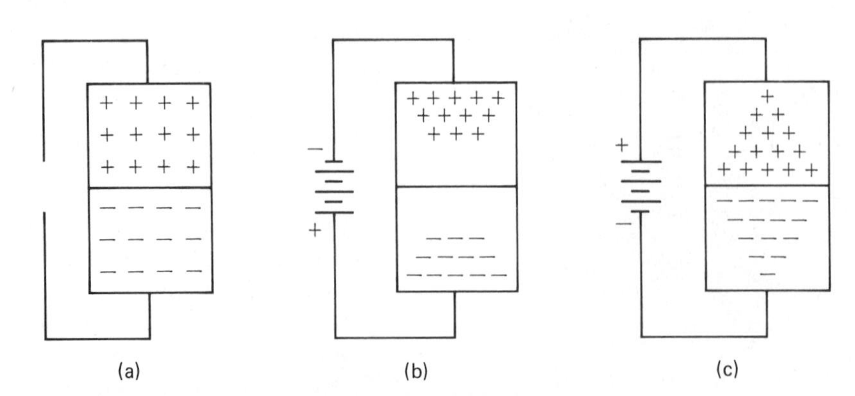

Fig. 12-5 Bias conditions: (*a*) no bias; (*b*) reverse bias; (*c*) forward bias.

12-4 CHARACTERISTIC CURVES

Characteristic curves are often used to explain the operation of semiconductors. The curve is divided into four quadrants on which the voltage and current are plotted. Figure 12-6 is the characteristic curve of a diode. Notice that the voltage is positive on the right side of the vertical line and negative on the left side of the vertical line. Current is plotted as increasing in the positive direction as the curve goes above the voltage axis. Current increases in value in the negative direction as it goes below the voltage axis. Where the curve intersects the axes, the voltage and current both have a value of zero. The characteristic curve of a diode indicates a small voltage in the forward direction (forward bias) will cause current flow. It takes a large voltage in the reverse-bias direction to cause current flow. The voltage point where current flows in the reverse direction is called the *avalanche* or *breakdown point.* In a regular diode when current flows in the reverse direction, it will likely destroy the diode. Some devices depend on the breakdown point for their operation. Study the characteristic curve and be sure you understand what this curve shows about the current-voltage behavior of diodes.

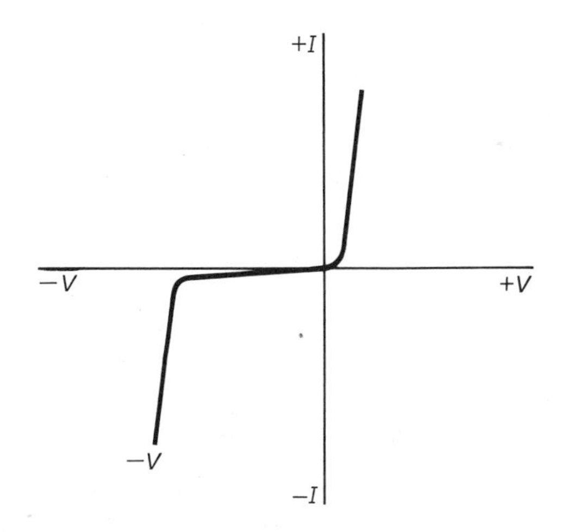

Fig. 12-6 Characteristic curve of a diode.

Diodes have many uses in electronic circuits. Their main use in industrial solid-state devices is for rectification, that is, changing alternating current to direct current. Rectifiers can provide half-wave or full-wave rectification. Another use for diodes is to block dc flow in specific parts of a circuit. They are used as steering diodes to steer signals through specific circuits. They can also be used to detect high-frequency radio signals and to separate the audio signals or low-frequency signals from radio frequency carriers. Programmable controllers contain many diodes.

12-5 ZENER DIODES

The characteristic curve of a diode (Fig. 12-6) has a point called the *avalanche* or *breakdown* point. The zener diode makes use of this point. The zener diode has the same junction schematic as the diode previously discussed but the symbols for the zener diode are somewhat different than those for other diodes (Fig. 12-7).

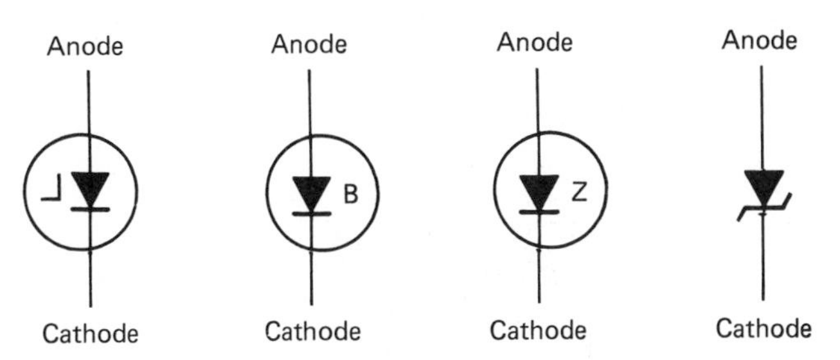

Fig. 12-7 Different symbols for zener diodes.

The avalanche effect, or breakdown, occurs when the anode voltage becomes high enough in reverse polarity that the diode starts to conduct in the reverse direction. This effect is undesirable in a regular diode used for rectification, but it is desirable in the zener diode. The zener diode is designed with the avalanche effect as its basis of operation. The characteristic curve for a zener diode is given in Fig. 12-8.

Zener diodes and regular diodes have similar characteristics curves. The principal difference is in their avalanche points. Both diodes conduct easily when a positive voltage is applied to the anode. But zener

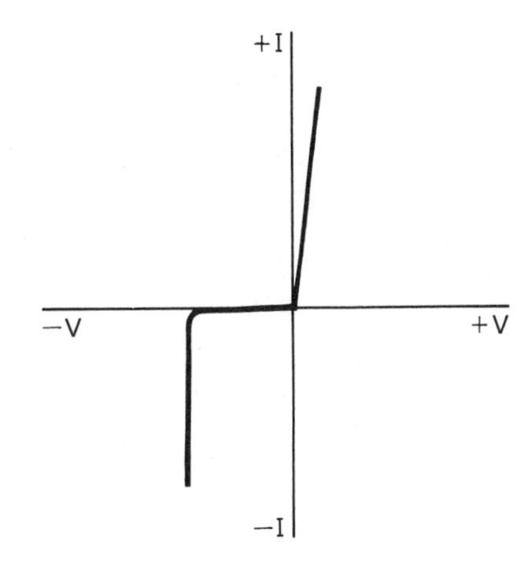

Fig. 12-8 Characteristic curve of a zener diode.

diodes have a controlled avalanche point, which can be very low, whereas regular diodes break down at a random voltage level well above their normal blocking voltage operating range.

Zener diodes are used for regulating, clipping, and limiting voltage. They are also used in voltage reference circuits. Voltage reference circuits in feedback-control systems use zener diodes to provide a fixed level of reference voltage. The voltage signal from the system under control is compared with the reference voltage to provide required control action.

A zener diode used to regulate voltage output is shown in Fig. 12-9. This figure shows a zener diode in

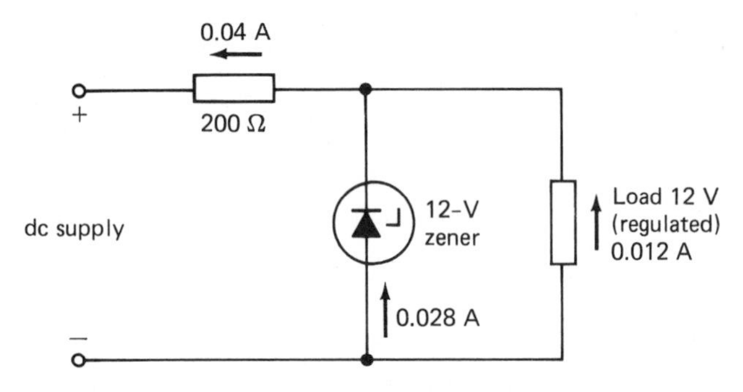

Fig. 12-9 Zener diode circuit providing a regulated voltage output to a load resistor.

series with a 200-Ω resistor. The purpose of the 200-Ω zener resistor is to produce a voltage drop equal to the difference between the supply voltage and the regulated voltage across the output load. If the load current varied in the output load resistance, the current through the 200-Ω resistor would also vary to maintain the regulated 12 V across the output load. If the load resistance causes the output voltage to drop, the current flow through the zener and the 200-Ω zener resistor will decrease slightly. The supply voltage divides between the 200-Ω resistor and the output load in such a way that the voltage across the load returns to its 12-V regulated value. If the output voltage across the load were to increase, the zener current would increase. The current in the 200-Ω zener resistor would increase. The voltage drop across the 200-Ω load resistor would increase and the voltage across the output load would drop until it reached its 12-V zener regulated value. The circuit will provide a 12-V regulated output as long as the current in the zener does not exceed a minimum or maximum value. The characteristic curves for zeners indicate there is a range of current flow through the zener where it will provide a constant voltage across it. The zener characteristic curves can be used to determine the size of zener resistors needed to supply proper operation to regulated circuits.

12-6 THYRECTORS

An ac thyrector diode consists of two selenium zener-type diodes connected in inverse series. Thyrector schematic symbols are shown in Fig. 12-10*a*.

The characteristic curve for an ac thyrector is shown in Fig. 12-10*b*. This shows that as the voltage increases above a certain value in either direction, the thyrector breaks into conduction and current flow increases rapidly. The breakdown is not as sharp as it is in the zener diode. Hence, the thyrector diode is not suitable for continuous clipping or voltage regulation, the ac thyrector operates the same way in both quadrants to limit excess voltage in a circuit.

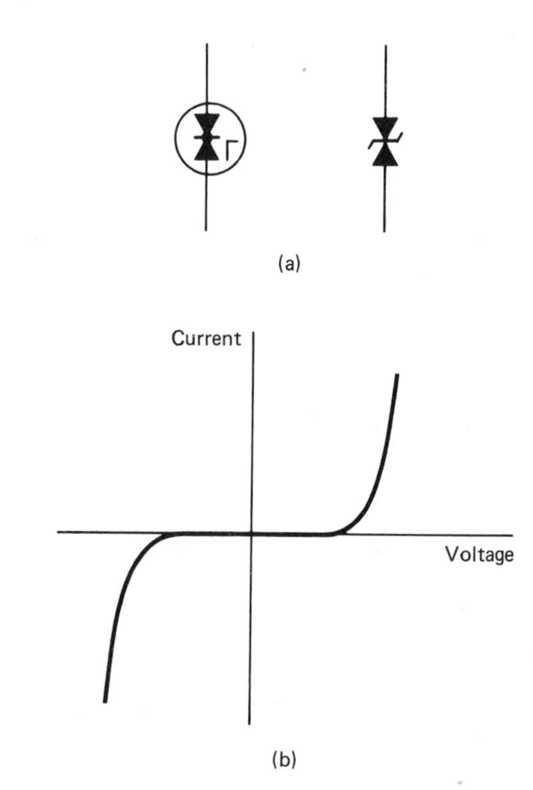

Fig. 12-10 (*a*) Thyrector schematic symbols; (*b*) thyrector characteristic curves.

One of the main uses of thyrectors is for arc suppression. When placed across the coil of a relay or other inductive device it discharges the inductive energy of the coil when coil current is interrupted. This helps lengthen the life of the switching contacts by eliminating arcing at the contacts.

Thyrectors are also used for transient-voltage suppression. They are placed across the ac line to shunt transient peaks of voltage coming into the system or generated within the circuit itself. This protects against damaging semiconductor diodes or other voltage-sensitive components connected to the system.

The thyrector was developed to make silicon semiconductor diodes acceptable in the power field. A typical silicon rectifier bridge, for instance, would short-circuit if a large transient voltage peak hit it. Selenium thyrectors were developed to solve this transient-voltage problem with germanium and silicon semiconductors.

A major use of thyrectors is in a circuit similar to the one shown in Fig. 12-11. Thyrectors are placed across the ac power input line to suppress incoming transient voltage peaks that could, if they were allowed to appear across the rectifiers, cause a considerable amount of damage to them.

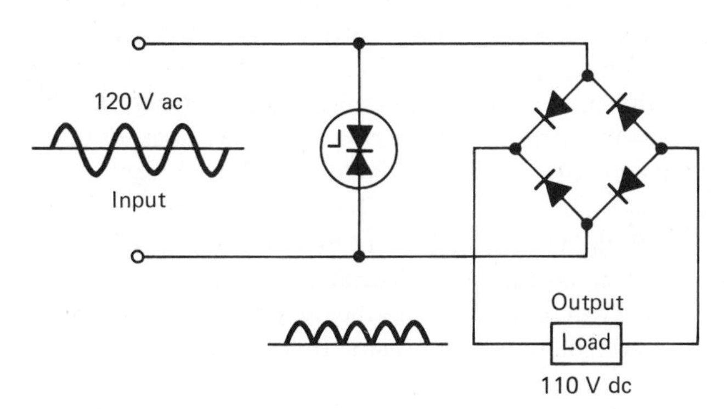

Fig. 12-11 Full-wave bridge rectifier circuit with a thyrector across the ac lines to clip the harmful voltage spikes that may enter the circuit.

The leads of a thyrector are labeled positive electrode and negative electrode instead of positive anode and negative cathode as in the junction schematic of a regular diode. *Cathode* and *anode* are terms used for devices such as rectifier diodes and zener diodes, but not for thyrectors.

12-7 TUNNEL DIODES

The circuit symbol for the tunnel diode is somewhat similar to that for the regular diode (Fig. 12-12). Tunnel diodes take their names from the *tunneling effect*—a current conduction process with the characteristics shown in Fig. 12-13. The tunneling effect occurs in PN junctions which are very thin and have high impurity levels. This means very many free charges exist in the semiconductor material. As the positive anode voltage increases from zero, the current increases as well. Notice in Fig. 12-13 what happens when the anode voltage increases beyond point V_P on the curve.

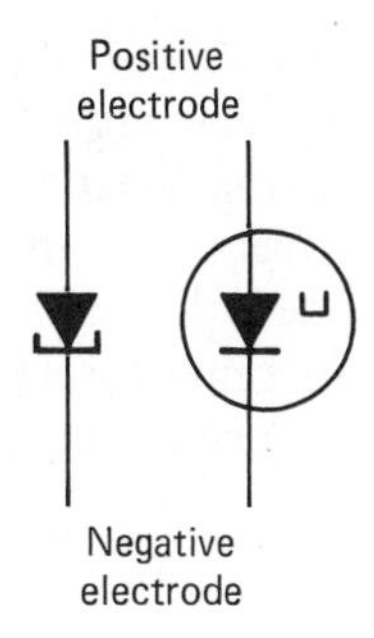

Fig. 12-12 Circuit symbols for the tunnel diode.

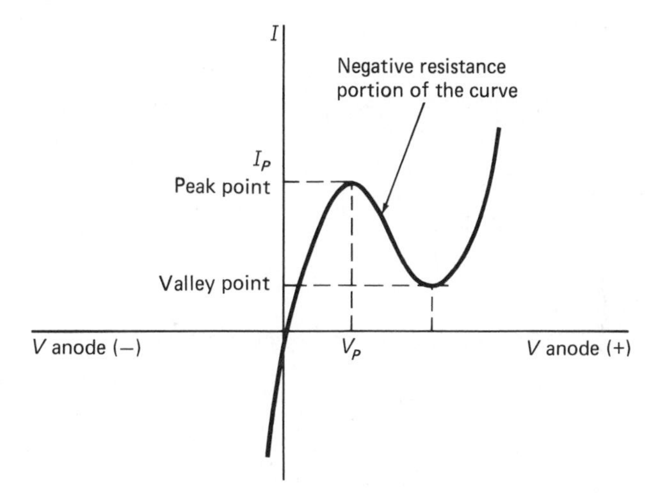

Fig. 12-13 Tunnel diode characteristic curve.

As the voltage increases to point V_P, the characteristic curve indicates that the current also increases. As the voltage increases beyond point V_P, the current decreases until the voltage reaches the valley point. As the voltage moves above the valley point, the current again increases rapidly. Notice between the V_P point and the valley point the tunnel diode exhibits a negative resistance. The current decreases with an increase in voltage.

This negative resistance is the key quality that makes the tunnel diode useful in several different types of circuits. Current through a positive (i.e., conventional) resistance causes it to dissipate power, that is, to heat up. A *negative* resistance, however, makes this power available to any circuit connected to it. A tunnel diode biased with negative-resistance will supply power to the rest of the circuit. Because of the extremely short distance between the N and P material in the diode, the speed at which this power can be supplied is very high, making the diode useful in ultra-high-frequency (UHF) circuits.

Because of its efficiency at high frequencies the tunnel diode can perform a few major functions well that transistors can't. Ultra-high-frequency circuits or microwave circuits often use tunnel diodes.

A back (or backward) diode is similar to a tunnel diode. It is used for the same functions regular diodes are, namely, for rectifying, detecting, and mixing signals, but those of extremely high frequencies. The symbol for a back diode is the same as that for a regular diode, but it has a different characteristic curve. Figure 12-14 is the characteristic curve for a back diode.

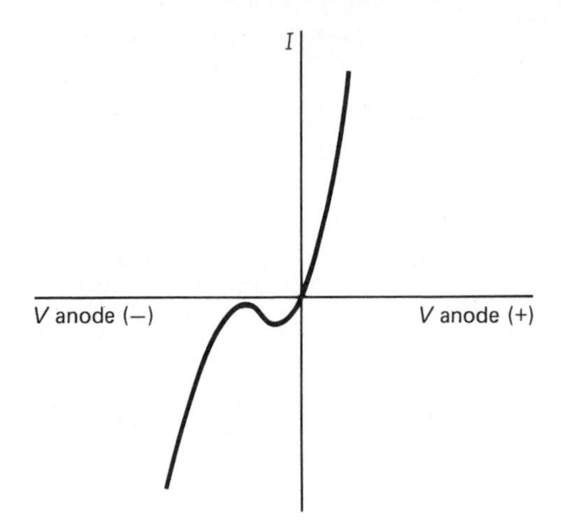

Fig. 12-14 Back diode characteristic curve.

12-8 NPN AND PNP TRANSISTORS

The NPN transistor is illustrated in Fig. 12-15*a*. Notice that the NPN transistor consists of a "sandwich" of N- and P-type semiconductor materials. The three leads of the transistor are the *emitter, base,* and *collector.* In the schematic symbol, the emitter is the lead with the arrow. The arrow always points toward the N-type material and away from the P-type material.

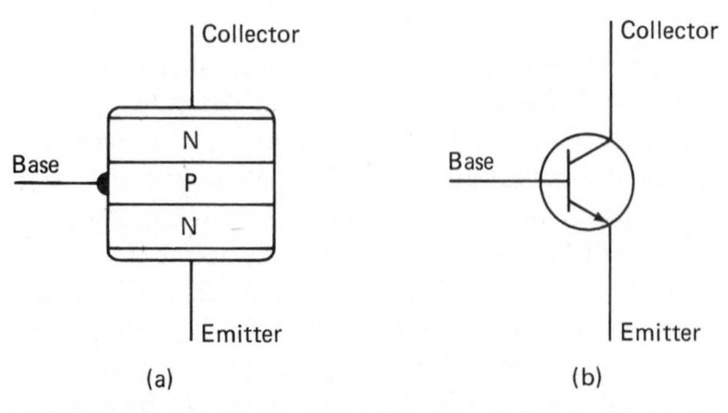

Fig. 12-15 NPN transistor: (*a*) junction diagram; (*b*) schematic symbol.

The NPN transistor can be turned on and off by a voltage applied across the emitter and base. Notice the emitter and base form an NP-junction diode. Forward bias turns on the NP junction. Notice then that if the base is positive with respect to the emitter, the emitter base junction is forward-biased and the transistor is on (providing the collector has its voltage more positive than the base voltage). Reverse-biasing between the emitter-base junction will turn the transistor off.

The transistor can be made to act like a switch by forward- and reverse-biasing the emitter-base junction. What would we have to do to make the transistor into an amplifier? If we take a look at the characteristic curve for the NPN transistor in Fig. 12-16, collector current I_C is represented by the vertical axis and emitter-to-collector voltage V_{CE} is represented by the horizontal axis. Voltage increases as we move to the right. The curve shown represents the I_C-V_{CE} characteristic at a particular value of base current I_B.

The collector family of curves in Fig. 12-17 shows how V_{CE} affects collector current. The various values of I_B increase from I_{B1} to I_{B5}. Note that a small change in the base current will cause a large change in the collector current. To amplify a signal a transistor must have a load resistor to develop the output signal. By using the curves in Fig. 12-18 and the circuit of Fig. 12-19 the voltage gain of a transistor amplifier for any input signal can be determined.

We will begin our amplifier analysis by determining the size of the load resistor in Fig. 12-19. We will

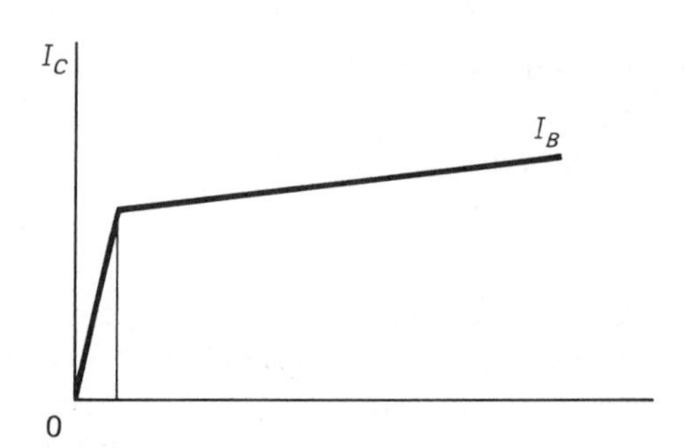

Fig. 12-16 I_C-V_{CE} characteristic curve for an NPN transistor.

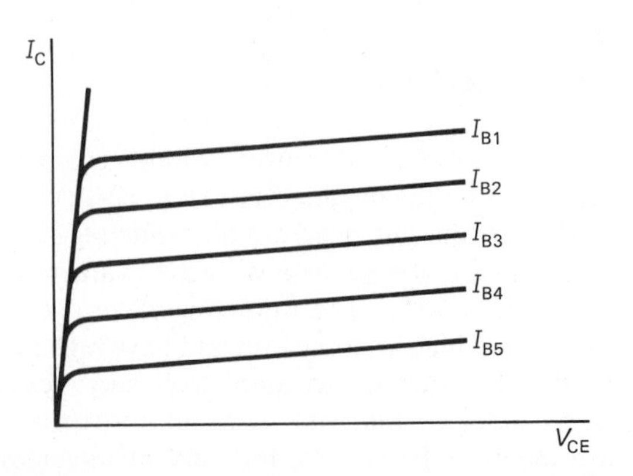

Fig. 12-17 Collector family of curves for an NPN transistor.

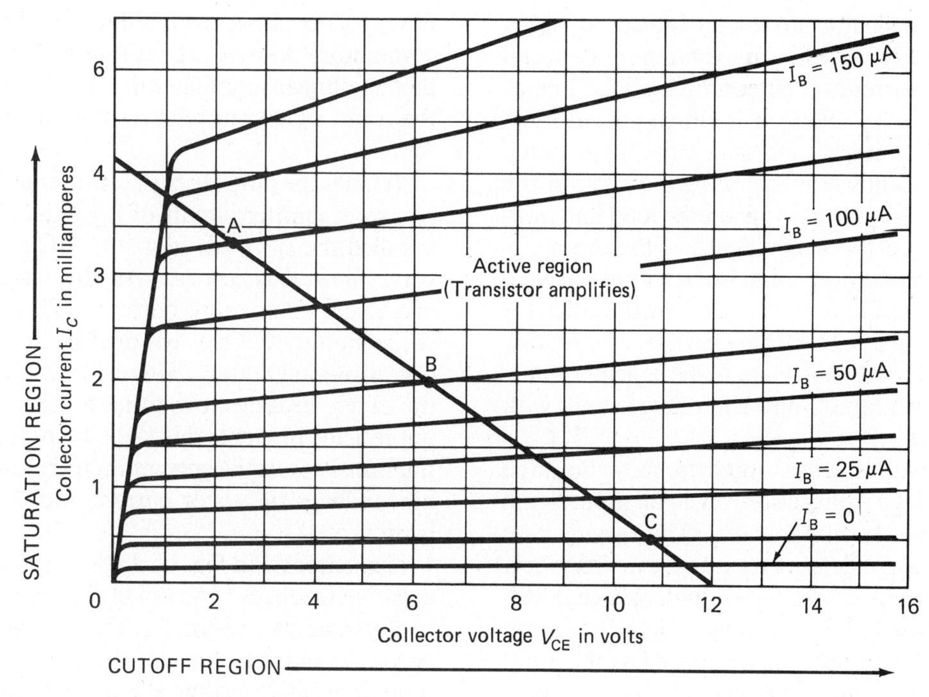

Fig. 12-18 Transistor characteristic curves and load line.

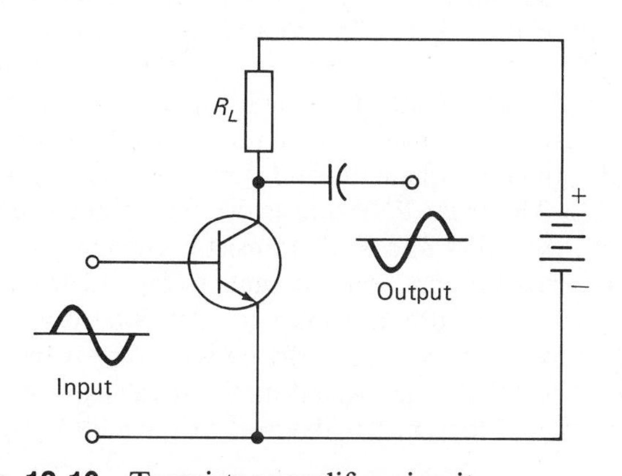

Fig. 12-19 Transistor amplifier circuit.

assume that the NPN transistor family of curves in Fig. 12-18 is for the transistor in the amplifier. Notice in the family of curves the straight line passing through the points A, B, and C. This is called the *load line* for the amplifier. The position of the line is determined by the value of the amplifier load resistor.

If we look at the transistor and the load resistor as being in series with a 12-V battery, we can see that the voltage will divide across the load resistor and the transistor. If the transistor acted like an open circuit in the amplifier circuit, the supply voltage would be measured across the transistor. With the transistor open, the current in the circuit would be zero. Notice on the transistor characteristic curves that at $I_C = 0$, $V_{CE} =$ 12 V. This is one end point of the load line. In Fig. 12-19 if the transistor were on and acted like a short in the circuit, the voltage would all be across the load resistor. The current in the circuit with the transistor acting like a short circuit would be determined by the size of the load resistor. Thus, using Ohm's law, we can find I_C when $V_{CE} = 0$. Assume the supply voltage is 12 V and the load resistor is 2800Ω.

$$I_C = \frac{12 \text{ V}}{2800\Omega} = 0.0043 \text{ A, or } 4.3 \text{ mA}$$

The other end of the load line is therefore at $V_{CE} = 0$ and $I_C = 4.3$ mA. Since the currents and voltages used to plot this load line are direct current, the load line is called a *dc load line* for this circuit. Similarly, using alternating current and voltage, an ac load line can be constructed. Figure 12-20 shows how a varying input

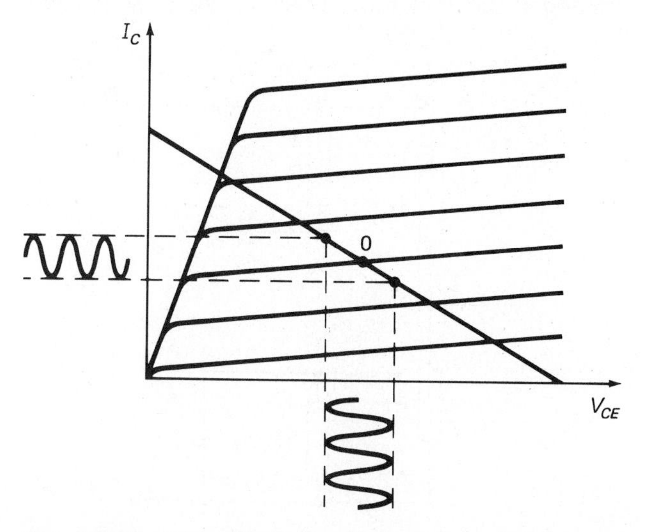

Fig. 12-20 The effect of a varying input signal.

signal affects the collector current and the output signal of the circuit in Fig. 12-19. The varying input signal causes the base current to vary along the load line as shown in Fig. 12-20. Point o is the *quiescent point* because it is the point around which the input signal varies. It is also known as the *operating point* of the amplifier. The two other points on the load line indicate the positive and negative peaks of the input signal. By following the horizontal broken lines across to the collector current axis we can see how the collector current varies in step with the base current. By following the vertical dotted lines down to the collector-emitter voltage axis, we can see how the voltage varies with the collector current.

Graphically we have attempted to show how the transistor provides amplification. A small input signal will cause the base current to vary. The varying base current causes the collector current to vary. The varying collector current causes the output voltage developed across the load resistor to vary. The voltage gain of the circuit is the output voltage divided by the input voltage. Measuring the input signal voltage we now have all we need to know to determine the gain of the amplifier.

The amplifier circuit presentation will be nonmathematical since we want to apply troubleshooting procedures to our presentation rather than design considerations.

Figure 12-21 shows a common-emitter linear amplifier. The circuit consists of two resistors R1 and R2 which are used to bias the transistor at its operating point (such as point B in Fig. 12-18). These two resistors also develop the input signal that is coupled into the circuit through the coupling capacitor C1. The input signal is applied to the base of transistor. As the signal goes positive there is an increase of emitter-collector current. Increase current flow through the load resistor causes the collector voltage to become less positive (more negative). When the input signal of the transistor becomes more negative, less base current flows. This causes the emitter-collector voltage to become more positive. Examining Fig. 12-21 indicates that the linear amplifier circuit gives amplification to the input signal but reverses the phase relation of the signal.

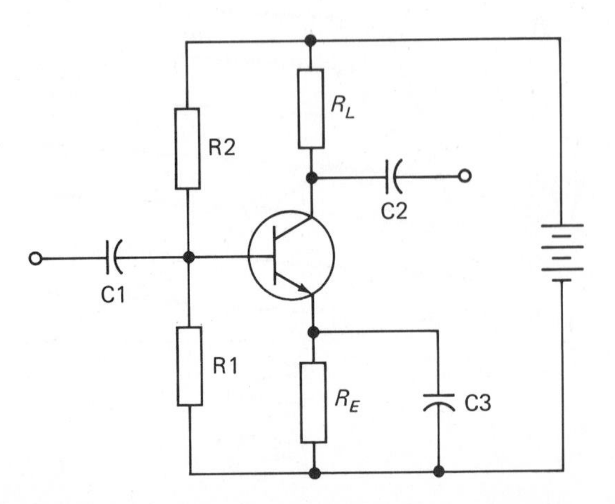

Fig. 12-21 Common-emitter linear amplifier.

What is the purpose of the emitter resistor (R_E) in the common emitter circuit of Fig. 12-21? This resistor is called the *swamping resistor.* Its purpose is to prevent *thermal runaway.* As the transistor becomes hot, it conducts more current. This causes it to become hotter, thereby conducting more current and becoming still hotter. This cycle would continue until the excess heat destroyed the transistor. The swamping resistor prevents this cycle by reducing the bias of the transistor as the current increases in R_E. Reduced bias reduces transistor current and thermal runaway is prevented.

Capacitor C3 in Fig. 12-21 prevents degeneration, or negative feedback. As the input signal causes current to fluctuate in resistor R_E, the voltage across R_E opposes the current through the transistor. Degeneration, or negative feedback, reduces amplifier gain. Capacitor C3 prevents this by bypassing the ac signal around resistor R_E. So C3 is in the circuit to prevent a reduction in the gain of the amplifier by negative feedback.

The transistor shown in the linear amplifier of Fig. 12-21 is a PNP transistor. A similar transistor is shown in the junction schematic and symbol diagram of Fig. 12-22. This is the PNP transistor. The difference between the NPN and PNP transistor can be seen by comparing the junction schematics (Fig. 12-15*a* and 12-22*a*). The NPN transistor requires a positive bias on its base with respect to the emitter to provide forward bias. The PNP transistor requires a negative bias on its base with respect to its emitter to provide forward bias.

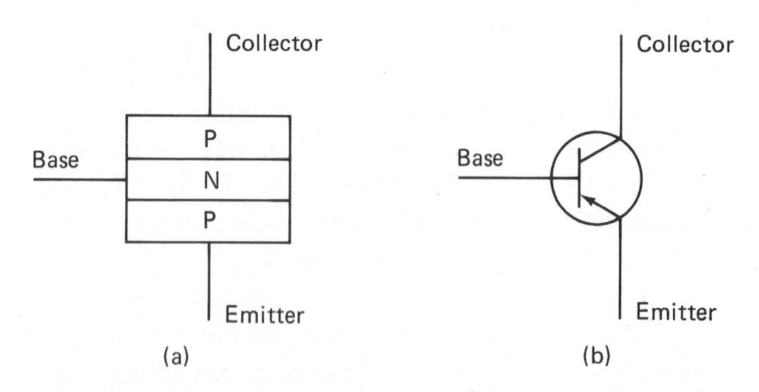

Fig. 12-22 A PNP transistor: (*a*) junction schematic; (*b*) symbol.

Notice that the arrow on the emitter of a NPN transistor is Not Pointing in (NPN). The arrow on the PNP transistor emitter is Pointing in (PNP). This is an easy way to remember the symbol for each transistor.

The arrow of a transistor always points to the negative power source. In the linear transistor amplifier of

Fig. 12-21, if the transistor were changed to a PNP, we could use this rule to determine the polarity of the power source. Notice all we would have to do is to reverse the battery terminals (+ and −) and the circuit would work with the PNP transistor.

12-9 UNIJUNCTION TRANSISTORS

Another common semiconductor device is the *unijunction transistor* or UJT. The junction schematic and symbol for the unijunction transistor are shown in Fig. 12-23.

The characteristic curve for the unijunction transistor is shown in Fig. 12-24. The vertical axis shows the amount of the voltage applied between the emitter and base 1. Emitter current is shown on the horizontal axis.

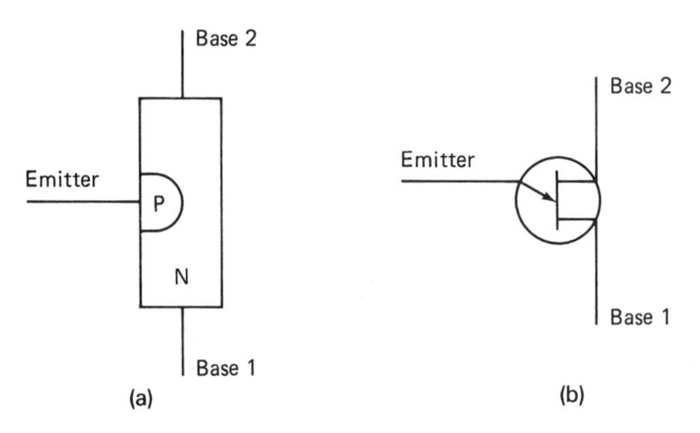

Fig. 12-23 Unijunction transistor: (*a*) junction schematic; (*b*) symbol.

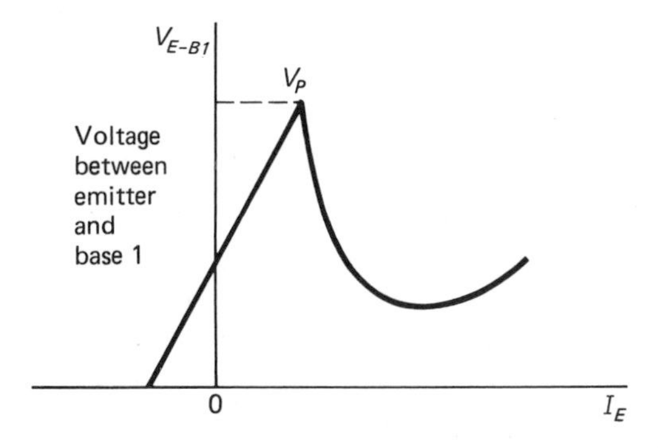

Fig. 12-24 Characteristic curve for a unijunction transistor.

The emitter allows only a negligible current flow until its voltage reaches V_P on the characteristic curve. At this point the UJT switches to a high emitter current as it moves into the negative resistance region on the curve.

A unijunction can be used as a *pulse generator* (also called a *relaxation oscillator*) as in Fig. 12-25.

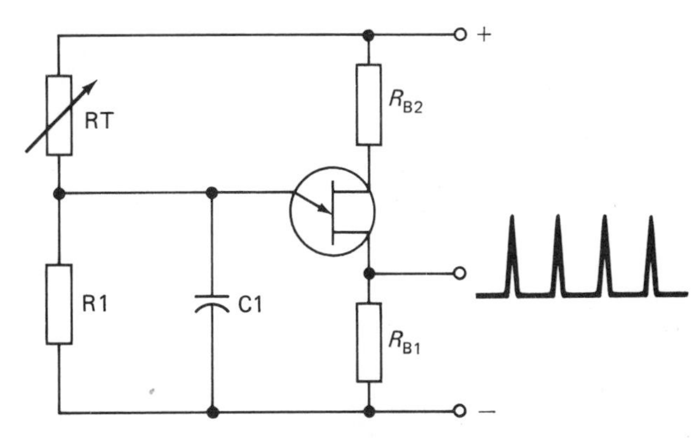

Fig. 12-25 UJT pulse generator, or relaxation oscillator.

An explanation of the UJT pulse circuit is as follows. The timing capacitor C1 charges through the timing resistor RT. As the top side of the capacitor becomes more positive, it will reach the standoff voltage of the unijunction transistor (point V_P on Fig. 12-24). At this point emitter E and base 1 become forward-biased. The emitter and base 1 become a low-resistance path for the capacitor to discharge. As the capacitor discharges through the UJT, a pulse of voltage develops across the resistor R_{B1}. As soon as the capacitor discharges, the process repeats. Figure 12-25 indicates the output of the UJT pulse generator is a series of pulses. The frequency of pulses is determined by the resistance of RT (or size of C1). With RT an adjustable resistor, the frequency is determined by the setting of RT. Resistor R_{B2} is in the circuit to prevent thermal runaway of the UJT.

The UJT is often used to produce trigger or gating pulses for such power-switching devices as triacs and SCRs, in a family of semiconductor devices known as *thyristors*. Thyristor devices are power-switching semiconductor devices. We will study the thyristor family next.

12-10 SILICON-CONTROLLED RECTIFIERS

A silicon-controlled rectifier (SCR) is a four-layer semiconductor device with alternating layers of P- and N-type material. An SCR has three leads, one of which is a gate connected to the P-type layer nearest the bottom N-type layer. The other leads are the normal anode and cathode.

A junction schematic and symbol are shown in Fig. 12-26 on the next page. The SCR symbol is identical with the common diode symbol without the gate lead.

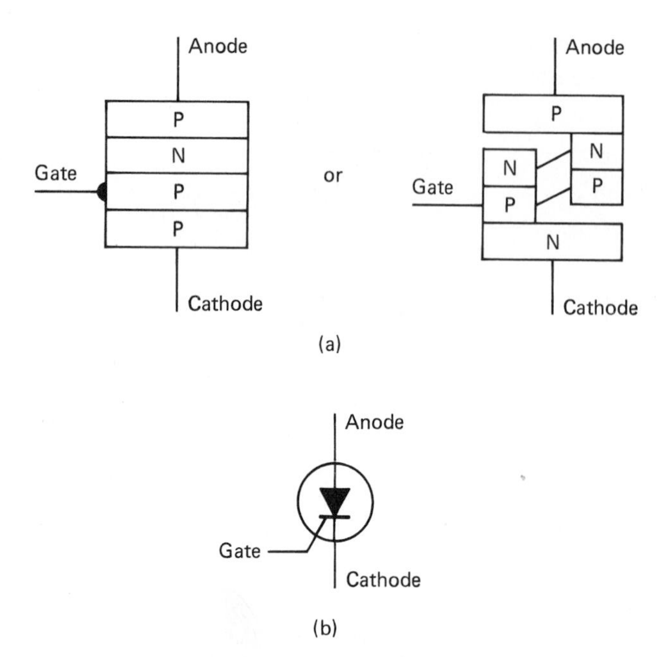

Fig. 12-26 Silicon-controlled rectifier: (*a*) junction schematic; (*b*) symbol.

The characteristic curve for an SCR is shown in Fig. 12-27. From Fig. 12-27 it can be determined that the SCR acts like a regular diode in that it conducts in one direction only. It conducts current from the cathode to anode only. It blocks current flow in the reverse direction from anode to cathode. In this way the SCR is like a regular diode. It differs from a diode in that it will not conduct until a trigger pulse at the gate causes gate current. The trigger pulse at the gate reduces the breakover voltage V_{BR}. When turned on, the gate loses all control for the SCR. It will not turn off until the anode current falls near zero. Notice where V_{BR} is located on the SCR characteristic curve of Fig. 12-27.

There are many uses for an SCR in industrial control circuitry. The SCR's ability to be controlled by injecting a pulse in its gate lead makes it possible to apply the SCR to such applications as power switching, phase control, inversion (direct current to alternating current), and chopping.

The SCR has many uses in circuits where current control is desired. Because of the ability of the SCR to maintain conduction of anode current after a pulse of current is applied to the gate and yet block the flow of anode current in the reverse direction, the SCR can function as a *controlled* rectifier. Current I_H is the point above which current must be maintained to keep the SCR in conduction.

One of the applications of an SCR is for an inverter circuit. An inverter circuit is a circuit that converts a dc voltage to an ac output voltage. This is illustrated in Fig. 12-28. When the dc voltage is applied, nothing happens in the circuit. The small circle at the end of the gate leads means that there is a trigger circuit connected to the gates, but for reasons of simplicity the trigger circuits themselves are left out of the drawing. The trigger circuits alternate the firing of the SCRs at any desired frequency, and C acts with L to reverse bias, or turn off, each SCR in turn when the other is turned on. The resulting pulsating dc voltage on the primary winding of the transformer sets up an ac voltage in the transformer secondary winding.

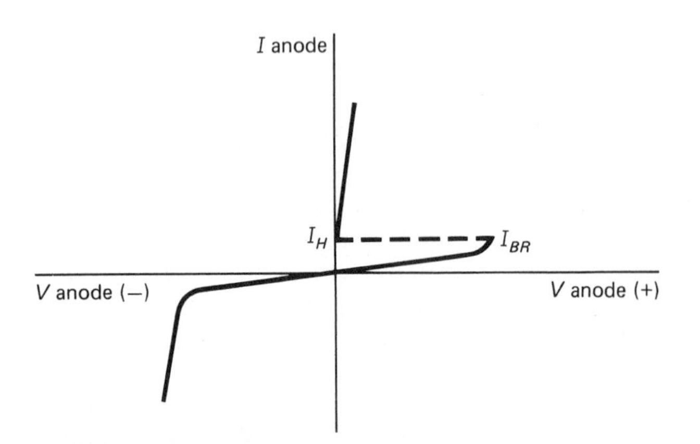

Fig. 12-27 SCR characteristic curve.

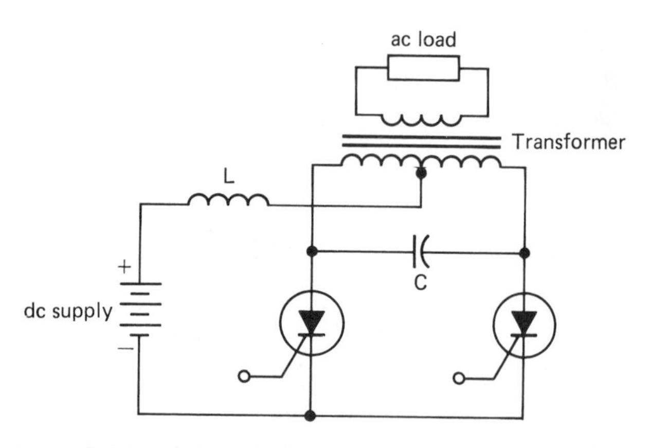

Fig. 12-28 Inverter circuit.

The SCR is used in many industrial applications to give what is called *phase control*. By controlling a portion of ac cycle, control similar to that given by a variable transformer or a variable resistor can be obtained. The SCR's voltage can be controlled by switching the SCR on during portions of the ac cycle.

Phase control is a very effective method of obtaining motor speed control, light dimming, and temperature control, or any condition where it is desirable to apply only a certain portion of an alternating current cycle to a load.

Figure 12-29 is a half-wave circuit, since only the positive half-cycle of the ac supply can be applied to the load. Other circuit arrangements allow both negative

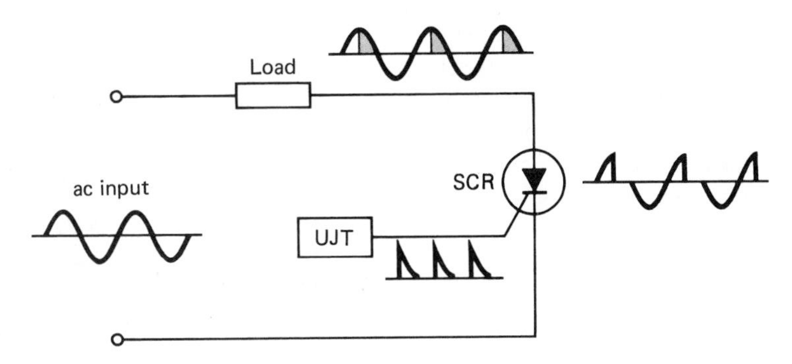

Fig. 12-29 Half-wave phase-control circuit.

and positive half-cycles to be applied to the load. These are called *full-wave circuits.* The shaded portions of the waveform next to the SCR show when voltage is blocked by the SCR. (We only want 25 percent of the total average voltage applied to the load.)

The shaded portion of the waveform next to the load shows when current is flowing through the load (when voltage is applied to the load). Note that current through the load and SCR always stops flowing at the exact instant when it tries to reverse through the SCR.

Current can be made to start through the SCR and load whenever the voltage waveform is positive by triggering the SCR gate at the proper instant.

Many times, especially in the use of motors operating from a dc supply, such as a battery-operated lift truck, speed is controlled by a dc *chopper circuit.*

With an SCR circuit switching many times per second, the on-off time ratio may be controlled so that, for example, only 50 percent of the 48-V battery voltage is applied to the motor. We do this by applying the full 48 V to the motor 50 percent of the time. This is illustrated in Fig. 12-30.

Let's say that you can slow the truck to one-half its full speed. The dotted line on the waveform in Fig. 12-31 shows how the dc voltage should be "chopped" to produce a speed one-quarter of the truck's full speed.

What is needed in an SCR circuit to make it control the direct current? We already have a way to turn the SCR on, but we also need a method to turn the SCR off.

In a dc system like this, it is easy to turn on the SCR with a gate trigger signal from a UJT oscillator, for instance, but we need a way to turn the SCR off at the end of each conduction period.

In the phase-controlled circuit, the SCR is turned off when the ac supply reverses the load current flow at the end of the positive half-cycle.

In our inverter circuit, which operates from a dc supply, the L and the C act as a commutation circuit

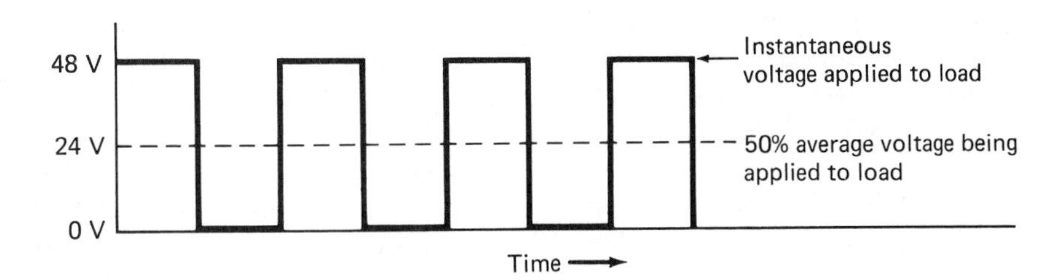

Fig. 12-30 Voltage pulses.

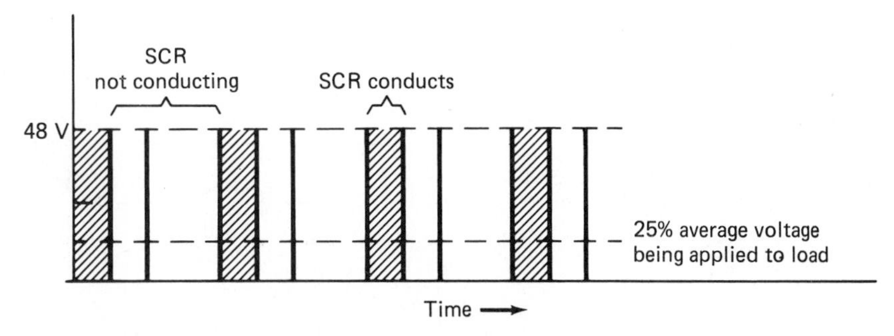

Fig. 12-31 Conduction of SCR affecting average voltage to load. In this case, the SCR conducts only 25 percent of the time.

which forces the SCRs to turn off by backing current momentarily against the normal direction of current flow through the SCRs. In choppers, which operate from dc supplies also, we also need a "commutation circuit" of some kind to turn off the SCR since no natural reversal of the supply voltage occurs, as happens when an SCR is operating from an ac supply.

12-11 LIGHT-ACTIVATED SILICON-CONTROLLED RECTIFIER

Silicon-controlled rectifiers can be triggered into conduction in many ways. The only way that a regular SCR cannot be triggered on is with a source of light. A variation of the SCR that will turn on with a light source is called an *LASCR* which stands for light-activated silicon-controlled rectifier.

Figure 12-32 illustrates the circuit symbol for the LASCR. The figure also indicates the leads found on the LASCR. The arrows that point in toward the symbol indicate that the device is light-activated.

The LASCR has the same operating characteristics as the SCR. It can be turned on with a gate pulse or it can also be turned on by light passing through a window to the LASCR's enclosure. To activate the LASCR you can either place an electrical trigger on the gate lead, increase the anode voltage beyond V_{BR}, or shine a sufficiently intense light on the gate junction. The LASCR and SCR both have the same characteristic curves. The characteristic curve for an LASCR is shown in Fig. 12-33.

A LASCR has the same major function as an SCR: to act as a switch. It can be used for relay replacement, position control, photoelectric detection, slave photo flashes, and many other functions where light is involved.

Like a rectifier, the LASCR will conduct current only in one direction. In Fig. 12-32 a bridge-rectifier circuit is shown that rectifys alternating current into direct current. To turn the LASCR on (make it conduct), shine the light source on the window of the device. This triggers the LASCR to conduct and the load will have power delivered to it. Remove the light, and the LASCR opens on the following half-cycle when the voltage dips momentarily through zero, thereby removing current from the load. Resistor R1 helps the LASCR block voltage when no light is shining on it. What if the light should burn out? Is there any other way you can make the LASCR operate? Yes. Apply a gate trigger to the gate of the LASCR.

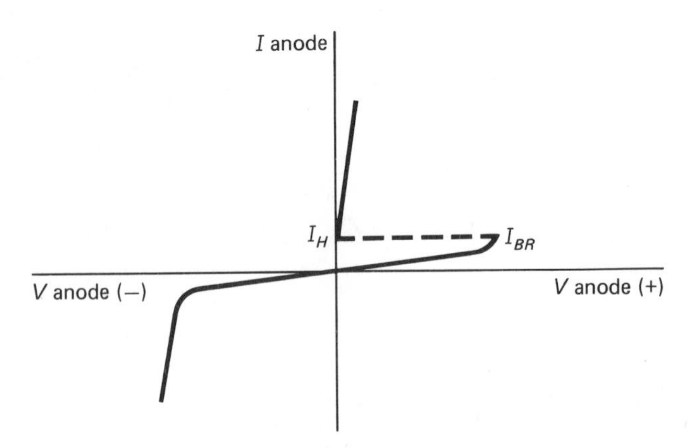

Fig. 12-33 Characteristic curve for light-activated SCR.

12-12 SILICON-CONTROLLED SWITCH

Like the SCR, a silicon-controlled switch, or SCS, is a four-layer device with alternating layers of P- and N-type material. The SCS is a four-lead device with a lead labeled *anode gate* and another labeled *cathode gate* as well as the leads labeled *anode* and *cathode.* The SCS junction schematic is shown in Fig. 12-34.

The SCS has a circuit symbol similar to the SCR. There is one addition to the SCS that is not found on the SCR and that is a lead coming from the *anode* side of the diode. Figure 12-35 shows the circuit symbol and the lead identification.

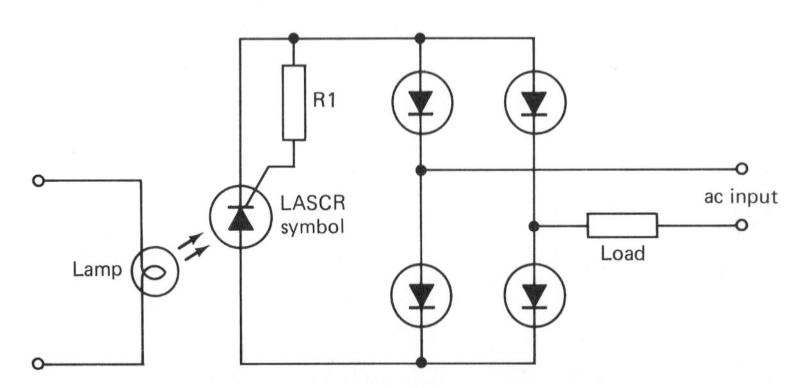

Fig. 12-32 LASCR circuit.

The characteristic curve for the SCS is the same as the curve for the SCR and the LASCR. The SCS can be triggered to conduct either by a positive signal on the cathode gate or by a negative signal on the anode gate. Figure 12-36 illustrates the characteristic curve for the SCS.

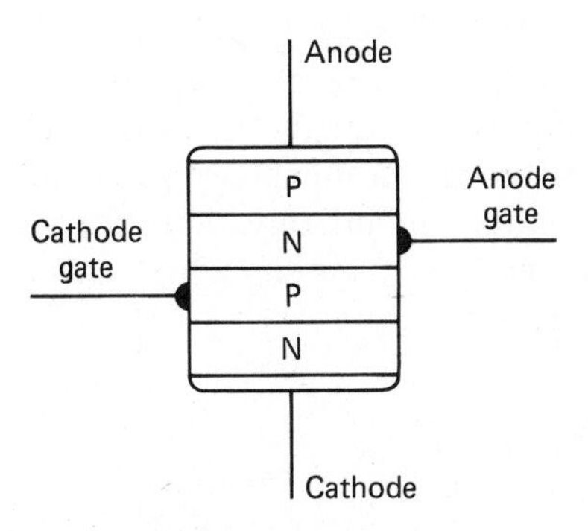

Fig. 12-34 SCS junction schematic.

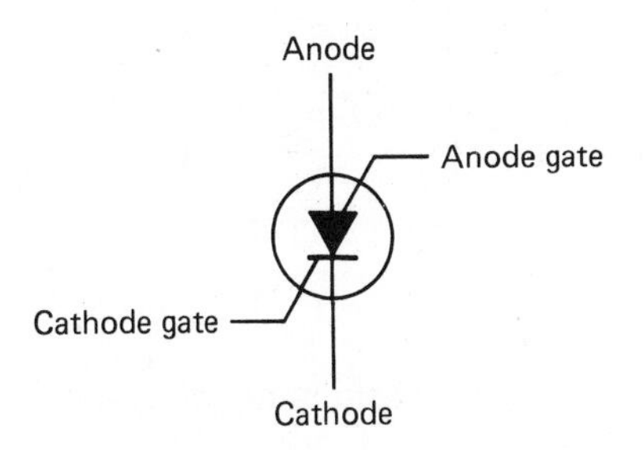

Fig. 12-35 SCS circuit symbol.

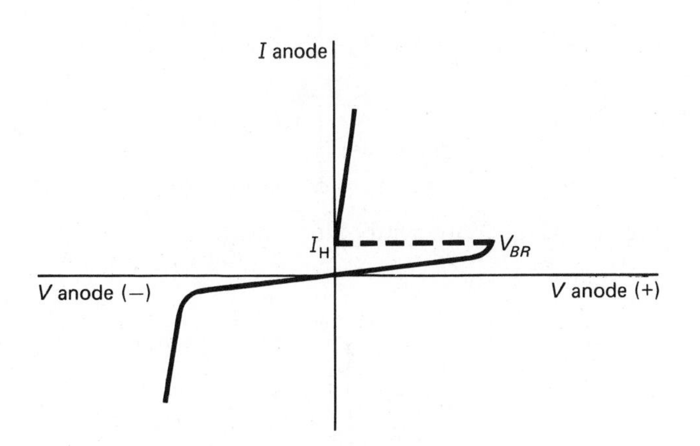

Fig. 12-36 LCS circuit symbol.

The major function of the SCS is to act like a switch. It is different than the SCR in that it has an extra gate. An SCR could be triggered to conduct by applying a positive signal to the cathode gate, thereafter staying in conduction as long as the anode current did not fall below the holding current (I_H).

The SCS can do the same thing. Its extra anode gate can trigger the SCS into conduction when a negative signal is applied to the anode gate with respect to the anode. In other words, the SCS can be triggered into operation either by a *negative signal* on the anode gate or by a *positive signal* on the cathode gate. Could an SCS take the place of an SCR? And, could an SCR take the place of an SCS?

The answer is that an SCS can replace the SCR of equal size because the SCS can be triggered by a positive signal to the cathode gate. An SCR can sometimes replace an SCS. If the SCS were connected for a positive gate signal, it could be replaced with an SCR. If the SCS were connected for a negative anode gate, it could not be directly replaced with an SCR.

The SCS can be used in many of the same jobs that the SCR is used for. The SCS can also be used in such other devices as logic circuits, counters, lamp drivers, controllers, and power switches.

12-13 GATE TURN-OFF SWITCH

The gate turn-off switch, or GTO, is a device that can be switched on or off by a gate signal. Large transistors have similar on-off control characteristics so in many circuits large transistors are taking the place of GTOs. Figure 12-37 is the junction schematic of a GTO.

The circuit symbol (Fig. 12-38) for the GTO looks very much like an NPN circuit symbol. A GTO, how-

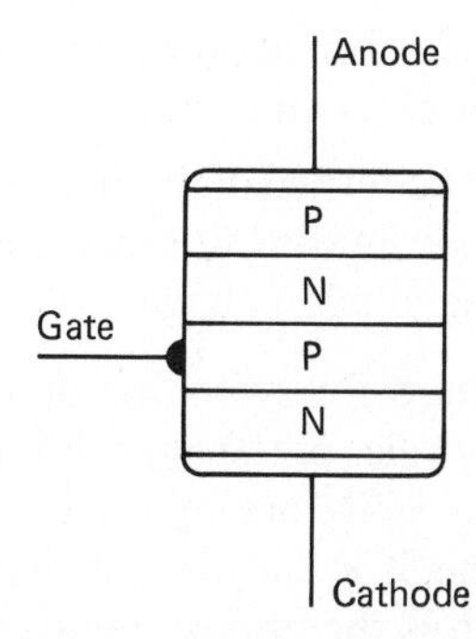

Fig. 12-37 Gate turn-off switch junction schematic.

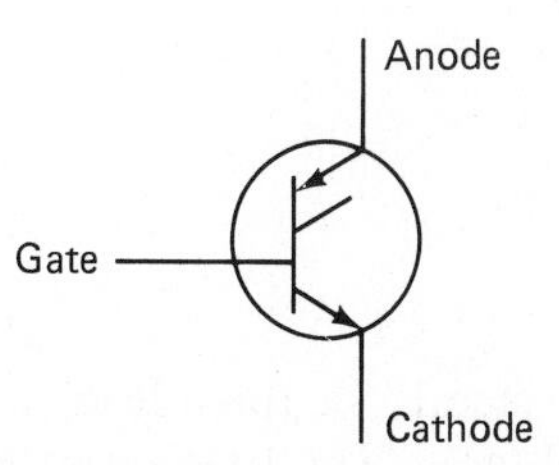

Fig. 12-38 Gate turn-off switch (GTO) circuit symbol.

ever, has a cathode that looks like the emitter for an NPN transistor, and an anode that looks like an emitter for a PNP transistor. The gate lead looks just like the base lead for either a PNP or an NPN transistor. There is an additional short diagonal line that extends from the gate lead. Figure 12-37 shows the leads, labeled *anode, cathode,* and *gate.*

The GTO characteristic curve is again similar to that of the SCR. The GTO has one feature that makes it different from the SCR, however. A negative gating pulse can turn off the GTO, whereas the SCR must be turned off by some external means that reduces anode current below the I_H, or holding current, level. Since GTOs are used for switching circuits on and off, GTOs are used in dc switching circuits, pulse-generating circuits, and chopping circuits. How the SCR, LASCR, SCS, and GTO can be turned on and off can be summarized as follows:

The SCR can be turned on by a positive trigger on the gate lead or exceeding V_{BR} on the anode.

The SCR can be turned off by decreasing the anode current below I_H or by reversing the anode voltage.

The LASCR can be turned on with a positive trigger on the gate lead or a light shining on the PN junction or by exceeding V_{BR}.

The LASCR can be turned off by decreasing anode current below I_H or by reversing anode voltage.

The SCS can be turned on with a positive trigger on the cathode gate or a negative trigger on the anode gate or by exceeding V_{BR}.

The SCS can be turned off by decreasing anode current below I_H or by reversing anode voltage.

The GTO can be turned on with a positive trigger to the ON gate or exceeding V_{BR}.

The GTO can be turned off with a negative trigger to the ON gate or decreasing the anode current below I_H or by reversing anode voltage.

The SCR is used for switching dc and ac circuits. When both half-cycles of the alternating current are to be switched, two SCRs are required. They are connected so that one SCR controls one half-cycle and the other SCR controls the opposite half-cycle. Can one semiconductor device be used to control both half-cycles of the alternating current? Yes, the name of the device is a *triac.*

12-14 TRIAC

The term *triac* stands for three lead (tri) alternating-current control (ac). The triac operates in the same way the SCR does, except that it can be triggered into anode conduction in either direction by a gate signal of either polarity. The triac is therefore ideal for switching ac circuits.

To perform this kind of operation, the triac has a complex structure, as is shown by its junction schematic in Fig. 12-39. The triac circuit symbol with lead names is shown in Fig. 12-40.

Figure 12-41 illustrates the characteristic curve for a triac. Characteristic curves for the triac will be much the same as the curves for the SCR, LASCR, and SCS. The same conduction effect found in the upper right-hand quadrant (first quadrant) in the SCR curve is found in the first quadrant of the triac curve. Since the triac can conduct in both directions, the same effect is present in the lower left-hand triac quadrant (third quadrant) as well.

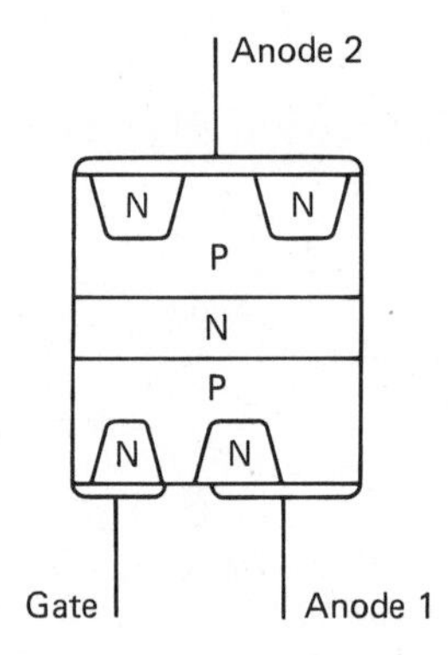

Fig. 12-39 Junction schematic of a triac.

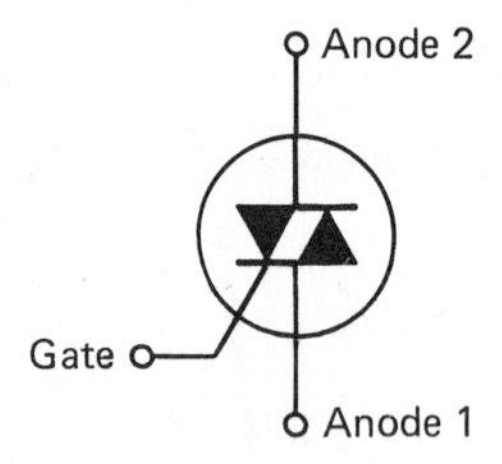

Fig. 12-40 Circuit symbol for a triac.

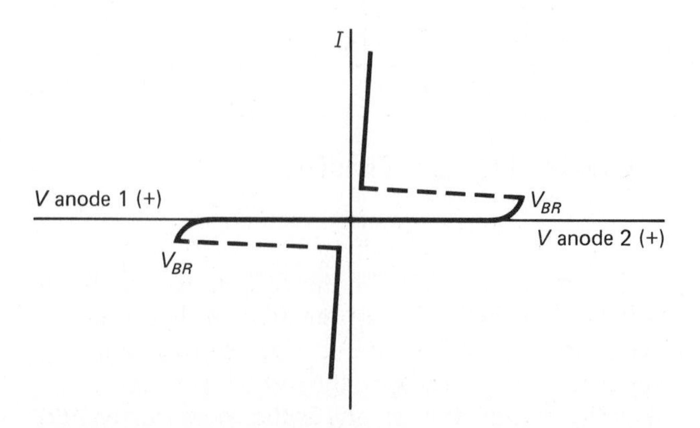

Fig. 12-41 Characteristic curve of a triac.

The triac, with its three connections, can perform the same operations that two SCRs connected in inverse parallel can perform. The triac is used primarily as a latch-type switching device.

The triac functions as a power-switching device that conducts in either direction when an ac voltage is applied as a gate signal of either polarity. It is used in ac switching circuits.

Figure 12-42 shows the triac being used in a circuit to switch an ac circuit on and off. This circuit shows how the triac works as a simple ac switch. When S1 is closed, current flows into the gate lead of the triac, triggering it into conduction between its anodes. Alternating-current power is delivered to the load.

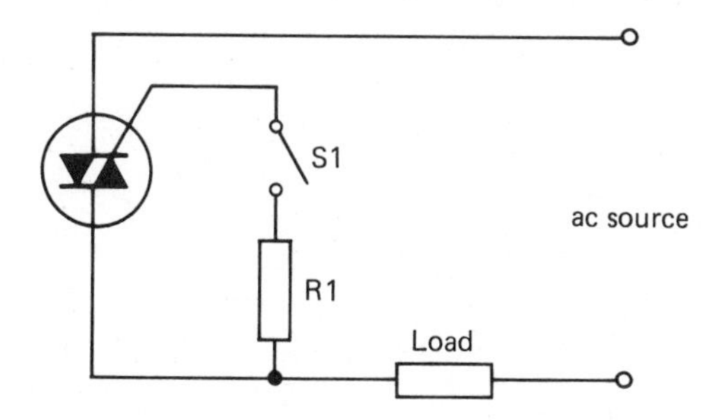

Fig. 12-42 Switching a triac.

When the switch S1 is opened, the triac turns off because it does not have a path for gate current. A gate signal is required to turn the triac back on.

The triac can also be used to provide phase control. The circuit of Fig. 12-43 provides phase control to a lamp. This circuit is essentially the same as found in lamp dimmers for home lighting. Notice the device called a *diac* in the circuit. Don't worry about the diac for the moment; we'll be getting to it shortly. The resistor R1 controls the time it takes for the capacitor C1 to charge. When C1 reaches a sufficient voltage, it will discharge through the *diac,* which triggers the triac. Action of this kind on subsequent negative and positive half-cycles will produce the waveforms shown. By increasing the resistance of R1, C1 will take longer each cycle to reach the trigger point and the amount of alternating current delivered to the lamp can be decreased, thereby making the lamp glow less brightly. In the previous circuit the triac was simply functioning as a switch, while here it is being used for phase control.

12-15 DIACS

The diac has a characteristic curve that begins in a manner similar to the curve for the triac. The diac, however, does not switch back to as low a voltage as the triac. By looking at the curve, you can see that the conduction begins when voltage in either direction reaches V_{BR}, approximately 35 V, but then drops to a value approximately 10 V down from V_{BR}. This is highly useful when the diac is used as a trigger for the triac.

The diac is a good device for triggering the triac, since it is used in ac circuits which usually need triggering on both the positive and negative half-cycles. The diac can be used to trigger the SCR, LASCR, SCS, and GTO, also. But it generally finds its main use with triacs.

As its curve shows, the diac can switch down. The diac turns on at, say, 35 V and drops to around 10 V, allowing the diac to discharge a capacitor, and this develops a trigger voltage. The diac stops conducting when C1 discharges below the breakdown voltage of the diac.

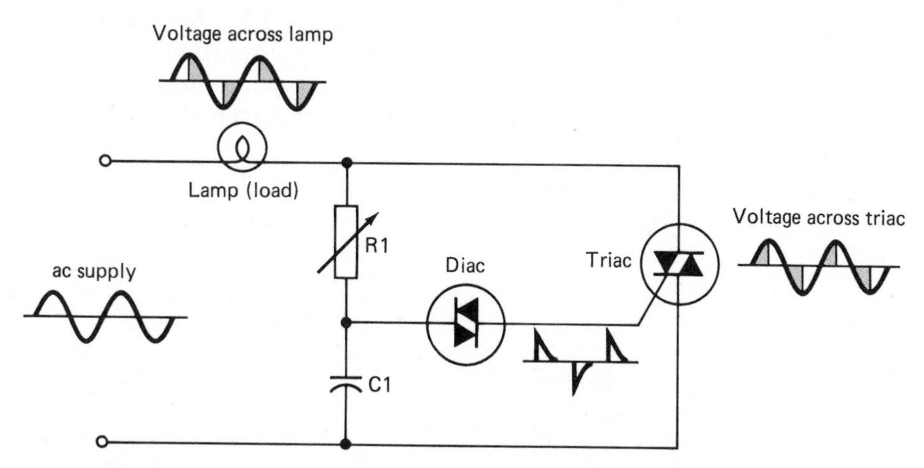

Fig. 12-43 Phase control with a triac.

Figure 12-44 indicates the junction schematic for a diac. The term *diac* stands for two lead (di) alternating-current control (ac). The diac junction schematic is the same as that for an NPN transistor except that the diac does not have a base lead. It breaks into conduction in either direction at a certain value of voltage.

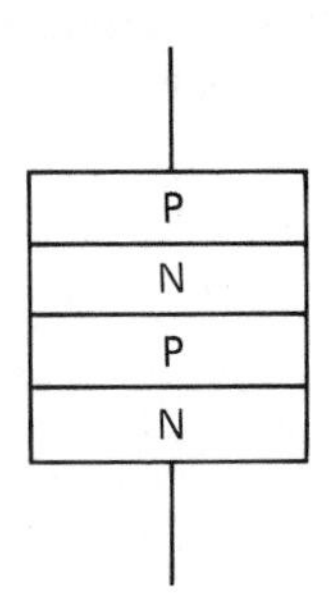

Fig. 12-44 Junction schematic for a diac.

As illustrated in Fig. 12-45, the diac has two circuit symbols.

There are several other solid-state devices whose mode of operation can be determined by their characteristic curve. Because the junction schematic for these devices will be of little value, we will use only their circuit symbol and characteristic curve.

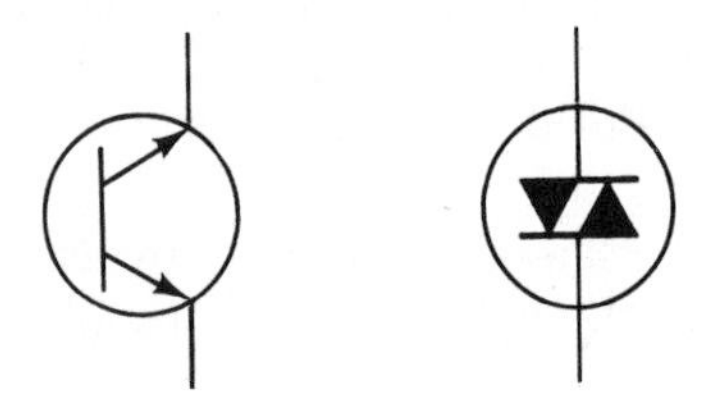

Fig. 12-45 Circuit symbols for diacs.

12-16 PHOTODIODE AND LIGHT-EMITTING DIODE

In the diode family there are two diodes we haven't discussed. One is the *photodiode,* and the other is the *LED,* or light-emitting diode. The symbols for these diodes are found in Fig. 12-46. A light source will make the photodiode act like a low resistance. It can be used to turn on a circuit with a light source.

The LED passes current in only one direction. When it is on, it glows, or produces light. It can be used as an indicating light or in displays.

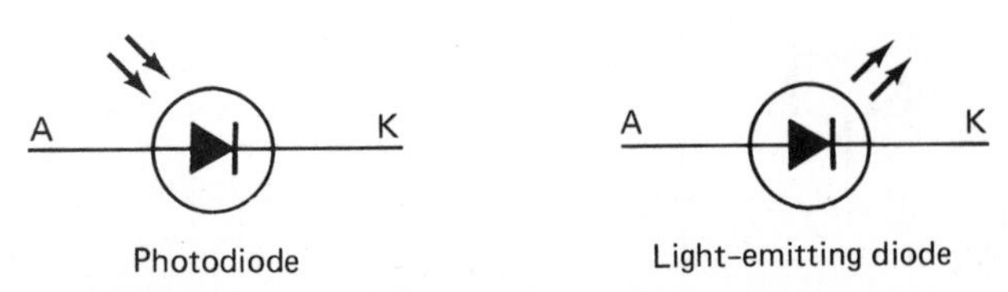

Fig. 12-46 Photo diode light-emitting diode.

12-17 DARLINGTON AMPLIFIER

The Darlington amplifier (Fig. 12-47) provides very high amplification. It is turned on when the emitter-base connection is forward-biased. The light-sensitive Darlington photoamplifier can be turned on with a light source or emitter-base bias.

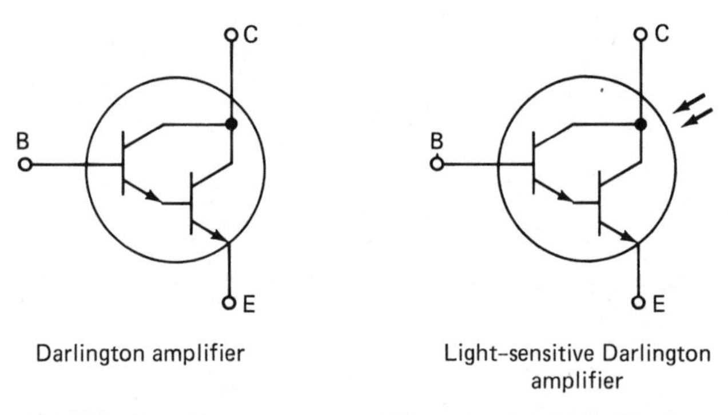

Fig. 12-47 Darlington and light-sensitive Darlington amplifier.

12-18 PROGRAMMABLE UNIJUNCTION TRANSISTOR

Another device that is used as a trigger device is the PUT, or programmable unijunction transistor. Figure 12-48 illustrates the PUT and LAPUT, or light-activated PUT. The PUT turns on when the anode becomes more positive than the gate. It turns off when the anode becomes less positive than the gate. The PUT can turn on with a light source or a gate signal.

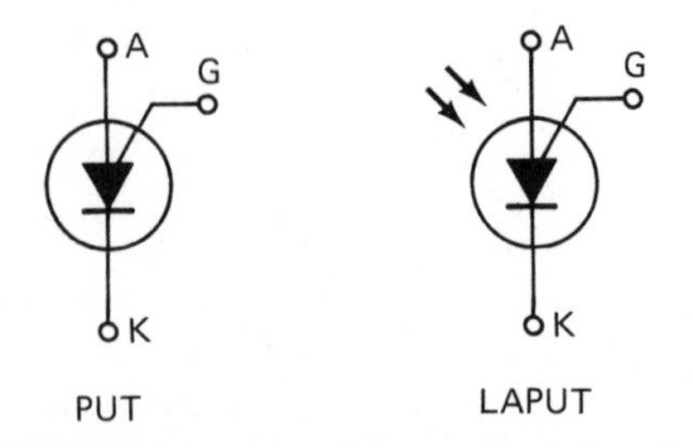

Fig. 12-48 PUT and LAPUT circuit symbols.

12-19
SUS, SBS, AND ASBS

Three more trigger devices are the SUS, or silicon unilateral switch; the SBS, or silicon bilateral switch; and the ASBS, or asymmetrical silicon bilateral switch. The SUS turns on at some value of voltage in one direction only. The SBS turns on at a certain voltage with voltage applied in either direction. The ASBS turns on with voltage applied in one direction and drops less voltage across it than it does with voltage applied in the opposite direction. The above statements can be verified by looking at the characteristic curves for these devices. Figure 12-49 illustrates the symbols for the three trigger devices and their characteristic curves.

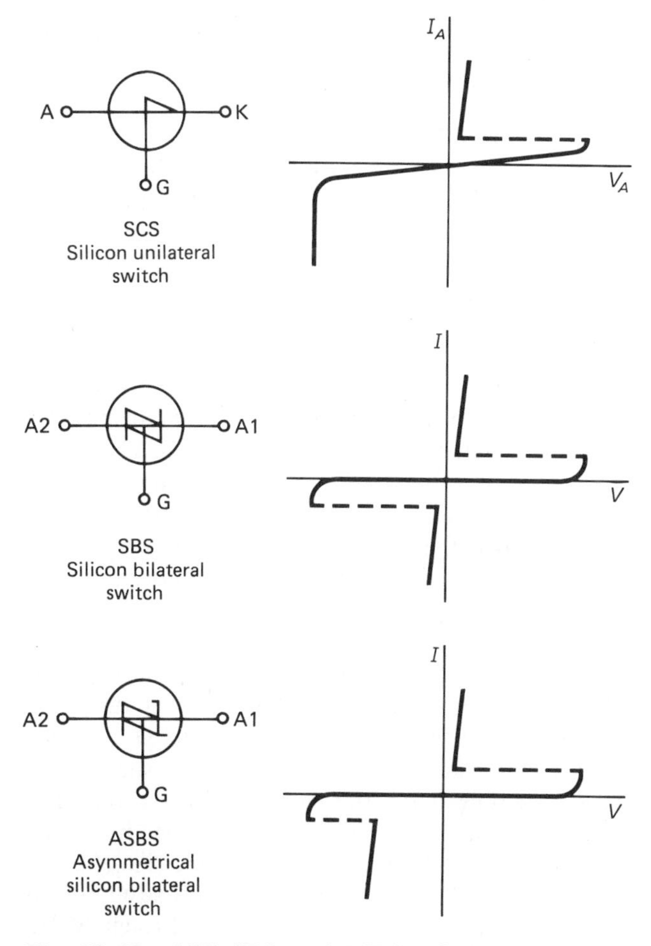

Fig. 12-49 SCS, SBS, and ASBS switches.

12-20
SOLID-STATE MOTOR CONTROL WITH TIME-DELAY TURN-OFF

We have now reviewed all the different solid-state devices used in motor controls and can identify them by name and symbol. Now let's see if we can put a few of them to use in controlling motors. Let's start with a problem for an ac motor that requires the motor to be started by a pushbutton. After the motor starts, the control action will cause it to run for a period of time then automatically stop. The circuit shown in Fig. 12-50 will give us this control action.

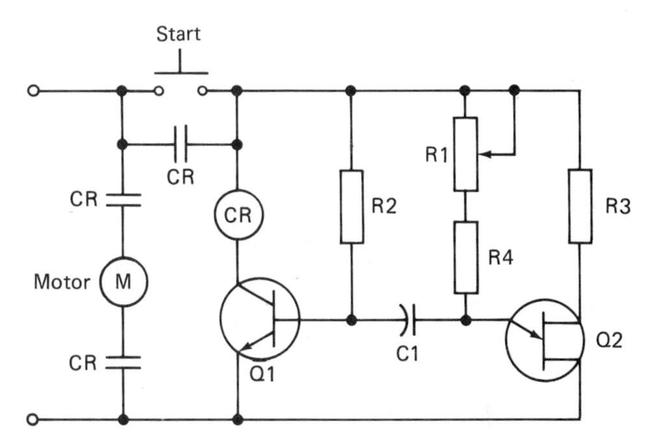

Fig. 12-50 Motor control circuit using NPN transistor and UJT for timing.

When the START switch is pushed, there is a path for the transistor base current that turns on the NPN transistor. Current flowing through the CR relay coil energizes the relay, closing the CR contacts.

The CR contact parallel with the START button bypasses the START pushbutton so that the pushbutton can be released. The transistor continues to have a source for its base current through the R2 and CR contacts. When CR is energized, a CR contact closes in the motor starter, energizing the starter, which in turn starts the motor.

The capacitor C1 starts to charge through R2. The charging rate of the capacitor C1 is determined by the setting of adjustable resistor R2. The higher the resistance, the longer the time delay to charge C1. When C1 charges to the stand-off voltage of the unijunction transistor, the UJT turns on. This occurs when the emitter of the UJT becomes sufficiently positive with respect to the base. When the UJT turns on, it acts like a closed switch across the capacitor. The capacitor tries to discharge back through the transistor. This condition will reverse-bias the transistor and turn it off, and the CR relay will then de-energize. The motor starter de-energizes when the CR contacts open and the motor stops. The electronic circuit can be packaged as a black box and called a solid-state time-delay relay. Notice that the solid-state circuit has the equivalent action of a time-delay relay with an instantaneous and a time-opening contact.

12-21 SOLID-STATE MOTOR CONTROL WITH OVERLOAD PROTECTION

The circuit in Fig. 12-51*a* illustrates a regular motor starter that will start and stop a dc motor as well as provide instantaneous trip overload protection for the motor. The solid-state equivalent of the motor starter circuit with instantaneous overload protection is shown in Fig. 12-51*b*.

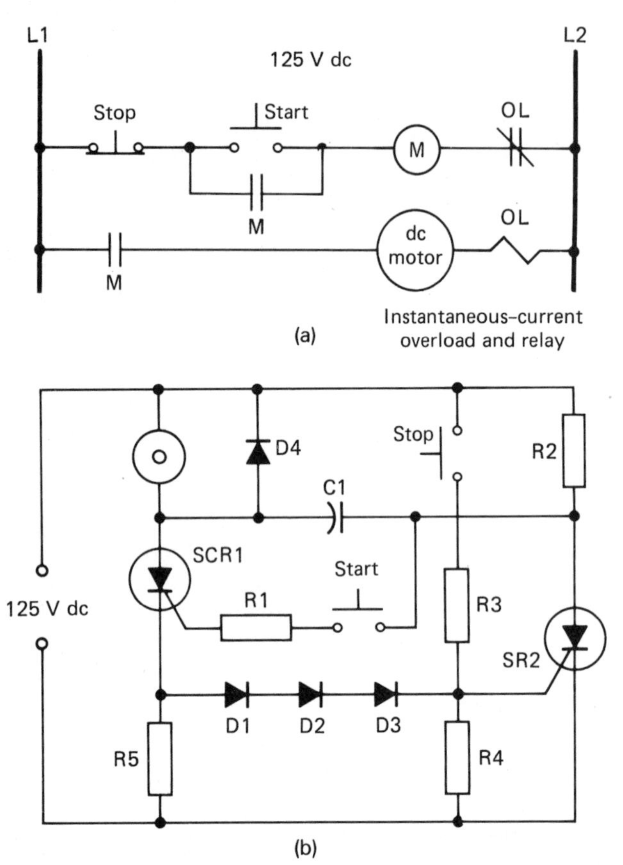

Fig. 12-51 (a) Dc motor starter using conventional motor starter. It gives start, stop, and overload protection. (b) Solid-state motor starter with stop, start, and overload protection.

In the solid-state motor starter of Fig. 12-51, when the START button is pressed, there is a path through R1 and R2 for the gate current and SCR1 turns on. Once SCR1 is on, the pushbutton can be released and gate current is no longer required because the SCR will remain on without a gate signal.

Current flowing through the SCR from cathode to anode turns on the motor. With SCR1 on, the capacitor C1 will charge to the voltage that is across the motor. Now the motor is running with SCR1 on and capacitor C1 charged.

When stopping the motor is desired, all we have to do is press the STOP button. Notice that this is a normally opened pushbutton. When the STOP button is pressed, current flows through R4 and R3. The voltage drop across R4 provides the positive gate signal to turn on SCR2. When SCR2 is on, the charged capacitor C1 is effectively across two closed switches and wants to discharge. Because of its charge it will attempt to discharge down through SCR1 from anode toward the cathode. When in conduction, the voltage drop across SCR1 is very low. The charge on C1 is practically line voltage. Capacitor C1 attempting to discharge through SCR1 reverses the bias on the cathode anode and turns the SCR off. Even though the STOP button is released, SCR2 remains on. Capacitor C1 charges to line voltage in the opposite direction. Diode D1 is in the circuit to discharge any inductive kick from the motor that may produce a high enough voltage to damage SCR1.

When the START button is pressed, SCR1 turns back on and the motor starts. Capacitor C1 now attempts to discharge through SCR2. This reverse-biases the anode-cathode circuit and the SCR2 turns off. Capacitor C1 reverses its charge and now we are back in the "on" condition for the motor.

How is overload protection accomplished? There are three diodes D1, D2, and D3 in the circuit. Examining the characteristic of a diode indicates that a voltage of, say, 0.7 V will forward-bias the diode, turning it on. (The type of diode will make a difference in the "on" voltage.) With three diodes in series, a voltage of 2.1 V will be required to turn on the diodes. The source of this voltage is the voltage drop across R5. If the size of R5 were 0.1 Ω and the motor current were 10 A, R5 would have a 1-V drop across it. The diodes are off because they require 2.1 V to turn on. If the motor current increased to 20 A, the diodes would still be off because they would have only 2 V across them and so they wouldn't turn on. At a motor current of 22 A, the voltage drop across R5 is 2.2 V. This is enough to turn on the diodes. Resistor R4 develops a voltage drop with the top of it becoming positive. This makes the gate positive on SCR2, and it turns on. Capacitor C1 then attempts to discharge through SCR1, from anode to cathode. This reverse-biases SCR1, turning it off and stopping the motor. The value of overload protection can be changed by adding more or fewer diodes or by changing the size of R5.

The current through SCR1 has to be large enough to match the motor current. The current through SCR2 can be very small because it is not required to pass very much current in turning off SCR1.

The previous solid-state motor starter requires a dc motor and dc power source. The starter provides only on-off control with overload protection.

There are many solid-state circuits that provide starting, stopping, reversing, and overload protection for ac and dc motors. Many solid-state starters also provide adjustable-speed control. Such a system is generally called a *solid-state adjustable-speed drive.*

12-22 SOLID-STATE AC MOTOR SPEED CONTROL (FHP PARAJUST)

Let us look at solid-state adjustable-speed drive circuits for ac motors first. We will start by looking at FHP's ParaJust® ac motor speed control. This ac motor speed control converts fractional-horsepower, three-phase ac motors into variable-speed drives. This controller is built by Parametrics under the FHP ParaJust registered trade name and is a fractional-horsepower version of the highly successful 1- to 10-hp ParaJust control. The FHP ParaJust control is simple to use, inexpensive, and versatile.

The FHP ParaJust ac motor speed control utilizes solid-state circuitry to convert plant power to variable-frequency power. Any three-phase motor connected to an FHP ParaJust control becomes a variable-speed motor when operating on the variable frequency from the FHP ParaJust control. There is no loss of motor torque as the motor speed changes.

Figure 12-52 illustrates an enclosure with pushbutton switches, pilot light, and speed control for the FHP ParaJust drive. Figure 12-53 illustrates the chassis with the solid-state components and wiring for the variable-speed ac motor controllers.

FHP ParaJust controls can be used with a three-phase ac motor or a gear motor of up to ¾ horsepower. First, wire the motor for 230 V, and connect it to the output of the ParaJust. The motor's nameplate must say 3 A or less. No motor starter is required. Next connect single-phase, 230-V input power to the FHP ParaJust. START and STOP/RESET pushbuttons and a speed-control potentiometer give you full control over your motor. The combination of FHP ParaJust's solid-state construction and a reliable three-phase ac induction motor eliminates brushes, commutators, belts, or clutches. There is nothing to wear out or require maintenance. Modifications and servicing are simple, too, because of the control's modular design.

Fig. 12-52 FHP ParaJust variable-frequency, variable-speed ac motor drive.

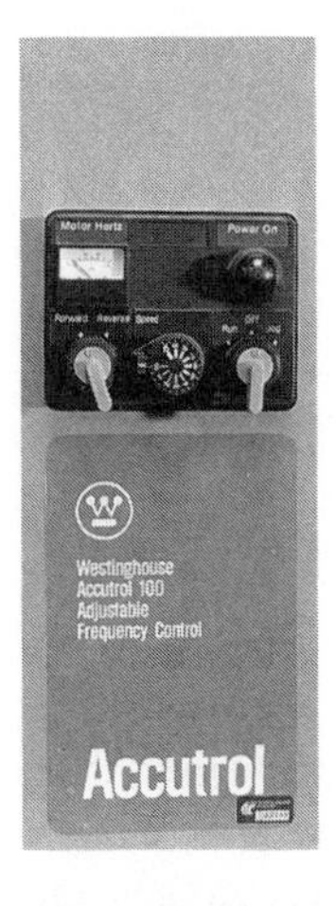

Fig. 12-53 FHP chassis for 1- to 10-hp motors. The solid-state components provide variable frequency, giving variable speed to the ac motor. *(Westinghouse)*

The solid-state speed control and a standard ac motor are priced competitively with both dc (SCR) drives, eddy-current drives, and enclosed mechanical variable-speed drives. Solid-state control saves hidden installation costs that accompany other drives, too.

Freedom from maintenance makes the solid-state control less expensive to operate than alternative systems. The efficiency of a solid-state controller and a three-phase motor means a saving in power costs.

A solid-state control (FHP ParaJust) is versatile. There's no need to match your motor to the solid-state control. Use with any type of 230-V, three-phase motor, including one that is open, explosion-proof, encapsulated, built-in domestic, or imported.

The FHP ParaJust control can be located any distance from a motor, and is available in standard duty (NEMA 1), oil-tight, wash-down (NEMA 4), or chassis versions. ParaJust controls can operate either manually or automatically from a wide variety of external signals. The operator's controls can be located either on the ParaJust or remotely from it.

Frequent starting is not a problem when solid-state variable-speed drives are used to give speed control to ac motors. This is because controllers accelerate three-phase motors instead of "starting" them, and the normal limitations on frequent motor starting do not apply. The motor can be "started" up to 20 times per minute without being overheated. The FHP ParaJust controller can require up to 10 s between a stop and a "start" for its internal circuitry to discharge. Discharge time will be less if the speed (frequency) at which it was running when stopped is less than maximum. If the FHP ParaJust controller is "decelerated" rather than "stopped," it may be immediately accelerated or jogged without a pause for discharging.

EXTENDED OR SHORTENED ACCELERATION/ DECELERATION TIMES

The FHP ParaJust is furnished with 1.5- to 15-s acceleration/deceleration adjustments. For longer or shorter adjustments, circuitry is changed on the plug-in control board. Any 10-to-1 range of adjustment can be furnished, such as 0.5 to 5 s or 10 to 100 s, up to maximum 25 to 250 s.

EXTENDED FREQUENCY

The standard FHP ParaJust control produces a 0- to 60-Hz output frequency whether the input frequency is 50 or 60 Hz. For other output frequency ranges, circuitry is changed on the plug-in control board.

ISOLATION TRANSFORMERS

These are required or suggested for solid-state drives used in certain environments. The FHP ParaJust control requires 230 V ac input power. They can be operated from 460- or 575-V plant power by inserting a transformer between the power line and the ParaJust input. Not only do they reduce the voltage to 230 V, but they also protect the control from failures due to grounding faults.

12-23 SOFT STARTS FOR POLYPHASE MOTORS WITH SOLID-STATE MOTOR CONTROL

Electrical South Inc. has designed a sophisticated electronic circuit to provide electronic soft starts for three-phase motors.

Softron® is a solid-state polyphase-motor control which, when connected in series with an existing motor starter, controls the power applied to standard squirrel-cage motors. It provides smooth acceleration to full speed without the large power surges associated with other types of starting. Obvious advantages of this method of starting are prolonged motor life, reduced power surges, and elimination of shock to mechanical coupling and gears. Operator adjustments are provided to set the desired acceleration time and initial starting torque for each individual requirement. Softron motor controllers offer a solution to controlled motor starting, eliminating the need for special dual-wound motors and electromechanical reduced-voltage starters. Electrical hookup is simple since the Softron is connected in series with the motor power leads (six wires total—three from the existing starter and three out to the motor). The Softron is available in 5- 400-hp versions up through 600 V. The NEMA 12 enclosure is standard in most sizes.

PRINCIPLES OF OPERATION

Softron operates automatically when the ac motor is energized. Utilizing a solid-state SCR assembly, Softron gradually increases the voltage applied to the ac motor by controlling the SCR conduction time. An electronic circuit in Softron directly controls the SCR conduction and includes operator adjustments for setting both initial starting torque (10 to 100 percent) and acceleration time (3 to 30 s). The typical adjustment procedure is to set the initial starting torque to the point where the motor shaft begins to turn and to set the acceleration time to smoothly bring the motor to full speed (normally 20 to 30 s). Certain applications may require more than 30 seconds.

Figure 12-54 provides connection diagrams of various types of Softron motors.

12-24 SUPER START SOLID-STATE CONTROLLER

Super Start® is a trade name for Electrical South, Inc.'s solid-state reduced-voltage motor starter. Solid-state starters such as the Super Start are making non-electronic reduced-voltage starters and dual-wound motors obsolete.

The block diagram in Fig. 12-55 on page 150 illustrates the circuitry required to provide control for solid-state reduced-voltage starting for ac three-phase motors. This block diagram indicates the special circuits used by Electrical South, Inc. in its Super Start controller.

The operation of Super Start can best be explained by referring to the block diagram. Super Start provides smooth, stepless acceleration by phase-controlling three sets of inverse-parallel SCRs connected in a three-phase-switch configuration. Three-phase voltage and current to the motor are controlled by varying the portion of the three-phase current sinewave conducted by each SCR. Three-phase current sensing is used for overload sensing and as the feedback signal necessary for stable control of current-ramp and current-limit starting.

The three-phase line voltages are sensed on the line side of Super Start. The signals are attenuated and used as inputs to the phase-unbalance, phase-loss, and phase-rotation protection circuits. The SCR control circuitry is synchronized to phase A and the phase-to-phase conduction balance is precisely determined with quartz crystal accuracy and rugged CMOS digital circuitry. Current is sensed on each input phase by a

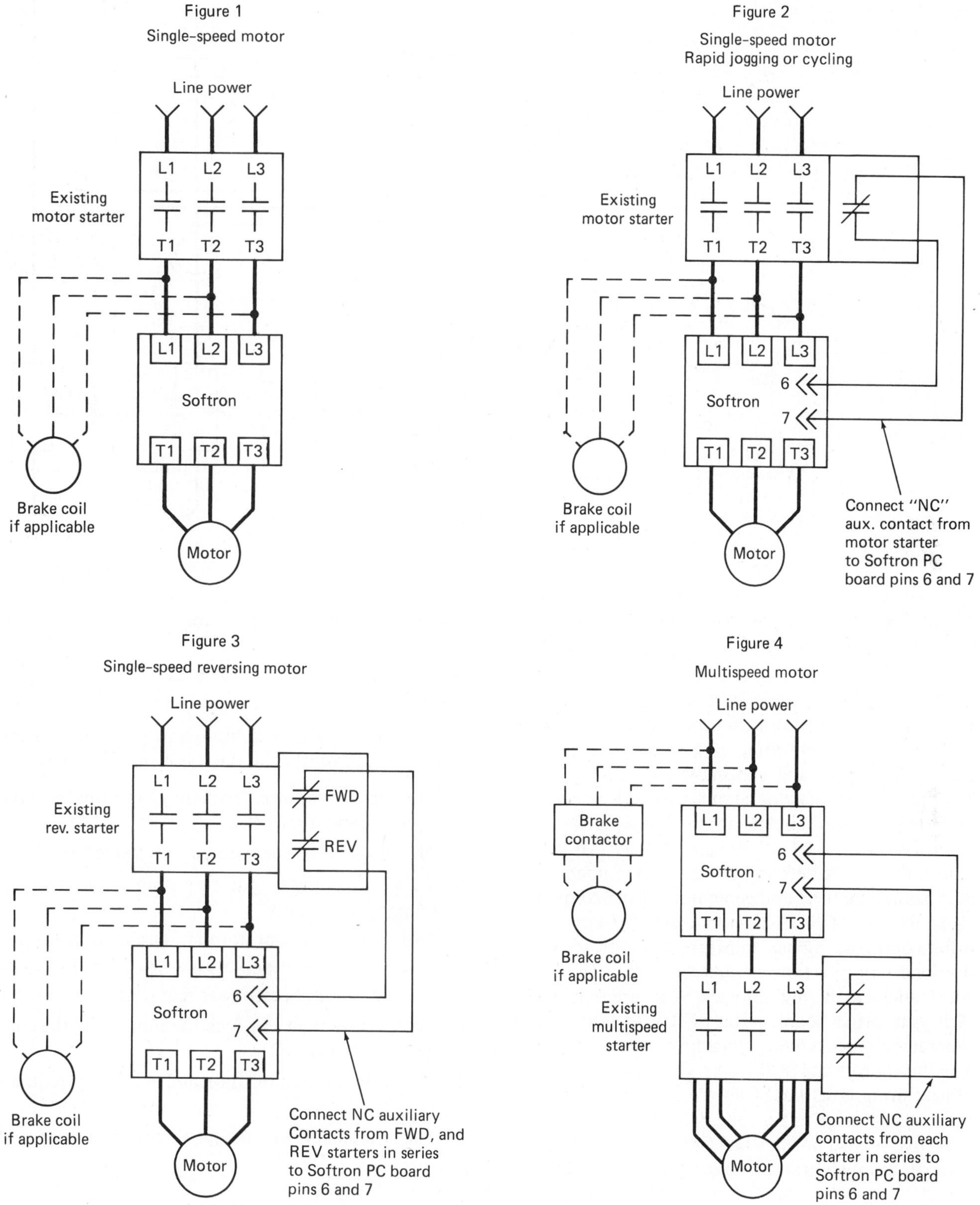

Fig. 12-54 Typical Softron connection diagrams. (*Courtesy of Electrical South Inc.*)

current transformer (CT). These signals are then converted to a true rms value. By converting the signals to true rms values, an accurate measurement of the current can be made. The true rms current signal is used as an input to the overload circuits and as a feedback signal for the starting circuit. The starting circuit controls current-ramp and current-limit starting by continuously comparing the true rms current signal to reference signals within the starting circuit and sending a phase-control command signal to the SCR firing con-

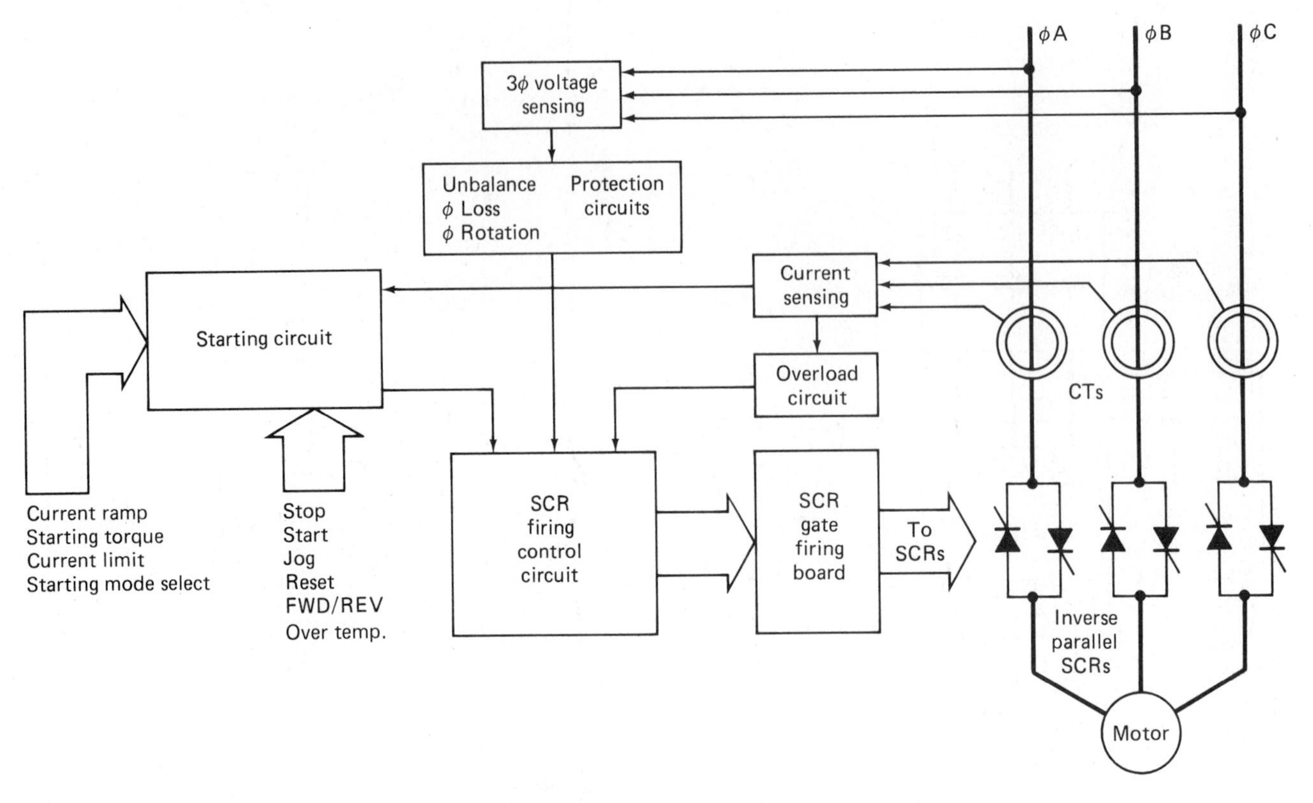

Fig. 12-55 Block diagram for a solid-state, reduced-voltage starter for a three-phase ac motor.

trol circuit. The starting circuit receives inputs from the control switches (stop, start, jog, reset, forward/reverse), the overtemperature sensor, and the starting potentiometers (current-ramp pot, starting-torque pot, current-limit pot, and starting-mode select switch). The output of the starting circuit, the overload circuits, and the protection circuits are the inputs to the SCR firing control circuit. Through digital CMOS circuitry, the circuit determines the exact moment each SCR should be fired and generates a firing command signal for each SCR. This circuit will also remove the firing command signals and turn off the SCRs if the overload or protection circuits tell it to. The firing command signals (one for each SCR) are sent to the SCR gate firing board. The SCR gate firing board generates a "picket fence"-type firing signal and sends it to the appropriate SCR gate. The picket fence-type of gate firing gives the SCRs a firing signal for more than enough time to turn the SCRs on with a highly inductive load, but does not force the SCR gates to dissipate as much heat as a "block firing"-type of signal would.

12-25 SOLID-STATE MOTOR CONTROL BY WESTINGHOUSE

The Guide to Solid-State Motor Control by Westinghouse Electric Corporation explains their solid-state ac and dc motor controls in the summaries that follow.

WHAT YOU MUST KNOW TO SELECT CONTROLLERS OR MOTORS

To select controllers and motors properly for any application, you must have the following information:

1. The motor's voltage, frequency, and phase of the power supply to be used
2. The motor's type or types of duty cycles
3. The driving or gear motor's output speed or speed-range requirements
4. The motor's starting torque, acceleration torque, and/or running torque for each step of the duty cycle
5. The maximum horsepower required of the motor
6. The environment in which the motor and the control are to operate
7. The motor's maintenance and reliability requirements
8. The type and NEMA design of the motor to be used

TYPES OF SOLID-STATE AC MOTOR CONTROLLERS

The chart of Fig. 12-56 shows how solid-state ac motor controllers vary in design, function, and available ratings. The descriptions of solid-state ac motor controllers are as follows:

1. Reduced-voltage starting controls (Westinghouse's Startrol and Startrol Power Miser) are used in conjunction with electromechanical starters or contac-

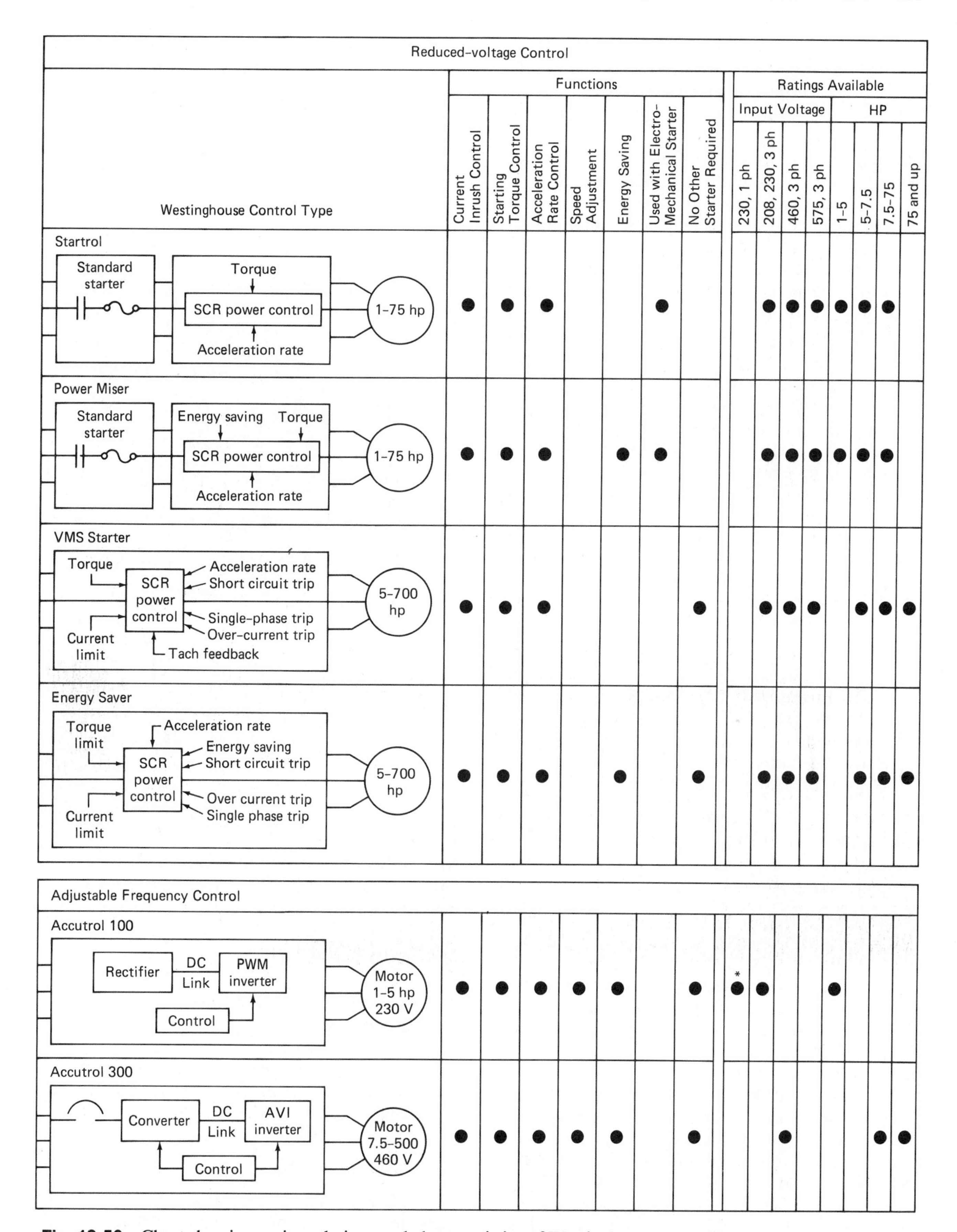

Westinghouse Control Type	Current Inrush Control	Starting Torque Control	Acceleration Rate Control	Speed Adjustment	Energy Saving	Used with Electro-Mechanical Starter	No Other Starter Required	230, 1 ph	208, 230, 3 ph	460, 3 ph	575, 3 ph	1-5	5-7.5	7.5-75	75 and up
Reduced-voltage Control															
Startrol	●	●	●			●			●	●	●	●	●	●	
Power Miser	●	●	●		●	●			●	●	●	●	●	●	
VMS Starter	●	●	●				●		●	●	●		●	●	●
Energy Saver	●	●	●		●		●		●	●	●		●	●	●
Adjustable Frequency Control															
Accutrol 100	●	●	●	●	●		●	●*	●			●			
Accutrol 300	●	●	●	●	●		●			●				●	●

(Functions: columns 2–8; Ratings Available — Input Voltage: columns 9–12; HP: columns 13–16)

Fig. 12-56 Chart showing various designs and characteristics of Westinghouse controllers.

tors or both. These controls provide soft starts or stops, or both, along with reduced current inrush. These controls also have starting torque and acceleration time settings that can be adjusted easily with a screwdriver.

2. Solid-state, reduced-voltage motor starters (Westinghouse's Vectrol VMS) do not have moving parts or current-interrupting contacts. The Vectrol VMS starter provides not only more motor-protection functions than a standard electromechanical starter, but also smooth starts, and adjustable acceleration and deceleration.
3. Energy-saving, solid-state motor controllers and starters (Westinghouse's Startrol Power Miser and Vectrol Energy Saver) are similar to reduced-voltage motor starters. They do, however, have additional energy-saving circuitry for more economical operation when load conditions vary widely from the full-load rating of the motor. The advanced design of these controls incorporates an automatic voltage-adjusting circuit. This circuit delivers true kilowatt savings on applications that require oversized motors for starting or on general-purpose machines that rarely use the full-rated capacity of the motor. The Vectrol Energy Saver starter and the Startrol Power Miser controller are advanced applications of the (NOLA/NASA energy-saving concept).
4. Reliable, solid-state, adjustable-frequency motor controllers (Westinghouse's Accutrol) provide for an easily controlled, stepless range of output-speed ac motors. The Accutrol inverter not only makes the motor into a variable-speed drive with up to a 2 to 1 speed range, it also controls starts, stops, reverses, accelerates, and decelerates.

Figures 12-57, 12-58, and 12-59 represent the physical appearance of solid-state motor controllers.

Fig. 12-57 Westinghouse Electric Corporation's Startrol for reduced-voltage starting.

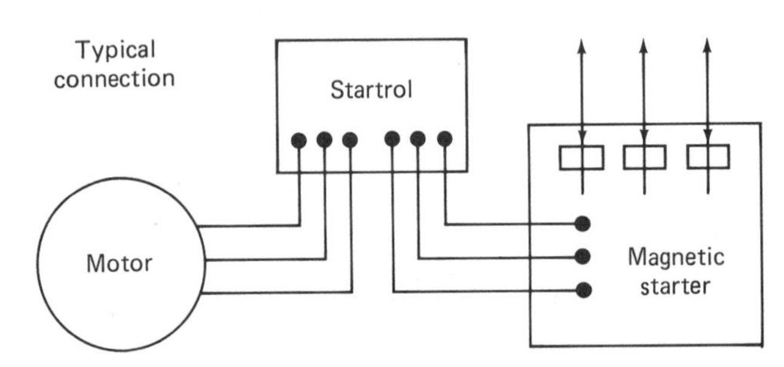

Fig. 12-58 The VPAC 300 three-phase series of power controllers.

GENERAL INFORMATION

The VPDC 500 Series "RED-PAC" unit is a high-performance compact ac-to-dc converter. Standard units, available in 120- to 480-V ratings, feature a full-wave bridge rectifier circuit consisting of two SCRs, two bridge diodes, and a freewheeling diode. The SCR trigger and logic circuitry can be rewired separately from the bridge circuit if control on the secondary side of a transformer is required.

This unique RED-PAC design provides for the incorporation of numerous options. This permits operation from standard milliampere dc signals, manual control, and dc voltage, or from current feedback signals for closed-loop control. The VPDC 500 Series provides a complete range of protection features, including high PIV-rated SCRs, hard-firing gate pulses, fast blow fuses and a voltage transient clipping network.

DESIGN FEATURES

RED-PAC units are designed with such protective features as high-PIV SCRs. A subcycle I^2t fuse is optional. A reliable and dependable voltage clipping network offers protection against transient and unstable voltages. Hard-firing gate pulses ensure accurate and dependable triggering of SCRs.

TYPICAL APPLICATIONS

The RED-PAC unit can be used as a battery charger with constant-voltage or constant-current control, or both. It can be used for dc motor control with armature or tachometer feedback. The unit can also be

Fig. 12-59 Westinghouse's SCR power controller.

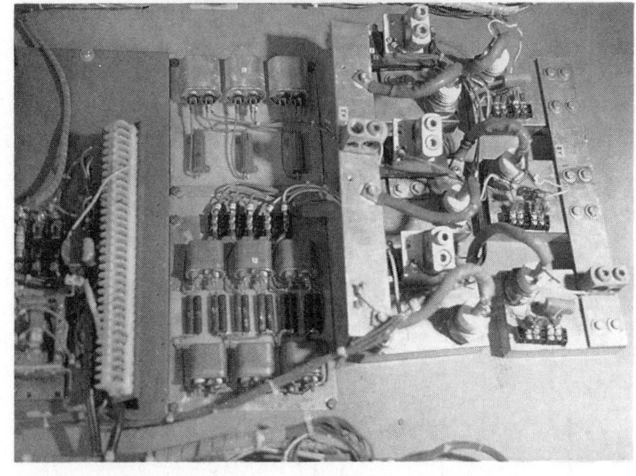

Fig. 12-60 General Electric's GP-100 three-phase 5- to 60-hp adjustable-speed drive.

used for temperature control with dc saturable reactor control and can be made to provide dc power with constant voltage and current and feedback control from the output voltage and a 50-mV shunt.

12-26 SOLID-STATE MOTOR CONTROL BY GENERAL ELECTRIC

HOW THE SCR DRIVE WORKS

Figure 12-60 is a picture of General Electric's GP-100 adjustable-speed drive. Alternating-to-direct current power conversion is effected by this controller.

The power conversion circuit of the SCR drive consists of a three-phase, full-wave rectifier bridge (Fig. 12-61). The negative legs of the rectifier bridge contain silicon-controlled rectifiers, while the positive legs contain diodes or uncontrolled rectifiers. The SCR conversion unit has a two-fold function: to rectify the ac voltage to a dc voltage and to control the dc voltage level.

A diode rectifier will conduct current in only one direction, blocking the voltage in the opposite direction. A rectifier bridge consisting of only diode rectifiers will convert an ac voltage into a dc voltage having an average value of 1.35 (for a three-phase bridge) times the rms value of the ac voltage. The rectifiers block current flow in the reverse direction such that it is impossible to regenerate power back into the ac line.

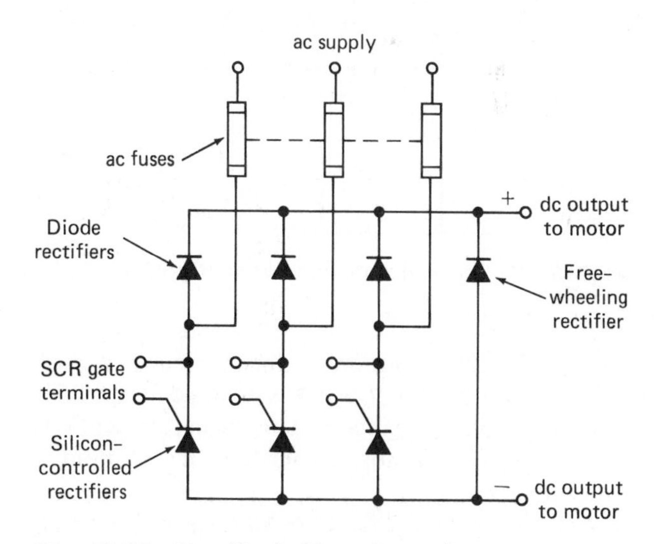

Fig. 12-61 Rectifier bridge schematic.

The function of controlling the dc voltage level is obtained by using silicon-controlled rectifiers in the negative legs of the rectifier bridge. An SCR is a solid-state, semiconductor which is basically a rectifier, but can also block voltage in the forward direction (the anode positive with respect to cathode) until fired by a gate signal. It then switches to a highly conductive state having a very low forward voltage drop. The SCR remains in the conductive state even after the gate signal is removed, until the forward voltage is removed. It then reverts to a blocking state in the forward direction. The sinusoidal ac voltage waveform satisfies these conditions in that each alternate half-cycle causes the anode to act positive with respect to the cathode to permit conduction, and the other half-cycle reverses this voltage to turn the SCR off. A positive firing signal is applied to the gate of the SCR at the proper point in time to turn the SCR on. Controlling the point in time (with respect to the ac supply voltage) when this gate signal is applied to the SCR controls the output voltage.

Figure 12-62 shows the output voltage and current waveform of a three-phase SCR conversion unit. It can be seen that as the firing pulses are advanced toward the zero time point, the average output voltage of the SCR conversion unit is increased.

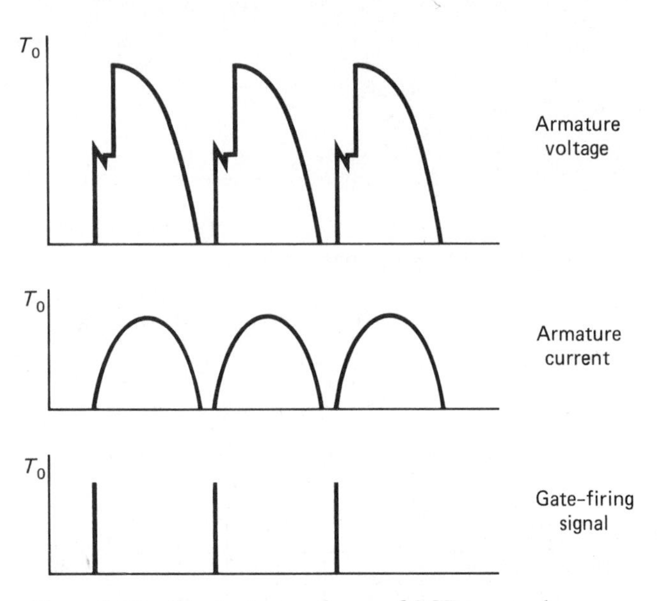

Fig. 12-62 Typical waveform of SCR control.

An SCR has a minimum forward current level, called a *holding current,* which is required for the SCR to stay in a conducting state. This means that the SCR conversion unit cannot be controlled if its output is open-circuited.

A diode rectifier, called a *commutating* or *freewheeling rectifier,* is connected across the dc terminals of the power-conversion bridge. This diode provides a freewheeling path for current produced by the induced voltages in the dc motor armature during phased-back operation. It prevents this freewheeling current from flowing through a controlled-rectifier leg. If current were allowed to flow, the SCR would fail to turn off when the supply voltage went negative and control would be lost. The freewheeling rectifier also acts to reduce motor-armature ripple current.

The conversion unit contains silicon-controlled rectifiers, together with transient-voltage protective circuits. The SCRs and diodes are mounted on aluminum cooling fins. These fins serve to transmit the heat produced inside the rectifiers to the surrounding air. The conversion unit is convection-cooled or force-cooled depending on the horsepower rating of the drive. Standard current-limiting fuses are placed in the ac supply lines to the rectifier bridge.

Thyrectors and other transient-voltage protective components are contained within the conversion unit. A thyrector is a special selenium rectifier which breaks down to limit high-voltage transients, in a way similar to, but more sharply than, a thyrite resistor. It "passes" the energy of the transient-voltage rating of the SCRs and diodes. Series resistor-capacitor networks are connected by shunting across each SCR and across the dc output. These provide a low-impedance path parallel to the rectifier to further limit transient voltage spikes. Noise-suppression capacitors are connected across the gate to the cathode of each SCR to prevent misfiring from extraneous signals.

The motor shunt field is connected between one ac line and the positive side of the full-wave rectifier bridge. The positive side of this bridge contains three diode rectifiers. Two diode rectifiers furnish two-thirds power to the motor shunt field, while the other diode rectifier acts as a commutating, or freewheeling, diode. This diode provides a freewheeling path for the current produced by the induced voltages in the motor shunt field.

12-27 REGULATOR

Figure 12-63 is a block diagram of the SCR drive regulator. There are two feedback signals used in this regulator. The primary signal is from the dc output voltage of the SCR conversion unit, and the secondary is an IR compensation signal. This IR compensation signal is obtained from the voltage drop across the armature resistor DRS, and is proportional to armature current. The magnitude of the compensation signal is established by the IR Comp potentiometer.

When these two feedback signals are combined, their difference is a measure of motor counter-emf, which is very closely related to motor speed. Thus, the combination of these two signals provides a feedback which is a good approximation of motor speed.

The timed-acceleration circuit consists essentially of a constant-current source and a capacitor. When this current source is used to charge the capacitor, its voltage will increase approximately linearly with time. The voltage appearing on this capacitor is the regulator reference.

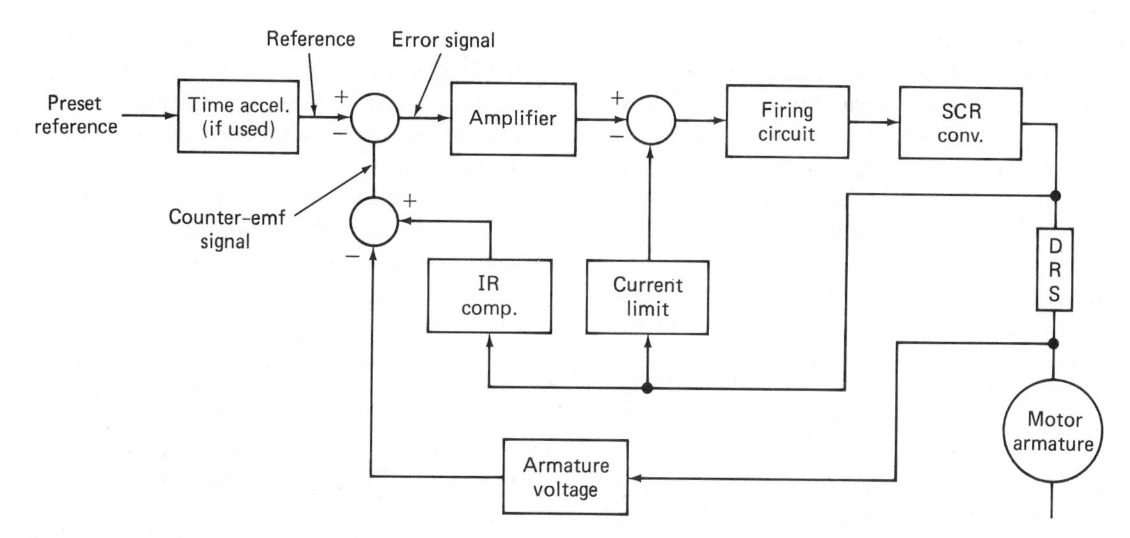

Fig. 12-63 Regulator block diagram.

The total accelerating time of the motor is thus the charging time for the capacitor and is determined by the setting of the Time Accel potentiometer, which controls the value of capacitor voltage (regulator reference).

Whenever the preset speed reference is at a higher value than the regulator reference, the timing circuit becomes operative. If the value of preset reference is reduced below the regulator reference voltage (by turning the Speed potentiometer down), the capacitor in the Time Accel circuit will discharge at a fixed exponential rate, reducing the regulator reference voltage.

Under these conditions, the drive will decelerate to the new preset speed at this rate, assuming that this rate is *slower* than the normal coasting time of the drive and its connected load.

When the STOP button is pressed, the CR1 relay is de-energized and the capacitor in the Time Accel circuit is discharged. This "resets" the circuit, ensuring that the next start will find the reference voltage at zero.

When this "net" feedback signal is compared to the regulator reference voltage, the resulting signal is called the *error signal.* The level of the error signal is very small and requires amplification (by the amplifier) before it can be used to control the firing circuit.

The output of the amplifier controls the value of current delivered by the constant-current source in the firing circuit. As the error signal increases, the magnitude of the capacitor constant-charging current increases. With a rapid charging rate, the capacitor is charged quickly and the firing pulse occurs early in the cycle. This results in turning the SCRs "on" at an earlier point in the ac cycle, increasing the average dc output voltage.

TIMED ACCELERATION

The timed-acceleration function is inserted between the preset reference voltage and the regulator reference, as shown in Fig. 12-63. This function is inoperative except during the acceleration or deceleration of the drive. When the START button is pressed, the F (or R) contactor applies the preset reference to the timed-acceleration circuit. This circuit allows the regulator reference voltage to increase smoothly at a rate established by the setting of the Time Accel potentiometer in a range of 2.5 to 10 s.

CURRENT LIMIT

The current-limit signal is obtained from the voltage across the DRS resistor. This voltage is proportional to the motor armature current. The magnitude of the current-limit feedback signal is adjusted by the Current Limit potentiometer.

12-28 SCR FIRING CIRCUIT

The firing circuits in the conversion unit provide pulses to the gate circuits of the three controlled rectifiers to "fire" them at the correct phase of the cycle determined by the signal from the amplifier.

Each firing circuit consists of a pulse transformer, a unijunction transistor, a capacitor, and a constant current supply. The firing circuit is controlled by a unijunction transistor. A unijunction is a special type of transistor that behaves similar to a switch; it is either fully "on" or fully "off." The unijunction transistor can be turned "on" by applying a voltage signal of sufficient magnitude to its input.

FIRING-CIRCUIT OPERATION

The amplifier determines the level of a constant current supply. The constant current linearly charges the capacitor, which is connected to the input of the unijunction transistor. When the voltage across the capacitor is high enough, the unijunction transistor turns "on." The capacitor discharges through the unijunction transistor into the primary of the pulse transformer. The pulse of power is passed on to the secondary winding of the transformer, which is coupled directly to the SCR gate circuit.

If the amplifier output is decreased, the magnitude of the constant current supply is reduced, the capacitor takes a longer time to charge and the unijunction transistor turns on later in the cycle. Hence, the SCRs fire closer to the end of the cycle and the average dc output voltage is decreased.

The firing circuits are synchronized with the ac supply such that the firing pulses to the three SCRs are spaced a third of a cycle apart. The SCRs will remain in the conducting state after the signal is removed, until the anode-to-cathode voltage is reversed.

MOTOR SHUNT-FIELD EXCITATION

Excitation for the motor shunt field is obtained from a part of the full-wave rectifier bridge used to control the motor armature circuit. When the armature current feedback signal increases to the limit setting, it acts to retard the charging rate of the capacitor in the firing circuit. The unijunction transistor fires later and the SCRs are phased back (turned on later in the cycle) until the armature current is reduced to the level determined by the setting of the current limit potentiometer.

STATIC INSTANTANEOUS OVERCURRENT CIRCUIT

A static instantaneous overcurrent (IOC) circuit senses armature current. When current exceeds approximately 300 percent, this circuit prevents the SCRs from firing and so reduces output voltage to zero in one-third of a cycle. This circuit is reset by pressing the STOP button on the operator's station.

MAGNETIC FUNCTIONS

Pressing the START pushbutton energizes the CR1 relay, which removes the current limit clamp and initiates the timed-acceleration circuit. The F (or R) contactor, also energized by the START pushbutton, connects the motor-armature circuit to the conversion unit and applies preset reference voltage.

Pressing the STOP pushbutton drops out relay CR1 and the F (or R) contactor. Relay CR1 applies the current limit signal and resets the timed-acceleration circuit. The F (or R) contactor disconnects the motor-armature circuit from the conversion unit and removes the preset reference voltage.

Relay CR2 provides loss of phase protection (L1) by removing the 115 V of ac control supply from the magnetic circuitry. Loss of phase protection for L2 and L3 is provided by the loss of excitation of the control power transformer, or CPT.

When dynamic braking is furnished, the F (or R) contactor is used to connect the dynamic-braking resistor to the motor armature.

On reversing drives, an antiplugging relay AP is connected across the motor armature. Relay AP locks out the start circuit as long as the armature voltage is above 20 V dc for 230-V ac drives and 40 V dc for 460-V ac drives.

The jog selector switch actuates the jog relay and connects the jog reference to the regulator. It also disconnects the seal circuits to the F (and R) contactors. Pressing the START pushbutton jogs the motor. Releasing the START pushbutton stops the motor.

TACHOMETER FEEDBACK MODIFICATION

The tachometer generator provides a voltage-feedback signal proportional to motor speed. This voltage-feedback signal is fed into a full-wave rectifier bridge, which makes the feedback signal insensitive to the polarity of the tachometer-generator output voltage. A resistor bridge, located immediately after the full-wave rectifier bridge, reduces the tachometer-generator output voltage to a suitable value. The output of the resistor bridge is then fed to the regulator.

The armature-voltage feedback is disconnected by removing a jumper thus permitting the tachometer generator to provide an accurate speed signal. The IR Comp circuit is not required and IR Comp is set to zero (CCW).

Figure 12-64 represents an elementary diagram of the General Electric adjustable-speed drive in Fig. 12-60.

SUMMARY

The dc motor offers many advantages. It can provide a wide range of adjustable speeds, it can provide directional control, and it can provide torque limit, dynamic braking, soft starts and stops, acceleration, and deceleration control. The one disadvantage of the dc motor is that it requires a dc power supply. However, most power-generating systems in the United States produce ac power. Until recent years, using dc motors in industry required large M-G sets or rectifier systems. Direct current motors presented problems because of the difficulty of obtaining dc power.

With the advent of solid-state devices, particularly the silicon-controlled rectifier (SCR), a dc motor can be operated from an ac power source with few problems. Solid-state drives now control much of the variable-speed and specialized machines used in the manufacturing industries.

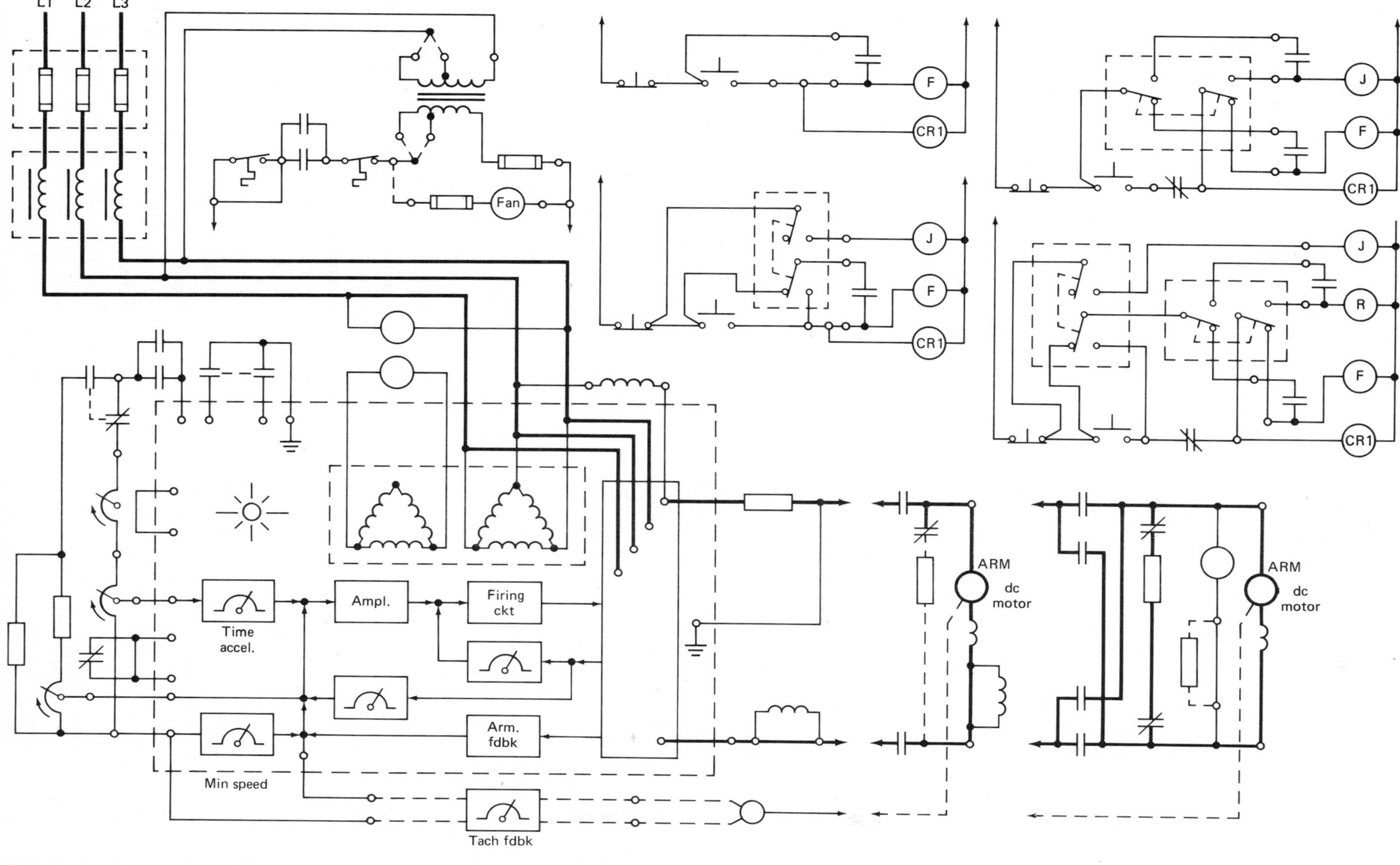

Fig. 12-64 Elementary diagram of G.E. Adjustable speed drive (Fig. 12-60).

Alternating current motors are normally considered to be either constant-speed or limited multiple-speed machines. Other than the physical characteristics of the motor, speed is determined by the frequency of the power source. With solid-state motor control, ac motors can be operated at variable frequencies. This allows a wider range of speed control for ac motors.

The diode and rectifier have been around for many years. These devices have many uses. Their main use is for changing ac to dc, steering and blocking circuits, and similar applications. Zener diodes provide reference-voltage and regulated-power supplies. A thyrector, which is two zener diodes connected back-to-back, is used for transient-voltage suppressors and arc suppressors. These devices are used to protect a large number of electrical components in solid-state motor control.

There are many solid-state devices that are trigger devices used to switch or turn on the power-switching thyristors in motor control. The UJT (unijunction transistor), the PUT (programmable unijunction transistor), the diac, the SUS (Silicon Unilateral Switch), the SBS (silicon bilateral switch), and the ASBS (asymmetrical silicon bilateral switch) are some of the names for trigger devices used in solid-state motor control.

The power switch devices used in solid-state motor control systems are the SCR (silicon controlled rectifier), the triac (bidirectional triode thyristor), the SCS (silicon controlled switch), and light-activated versions of these thyristors.

In just a few short years, solid-state devices have revolutionized industrial motor control. They have also changed the knowledge and skills required by the industrial electricians and maintenance technicians who install and maintain solid-state drive systems.

REVIEW QUESTIONS

1. Why are dc motors preferred for adjustable-speed drives in industrial control?
2. What are some of the characteristics offered by dc motors?
3. How was a dc power source obtained in "the past" when only an ac source of power was available?
4. What is a modern method for obtaining direct current when only an ac power source is available?
5. How is speed and directional control of dc motors obtained using solid-state devices?
6. What is the advantage of solid-state drives in the mining industry?
7. Is solid-state motor control limited to small motors? How large are some of the motors found on industrial solid-state drives?
8. What is a hole as it pertains to semiconductors?
9. What is accomplished by the doping process?
10. What types of materials are required to make semiconductor devices?
11. How is a diode constructed using semiconductor materials?
12. Define forward bias.
13. Define reverse bias.
14. How can a diode be made to act like an ON and OFF switch?
15. What direction does current flow in semiconductor devices?
16. If a diode is operated at its avalanche or breakdown voltage, what will happen to the diode?
17. Describe the zener diode.
18. What are some applications for zener diodes?
19. What is the purpose of the load resistor in a zener-diode voltage regulator?
20. What is a thyrector?
21. What are the industrial applications for a thyrector?
22. How are the leads of a thyrector labeled?
23. What is a back diode?
24. What are some of the applications of a back diode?
25. How is an NPN transistor biased to act like an open switch? A closed switch?
26. What can be determined from the characteristic curves of a transistor?
27. How is the size of a load resistor for a transistor common emitter amplifier determined from the transistor's characteristic curves?
28. What is the phase relation of the input and output signal of a common emitter transistor amplifier?
29. What is the purpose of R_E in Fig. 12-21?
30. Define degeneration? What causes it?
31. How can the arrow on the emitter be used to determine if a transistor is an NPN or a PNP?
32. What is a unijunction transistor used for?
33. What is an SCR?
34. How does an SCR differ from a diode?
35. What are the industrial applications of SCRs?
36. What is an inverter circuit used for?
37. What is meant by phase control?
38. What is a commutation circuit used for when applied to SCRs?
39. How is an SCR turned on? How is it turned off?
40. How is an LASCR turned on and off?
41. What are LASCRs used for?

42. What is an SCS?

43. How is an SCS turned on and off?

44. Why can an SCS be used for tasks that an SCR cannot be used for?

45. What is a GTO?

46. What are some of the applications for GTOs?

47. What are some of the applications for a triac?

48. What polarity gate signal is required for a triac?

49. What is a diac used for?

50. What are some of the applications of a photodiode?

51. How is an LED turned on and off?

52. What are some of the applications of LEDs?

53. What is the advantage of a Darlington amplifier?

54. What is a PUT?

55. How is a PUT turned on?

56. What is a LAPUT?

57. How is a LAPUT turned on?

58. What are SUS, SBS, and ASBS used for?

59. What part of the circuit in Fig. 12-51 provides the time delay?

60. What part of the circuit in Fig. 12-51 provides overload sensing?

61. Define a solid-state adjustable-speed drive.

62. How does an FHP ParaJust ac motor control provide speed control?

63. What is the advantage of solid-state motor control?

64. Why is frequent starting not a problem with FHP ParaJust controllers?

65. Why are isolation transformers required with FHP ParaJust controllers?

66. What is the advantage of providing soft starts to polyphase motors?

67. What is the function of a Softron controller?

68. What size motor can a Softron controller be used with?

69. What are Super Start controllers making obsolete?

70. How does a Super Start controller provide smooth, stepless acceleration to a three-phase ac motor?

71. What does a Super Start controller use for a true rms current signal?

72. Where is a "picket fence" firing signal used?

73. What is meant by the forward direction of an SCR?

74. What is holding current in an SCR?

75. How are SCR conversion units cooled?

76. What are thyrectors used for in an SCR conversion unit?

77. What is the purpose of noise-suppression capacitors in the SCR conversion unit?

78. What do the "firing" circuits in the SCR conversion units do?

79. What components are found in the firing circuit?

80. What turns the unijunction circuit on in the firing circuit of the SCR conversion circuit?

81. How are the firing circuits synchronized with the ac power supply?

82. How is excitation for the shunt field of the motor obtained in an adjustable-speed drive?

83. In the shunt-field circuit of the adjustable-speed drive, where does the current that flows in the freewheeling diode originate from?

84. What does an antiplugging circuit do on a reversing drive?

13
Basic Concepts of Static-and-Logic Control

This chapter is concerned with digital control systems that are used in the decision-making circuits of a motor control circuit. Emphasis is on the static-and-logic components that perform decision-making functions.

Just as the magnetic starter liberated the machine from the line shaft, static-and-logic control is liberating the machine and operator from servitude to the slow, failure-prone, magnetic relay and contactor. The advent of static-and-logic control opens a new field of possibility for fully automated machines and processes.

13-1 BASIC CONCEPTS OF STATIC-AND-LOGIC CONTROL

The previous chapters of this book have been concerned with magnetic control, that is to say, control by means of moving contacts and magnetic cores. This chapter will present a new kind of control. We may define the term *static* as "pertaining to or characterized by a fixed or stationary condition." This definition gives us the key to the meaning of *static control,* that is, "control by means of devices without moving parts." The term *logic control* refers to "the use of circuits to express the state, or condition, of control devices."

The ever-present problem of magnetic control has always been the failure of components. Magnetic switching devices, such as relays and contactors, have coils which require relatively large currents to operate the mechanical linkages attached to the contacts. These coils tend to burn out, and the linkage parts are constantly subject to wear. The contacts themselves are often victims of dirt, grease, and other foreign matter which cause arcing and burning or pitting of their surfaces.

For many installations with relatively simple control functions, magnetic controls may still be a more practical and economical approach to control. However, when the demands made on the circuit amount to a significant number of control functions, when rapid switching becomes a significant requirement, and when long life is essential, static switching becomes not only economically feasible but almost mandatory. One other factor which must always be considered when selecting a control system for a machine or process is the space available for control components. Magnetic control for a complex system puts serious demands on the available space. In contrast, a fraction of that space is required by static switching devices. The environment in which the machine operates may also provide sufficient reason for using static control.

Solid-state static-and-logic switches operate at low dc voltages, usually 5 to 20 V, and very low current. They have no moving parts which would be subject to wear or require adjustment. They also have no contacts to arc or burn or to collect dirt and other foreign matter; therefore no cleaning is required.

Static-and-logic control offers several advantages over magnetic control. The first and very important advantage is the increased reliability of the circuit. A static system has a much greater ability to produce a signal output when and only when an output is called for. The long life of static switches, which is completely independent of the number of operations performed, makes them almost indispensable for automated control systems. Static-and-logic switching provides a much higher speed of operation, which is often required by modern machines and processes. Many control functions must be performed in environments where magnetic control devices would be destroyed or at least limited to a short life, for example, by environmental factors such as corrosive or explosive atmospheres. The condition of the operating environment is generally not a major consideration when static switching devices are used. Static switching also requires a much simpler circuit design than magnetic switching. A static switch is a multiple-input and single-output device, as contrasted with a relay or contactor, which is inherently a single-input, multiple-output device. The single output of the static switch may be used to provide inputs to many other static-and-logic control switches; this property is referred to as *fan-out.*

Semiconductors, solid-state devices, and integrated circuits (ICs) are used in industrial control circuits to perform both digital and analog functions. This chapter is concerned with digital control or, in more familiar terms, switching-type control.

13-2 ESSENTIALS OF STATIC-AND-LOGIC CONTROL

The language of logic control consists of only a few words—five to be exact. These five words are: AND, OR, NOT, MEMORY, and DELAY. There are also a few derivatives and combinations of the basic words such as NOR, which is really a combination of OR and NOT and is sometimes called an OR-NOT.

If you feel slightly confused at this point, you have fallen victim to the limitations of a five-word language. The very simplicity of the language makes it sound like double-talk if you are not careful. Students who do not let themselves get confused by the simplicity will have no real trouble with this new but astonishingly useful digital *logic.*

Consider the possibilities of a control system in which even the most fantastically complicated specifications can be met by the use of a handful of basic building blocks in the proper combinations. This is static control. Each of the words in static control represents a basic building block called a *logic function* or *logic element.* Each logic function has a symbol which is used in what is known as a *logic diagram.*

The first word in static control language is AND. To understand the meaning it is necessary to remember that all logic elements have multiple inputs and only one standard output. This is just the opposite of a relay, which can have only one input, the coil, and may have multiple outputs, the contacts.

Consider the basic logic symbol of Fig. 13-1 on the next page. This is the common form of the AND symbol. The requirement of an AND element is that all inputs must be present in order to have an output. Table 13-1 shows the NEMA forms of symbols used by several different companies in their logic control diagrams. The NOT symbol is shown in row 1 in the table; the AND symbol in row 2; the OR symbol in row 3; the time DELAY symbol in row 4; and the MEMORY symbol in row 5.

In the AND circuit, inputs *A*, *B*, and *C* must be present in order to have an output. The output in Fig.

Logic Function	ANSI Standard	NEMA Standard	Relay (Ladder) Diagram
NOT	A — $\bar{A}$	A — $\bar{A}$	A, CR; CR, $\bar{A}$
AND	A, B, C — Y	A, B, C — Y	A, B, C, CR; CR, Y
OR	A, B — Y	A, B — Y	A, B, CR; CR, Y
Delay	A — Y	A — Y	A, TD; TD, Y
Memory	A (S), B (R) — Y	A, B — Y	B, A, CR; CR; CR, Y

Table 13-1 NEMA Forms of Symbols

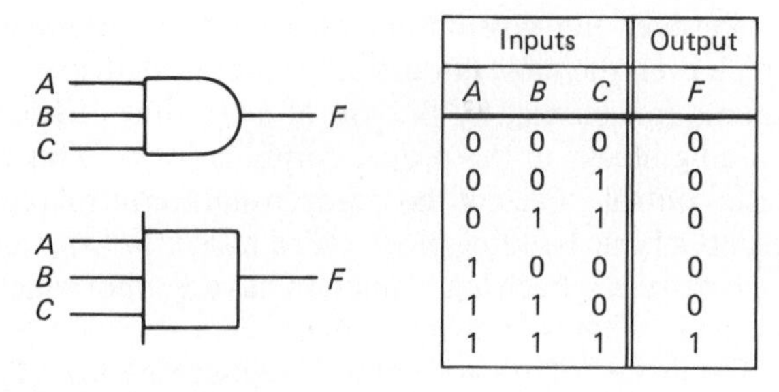

Inputs			Output
A	*B*	*C*	*F*
0	0	0	0
0	0	1	0
0	1	1	0
1	0	0	0
1	1	0	0
1	1	1	1

Fig. 13-1 The AND symbol.

13-1 is at F. It should be remembered that the loss of any one input will turn off the output. This is indicated by the table in Fig. 13-1, which is called a *truth table.* A truth table is a means of visualizing all the combinations of inputs and outputs to a logic circuit. Across the top of the table are the designations for each input and the single output. In Fig. 13-1, the inputs are *A*, *B*, and *C*, and the output is *F*. The 1s and 0s in the body of the table represent the state of the input and the output. A 1 denotes ON or high or closed; a 0 denotes OFF or low or open. The truth table will be discussed further in the following sections.

At this point do not be concerned about what an input or an output is or how the logic element performs its functions. Just learn to read the language in symbolic form.

The second word in the language of logic is OR. The common forms of the OR logic symbol are shown in Fig. 13-2. The requirement of an OR logic element is that it will have an output when any one or more of its inputs are present. There will be an output at *F* if input *A* or *B* or *C* or any combination of these inputs is present.

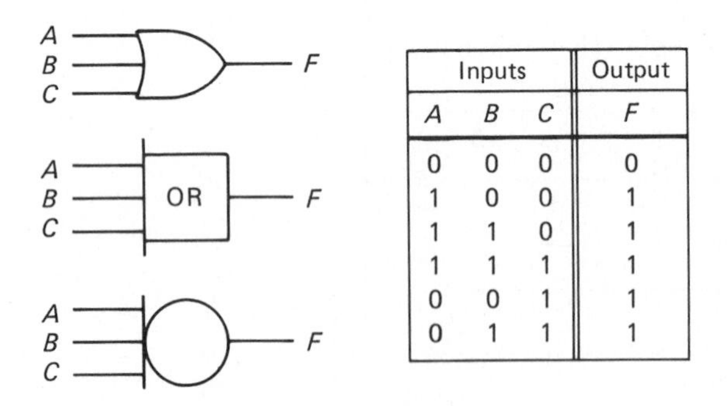

Inputs			Output
A	*B*	*C*	*F*
0	0	0	0
1	0	0	1
1	1	0	1
1	1	1	1
0	0	1	1
0	1	1	1

Fig. 13-2 Forms of the OR symbol.

The third word in static language is NOT. The symbols for NOT are shown in row 3 of Fig. 13-3. The requirement of a NOT is that it will produce an output when an input is not present, and no output if an input is present.

The one input inverter has the logic function of NOT. The NOR is merely a multiple-input NOT and will

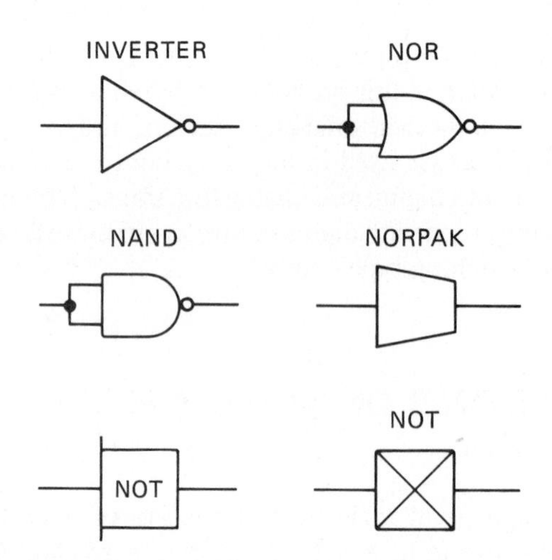

Fig. 13-3 The NOT symbol and circuits that give the NOT logic function.

have an output only when all its inputs are not present. When any one or more inputs to the NOR element are present, the output is turned off.

A third form of a negated-input logic element similar to the basic NOT or NOR is the NAND element. Figure 13-3 and 13-4 shows the NAND symbol. The requirement of a NAND is that it will have an output if no inputs are present. When all inputs to the NAND element are present, it will have no output.

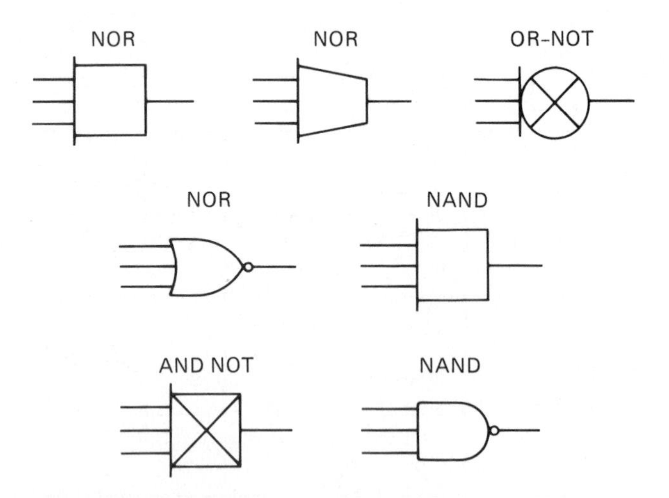

Fig. 13-4 Multi-input symbols that give inverted outputs.

The fourth word in the static control language is MEMORY. The symbols for the basic form of MEMORY elements are shown in Table 13-1 and Fig. 13-5. A momentary input to the ON side of the memory turns the memory on. It will remain on until there is a momentary input to the OFF side. The memory is then in the "off" condition. The MEMORY element re-

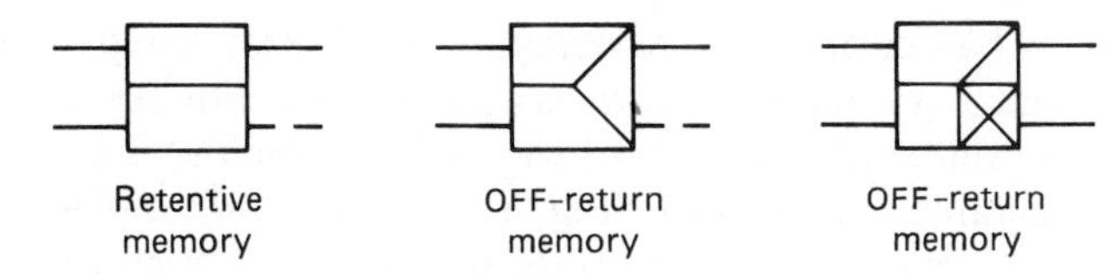

Fig. 13-5 Memory symbols.

members the condition of its output as long as the power remains on. The *retentive* MEMORY element "remembers" the state of its output even after the power is turned off. Its behavior resembles that of a manual switch or latching relay which mechanically remembers its last position.

The OFF-return MEMORY remembers the state of its output until the power is turned off and then always goes to the OFF condition. This is similar to the action of the magnetic starter and a three-wire control circuit.

The fifth word in static control language is DELAY. Symbols for DELAY elements are shown in Table 13-1 and Fig. 13-6. The function of the DELAY element is to provide an output after a specific delay following the application of an input. The above function would be known as *time delay on energizing.* DELAY elements can also be built to provide delay upon the de-energizing of the circuit. The symbols in Fig. 13-6 illustrate the four common forms of DELAY logic elements.

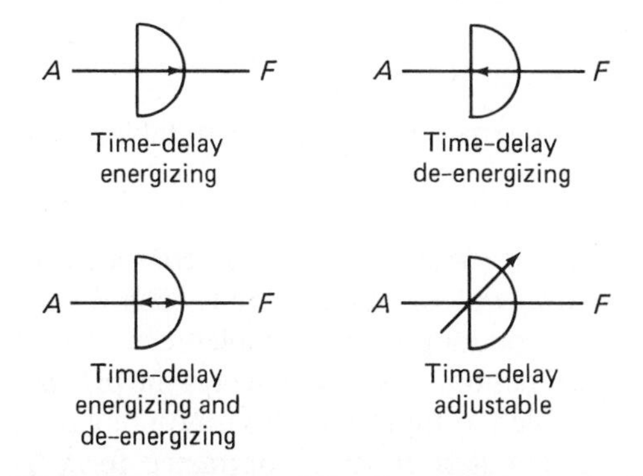

Fig. 13-6 Four forms of the time-delay control.

At this point it would be well to test your understanding of this new language. Consider a wall with three pushbuttons mounted below a lamp. You walk up and press one button at a time; nothing happens. You then press the buttons in pairs; nothing happens. You then press all three buttons at once. The lamp lights. What type of logic element are the pushbuttons and lamp connected to? (The answer is a logic AND.)

What type of logic element would necessarily have been used if the lamp came on when you pressed any one or any combination of buttons. (It would have been an OR element.)

You supplied the input by pressing the buttons, and the lamp indicated an output by turning on. The other logic elements can be as easily understood by applying the same reasoning. If the lamp had been on and was turned off by pressing all three buttons at once, the logic element would have been a NAND.

All logic-function elements operate at very low power levels; therefore, they must be followed by an amplifier in order to bring the level of power high enough to operate the device which is to be controlled. Figure 13-7 shows the symbol for an amplifier. It is used regardless of the physical or electrical makeup of that amplifier.

Fig. 13-7 Two forms of the amplifier symbol.

13-3 DEVELOPMENT OF LOGIC CIRCUITS

There are three methods used by logic circuit designers to develop logic circuits. One method is the direct conversion method. With this method, an electric relay ladder circuit would be converted to a logic diagram by taking each rung of the relay ladder circuit and developing it into an equivalent logic circuit. After each line of the ladder diagram is converted into its equivalent logic circuit, the circuits are combined to complete the conversion. The symbols, style of drawing, and numbering and wiring information (for the logic components you use in your diagram) can be determined by referring to the manufacturer's data sheets.

The second method of developing logic diagrams is called the *English method;* that is, you talk to yourself concerning the requirements of a circuit. As you orally describe the needs of a circuit, a logic diagram can be developed from your description.

The third method of making a logic diagram is to convert the electric circuit from a ladder diagram to boolean algebra expressions. Each line of the circuit will have an equivalent expression. The logic circuit is developed one boolean algebra expression at a time. Combining these expressions will result in a complete logic diagram. The symbols used for any logic system can come from any of the many systems available from logic-control manufacturing companies.

Boolean algebra is used a lot in programming controllers. Such controllers are called *programmable controllers* or *P.C.s.* Inasmuch as boolean algebra is being used widely in digital and logic control, we will start our discussion with a brief description of its theory and application in developing logic circuits. In a later chapter we will see how it is used to program P.C.s from relay ladder diagrams.

13-4 SWITCHING LOGIC AND BOOLEAN ALGEBRA

In light of the existence of digital control systems, P.C.s, microprocessor-based control system, computers, and automatic data-processing equipment, manually operated switching networks are a rather impractical way of applying logic functions to control. Solid-state devices are much more capable of meeting the requirements of size, weight, power consumption, speed of operation, etc., of industrial controllers. Solid-state devices may include circuit boards with discrete components such as capacitors or inductors, or integrated packages, such as integrated circuit (IC) chips, in which the components are formed as part of the interconnecting circuitry. In either case, the term *components* includes diodes, transistors, resistors, capacitors, and the like.

A thorough knowledge of the operation of these components is essential to an understanding of the design and operation of logic circuits. However, for practical applications, if the maintenance electrician understands the logic of a package of electronic components and how to check for inputs and outputs, he or she can successfully keep the logic control systems in operation without an electronic background. Replacing logic elements instead of repairing plug-in modules or printed circuit boards is the accepted procedure for maintaining logic control system. Most industrial logic systems are designed for ease in troubleshooting and replacing defective units.

The basic logic functions AND, OR, NOT, NAND, and NOR, and exclusive OR may be combined in various ways to perform any logical operation required for electronic digital control circuits. These logic functions are considered to be the basic elements, or building blocks, for digital logic control systems. Most industrial logic control systems can be constructed from only three basic logic functions: the AND, OR, and NOT functions. An analysis of other logic functions will show that they are formed by combining one or more of these basic functions.

To construct truth tables for logic functions, such as those utilized in digital circuits, it must be assumed that these functions exist in one of two possible states; that is, a function may be either "on" or "off," "true" or "false." The states of a function may be represented numerically by assigning a value of 1 to the function when it is in its normal "on" or "true" state and the value of 0 when it is in the complementary form, that is, "not," or "false," state. All possible states of a function having one or more terms may thus be tabulated in what is commonly known as a *truth table.* For example, the truth table and logic symbols for two logic diagrams are shown in Fig. 13-1. The "on" inputs (represented by a 1) that are required to give an on output (represented by a 1) can be determined from these tables.

A logic symbol is a graphic method of representing a given logic function on a logic diagram. It can take a distinctive shape, as will be illustrated in the figures that follow, or it can be in the form of a block with the logic function identified in some manner within the block. Note the graphic symbols for logic diagrams used in Fig. 13-8. This represents the industrial and electronic symbols used in modern control systems. A truth table for the logic circuits is also illustrated in the figure.

The NAND function illustrated in Fig. 13-8 functions as follows. If either (or both) A and B are open (logic 0), then F will be equal to logic 1. However, should A and B both be closed, then F will be equal to logic 0. These circuit conditions are illustrated by the truth table in the figure.

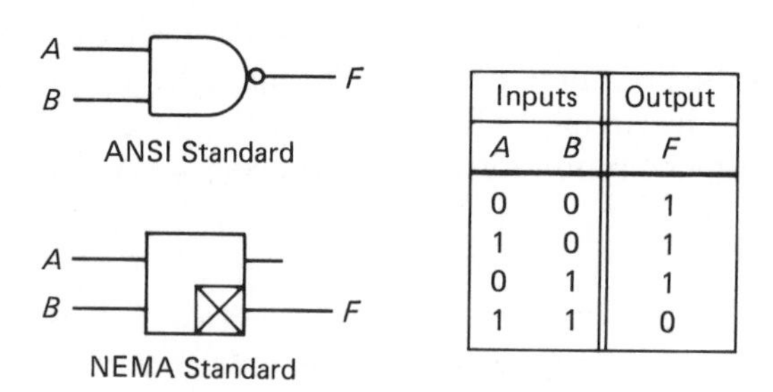

Inputs		Output
A	B	F
0	0	1
1	0	1
0	1	1
1	1	0

Fig. 13-8 NAND symbols and truth table.

The NOR function, or operation, combines the NOT and OR functions or operations, is such a way as to produce, at the output, the complement, or negation, of the OR function. Then NOT-OR is the proper designation for such as function. Common practice, however, uses the shortened or contracted form to read NOR.

The circuit for the NOR function is illustrated in Fig. 13-9. In this circuit if the input switches A and B are

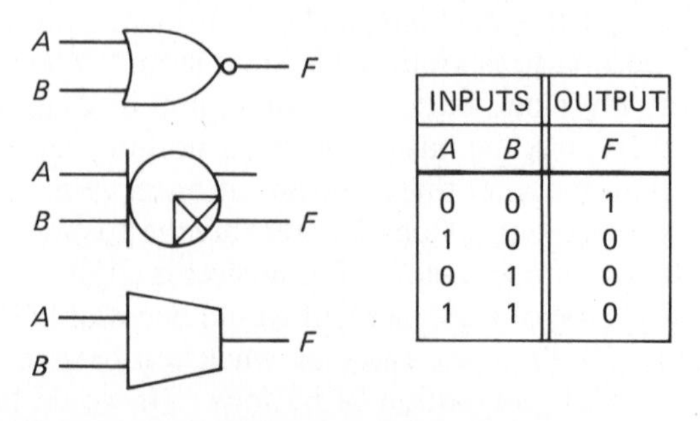

INPUTS		OUTPUT
A	B	F
0	0	1
1	0	0
0	1	0
1	1	0

Fig. 13-9 NOR symbols and truth table.

both open, then F will be equal to logic 1. However, should either or both A and B be closed, then F will be equal to logic 0 and will remain at 0. These circuit conditions are shown by the truth table in the figure.

The exclusive OR function, or operation, may be represented by the diagram in Fig. 13-10. It may be expressed mathematically by the logic statement, $F = A\bar{B} + \bar{A}B$, where F represents a function of A and B which will be true if either A or B, but not both, are true. The bar over the letter indicates the complement of that letter. Thus A is the complement of $\bar{A}$. If $A = 0$ then $\bar{A} = 1$ and if $A = 1$, $\bar{A} = 0$. The OR function, then, will accept only one input at a time; hence the term exclusive OR.

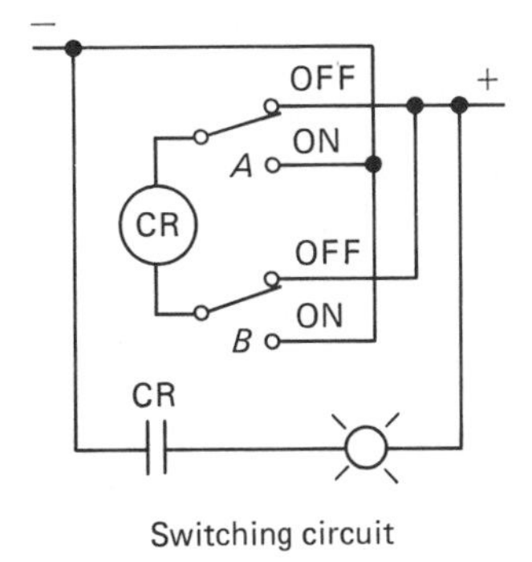

INPUTS		OUTPUT
A	B	$F = A\bar{B} + \bar{A}B$
0	0	0
1	0	1
0	1	1
1	1	0

Truth table

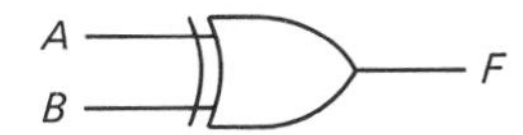

Fig. 13-10 The exclusive-OR function.

The exclusive OR function is represented by the logic symbol in Fig. 13-10. If in this circuit either switch A or switch B is closed, then F will also be equal to 1 and there will be an "on" output. If, however, switches A and B are both open or both closed, then F will be equal to 0. These conditions are illustrated by the truth table in the figure.

Truth tables for more complex logic circuits are illustrated in Fig. 13-11 on the next page. See if you can trace logic 1s and 0s through the circuits and verify each row of the truth tables. The circles at the inputs to the logic symbols are the same as inverters. For example, if the output of the previous logic unit is 0, the input at the circle would be 1.

13-5 LOGIC POLARITY

Digital circuits utilize a form of two-value logic which is easily adaptable to the binary numbering system. Most circuits or elements within such a digital control system will have two operating states, and these states may be assigned a logic value of 1 or 0. Actually, any two distinctive voltages or currents may be used to represent the logic values of 1 and 0 in logic control circuits. For example, a negative voltage could indicate 0 and a positive voltage could indicate 1, or vice versa. Likewise, current flow *into* the circuit could indicate 0, and current flow *out of* the circuit could indicate 1. Similar results could be obtained with amplitude control, using small and large voltages or currents. Pulses could also be used in a like fashion, with a negative pulse indicating 0 and a positive pulse indicating 1; or 0 could be indicated by the absence of a pulse, in which case the presence of a pulse could signify 1. Many combinations of logic expressions are possible; they can also be used interchangeably, since each logic element (circuit) or operating entity can actually function independently as long as the desired result is achieved. Mixed logic systems will normally not be utilized in large digital control systems, since this would lead to more complicated circuitry and added costs. Most present-day logic systems utilize polarity to define the circuit state, since positive and negative voltages are easily obtained and manipulated, regardless of whether the actual logic element (circuit) employs relays, switches, diodes, transistors, or ICs. According to polarity logic circuits can be divided into three general classes: positive, negative, and mixed logic. As employed on logic diagrams, a signal may assume either the "active" (or "true") state (logic 1) or the "inactive" (or "false") state (logic 0). The electrical signal levels used and a statement concerning whether positive or negative logic applies are usually specified explicitly on the individual logic diagrams by the manufacturer or the logic designer.

In practice, many variations of logic polarity are employed; for example, from a high-positive to a low-positive voltage, or from positive to ground; from a high-negative to a low-negative voltage, or from negative to ground; and mixed polarity, from a high-positive to a negative (below ground) potential, and vice versa. A brief discussion of the two general classes of polarity is presented below.

Positive logic polarity is defined as follows. When the logic 1 state is relatively more positive than the logic 0 state and the circuit is activated (operated) by the logic 1 signal, the logic polarity is considered to be *positive*. The following typical examples illustrate the manner in which positive logic may be employed:

Example 1: Logic 1 = +10 V
Logic 0 = 0 V

Example 2: Logic 1 = 0 V
Logic 0 = −10 V

In both examples, the logic 1 state is always more positive than the logic 0 state, even though in Example

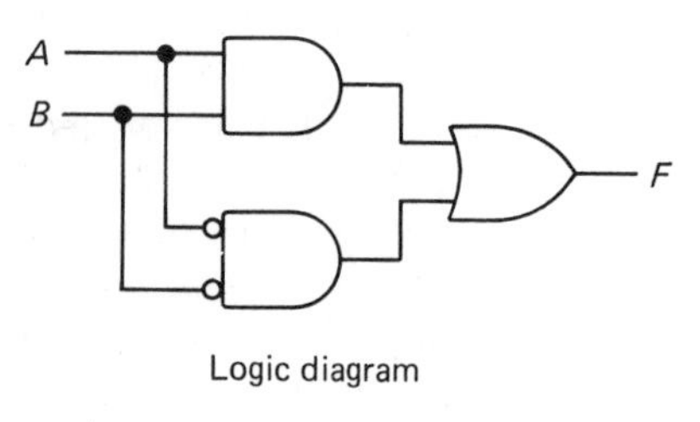

Inputs		Output
A	B	F
0	0	1
0	1	0
1	0	0
1	1	1

Truth table

(a)

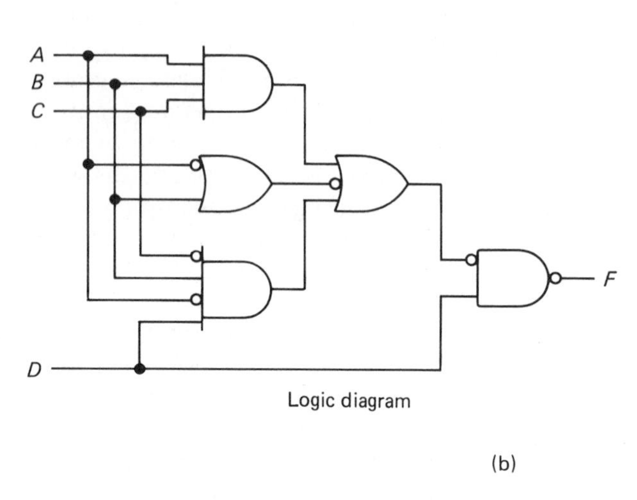

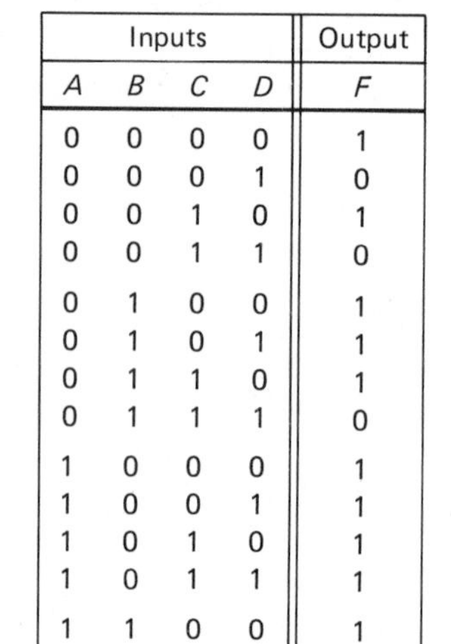

Inputs				Output
A	B	C	D	F
0	0	0	0	1
0	0	0	1	0
0	0	1	0	1
0	0	1	1	0
0	1	0	0	1
0	1	0	1	1
0	1	1	0	1
0	1	1	1	0
1	0	0	0	1
1	0	0	1	1
1	0	1	0	1
1	0	1	1	1
1	1	0	0	1
1	1	0	1	0
1	1	1	0	1
1	1	1	1	1

Truth table

(b)

Fig. 13-11 Example of truth tables for combinations of logic functions.

2 the logic 0 state is numerically greater (but negative). The previous statements and definitions are particularly appropriate for dc switching circuits, but apply to ac circuits as well. For example, a positive pulse can be used to simulate a positive voltage and a negative pulse can be used to simulate a negative voltage. However, such complexity is unnecessary, since the absence of a pulse can signify the logic 0 state and the original definition of positive polarity will still apply. That is, the logic 1 state is more positive than the "no" signal or the logic 0 state.

Negative logic polarity is defined as follows. When the logic 1 state is relatively more negative than the logic 0 state and the circuit is activated (operated) by the logic 1 signal, the logic polarity is considered to be negative. The following typical examples illustrate the manner in which negative logic can be employed:

Example 1: Logic 1 = 0 V
Logic 0 = +10 V

Example 2: Logic 1 = −10 V
Logic 0 = 0 V

At present, however, it is unimportant which type of logic polarity (positive or negative) is used, since logic components and circuits are available for all types. In fact, by using a form of mnemonic (memory-aiding) notation, such as H for the high, or active, state (1) and L for the low, or inactive, state (0), logic design may be completed and circuitry devised without concern for the polarity or levels used. Once the logic design is completed, standard circuits of the proper type and polarity for the components and level to be used are selected, and the unit is constructed.

13-6 BOOLEAN ALGEBRA AND CIRCUIT SIMPLIFICATION

Boolean algebra was introduced in 1847 by the English mathematician George Boole. The purpose of the algebra was to find a shorthand notation for the system of logic originally set forth by Aristotle. Aristotle's system of logic dealt with statements that were considered to be either true or false, but never partially true or false. Boole's algebra was based on a single-valued function with two discrete possible states.

Boolean algebra is used as an efficient method for handling any single-valued function with two possible states. When it is applied to binary arithmetic, the two states are 0 and 1. When it is applied to electric circuitry such as switching circuits, the two values are "open" and "closed," or open circuit and continuity. The numerals 0 and 1 are often used to indicate these electrical conditions. They are simply used to indicate the presence or absence of a conducting path.

Boolean algebra uses special notations and methods to derive logical and simplified circuit design. It permits the manipulation and rearrangement of complex series-parallel networks into simpler circuit arrangements. Redundant contacts, which are not always apparent, are eliminated to give a circuit with a minimum number of components. Boolean algebra enables the user to simplify a circuit in a rapid, smooth manner.

CIRCUIT SIMPLIFICATION PROCEDURES

One of the chief purposes of using boolean algebra in motor control design is to reduce switching circuits to their simplest form, with a minimum number of switches and contacts. Five steps for simplifying switching arrangements are as follows:

1. Draw the logic diagram of the original switching circuit.
2. Derive the boolean expression for the diagram.
3. Using the laws of boolean algebra, reduce the expression to its simplest form. Draw a simplified circuit diagram anytime during this process when it proves helpful.
4. Convert the minimum boolean function into its equivalent logic diagram.
5. Use a truth table to verify that the simplified circuit is the equivalent of the original circuit.

BOOLEAN ALGEBRA RULES AND STANDARD TERMINOLOGY

The following rules are used in boolean algebra to simplify circuits:

1. Relays and their contacts are designated by letters.
2. Normally open contacts are denoted by letters such as A, B, or C.
3. Normally closed contacts are denoted by letters with a bar over the letter. Example: $\overline{A}$, $\overline{B}$, $\overline{C}$, etc.
4. Contacts connected in series are denoted by the multiplication symbol $\times$ or $\cdot$ such as in $A \cdot B \cdot C$, etc., or by writing the letters together with no symbol between them, as in AB and ABC. Contacts in series are also known as an AND circuit. Contacts in series may also appear as $A \times B \times C$ or just ABC.
5. Contacts connected in parallel are denoted by the plus sign. $A+ B+ C+$, etc. Contacts in parallel are known as OR circuits.
6. Normally closed contacts in parallel would be denoted as $\overline{A}$ (A bar or not A) plus $\overline{B}$ (B bar or not B) or $\overline{A} + \overline{B}$. Normally closed contacts are known as NOT circuits.
7. When a contact is closed, transmission can occur; a 1 is used to denote this condition. When a contact is open, transmission cannot occur or the circuit has infinite resistance. This is denoted by 0.
8. The equal sign, as is the case in conventional mathematics, represents a relationship of equivalence between the expressions on both sides of the sign.
9. The familiar signs of grouping ( ) and { } are used in the customary fashion to indicate that all terms so contained are to be treated as a unit.
10. Various letters are used to represent the variables under consideration, generally starting with A. Since the variables are capable of being in only one of two states, the numerals 0 and 1 are the only numbers used in boolean expressions.

13-7 FUNDAMENTAL LAWS AND AXIOMS OF BOOLEAN ALGEBRA

If $\overline{A} = 0$, then $A = 1$, and if $\overline{A} = 1$, then $A = 0$. These expressions, though simple in appearance, may be used to construct a boolean algebra system, determining all of the following relationships:

Logical addition (OR):	$0+0=0$
	$0+1=1$
	$1+0=1$
	$1+1=1$
Logical multiplication (AND):	$0 \cdot 0=0$
	$0 \cdot 1=0$
	$1 \cdot 0=0$
	$1 \cdot 1=1$
Rules of complement:	$\overline{0}=1$
	$\overline{1}=0$

The following laws and axioms should be memorized since they may be utilized to simplify any boolean expression. However, a problem exists in applying these laws and axioms: it is difficult to determine when a given expression is in its simplest form. To overcome this difficulty over methods of simplification have been devised and will be explained later in this chapter.

1. $A+\overline{A}=1$
2. $A \cdot \overline{A}=0$

3. $A + A = A$
4. $A \cdot A = A$
5. $A + B = B + A$
6. $A \cdot B = B \cdot A$
7. $A + (B + C) = (A + B) + C$
8. $A(BC) = (AB)C$
9. $A(B + C) = AB + AC$
10. $A + BC = (A + B)(A + C)$
11. $\overline{A + B} = \overline{A}\,\overline{B}$
12. $\overline{AB} = \overline{A} + \overline{B}$
13. $\overline{\overline{A}} = A$
14. $A(A + B) = A$
15. $A + AB = A$

THEOREMS USED TO SIMPLIFY BOOLEAN EXPRESSIONS

Law of Indentity

1. $$A = A$$

The law of identity states that a quantity is the equivalent of itself.

Complementary Law

1. $$A + \overline{A} = 1$$

2. $$A\overline{A} = 0$$

According to the complementary law, the logical addition of a quantity and its complement will result in a sum of 1 and the logical multiplication of a quantity and its complement will result in a product of 0.

Tautology Law

3. $$A + A = A$$

4. $$AA = A$$

The tautology law states that combining a quantity with itself either by logical addition or logical multiplication will result in a logical sum or product that is the equivalent of the quantity.

Commutative Law

5. $$A + B = B + A$$

6. $$AB = BA$$

The commutative law states that changing the order of the terms in an equation will not affect the value of the equation.

Associative Law

7. $$(A + B) + C = A + (B + C)$$

8. $$(AB)C = A(BC)$$

The associative law states that when combining three or more terms, either by logical addition or logical multiplication, the order in which the terms are combined will not affect the result.

Distributive Law

9. $$A(B + C) = (AB) + (AC)$$

10. $$A + (BC) = (A + B)(A + C)$$

The distributive law states that if a group of terms connected by like operators contain the same variable, the variable may be removed from the terms and associated with them by the appropriate sign of operation.

Law of Dualization (DeMorgan's theorem)

11. $$\overline{A + B} = \overline{A}\,\overline{B}$$

12. $$\overline{AB} = \overline{A} + \overline{B}$$

The law of dualization states the (1) the complement of the logical sum of two or more terms is equal to the logical product of the complements of the terms and (2) the complement of the logical product of two or more terms is equal to the logical sum of the complements of the terms.

Law of Double Negation

13. $$\overline{\overline{A}} = A$$

The law of double negative states that the complement of the complement of a term is the equivalent of the term.

Law of Absorption

14. $$A(A + B) = A$$

15. $$A + AB = A$$

By law of absorption the odd term will be absorbed when (1) a term is combined by logical multiplication with the logical sum of the term and another term and (2) a term is combined by logical addition with the logical product of the term and another term.

Axioms

1. $$A + 0 = A$$

2. $$A + 1 = 1$$

3. $$A \cdot 0 = 0$$

4. $$A \cdot 1 = A$$

These four expressions do not fall into the category of laws; however, as previously indicated, they are axiomatic expressions and are very useful in simplifying boolean equations.

13-8 EQUATION FORMS USED TO ELIMINATE REDUNDANCIES

In this discussion, consider the following equations:

$$F = AB + A\overline{B} + \overline{A}\,\overline{B} \qquad (1)$$

$$= A + \overline{B} \qquad (2)$$

Equation 1 is the sum of three boolean terms, each of which is the product of two variables (A and B). Each variable is represented in either its true or complement form. Equation 2 is the sum of two variables (A and B), B appearing in its complement form. Equation 1 is a *minterm* expression of the two variables. The proof of this is as follows:

$$F = AB + A\overline{B} + \overline{A}\,\overline{B}$$
$$= (\overline{A} + \overline{B})(\overline{A} + \overline{\overline{B}})(\overline{\overline{A}} + \overline{\overline{B}})$$
By negation and application of DeMorgan's theorem.

$$= (\overline{A} + \overline{B})(\overline{A} + B)(A + B)$$
By application of the law of double negation.

$$= A\overline{A}\,\overline{A} + \overline{A}AB + \overline{A}\,\overline{A}B + \overline{A}BB + A\overline{A}\,\overline{B} + AB\overline{B} + \overline{A}B\overline{B} + BB\overline{B}$$
$$= \overline{A}B$$
Collecting and combining terms.

Applying the four axioms results in the simplest minterm expression for Eq. 1.

$$= \overline{\overline{A}} + \overline{B}$$
By negation and application of DeMorgan's theorem.

$$= A + \overline{B}$$
Application of the law of double negation now results in Eq. 2.

Boolean algebra is an algebraic method that can be used to simplify logic circuits by spotting and eliminating redundancies. You substitute mathematical symbols with ANDS, ORS, and NOTS, simplify the resulting equations by factoring or other simple manipulation, then translate the results back into the logic symbols of the logic circuit.

To translate a logic design into a boolean algebra expression you use letters for inputs and outputs and use the algebraic symbols for AND and OR. Use parentheses to set off isolated expressions, as is done in ordinary algebra or in English. The importance of the correct use of parentheses can be seen in Fig. 13-12.

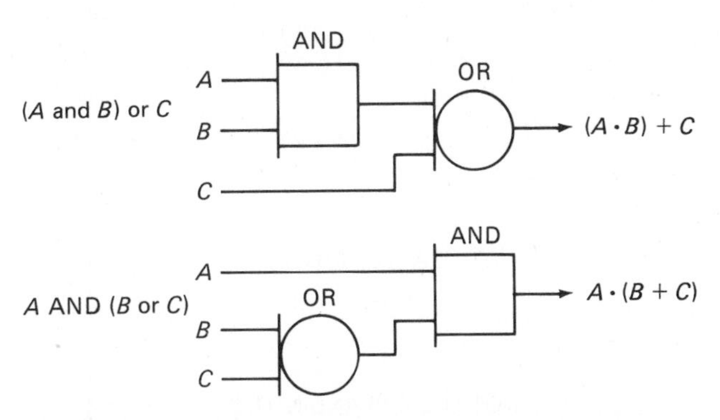

Fig. 13-12 Logic diagrams converted to boolean algebra expressions. Note that NEMA standard symbols are used.

Once the logic design is translated into boolean algebra, you can factor just as in ordinary algebra by treating the center dot (·) as the multiplication operator and the plus (+) as the addition operator. The logical meaning will remain the same.

Besides factoring, another form of manipulation to reduce complexity is to apply DeMorgan's theorem (Fig. 13-13). As shown in the figure, NOT *A* OR NOT *B* OR NOT *C* is the equivalent of NOT (*A* AND *B* AND *C*). In general, you take the bar off each term, change the connectives (AND to OR or OR to AND), and put the bar over the whole expression. The manipulation is reversible. Prove it to yourself by tracing signals through the two logic circuits in the figure. For instance, if you assume no input to the three NOTS in the OR circuit, you will have an output from the OR.

If you apply an input signal at *A*, you turn off the appropriate NOT, but the OR will still produce an output because it is receiving an input from NOT *B* and NOT *C*. Notice, in fact, that you'd have to apply an input to all three (*A*, *B*, and *C*) to turn off all three NOTS in order to turn off the OR, because, by definition, the OR will have an output if it receives a signal from any one or more inputs. Now, look at the AND circuit. If all inputs are off, the NOT is producing an output and will continue as long as an input is applied to all three (*A*, *B*, and *C*). This will then operate the AND, and send an output signal to the NOT, turning it off. This is exactly the same result we got in the OR circuit.

Figures 13-14 through 13-16 on pages 170 and 171 show the use of parentheses and brackets and the simplifying of logic circuits according to DeMorgan's theorem.

Notice that the number of sets of parentheses (or brackets) equals the number of connector logic elements (ANDS and ORS). When factoring to reduce the sets of parentheses, the aim is to reduce the complexity of the circuit.

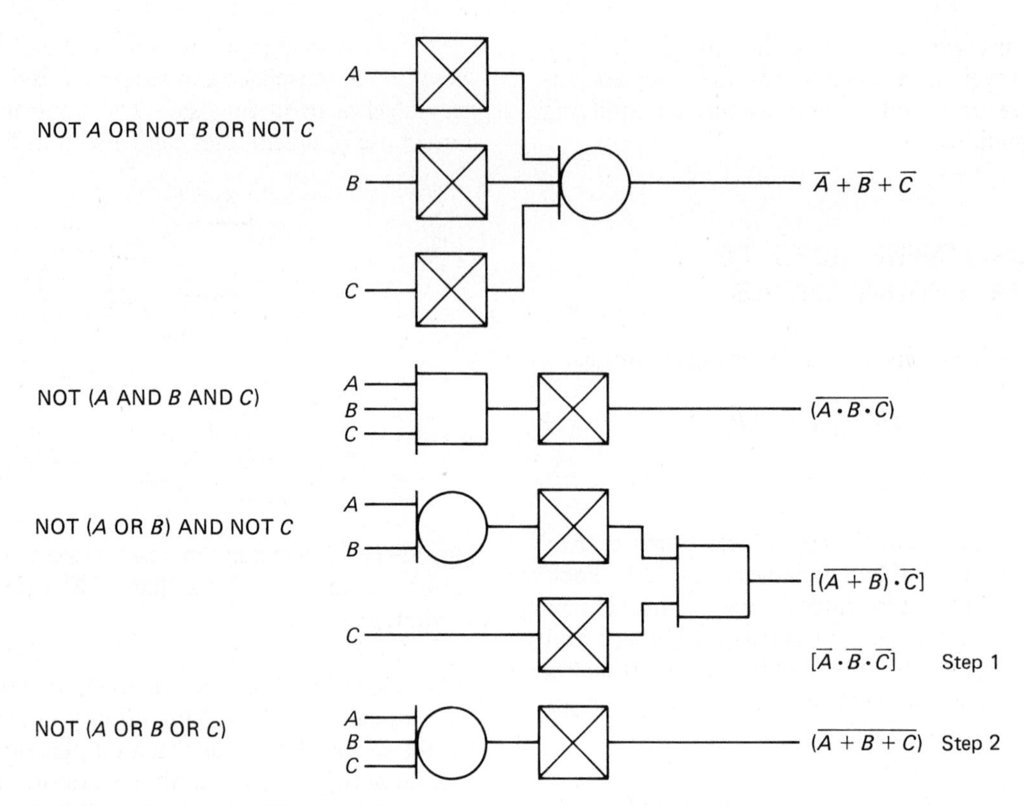

Fig. 13-13 DeMorgan's theorem illustrated.

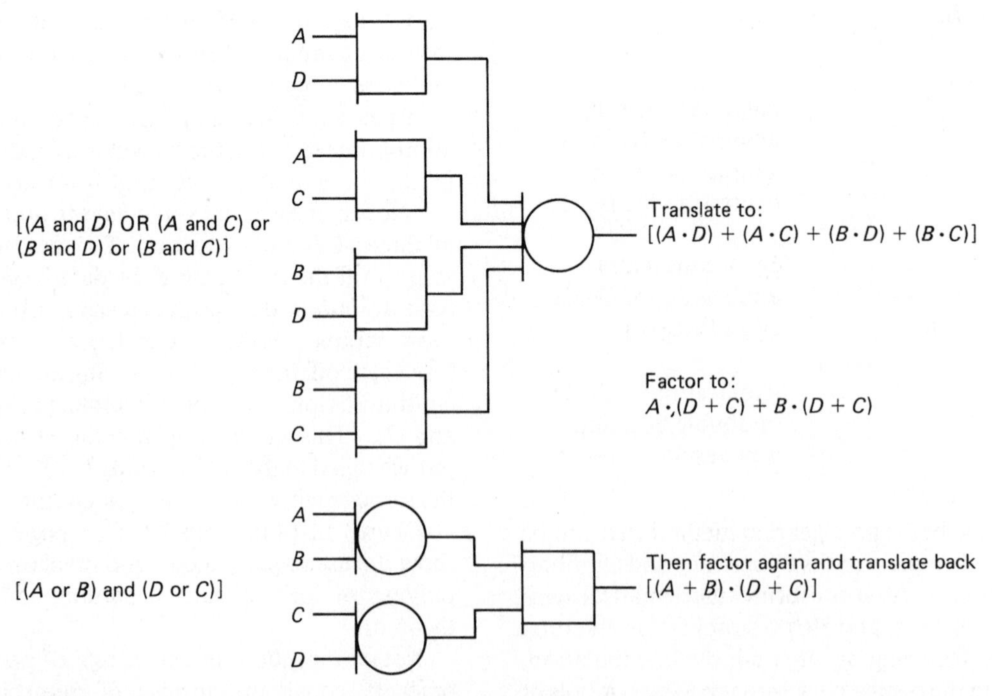

Fig. 13-14 Simplifying with DeMorgan's theorem showing the use of parentheses and brackets.

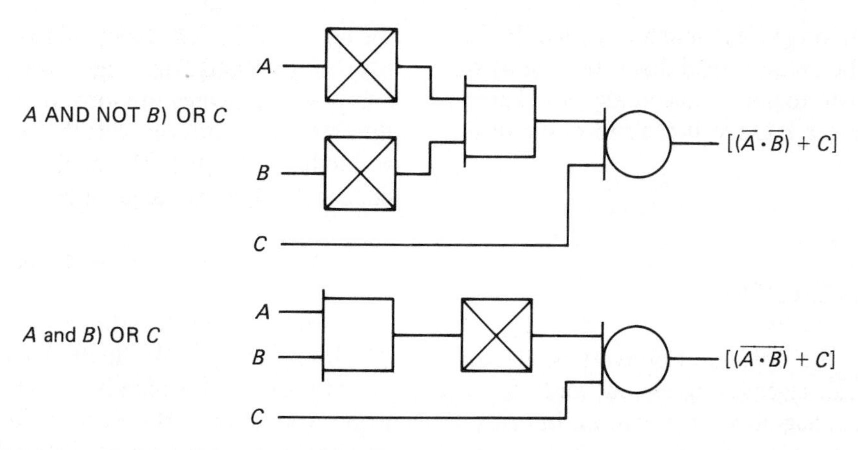

Fig. 13-15 Simplifying a circuit with NOTs using DeMorgan's theorem.

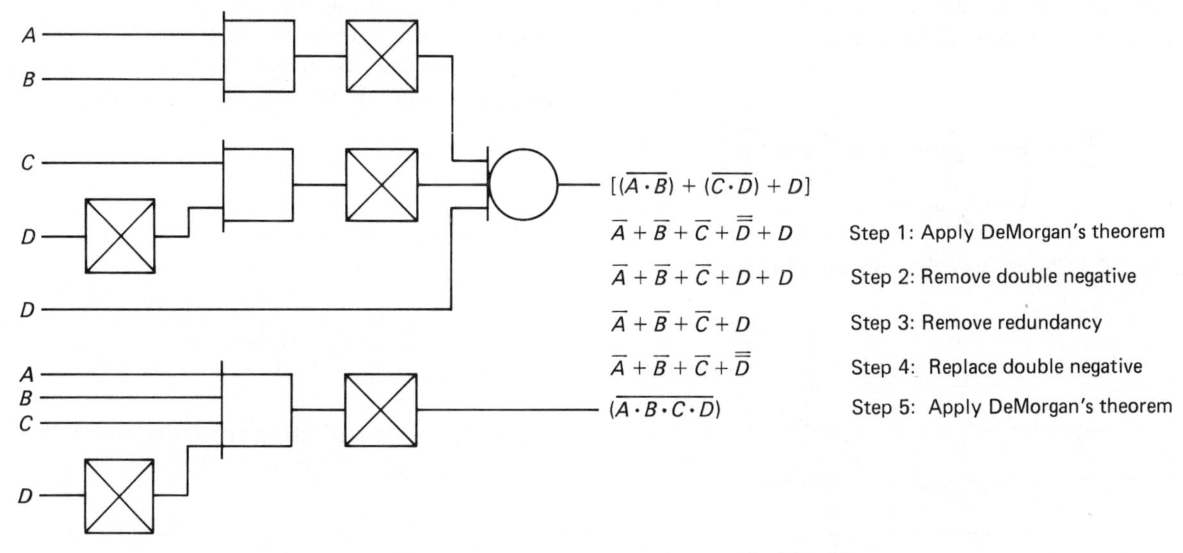

Fig. 13-16 Applications of DeMorgan's theorem using the theorem $\overline{A} \cdot \overline{A} = \overline{A}$.

The NOT is represented by a bar over the top of the letter identifying its input. The diagram in Fig. 13-16 shows how several NOTs are handled.

Another useful theorem says simply that NOT $\overline{A}$ or $\overline{\overline{A}}$ is the same as A. This is just another way of saying that double negatives cancel. Also (A AND A) or (A OR A) are redundancies and are equal to A.

Boolean algebra expressions find applications in P.C.s. For example, ladder diagrams for control circuits are converted to logic expressions, simplified with boolean algebra, and then converted to a program that operates a P.C. Figure 13-17 illustrates a ladder diagram converted to a boolean algebra logic expression and then converted to a program for a TI 510 P.C. This example indicates the value of boolean algebra in converting elementary line diagrams for industrial

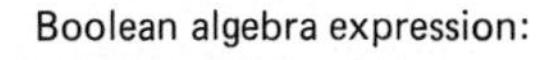
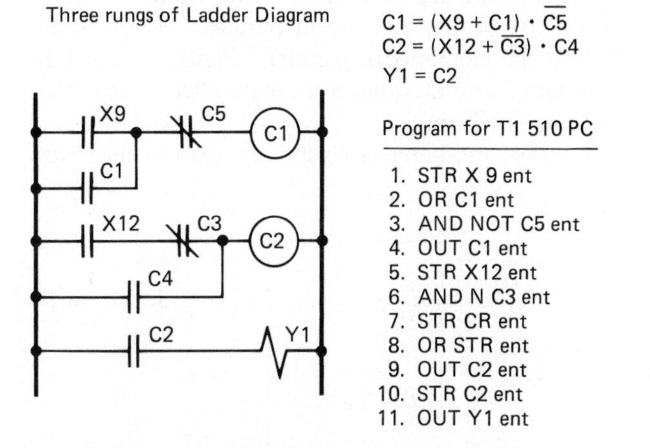

Fig. 13-17 Ladder diagram, boolean algebra expression, and program for TI 510 Programmable Controller.

control circuits into programs for some types of P.C.s. Not all P.C.s will be programmed this way. Many of the smaller P.C.s will require boolean algebra expressions while the larger P.C.s will require other techniques.

13-9 CIRCUIT CONVERSION

Let's look now at the process of converting electric circuits into boolean algebra expressions and the expressions into logic diagrams. We will use industrial logic symbols for our logic circuits.

Refer to Fig. 13-18. The first step in the conversion is to use the *circle-and-block* method of analyzing the circuit. Place a block around the circuit elements that are in series. As indicated in Fig. 13-19, these blocks indicate the AND logic of the circuit.

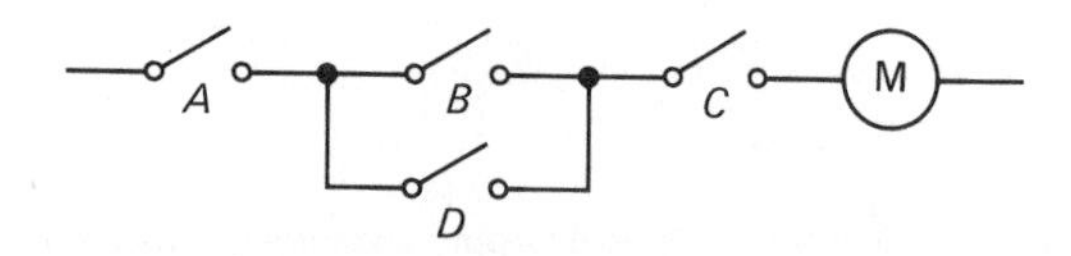

Fig. 13-18 Elementary diagram of an electric control circuit.

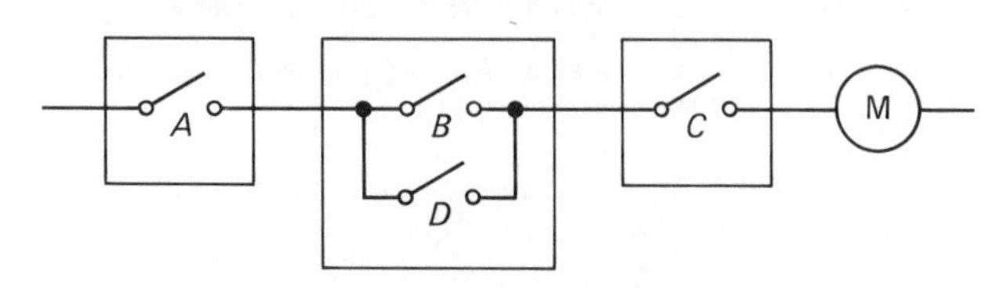

Fig. 13-19 Blocks representing components in series. Series circuits have the AND logic function.

The second step is to indicate the circuit elements within the blocks that are in parallel. Place circles around the elements in parallel. Figure 13-20 indicates that *B* and *D* contacts are in parallel and have the OR logic function.

To write the boolean algebra expression start with the circles within the block and work outward. Identify contact *B* and *D* in a logic expression as $B + D$. To complete the expression look at the blocks to see what they represent and complete the logic expression. This would be *A* and (*B* or *D*) and *C* equals *M*. This would appear in a boolean algebra expression as

$$A \cdot (B + D) \cdot C = M, \text{ or } A(B + D)C \doteq M$$

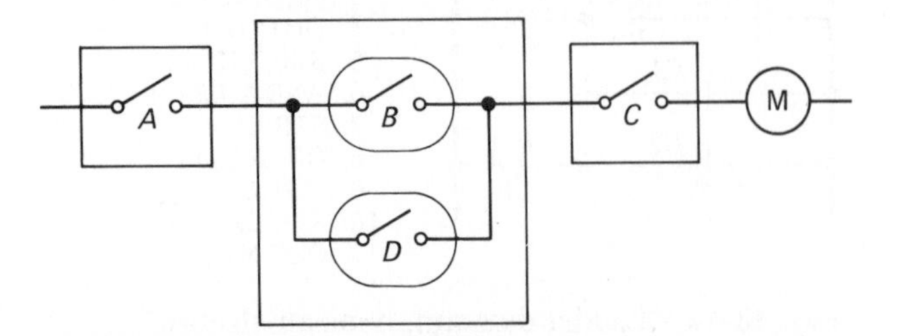

Fig. 13-20 Circles identify components in parallel. Parallel circuits have the OR logic function.

We will now convert the simple expression above to a logic expression. We do this by determining the output of the circuit, which is coil *M*. By examining the logic expression or the electric circuit with the blocks and circles, we can see that a logic AND is required as an input to the *M* coil. This is illustrated in Fig. 13-21. Note that an amplifier is driving the *M* coil. In an industrial logic system, the amplifier is required to raise the low voltage of the logic elements to a voltage sufficient to energize the load. Signal converters are used on input switches to change high-voltage inputs to logic voltages. Start at the output and work toward the input.

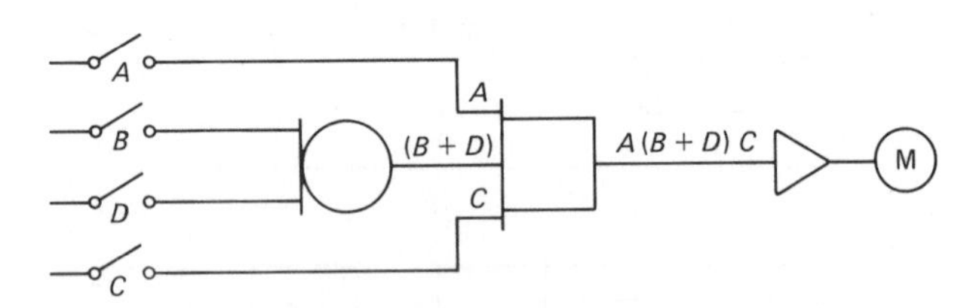

Fig. 13-21 Logic diagram converted to boolean algebra expression $M = A(B + D)C$.

In this problem a more complex electric circuit will be converted to a boolean expression and its equivalent logic diagram. Figure 13-22 illustrates our problem circuit.

The first step is to block out the circuit to determine the circuit elements in series. Note that the blocks are identified as 1, 2, and 3. At this point, the expression can be read as "block 1, AND block 2 AND block 3 equal coil *Z*."

Now if we break down block 1, it would look like Fig. 13-23, with the circles representing contacts in parallel forming the OR circuit.

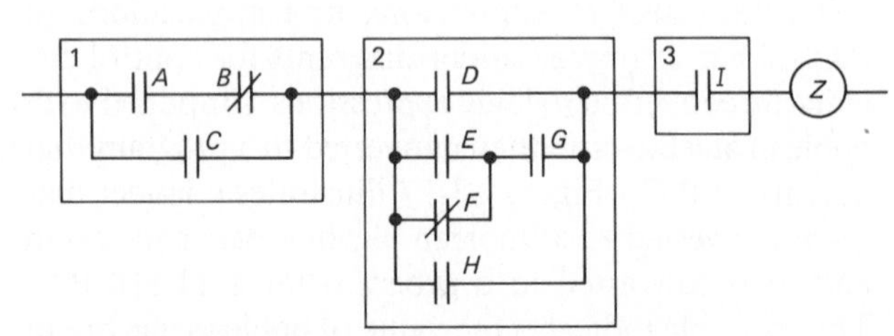

Fig. 13-22 Electric circuit for Example 2.

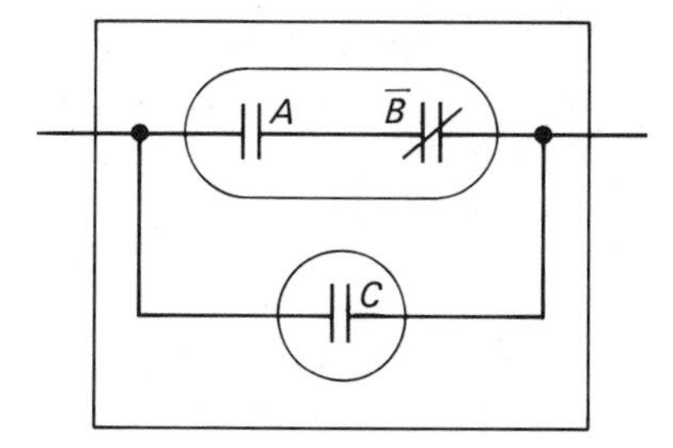

Fig. 13-23 First block of Fig. 13-22, identifying the OR function.

Continuing to break block 1 down will show that contacts A and $\overline{B}$ are in series ($\overline{B}$ represents a normally closed contact). Figure 13-24 illustrates the block broken down into ANDs and ORs, or boxes and circles.

Within the block, circle 1 shows we have $A \cdot \overline{B}$ and circle 2 has a C contact. The circles represent OR circuits denoted by a + sign. Block 1 will convert to the logic expression $A\overline{B} + C$.

Figure 13-25 illustrates block 2 of Fig. 13-22. Note how the block is broken down to derive the logic expression:

$$D + (E + \overline{F})G + H$$

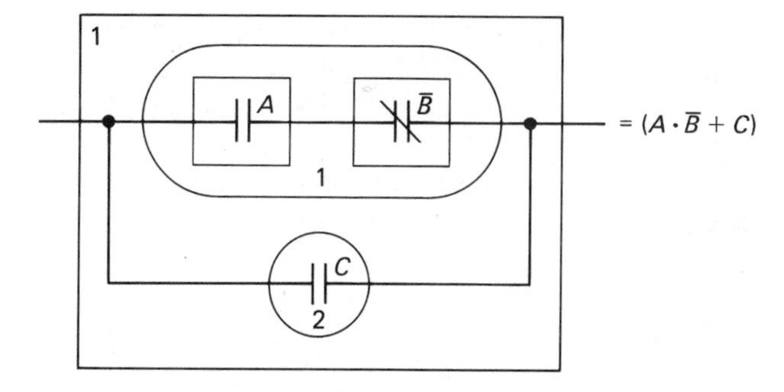

Fig. 13-24 First block of Fig. 13-23 with the AND function identified by blocks and the equivalent boolean expression.

The complete expression for the circuit in Fig. 13-22 has now been determined

$$(A \cdot \overline{B} + C) \cdot [D + (E + \overline{F})G + H] \cdot I = Z$$

13-10 VEITCH DIAGRAMS AND KARNAUGH MAPS

Veitch diagrams and Karnaugh maps provide a very quick and easy way for finding the simplest logic equation needed to express a given function. Veitch diagrams and Karnaugh maps can be developed for any number of variables but become difficult to use with

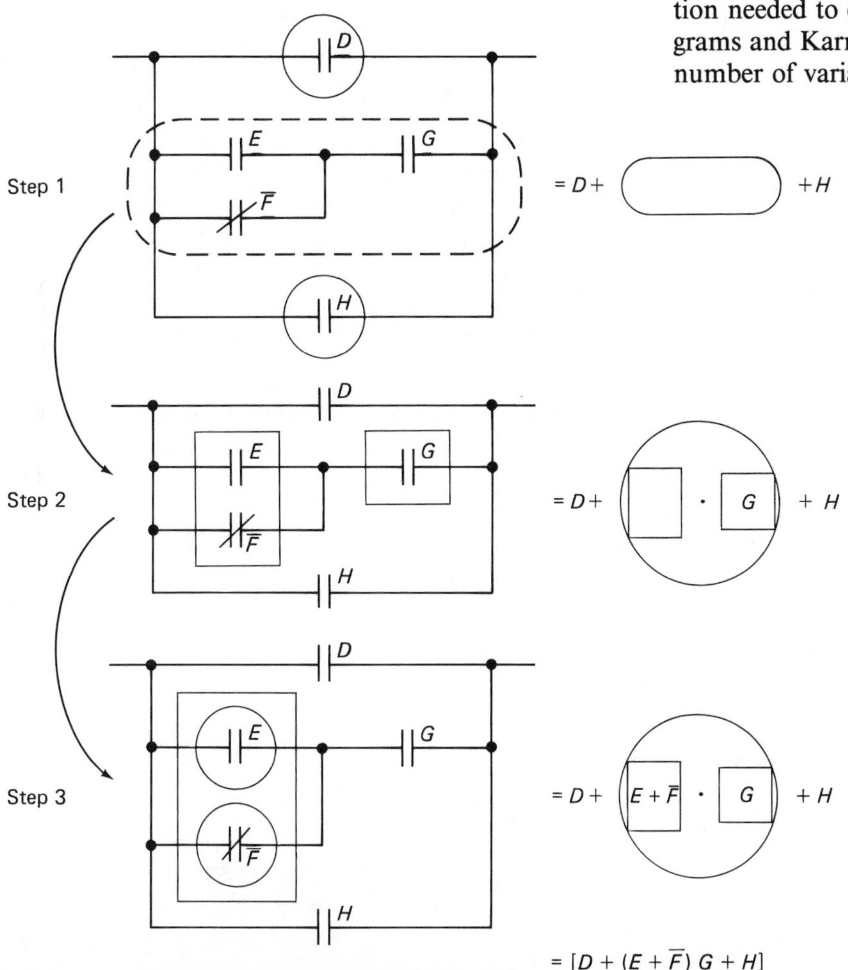

Fig. 13-25 Breaking down the center block of Fig. 13-22 and deriving the boolean expression.

more than four variables. However, when electric circuits are converted to their logic equivalents they rarely contain more than four variables. Boolean algebra equations rarely use more than four variables. We will therefore limit our discussion to four-variable Veitch diagrams and Karnaugh maps to avoid complicated problems.

An example of an electric circuit with many N.O. and N.C. contacts is illustrated in Fig. 13-26. It can be seen that there are also many redundant contacts. The unnecessary contacts can be eliminated by developing a logic expression for the electric circuit and simplifying with boolean algebra techniques. The circuit can also be simplified by using Veitch diagrams and Karnaugh maps.

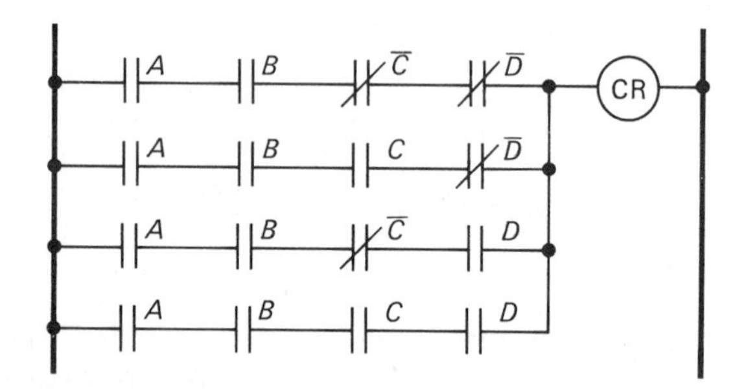

Fig. 13-26 Electric circuit with many redundant contacts.

The boolean algebra expression for the circuit in Fig. 13-26 is as follows:

$$AB\overline{C}\,\overline{D} + ABC\overline{D} + AB\overline{C}D + ABCD = CR$$

By using the boolean algebra principles explained earlier the circuit of Fig. 13-26 can be simplified to contain the minimum number of contacts. Veitch diagrams and Karnaugh maps can also be used to simplify the circuit of Fig. 13-26. The logic equation for Fig. 13-26 can be simplified much more rapidly by using Veitch diagrams and Karnaugh maps than by using the boolean equations and identities. Two contacts in series, A and B, will give the equivalent control that the 16 contacts in Fig. 13-26 give in controlling the energizing of the CR relay.

Figure 13-27 shows how a Karnaugh map and a Veitch diagram appear, with each square identified with letters, representing the variables A, B, C, and D. These variables represent the terms in four-variable logic equations. The letters for every combination of normally open and closed contacts are placed inside the squares of the Veitch diagram and Karnaugh map. A contact letter with a bar over it represents a closed contact, and when there is no bar over a letter, the letter represents an open contact.

For example, the variables $AB\overline{C}\,\overline{D}$ appear in the top left square of the Veitch diagram. This represents normally open A and B contacts in series with normally closed C and D contacts. The same square in the Karnaugh map does not have the same expression. In the Veitch diagram the top left square represents contacts $AB\overline{C}\,\overline{D}$. In the Karnaugh map the top left square represents $\overline{A}\,\overline{B}\,\overline{C}\,\overline{D}$.

By examining the Veitch diagram and Karnaugh map it can be seen that the two diagrams have the same variables but they are in different locations. The method you use can be either the Veitch diagram or Karnaugh map. The only difference between the two is the location of the terms. Whether you use the Veitch diagram or the Karnaugh map to simplify a logic equation, the results will be the same.

To demonstrate the difference between the Veitch diagram and Karnaugh map we will take the expressions for Fig. 13-26 and plot the location of each variable in Fig. 13-27. The logic expression for the circuit of Fig. 13-26 was indicated earlier to be

$$AB\overline{C}\,\overline{D} + ABC\overline{D} + AB\overline{C}D + ABCD = CR$$

Veitch diagram

	A		$\overline{A}$		
B	$AB\overline{C}\,\overline{D}$	$ABC\overline{D}$	$\overline{A}B\overline{C}\,\overline{D}$	$\overline{A}\,\overline{B}\,\overline{C}\,\overline{D}$	$\overline{D}$
B	$AB\overline{C}D$	$ABCD$	$\overline{A}BCD$	$\overline{A}B\overline{C}D$	D
$\overline{B}$	$A\overline{B}\,\overline{C}D$	$A\overline{B}CD$	$\overline{A}\,\overline{B}CD$	$\overline{A}\,\overline{B}\,\overline{C}D$	D
$\overline{B}$	$\overline{A}\,\overline{B}\,\overline{C}\,\overline{D}$	$\overline{A}\,\overline{B}C\overline{D}$	$\overline{A}\,\overline{B}C\overline{D}$	$\overline{A}\,\overline{B}\,\overline{C}\,\overline{D}$	$\overline{D}$
	$\overline{C}$	C	C	$\overline{C}$	

Karnaugh map

$AB \backslash CD$	00	01	11	10
00	$\overline{A}\,\overline{B}\,\overline{C}\,\overline{D}$	$\overline{A}\,\overline{B}\,\overline{C}D$	$\overline{A}\,\overline{B}CD$	$\overline{A}\,\overline{B}C\overline{D}$
01	$\overline{A}B\overline{C}\,\overline{D}$	$\overline{A}B\overline{C}D$	$\overline{A}BCD$	$\overline{A}BC\overline{D}$
11	$AB\overline{C}\,\overline{D}$	$AB\overline{C}D$	$ABCD$	$ABC\overline{D}$
10	$\overline{A}\,\overline{B}\,\overline{C}\,\overline{D}$	$\overline{A}\,\overline{B}\,\overline{C}D$	$A\overline{B}CD$	$A\overline{B}C\overline{D}$

Fig. 13-27 Veitch diagram and a Karnaugh map are used to simplify a boolean algebra expression.

In the Veitch diagram and Karnaugh map of Fig. 13-28 we have not indicated the letters inside the squares, as we have in Fig. 13-27, but have used 1s to indicate the location of the first four-variable term. The four-variable expressions representing the contacts in Fig. 13-26 will be used in Figs. 13-28 through Fig. 13-32. (See pages 176 and 177.)

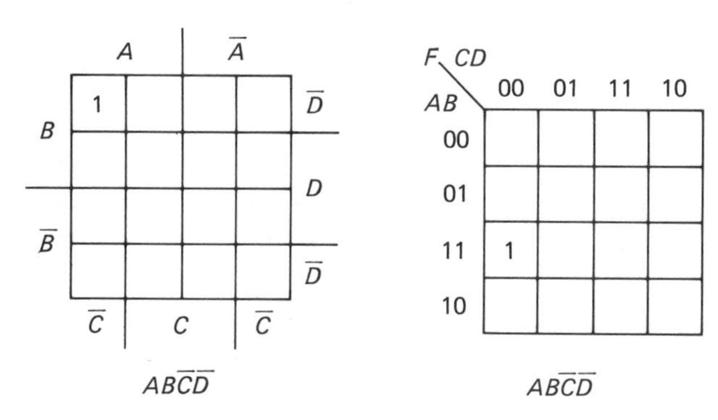

Fig. 13-28 Veitch diagram and Karnaugh map with the square for the expression $AB\overline{C}\overline{D}$ identified with a 1.

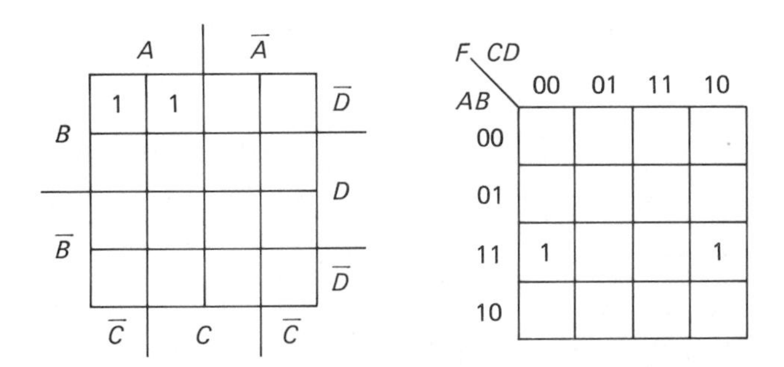

Fig. 13-29 Veitch diagram and Karnaugh map identified with 1s for expression $AB\overline{C}\overline{D}$ and $ABC\overline{D}$.

The squares with 1s marked in them in Fig. 13-28 represent the first term of the logic expression, $AB\overline{C}\overline{D}$. Notice its location on the Veitch diagram and the Karnaugh map. The next term in the logic expression is $ABC\overline{D}$. Notice the location of the 1 in Fig. 13-29. Next, notice the location of $AB\overline{C}D$ on the Veitch diagram and the Karnaugh map of Fig. 13-30. Finally, note the location of $ABCD$ in Fig. 13-31.

In Fig. 13-31 we have the complete diagram and map for the electric circuit of Fig. 13-26. What do we do now to simplify the circuit? The rules for simplifying logic equations using Veitch diagrams and Karnaugh maps are as follows:

RULE A. If 1s are located in adjacent squares or at opposite ends of any row or column, one of the variables may be dropped. The variable that changes in adjacent squares is dropped.

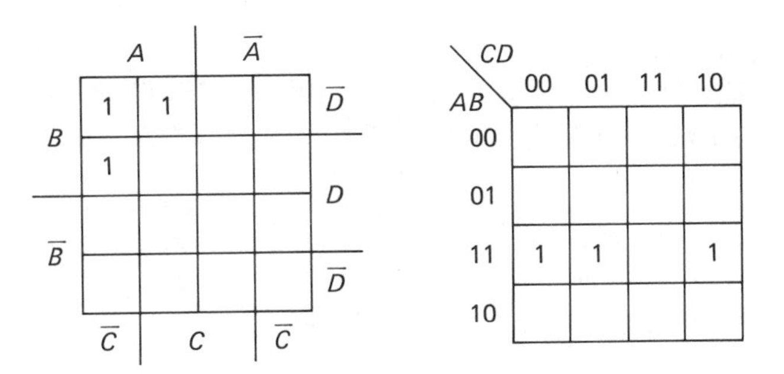

Fig. 13-30 Veitch diagram and Karnaugh map identified with 1s for the expression $AB\overline{C}D + AB\overline{C}\overline{D} + ABC\overline{D}$.

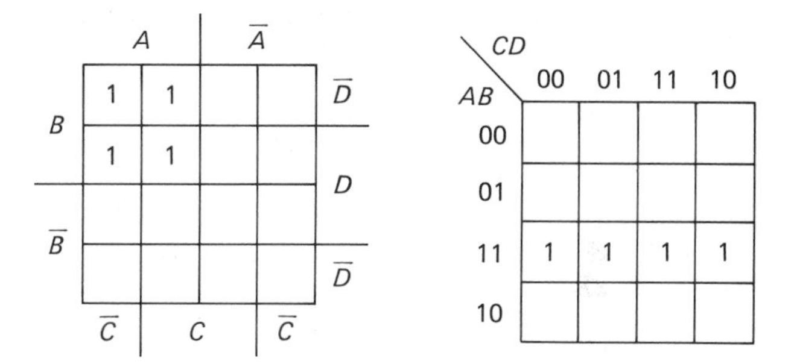

Fig. 13-31 Veitch diagram and Karnaugh map identified with 1s for the expression $AB\overline{C}\overline{D} \pm ABC\overline{D} + AB\overline{C}D + ABCD$.

RULE B. If any row or column of squares, any block of four squares, or the four end squares of any adjacent rows or columns, or the four corner squares are filled with 1s, two of the variables may be dropped. The variables that change in adjacent squares are dropped.

RULE C. If any two adjacent rows or columns, the top and bottom rows, or the right and left columns are completely filled with 1s, three of the variables may be dropped. The variable that does not change in adjacent squares is not dropped.

RULE D. To reduce the original equation to its simplest form, sufficient simplification must be made until all 1s have been included in the final equation. The 1s may be used more than once, and the largest possible combination of 1s in groups of 8, 4, 2, and 1 should be used.

In Fig. 13-32 Examples 1 to 14 represent the simplification rules. Note that two adjacent squares with 1s on either the Veitch diagram or Karnaugh map give a simplified circuit with three letters and four adjacent squares with 1s give a simplified circuit with two letters. Eight adjacent squares with 1s give a simplified circuit with only one letter.

13-11 SIMPLIFYING CIRCUITS WITH VEITCH DIAGRAMS AND KARNAUGH MAPS

The use of Veitch diagrams or Karnaugh maps for simplifying logic expressions with fewer than four functions requires the use of "I don't care" terms. These terms apply to Karnaugh maps, also.

To illustrate the use of these terms we will derive a logic expression from an electric relay circuit. From our expression, we will place 1s in each square containing the letters and the missing letters in the terms that would be required to make them each have four letters. These letters represent the boolean algebra expression for each line of the electric circuit. Once we simplify this expression with mapping techniques, the

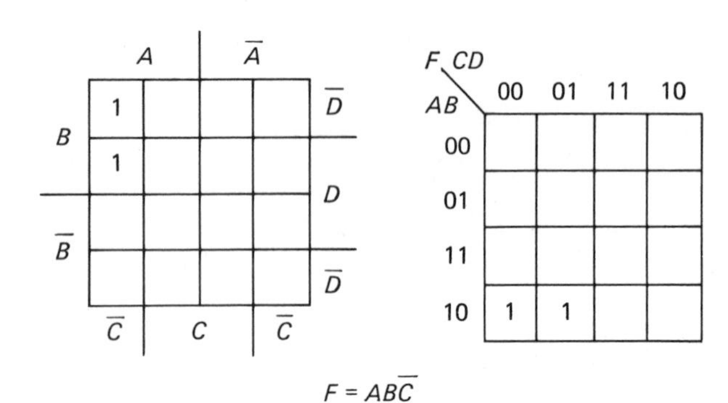

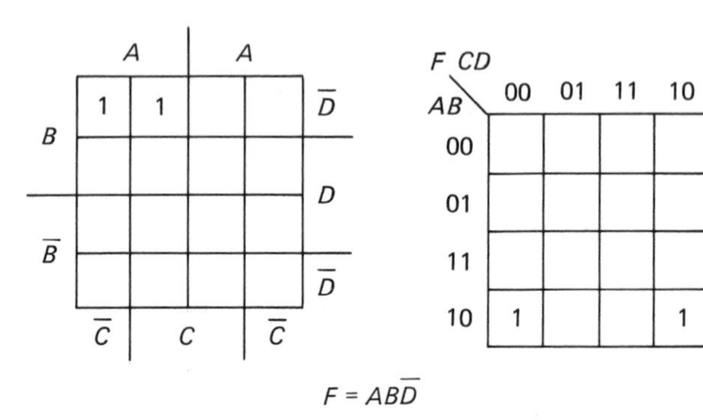

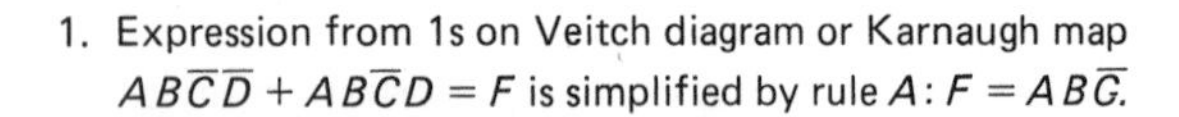

1. Expression from 1s on Veitch diagram or Karnaugh map $AB\overline{C}\overline{D} + AB\overline{C}D = F$ is simplified by rule A: $F = AB\overline{C}$.

2. Expression from 1s on Veitch diagram on Karnaugh map $AB\overline{C}\overline{D} + ABC\overline{D} + ABC\overline{D} = F$ is simplified by rule A: $F = AB\overline{D}$.

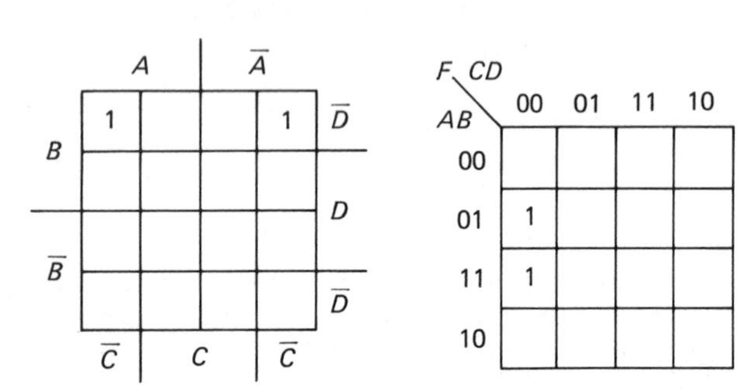

3. $AB\overline{C}\overline{D} + \overline{A}B\overline{C}\overline{D}$
Simplified: $F = B\overline{C}\overline{D}$

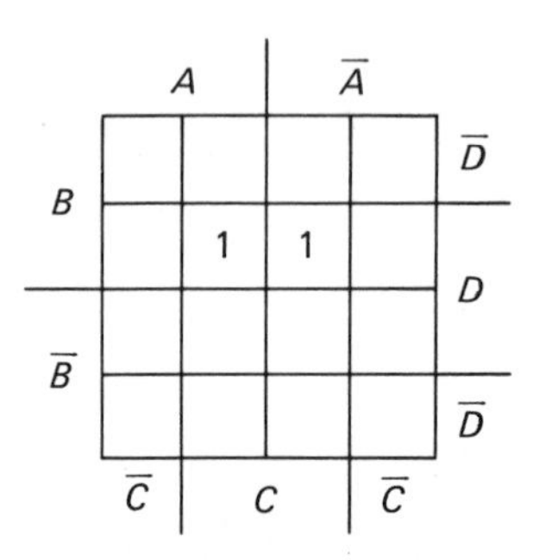

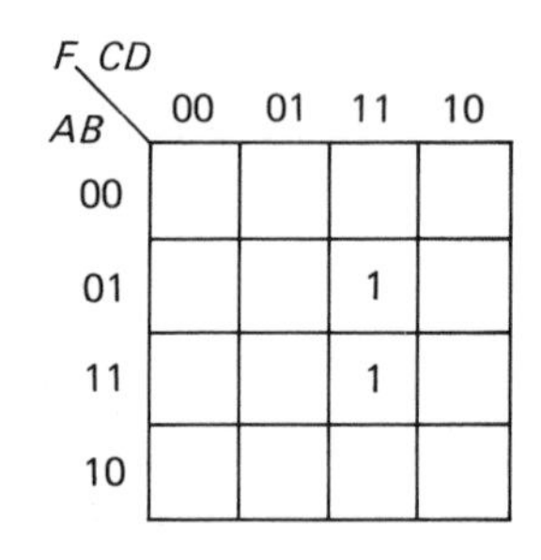

$F = BCD$

4. $ABCD + \overline{A}BCD = F$
Simplified: $F = BCD$

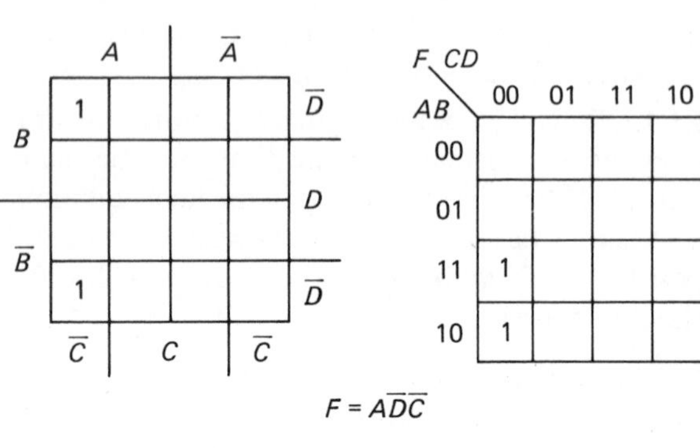

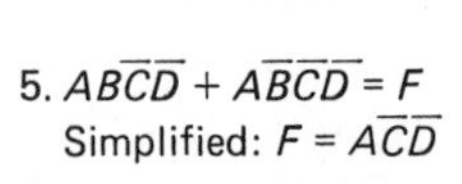

5. $AB\overline{C}\overline{D} + A\overline{B}\overline{C}\overline{D} = F$
Simplified: $F = A\overline{C}\overline{D}$

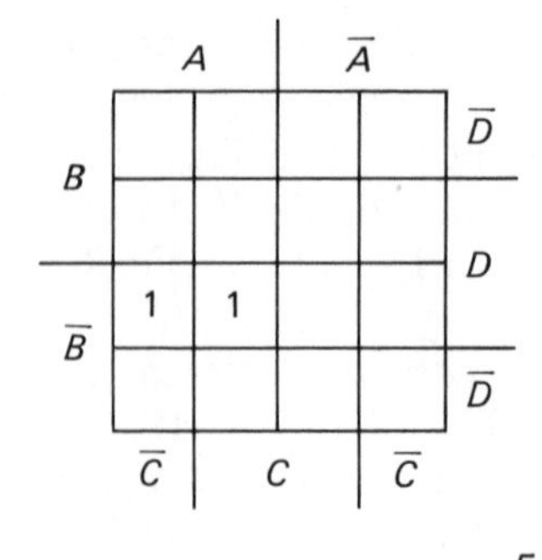

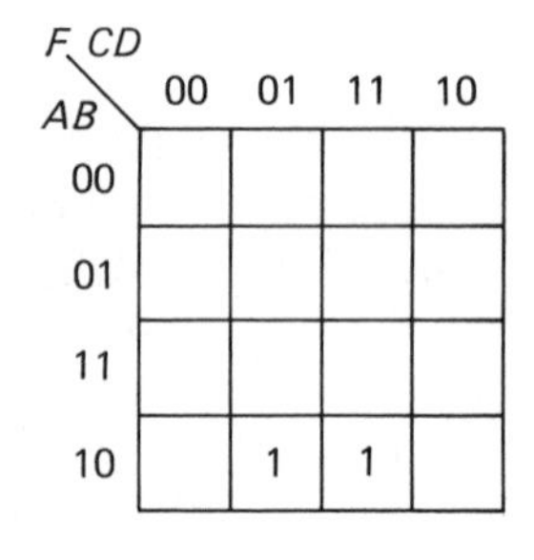

$F = A\overline{B}D$

6. $A\overline{B}\overline{C}D + A\overline{B}CD = F$
Simplified: $F = A\overline{B}D$

(a) Rule A

Fig. 13-32 Boolean algebra expressions simplified with Veitch diagrams and Karnaugh maps. (*a*) Application of Rule A.

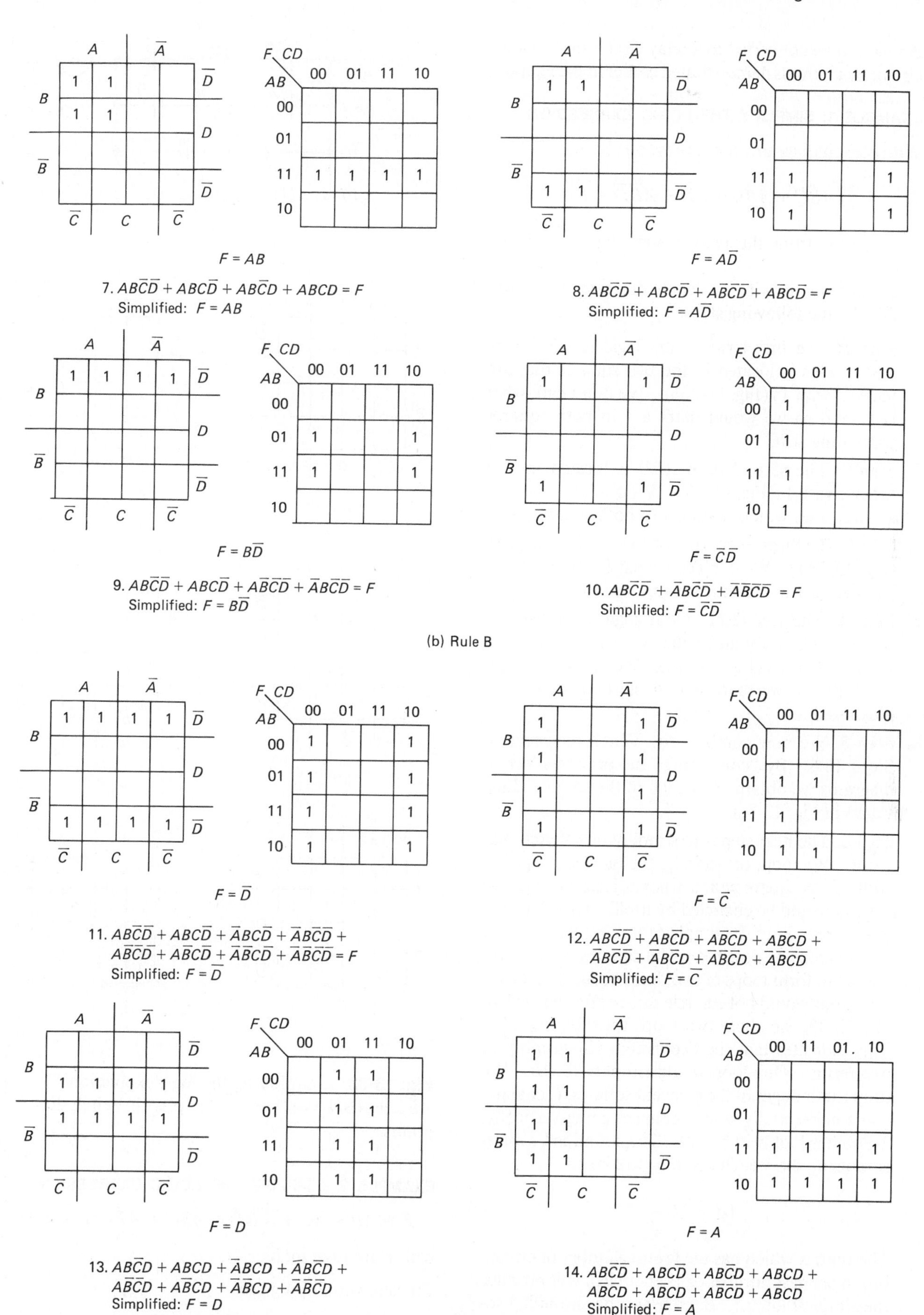

Fig. 13-32 *(cont.)* (*b*) Application of Rule B. (*c*) Application of Rule C.

circuit can be converted to a relay circuit or to a logic circuit by methods demonstrated earlier in the chapter.

EXAMPLE 1: SIMPLIFY THE LOGIC EXPRESSION

With Veitch diagrams the expression

$$A\bar{B}C + A\bar{B}D + \bar{A}D + A\bar{B}\bar{D} = F$$

was derived from the relay ladder diagram of Fig. 13-33.

Simplifying it with a Veitch diagram is accomplished by the following steps:

Step 1. The first term in the logic expression is $A\bar{B}C$. This is located in the two squares that are marked with 1s (Fig. 13-33*b*). We don't care what D is, and so we would mark a 1 in both squares containing $A\bar{B}C$.

Step 2. The second term $A\bar{B}D$ is located in two squares with 1s (Fig. 13-33*c*). We don't care what C is, and so we would mark a 1 in both squares.

Step 3. The third term, $\bar{A}D$, is located in four squares (Fig. 13-33*d*). We don't care what B or C is, and so we mark a 1 in the four squares, as illustrated.

Step 4. The last term of our logic expression is $A\bar{B}\bar{D}$. This is found in the two squares that are marked with 1's (Fig. 13-33*e*). We don't care what C is, and so we mark a 1 in the two squares, as illustrated.

Step 5. We now combine the Veitch diagrams in Steps 1–4. By "combining" the previous Veitch diagrams we obtain 1s in each of the squares illustrated in Fig. 13-33*f*.

Step 6. The next step is to simplify this Veitch diagram. We loop, or encircle, the blocks of squares with 1s. A square with a 1 not adjacent to squares with 1s would be encircled by itself. To reduce the logic expression to its simplest form, we try to obtain as few loops as possible. The squares with 1s are looped to form loops of two, four, eight, etc., terms. The loops would not encircle three or five terms. In Fig. 13-33*g* we chose two loops with four squares. This indicates that the expression will simplify to two terms. One loop would simplify to $A\bar{B}$. We drop the changes in the adjacent squares, C changes going horizontally, and D changes going vertically. The other loop will be $\bar{A}D$. The simplified expression using the Veitch diagram will be

$$A\bar{B} + \bar{A}D = F$$

The matrix which has the fewest number of circles, but the largest number of 1s encircled, will produce the simplest circuit. Sometimes the simplified solution can be simplified further by using boolean algebra.

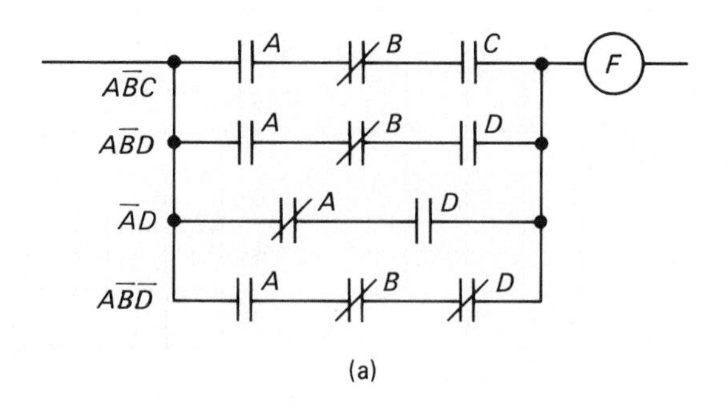

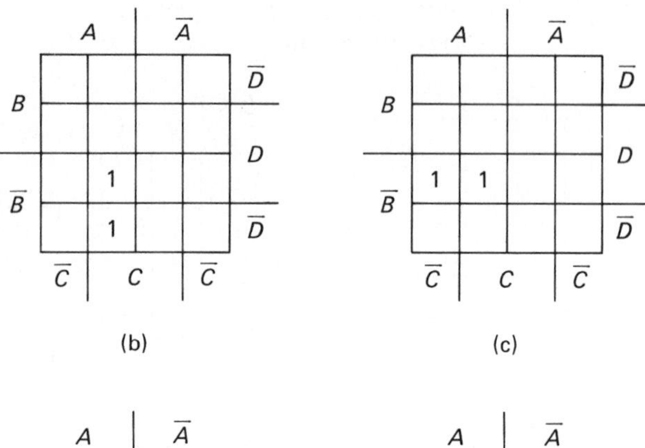

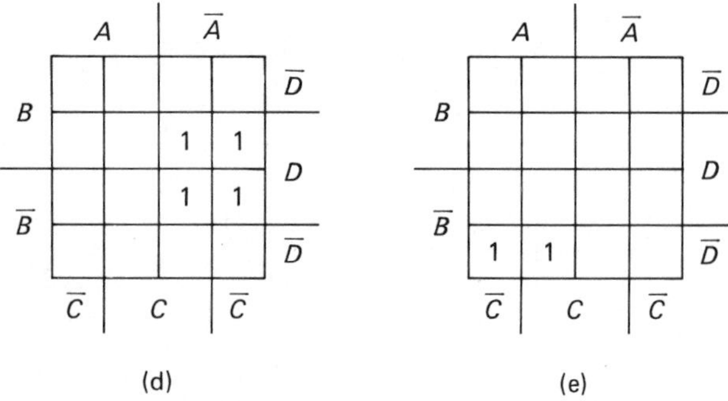

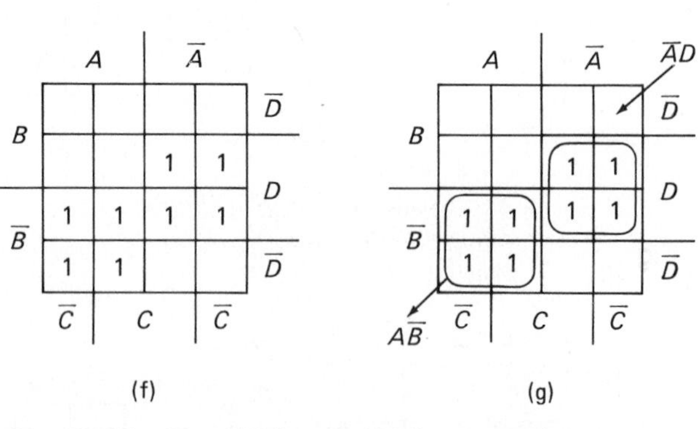

Fig. 13-33 Simplifying the logic expression $A\bar{B}C + A\bar{B}D + \bar{A}D + A\bar{B}\bar{D} = F$ with Veitch diagrams.

EXAMPLE 2: SIMPLIFY THE LOGIC EXPRESSION

$$F = AB + A\bar{C} + \bar{A}\bar{B}\bar{D} + ABC + \bar{A}\bar{B}C + ABC\bar{D}$$

with Karnaugh maps.

Steps in simplifying are:

Step 1. Refer to Fig. 13-34*a*. The squares that contain the first term, AB, are marked on the Karnaugh

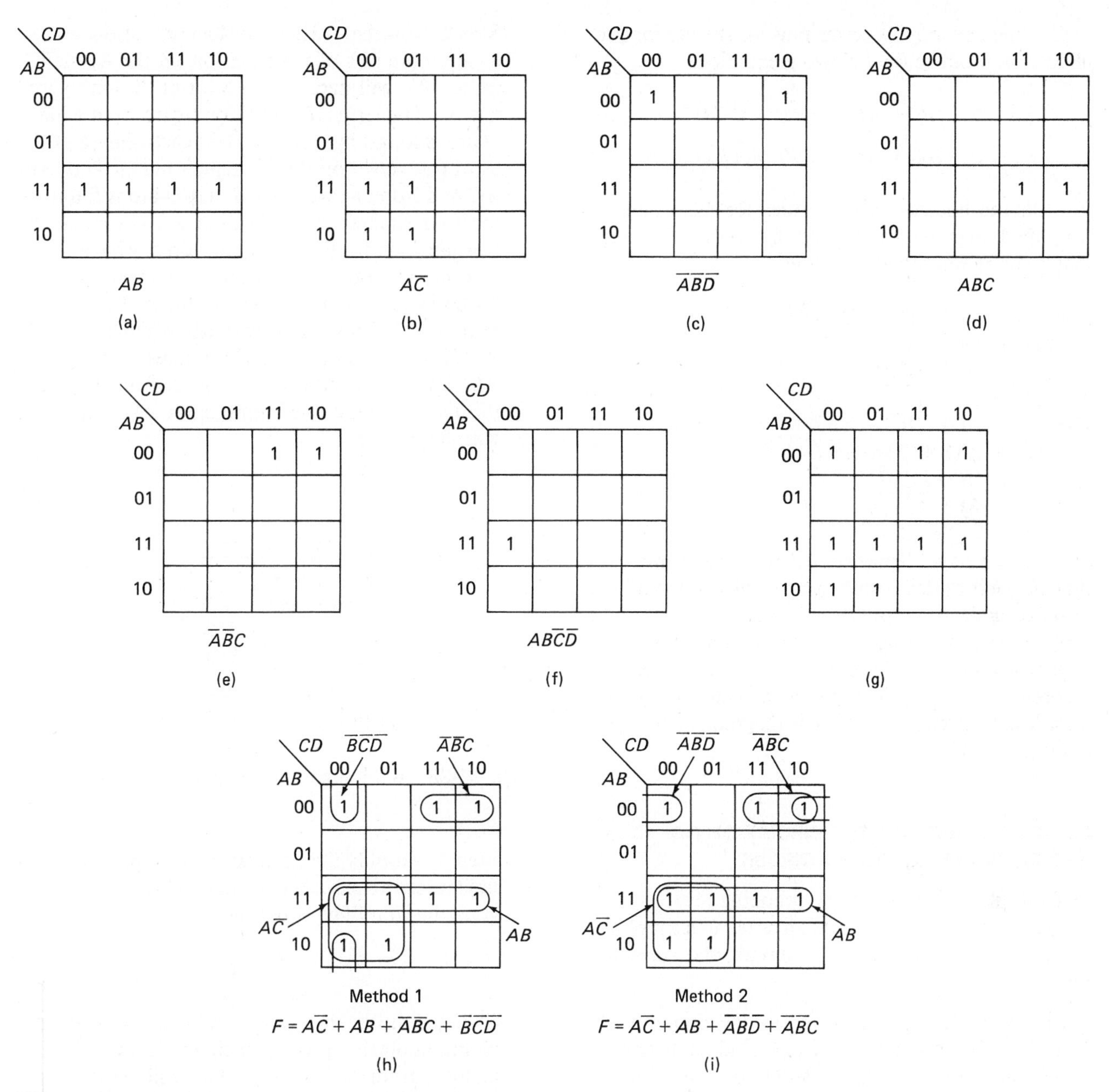

Fig. 13-34 Simplifying a circuit using Karnaugh maps.

map. Because the first term contains only *AB*, we don't care what *C* and *D* are. Notice that the third row down of squares all contain *AB*. We mark all squares with *AB* regardless of what *C* and *D* are.

Step 2. In Fig. 13-34*b* the squares that contain the second term, $A\overline{C}$, are now marked. We don't care what *B* and *D* are. The four squares in the lower left corner contain $A\overline{C}$. We mark 1 in all four squares. Because the term $A\overline{C}$ is found in these squares, they are marked with a 1 regardless what *C* and *D* are.

Step 3. The remaining terms are located as illustrated in Fig. 13-34*c* through *f*.

Step 4. Combining all the terms from the Karnaugh maps in Fig. 13-34, we would have 1s in the squares as illustrated in the Karnaugh map of Fig. 13-34*g*.

Step 5. Draw loops to encircle groups of 1s. Two methods are illustrated in the example, with a simplified expression for each (Fig. 13-34*h* and *i*). Either arrangement of loops will simplify the maps of Fig. 13-34*a* through *g* to the simplest circuits but with somewhat different results. Either result has the same number of contacts. Because complex circuits can always be simplified in several ways, the results of simplifying with Karnaugh maps will differ depending on how you do the looping of the squares containing the 1s.

Circling and looping terms plus using rules for simplification produce the following equations:

$$F = A\bar{C} + AB + \bar{A}\bar{B}C + \bar{B}\bar{C}\bar{D}$$

$$F = A\bar{C} + AB + \bar{A}\bar{B}C + \bar{A}\bar{B}\bar{D}$$

With the boolean algebra rules for factoring, the simplified expressions derived from the Karnaugh maps can be simplified as follows:

$$F = A\bar{C} + AB + ABC + \bar{B}\bar{C}\bar{D}$$
$$A\bar{C} + AB(1 + C) + \bar{B}\bar{C}\bar{D}$$
$$A\bar{C} + AB + \bar{B}\bar{C}\bar{D}$$
$$F = A(\bar{C} + B) + \bar{B}\bar{C}\bar{D}$$
$$F = A\bar{C} + AB + ABC + \bar{A}\bar{B}\bar{D}$$
$$A\bar{C} + AB(1 + C) + \bar{A}\bar{B}\bar{D}$$
$$A\bar{C} + AB + \bar{A}\bar{B}\bar{D}$$
$$A(\bar{C} + B) + \bar{A}\bar{B}\bar{D}$$

What we have demonstrated in these examples is that an electric relay circuit can be converted to a logic or boolean algebra expression. The expression can be simplified by mapping. It can be simplified further by factoring and using boolean algebra postulates and theorems. The simplified expression can be converted back to a relay circuit, to a logic diagram, or to a program for a P.C.

EXAMPLE 3: SIMPLIFY THE CIRCUIT OF FIG. 13-35 AND DRAW THE SIMPLIFIED CIRCUIT

In Example 3, Fig. 13-35, by mapping techniques, a relay circuit is simplified and converted to a simplified electric circuit, a simplified logic circuit, and a program for a P.C.

Step 1. Write a logic equation for the circuit in Fig. 13-35. The top line or rung of the ladder diagram is AB, the second rung is $A\bar{C}$, and so on. Writing this into a logic equation or expression, we have

$$F = AB + A\bar{C} + \bar{A}\bar{B}\bar{C} + ABC + \bar{A}BC + AB\bar{C}D$$

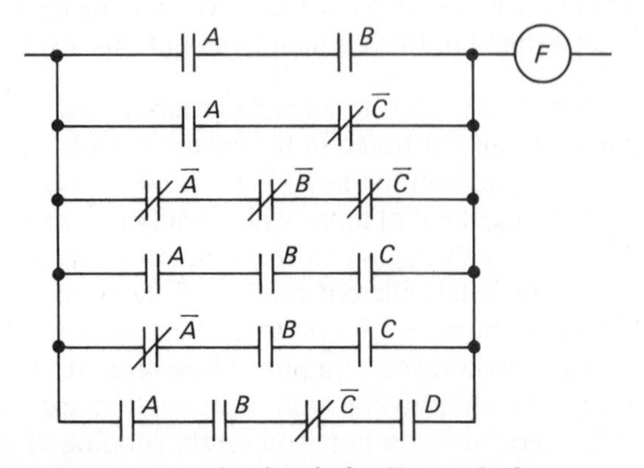

Fig. 13-35 Electric circuit for Example 3.

Step 2. Now that we have the logic equation, we can locate the terms of the equation on the Karnaugh map. We will need to use a lot of "I don't care" terms. The term AB and I-don't-care-what-C-and-D-are is found by placing a 1 in each square of the third row down on the Karnaugh map. The term $A\bar{C}$ and I-don't-care-what-B-and-D-are is found in the four squares in the bottom left corner of the Karnaugh map. When a square already has a 1 in it, no other 1s should be placed in it. The term $\bar{A}\bar{B}\bar{C}$ is found in the two left squares of the first row. The term ABC is found in the two right squares in row 3. $\bar{A}BC$ is found in the two right squares of row 2, and $AB\bar{C}D$ is found in the second square from left in the third row. Figure 13-36 represents the "completed" Karnaugh map.

AB \ CD	00	01	11	10
00	1	1		
01			1	1
11	1	1	1	1
10	1	1		

Fig. 13-36 Karnaugh map derived from Fig. 13-35.

Step 3. Simplify the expression by looping the 1s as illustrated in the Karnaugh map in Fig. 13-37. This gives the simplified equation

$$F = A\bar{C} + BC + \bar{B}\bar{C}$$

Step 4. Simplify the expression by factoring and applying boolean algebra principles. First step is to factor $\bar{C}$ from $(A\bar{C} + \bar{B}\bar{C})$. This will leave

$$F = \bar{C}(A + \bar{B}) + BC$$

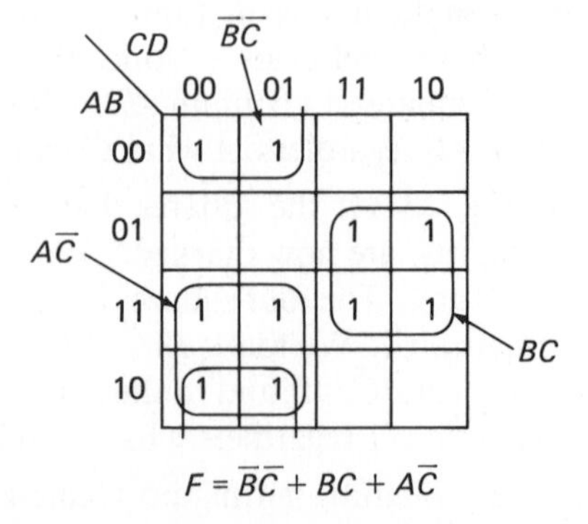

Fig. 13-37 Looping of 1s to give the simplified circuit for Fig. 13-35.

The simplified electric relay circuit is found in Fig. 13-38. The equivalent logic circuit for this simplified circuit is illustrated in Fig. 13-39.

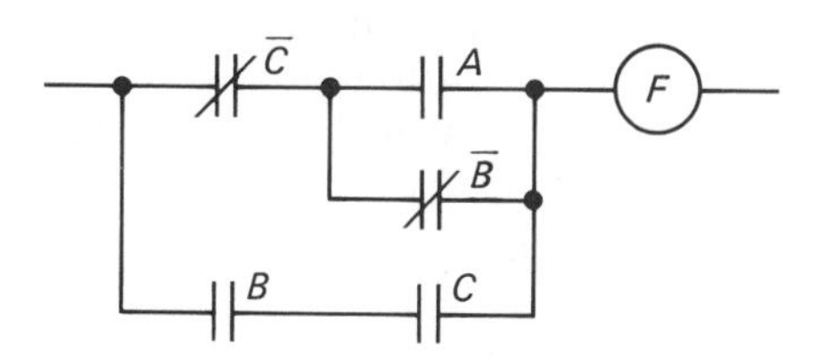

Fig. 13-38 Simplified equivalent electric circuit for Fig. 13-35.

13-12 DEVELOPING LOGIC CIRCUITS FROM CONTROL CIRCUIT REQUIREMENTS

Earlier in this chapter it was indicated that there were three methods used to develop logic control circuits. The use of boolean algebra has been explained, and now we are ready to examine how a logic circuit can be developed from the specifications of a machine. These specifications are called *English expressions.*

Any explanation of the control of a machine can be expressed in terms of the logic relationships of each function. Therefore, a control system breaks down into basic logic functions. The designer of conventional control circuits probably is not conscious of this, but examination of the progressive steps in designing a circuit will illustrate that the logic-function technique was actually one of the methods used to determine the circuit design.

Consider a circuit whose requirement is that a coil M be energized when either a pressure switch PS1 or a limit switch LS1 is closed. If this circuit were developed for magnetic control, it would be as shown in Fig. 13-40. The logic statement of this circuit is: "The coil will be energized when either PS1 or LS1 is closed." Therefore, the logic element required would be a two-input OR, as indicated in Fig. 13-41. Note the amplifier which must be inserted between the OR element and coil M to raise the power level sufficiently to energize M.

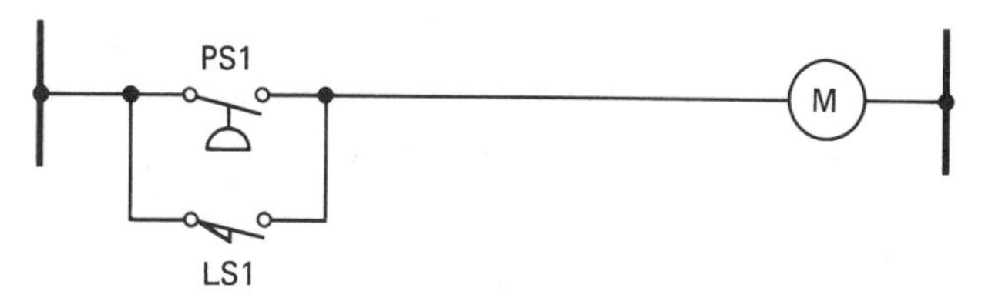

Fig. 13-40 Magnetic control circuit.

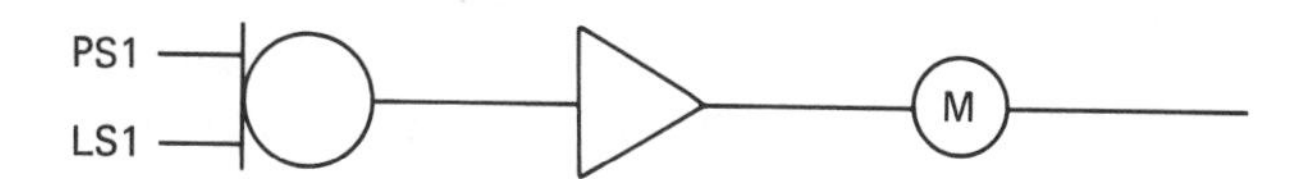

Fig. 13-41 Basic logic circuit for Fig. 13-40.

Possibly the greatest difficulty when studying static control for the first time stems from the difference between the circuit shown in Fig. 13-40 and that shown in Fig. 13-41. The conventional, or electromagnetic, circuit provides a means of tracing the flow of current from line 1 through the control devices, through the coil, and to line 2 is easy to understand. However, the logic diagram does not indicate the power circuit as such; rather, it is a block diagram of the control function of the circuit, and at this stage may tend to leave something to be desired from the student's point of view. As logic circuits become more complex, the desirability of omitting the actual wiring and using symbolic representation of control functions will become more understandable. The OR element of Fig. 13-41 could be made up of vacuum tubes, saturable reactors, or, more probably, solid-state devices such as transistors. The amplifier of Fig. 13-41 might well be a relay, a vacuum-tube amplifier, a saturable reactor, or a silicon-controlled rectifier used in a switching mode to provide the necessary power amplification. The logic diagram does not in any way indicate the actual circuitry or components used within the logic element. This information would be found in the circuit for each element and would vary from manufacturer to manufacturer. The student who feels it necessary to visualize a complete circuit with the logic diagram might be aided by considering that there is a common, or

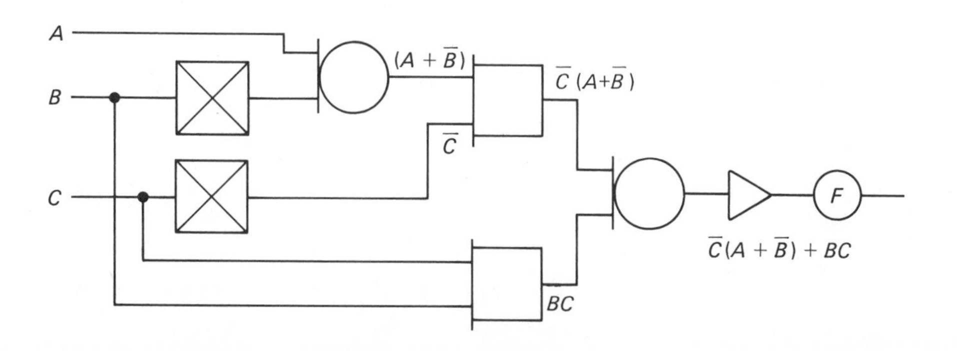

Fig. 13-39 Logic circuit developed from Fig. 13-38.

ground, bus which is not shown and all input and output voltages are taken from this common bus. Normal practice dictates that the common bus not be shown in order that the diagram will not be unnecessarily cluttered, since its presence adds nothing to the information given by the logic diagram.

Suppose that we add to the circuit of Fig. 13-41 a further specification that coil M will be energized by PS1 or LS1 only when contacts T1 and T2 are closed. Our circuit would need to be modified to that of Fig. 13-42. The logic function we have used could be stated: "The outputs of the OR and T1 and T2 must all be providing an input, or ON condition, before M is energized." This naturally indicates the use of a three-input AND unit, as shown in Fig. 13-42.

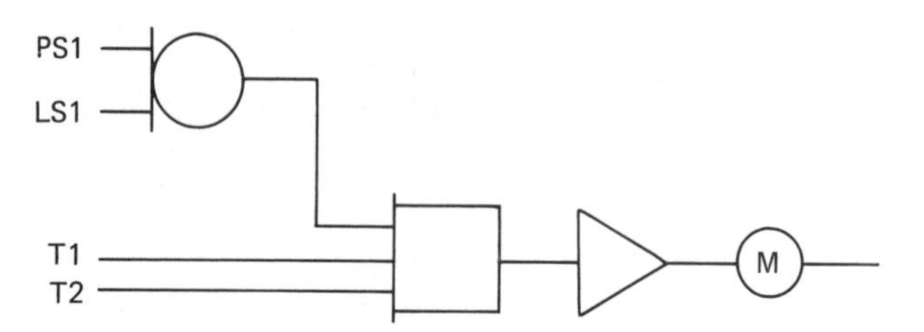

Fig. 13-42 First addition to the logic circuit.

A further development of the logic circuit of Fig. 13-42 might include the following as the total logic statement: "Coil M is to be energized when PS1 or LS1 and T1 and T2 are closed, and only if there is no input from T3 or T4." The logic requirements assigned to T3 and T4 would indicate the use of an OR-NOT element, since the presence of an output must be accompanied by the lack of an input from T3 and T4. The circuit for our new specifications is that given in Fig. 13-43. The two-input AND could be eliminated by using one four-input AND.

To read the diagram of Fig. 13-43 when no specifications are provided is relatively simple. The logic elements supply the key words to indicate the operation of the circuit. If we start at the top of the left-hand part of our circuit, the diagram will read as follows. "Whenever PS1 or LS1 provides an input to the OR element, there will be an output. When there is an output from the OR, it provides one of the three inputs to the first AND element, which requires an input from T1 and T2 in order that it have an output. When the above conditions are met, the first AND unit provides an input to the second AND unit, which must also have an input from the OR-NOT element. The OR-NOT will have an output only when there is no input from T3 or T4. When this condition is met, both inputs are present at the second AND element, thus providing an output. The output of the second AND element provides an input to the amplifier, which supplies the power amplification necessary to energize coil M."

At this point the student may well wonder what PS1, LS1, T1, T2, T3, and T4 are. The details of the field wiring or sensing devices are normally found on a separate diagram and not shown on the logic diagram. These might well be contacts of pressure switches, limit switches, and thermostats. The action part of the control circuit is represented in the logic diagram, and the power, or utilization, circuit normally is shown in a third diagram.

Consider the circuit of Fig. 13-43 from an operational standpoint. If PS1 is closed but LS1 is open, there will still be an output from the OR element. If T1 and T2 are closed, they will complete the necessary inputs to the AND element, thus providing one input to the second AND element. If T3 and T4 are open, there will be an output from the NOR element. The second AND element now has an output and the coil is energized. Consider what would happen if T3 closed, thus providing one input to the NOR element. The result of T3 providing an input to the NOR element is that it will lose its output. Since the second AND element now has only one input present, it no longer has an output. The amplifier has no input; therefore, it has no output and coil M is no longer energized. Consider that instead of T3 closing and providing an input to the NOR element, T2 opens and thus eliminates one input to the first AND element. The net result would be the same in both cases—coil M would not be energized.

What would the circuit of Fig. 13-43 look like if it were an electromagnetic circuit and T1, T2, T3, and T4 were thermostat contacts, and if we assumed that PS1

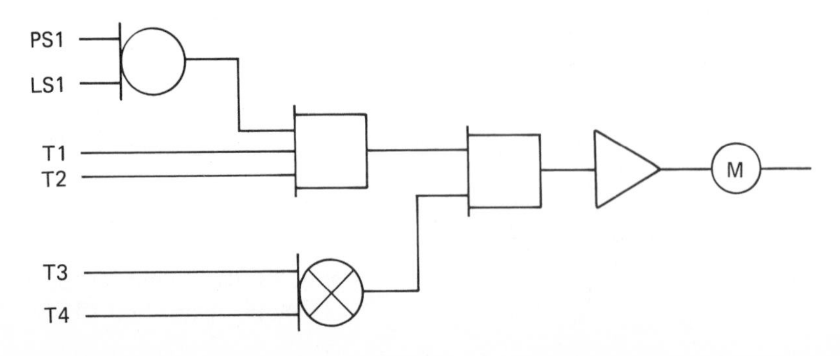

Fig. 13-43 Second addition to the logic circuit.

were a pressure switch and LS1 a limit switch? The circuit is shown in Fig. 13-44. This may seem to the student to be a much simpler representation of our circuit than the logic diagram of Fig. 13-43. But suppose now we add a further requirement to the circuit, that it provide complete isolation of coil M from the sensing devices and their contacts. Now the circuit becomes more complex and might possibly be wired as shown in Fig. 13-45. Neither the logic diagram nor the schematic diagram is that of a particularly complex circuit, but you should be able to visualize the simplicity of the logic diagram when applied to the complete automation of a machine or production line.

The three-wire control circuit which was used so frequently in the study of electromagnetic control circuits can be represented in the logic circuit by the use of feedback. The schematic for an electromagnetic circuit and its equivalent logic diagram are shown in Fig. 13-46. To understand the operation of the logic diagram, consider that one input is always provided to the AND element by the STOP button. When the START button is pressed, the second and final input to the AND element is provided and an output is produced. Once an output is produced, the internal feedback loop maintains an input even though the START button is released. This is equivalent to the action of the maintaining contacts parallel with the START button in the electromagnetic circuit. If the STOP button is pressed, the AND element loses its output, and there is no feedback to provide an input to substitute for the START button. The circuit is now in its original "off" state, and even when the STOP button is returned to the closed position, the START button must be pressed before there will again be an output from the AND element.

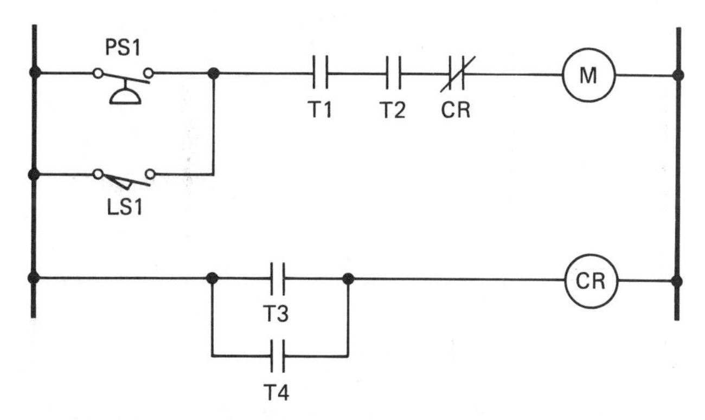

Fig. 13-44 Electromagnetic circuit equivalent of the final logic circuit.

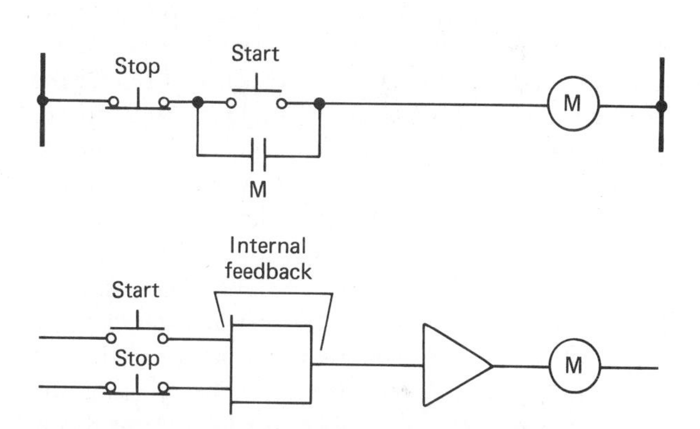

Fig. 13-46 Electromagnetic three-wire control circuit and logic equivalent.

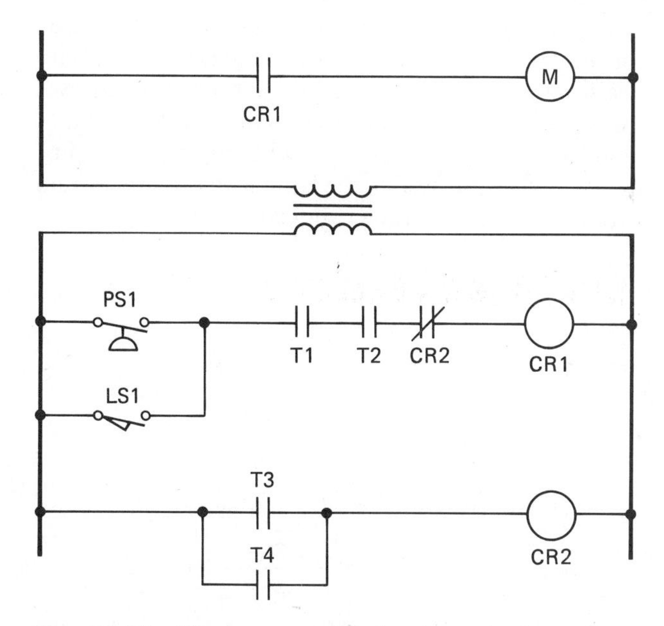

Fig. 13-45 Electromagnetic circuit equivalent of final logic circuit providing isolation of pilot devices.

13-13 APPLICATION OF STATIC-AND-LOGIC ELEMENTS

Any control system can be divided into three major sections (Fig. 13-47 on the next page). Sensing devices such as pushbuttons, limit switches, and thermostats constitute the information-gathering section. As the information is gathered, it must be interpreted by the decision section in order that a decision can be made about what the system should do. In electromagnetic circuits this section contains mostly relays. Having arrived at a decision based upon the information gathered, the system must take appropriate action. The action section of the control system consists of the final output device or devices, such as motor starters, indicating lamps, and solenoids.

Static control in the digital form of logic circuitry generally is applied to the decision section of the con-

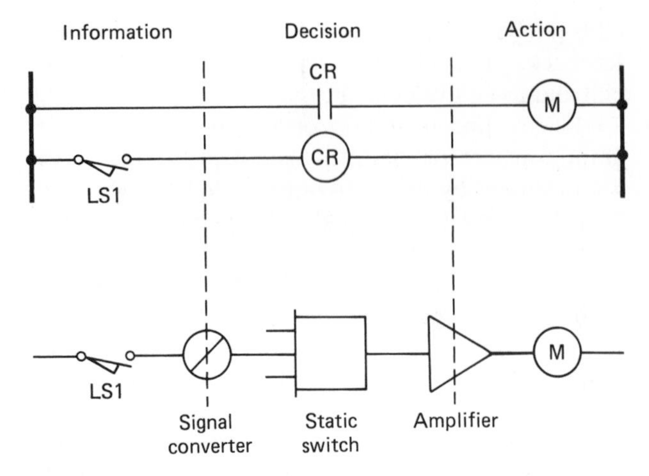

Fig. 13-47 The three major divisions of control circuits.

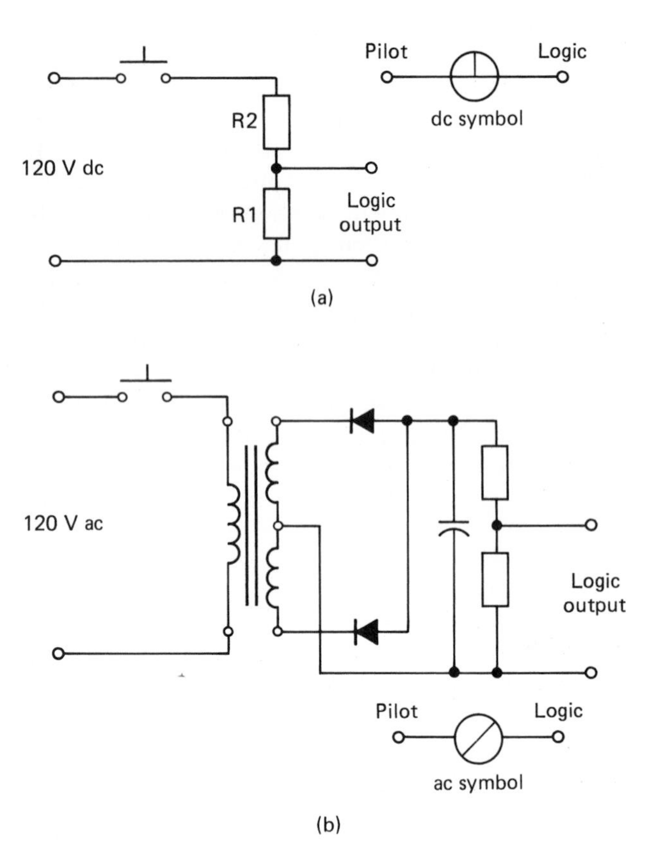

Fig. 13-48 Signal converters and their symbols.

trol system. Logic elements are low-voltage, low-power devices. They therefore require signal converters, sometimes called *original inputs,* to reduce the necessary high voltage of the sensing section to proper logic-signal values. They also require amplifiers to convert their low output power to the level of power required by the action section of the system.

The information section of the control circuit can be made entirely static by the use of static sensors such as the proximity limit switch. This section is much more likely to contain the familiar contact-type devices found in electromagnetic control. The opening and closing of contacts can be used as direct input to the logic elements, provided the proper value of low-voltage direct current is used through the contacts. When this is done, the contacts may become unreliable because the voltage is not high enough to overcome the resistance of dirty contact surfaces. The resistance of long leads can produce excessive voltage drop in the sensing circuits, which also will produce unreliable operation. Experience has shown that reliable operation of static logic fed from contact sensors requires a high-voltage contact circuit, usually 48 to 125 V dc, or ac if desired. This value of voltage used on contacts which have good wiping action will provide reliable input to the logic elements. The voltage used in the sensing section must be reduced and sometimes converted from alternating to direct current by means of an original-input signal converter (Fig. 13-48). Generally speaking, the signal converter should be mounted as close to the logic elements as possible to reduce electromagnetic interference, or EMI, problems, commonly referred to as *noise.*

The dc signal converter of Fig. 13-48*a* is a simple voltage divider used to lower the dc input to the proper value. This type of converter is the least expensive and probably the most used. This simple circuit can be improved by adding a transistor circuit to the voltage-divider network.

The circuit of Fig. 13-48*b* is a simple transformer and rectifier-supply ac signal converter which would be greatly improved by the addition of a transistor circuit to its output. Most manufactured signal converters are supplied with a pilot lamp to indicate the condition of the pilot device—on or off. A good practice with static switching is to wire each pilot device to its own original-input signal converter, even though it might be connected in parallel or series with other devices (Fig. 13-49). The purpose of the lamp is to provide instant trouble indication at the logic panel.

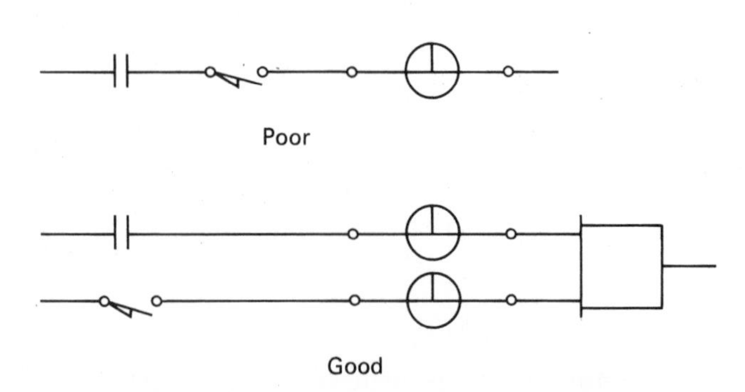

Fig. 13-49 Connections for signal converters.

Manufacturers' specifications should be followed very carefully in regard to noise suppression and the location of original inputs. The logic element operates at such low power levels that noise spikes can produce false switching if not properly suppressed.

The action section of the control circuit is where the work is done, that is, where power is consumed or switched. When the action required is to light a pilot lamp, the power may be a watt or two. When a large solenoid valve or starter must be actuated, the power requirements are much greater. The logic element is made to be a milliwatt device and therefore cannot switch such loads. Static amplifiers are used to directly switch loads up to several hundred watts or to activate relays and starters for larger loads. It is quite likely that future development in semiconductor devices will make it practical for all action devices to be static. Amplifiers using power transistors, reed relays, and SCRs are available. Manufacturers' specifications should be carefully followed, and a unit should be selected which fills the requirements of the load.

The decision section of the control circuit is the chief application of logic elements and therefore has been left until last. Remember, the input to the logic has been adjusted by the signal converter to optimum value, and the amplifier will convert the logic output to match the load. This gives the designer of the logic circuit freedom of choice of components limited only by the logic of the decisions to be made. The designer need not even know how each element works; nor is it necessary for the designer to know the internal circuitry.

13-14 DEVELOPMENT OF LOGIC DIAGRAMS

It should be apparent that the symbols used in this chapter are for one company only and are not the symbols used in logic circuits made by all manufacturers of logic components. Obtaining the manufacturers' data sheets for their logic components would be required to develop a logic diagram using a specific company's logic symbols and circuits.

The first step in the development of a logic diagram is to convert the specifications into simple English logic statements. Each logic statement will be the equivalent of one line on the schematic diagram studied in Chap. 7. The second step is to draw each sequence, or statement, in logic-symbol form. The third step is to integrate and interconnect the individual sequences where necessary to provide a complete control circuit. The fourth step is to examine the total circuit to see whether some functions can be combined to reduce the number of logic elements required. The fifth step is to review the circuit for possible conflicts between sequences and to check for correct overall circuit satisfaction of the specifications.

13-15 DEVELOPMENT OF A SOLENOID VALVE CONTROL CIRCUIT

The specifications for this circuit are that a solenoid valve (SOL) is to be energized whenever a normally open pushbutton PB1 is pressed, regardless of other inputs, or whenever pressure switch PS1 and thermostat T1 are closed and extreme-limit pressure switch PS2 and extreme-temperature thermostat T2 are both open.

The first step is to convert the specifications to logic statements. The first English logic statement is: SOL will be energized when PB1 or OS1 and T1 are closed. The logic diagram for this statement is shown in Fig. 13-50*a* on the next page.

The second English logic statement is: statement 1 will be true only if PS2 and T2 are not energized. This statement calls for the NOT output from an OR function, as shown in Fig. 13-50*b*. The first two steps in development are now complete. The third step is to combine the statements in the logic diagram, as in Fig. 13-50*c*.

The fourth step is to combine logic elements where possible and simplify the circuit if this is practical, which can be done in this case by using an OR unit with a built-in NOT output, as shown in Fig. 13-50*d*.

The fifth step is to analyze the circuit to be sure that it will perform the functions specified. We see from Fig. 13-50*d* that when PB1 is closed, it provides an input to the OR through its signal converter. The OR requires only one input to have an output; therefore, there is an input to the amplifier. The amplifier has been chosen to provide proper output for the solenoid when it is provided with an input; therefore, SOL is now energized and satisfies the specifications for PB1.

If PS1 is closed but T1 is open, there will be only one input to the AND: therefore, there will not be an output. If T1 closes while PS1 is closed, there are two inputs provided to the AND, but this will not produce an output unless there is an output from the NOR element. We must now examine PS2 and T2. If these two devices are open, there is no input to the NOR element, and therefore it will have an output and supply the third input to the AND element. When all three inputs are present at the AND, it will have an output. The output of the AND provides an input to the OR and energizes the solenoid through the amplifier.

13-16 DEVELOPMENT OF A THREE-STAGE AIR-CONDITIONING SYSTEM CONTROL CIRCUIT

This circuit is for a three-stage air-conditioning system. Machine 1 is the smaller of two machines and

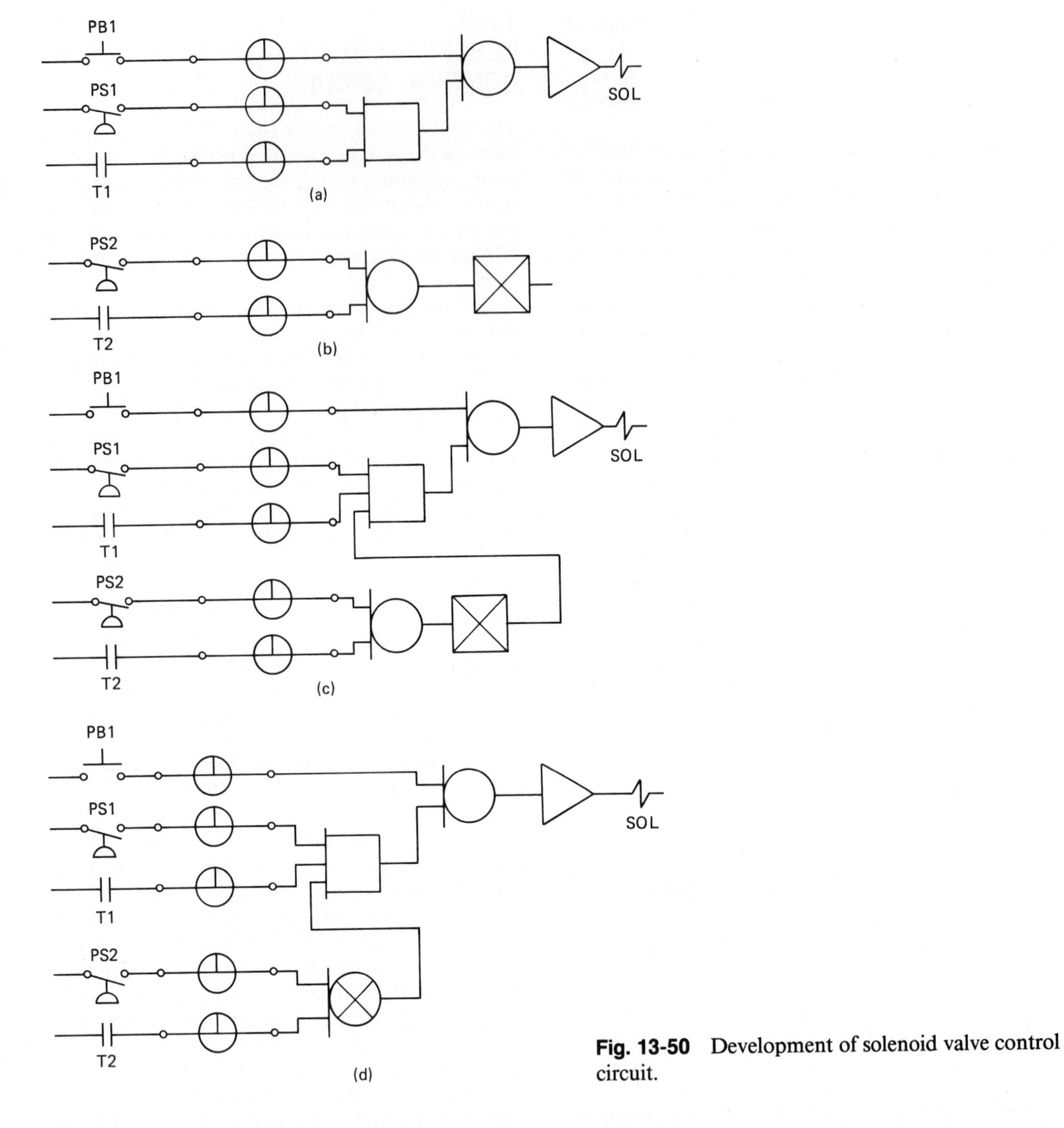

Fig. 13-50 Development of solenoid valve control circuit.

should run whenever the chilled-water flow switch FS1 and the main-control thermostat T1 are on and the second-level thermostat T2 is not on. This provides normal operation for most conditions.

When machine 1 cannot keep up with the load, the second-level thermostat T2 must first shut down machine 1 and then cause machine 2 to start, provided FS1 is still on. This provides a greater capacity to carry the load.

Whenever the load exceeds the capacity of machine 2, a third thermostat set at a higher temperature than that for T1 and T2 closes. This will keep machine 2 running and start machine 1. This provides a third capacity to carry large heat loads; in this case, both machines are used.

The first step in the development will be to reduce the specifications to logic statements. Then a logic diagram will be drawn for each.

The statement for machine 1 for all three stages of operation is: Run when T1 and FS1 are on and T2 is not on or when T3 and FS1 are on. The logic diagram for machine 1 is shown in Fig. 13-51*a*. The logic diagram for machine 2 is shown in Fig. 13-51*b*.

The third step is to combine the logic diagrams of the two machines into a complete circuit using only those components actually needed. This complete circuit is

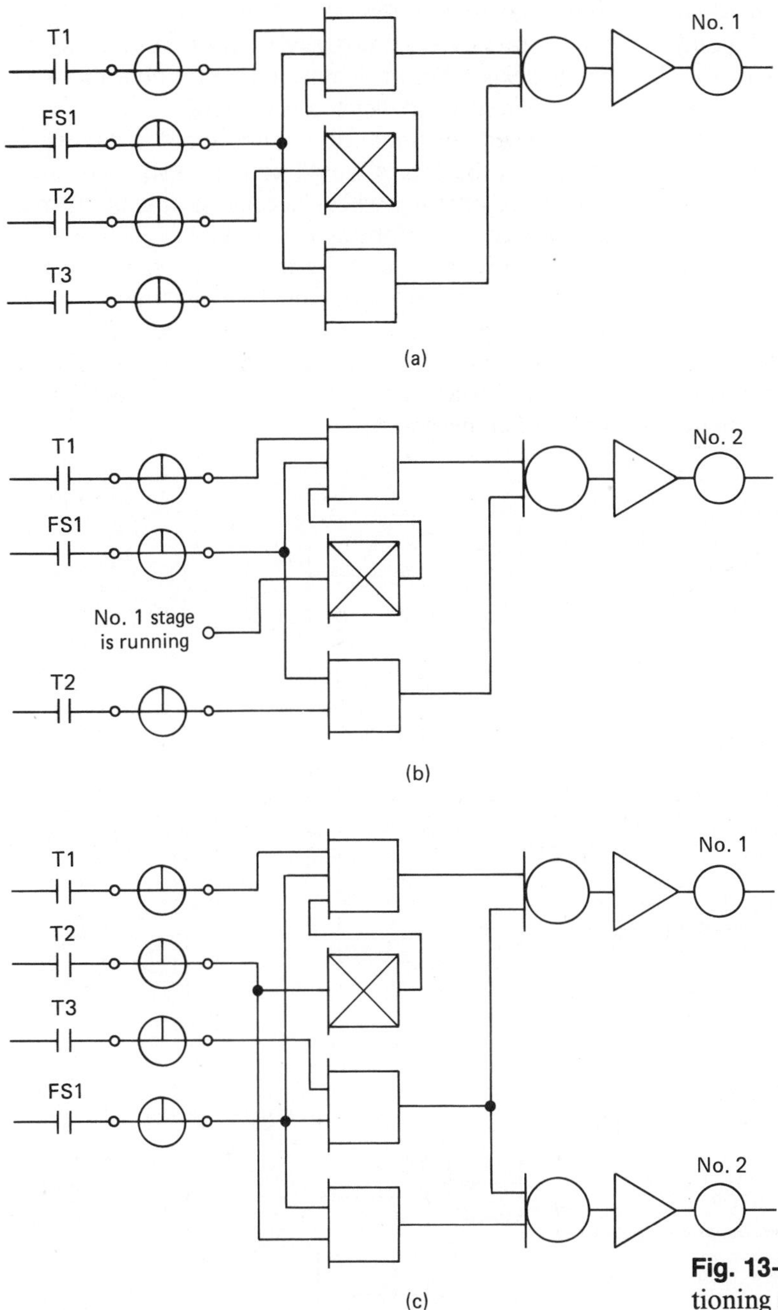

Fig. 13-51 Development of a three-stage air-conditioning system control circuit.

shown in Fig. 13-51*c*. Careful analysis of the final circuit will show that one of the AND units has been eliminated because one unit meets the need of the circuit. One NOT unit has also been eliminated for the same reason.

13-17 DEVELOPMENT OF A THREE-CONVEYOR CONTROL CIRCUIT

This is to be a circuit for three conveyors. There are two START buttons, one located at each end of the conveyor system. There are three STOP buttons, one located at each conveyor. Each conveyor is to be protected by a limit switch.

Pushing either START button will start all conveyors in sequence. Operation of any STOP button or limit switch will stop its conveyor and the conveyor immediately preceding it in the sequence.

The logic statement for conveyor 1 is: run when either START button 1 or START button 2 is pressed provided STOP button 1 and LS1 are closed and push-button 2 and LS2 are closed.

The logic statement for conveyor 2 is: run when conveyor 1 is running and PB2 and LS2 are closed and PB3 and LS3 are closed.

The logic statement for conveyor 3 is: run when conveyor 1 and conveyor 2 are running and PB3 and LS3 closed. The complete circuit is shown in Fig. 13-52.

13-18 DEVELOPMENT OF A SEQUENCE SPEED CONTROL CIRCUIT

This circuit is to provide sequence speed control with definite time delay. The motor must be started in its first speed by START button 1 and can then be raised to its second speed by means of START button 2, after a time delay to allow for acceleration. The motor may then be raised to its third speed by START button 3, after a time delay. The STOP button stops the motor regardless of the speed at which it is running.

The logic statement for the first speed is: run when the START button is closed and the STOP button is closed. A MEMORY will be needed because of the momentary contact of the START button.

The logic statement for the second speed is: run when the START button is closed and the time delay has lapsed since the first speed was energized. The completed circuit is shown in Fig. 13-53. The NOT outputs of the MEMORYS are used to shut down each lower speed when necessary.

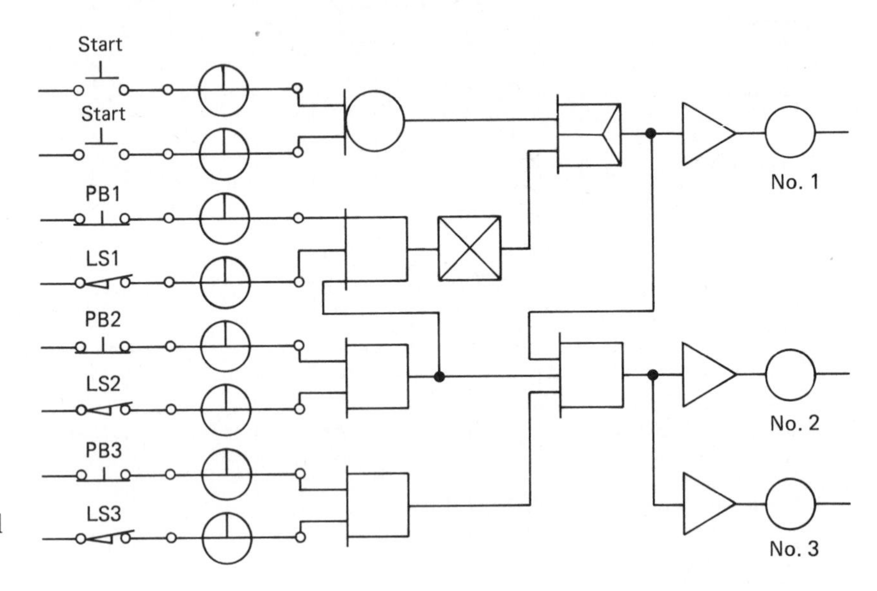

Fig. 13-52 Three-conveyor control circuit.

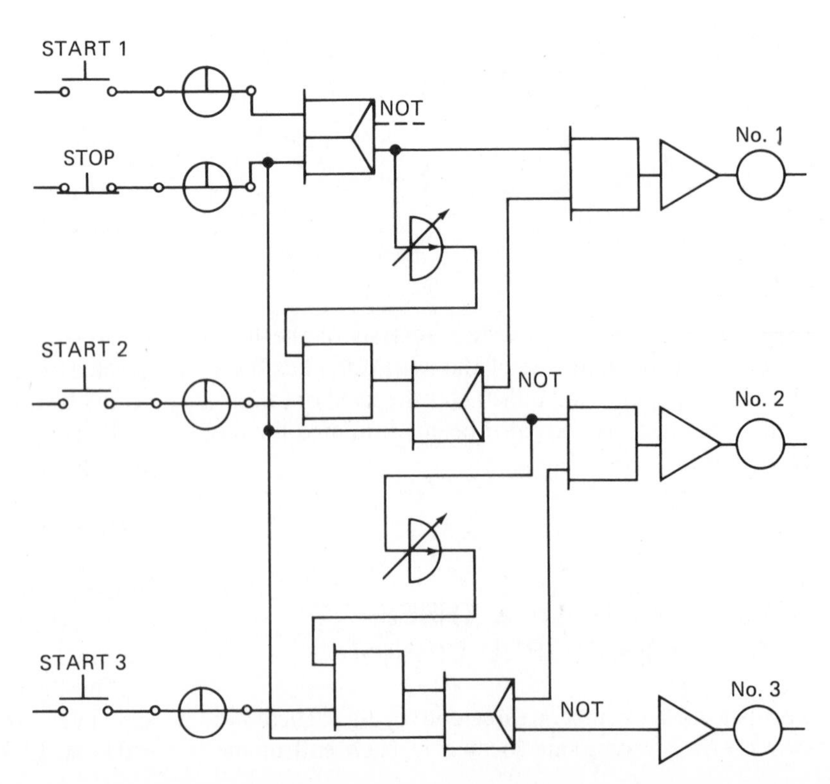

Fig. 13-53 Sequence speed control circuit.

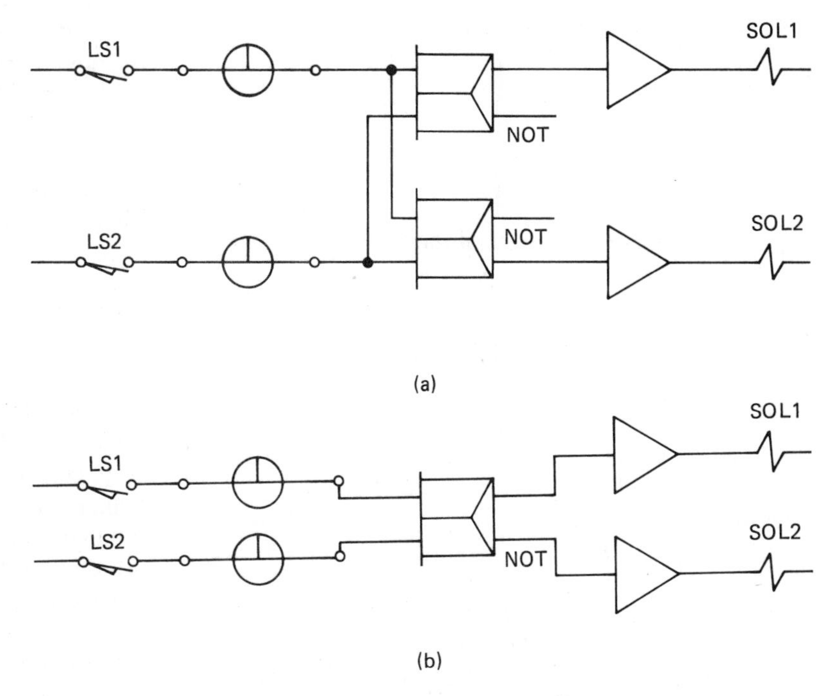

Fig. 13-54 Two-solenoid control circuits.

13-19 DEVELOPMENT OF A TWO-SOLENOID CONTROL CIRCUIT

There are two solenoids on the machine. The first solenoid, SOL 1, is to turn on when momentary-contact limit switch LS1 is closed. The solenoid must remain on until shut off by the closing of momentary-contact limit switch LS2. The closing of LS2 must also turn on a second solenoid, SOL 2, which will remain on until shut off by the momentary closing of LS1 on the next cycle of the machine.

Since the contacts of LS1 and LS2 are closed only momentarily and the solenoid must remember whether they were turned off or turned on, the circuit will require the use of MEMORY elements.

When LS1 is connected to the ON input of one MEMORY unit and the OFF input of a second MEMORY unit half of the requirements of the circuit will be satisfied.

When LS2 is connected to the OFF input of the first MEMORY and the ON input of the second MEMORY, the required circuit will be complete (Fig. 13-54*a*). This circuit can be simplified by utilizing the inverted, or NOT, output of only one MEMORY, as shown in Fig. 13-54*b*.

13-20 OFTEN-USED RELAY EQUIVALENTS

Any number of relay contacts connected in series (Fig. 13-55*a*) can be represented by equivalent AND circuitry (Fig. 13-55*b*). Sometimes it may be necessary to use more than one AND element to provide the required number of inputs. Figures 13-55*c* and 13-55*d* illustrate how nine inputs can be provided when only three AND elements are available.

The logic equivalent of paralleled normally open relay contacts is the OR (Fig. 13-56 on the next page).

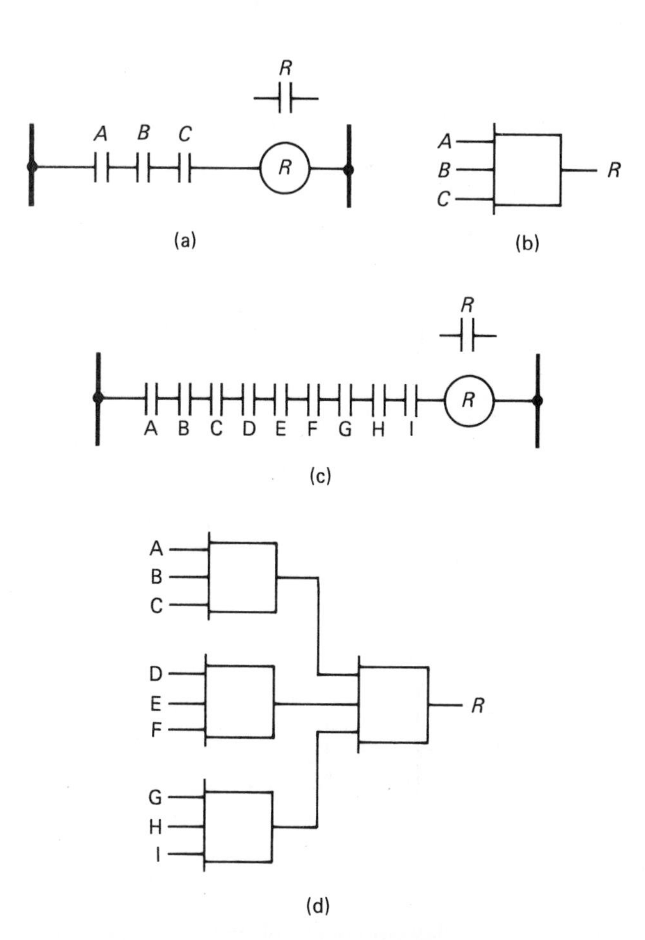

Fig. 13-55 Relay equivalent of logic AND.

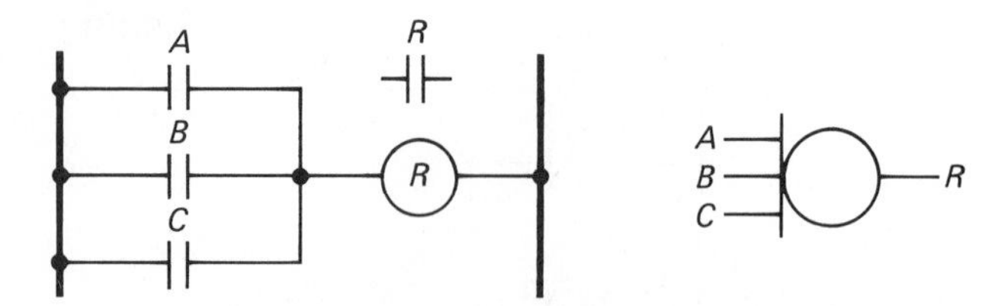

Fig. 13-56 Relay equivalent of logic OR.

Relays with normally closed contacts become a NOT (Fig. 13-57*a*), a NOR (Fig. 13-57*b*), or a NAND (Fig. 13-57*c*), depending upon the number of contacts and how they are connected.

The popular off-return of the magnetic control using three-wire control (Fig. 13-58*a*) can be duplicated in logic circuits by using an off-return MEMORY (Fig. 13-58*b*). The static circuit can also have the NOT, or inverted, output of the intended function shown by dotted lines on the symbol or by a NOT symbol within the MEMORY symbol. The circuit of Fig. 13-58*c* provides the same basic control using a sealed (feedback) AND and allows other inputs to be used.

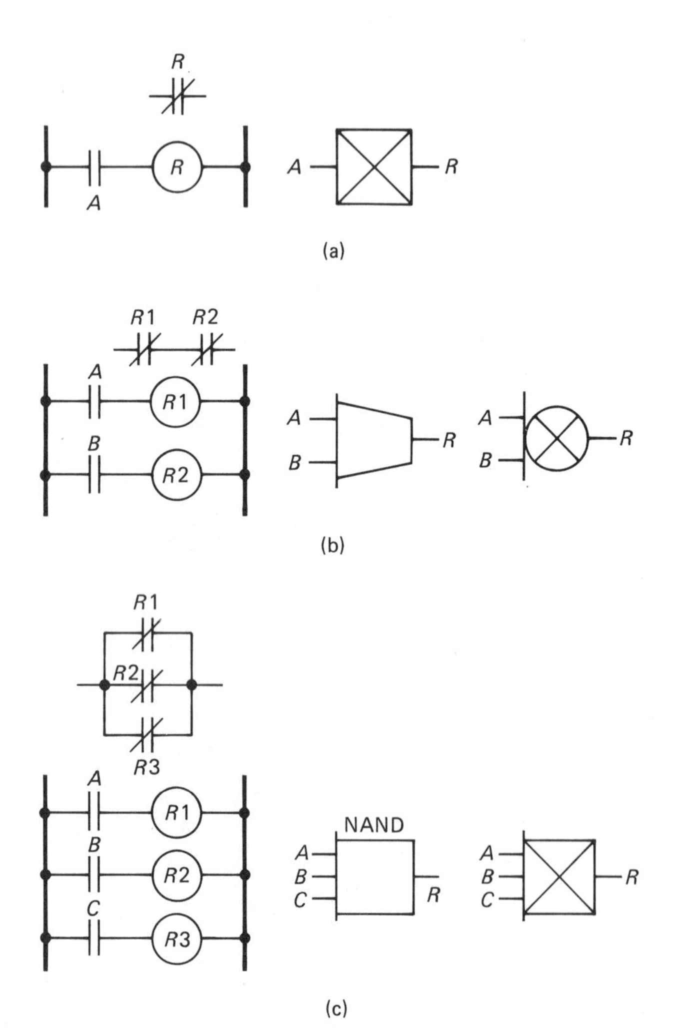

Fig. 13-57 Relay equivalent of logic NOT, NOR, and NAND.

Interlocking, which is so important in machine and process control, is easily achieved in logic circuitry by using the output of a logic function in one part of the circuit as one of the inputs to another logic element in a different part of the circuit (Fig. 13-59).

13-21 THE TRANSISTOR AS A STATIC SWITCH

The logic diagram gives complete information on the operation of the control system but tells nothing about the circuit of the logic element itself. When a system is installed, it must use the logic elements made by only one manufacturer, because two systems are not generally compatible. The voltage and current requirements for the logic elements differ among manufacturers.

Once a particular manufacturer's system is chosen, the installer can be confident that the individual logic elements will work when properly connected in accordance with the logic diagram. Power supplies are designed to supply all the correct voltages. Signal converters are designed to supply the correct input for the logic elements. The output of each logic element is designed to provide the proper input for other logic elements or amplifiers. Since the interconnection requirements are built into the logic elements, it is not absolutely necessary for the installer or servicer to fully understand the actual circuit of the logic element itself. General practice is to replace a defective element or return it to the factory if repair is to be attempted. The installer or servicer will need specific installation information on the individual system being worked on, and will be much better prepared to clear or prevent trouble if how the logic elements work is understood.

All commercially available logic components use the transistor as the basic switching device. During the late 1950s logic elements were designed around a saturable reactor; however, these are no longer manufactured, except for replacement in existing systems.

To understand transistorized logic components, you must first understand how a transistor responds to voltage and current. Figure 13-60 shows the symbols for NPN and PNP transistors with the necessary polarity for full conduction. The three external leads are identified by abbreviations: C for collector, B for base, and E for emitter. Conventional current flow is always used in transistor circuits and is indicated by arrows on the symbols.

The transistor is a semiconductor device which has two circuits; emitter to base and emitter to collector. The emitter-to-base circuit is generally referred to as the *base circuit*. The emitter-to-collector circuit is referred to as the *power* or *output circuit*.

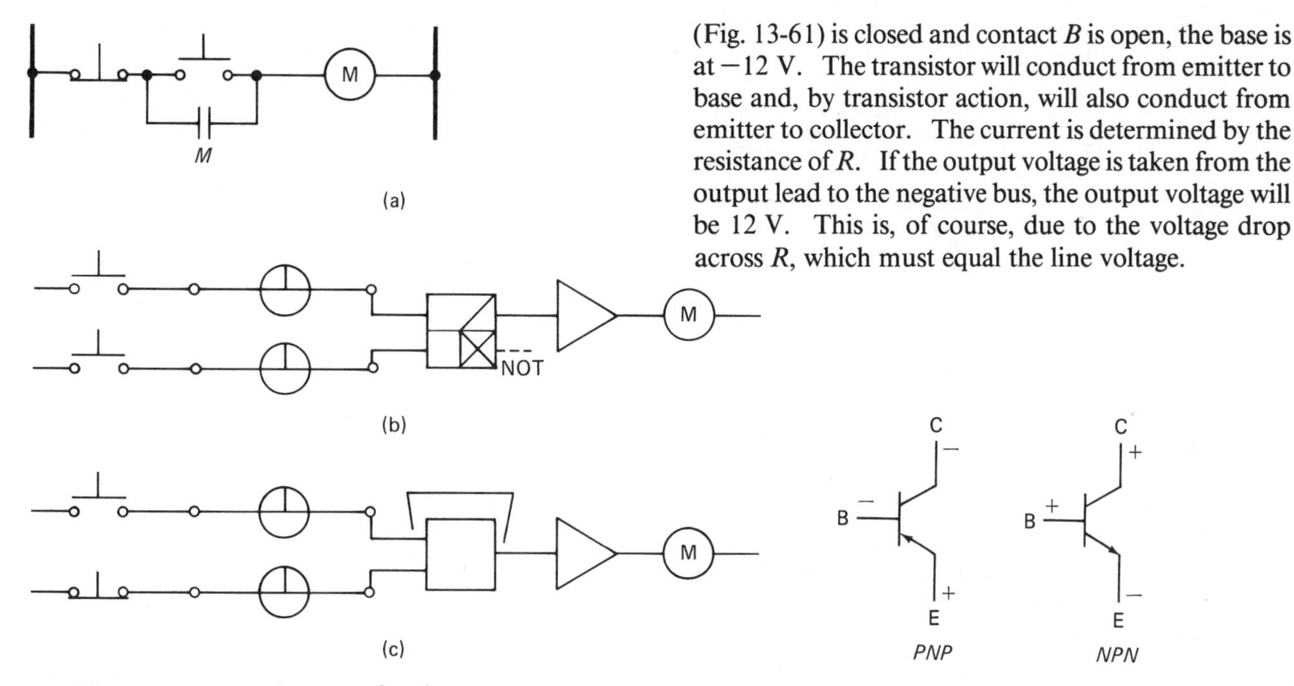

Fig. 13-58 Relay equivalent of logic OFF-return memory.

Fig. 13-60 Transistor symbols.

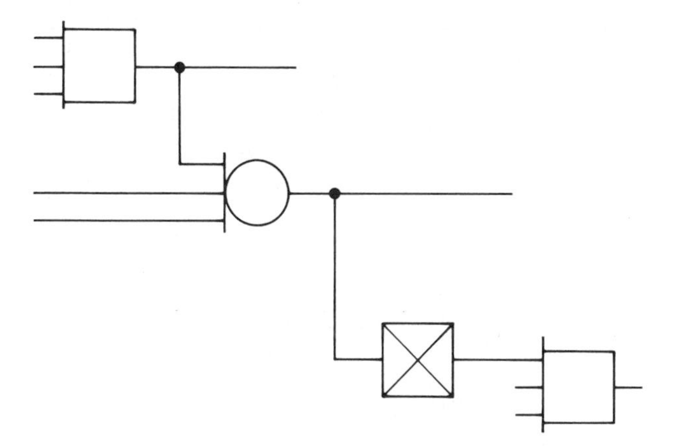

Fig. 13-59 Interlocking in logic circuits.

Fig. 13-61 Basic transistor switch.

Consider the circuit of Fig. 13-61. When contact B is closed, the base of the transistor is at the same potential as the emitter, and the transistor will not conduct from emitter to collector. For general circuit analysis, consider the emitter-to-collector circuit to be an open switch whenever the base of a PNP transistor is at zero voltage difference from the emitter. The same result will be achieved whenever the base is more positive than the emitter.

Whenever the base of a PNP transistor is made sufficiently more negative than the emitter, the transistor will conduct emitter to collector. The emitter-to-collector circuit acts as a closed switch, with practically no voltage drop across the transistor. When contact *A* (Fig. 13-61) is closed and contact *B* is open, the base is at −12 V. The transistor will conduct from emitter to base and, by transistor action, will also conduct from emitter to collector. The current is determined by the resistance of *R*. If the output voltage is taken from the output lead to the negative bus, the output voltage will be 12 V. This is, of course, due to the voltage drop across *R*, which must equal the line voltage.

The previous explanation should make the switching action clear. When the base of the transistor (Fig. 13-61) is positive, with contact *B* closed and contact *A* open, the output is at 0 V because there is no voltage drop across *R*. When the base is negative, with contact *A* closed and contact *B* open, the output is at 12 V because of the voltage drop across *R*.

Because of various design considerations, it is desirable that the switch be conducting when no signal is applied and stop conducting when a signal is applied. In a slight modification of the circuit (Fig. 13-62 on the next page), contact *A* is replaced by resistor *R*2 and resistor *R*4 is added. Resistors *R*2 and *R*4 form a voltage divider across the line voltage. Proper selection of *R*2 and *R*4 will provide the ideal value of negative

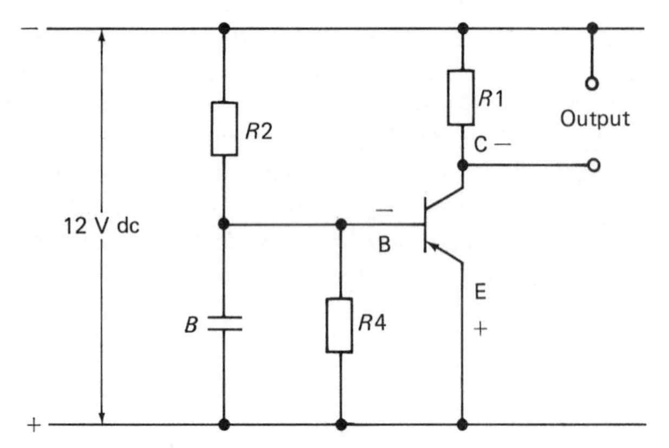

Fig. 13-62 Transistor switch as a NOT.

potential at the base in order to provide full conduction in the emitter-to-collector circuit. The transistor is now conducting and has an output voltage of 12 V. The closing of contact B will connect the base lead to the positive line and cause the transistor to cut off, thus reducing the output to 0 V. Consider contact B to be the pilot device used to control the switch. When the contact is closed, there is an input to the switch, but there is no output. When the contact is not closed, there is no input to the switch, but it does have an output. This action is known as NOT or *inverted logic.* This is the building block upon which all transistorized logic elements are built. Different manufacturers will use different circuit arrangements to achieve the NOT operation of the transistor. However, the analysis given above will suffice to explain the operation of the transistor switch itself.

When the NOT action of the transistor switch (inverted logic) is not desirable, a second transistor is added to the basic circuit. Figure 13-63 shows the new circuit. When contact B is closed, T1 will not conduct. The base of *T2* is then at a negative value in relation to its emitter, because of the circuit from emitter to base through *R*1 to the negative line. The negative potential on the base of *T2* will cause it to conduct, producing a current through *R*3 and an output of 12 V.

When contact B is open, *T1* conducts and brings the base of *T2* to the same potential as the positive bus; therefore, *T2* is cut off, and there is no output. The transistor switch is now a combination of two NOT basic circuits and produces an output when there is an input. When there is no input, there is no output.

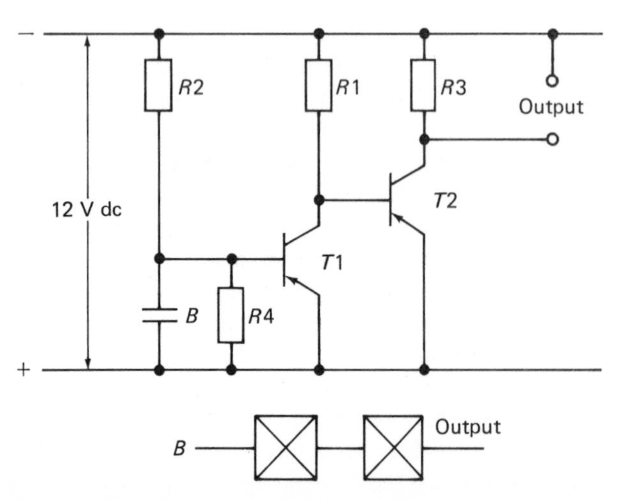

Fig. 13-63 Transistor switch using two transistors.

13-22 THE TRANSISTOR SWITCH AS AN AND ELEMENT

The basic transistor switch can be made to perform the AND function by proper input circuitry. The actual circuits will be considered in later chapters, but a symbolic representation here should improve understanding of the function.

Figure 13-64 shows the basic switch connected through contacts *A, B,* and *C* in series. The switch cannot have an output when the base of *T1* is connected to the positive bus. Resistance voltage dividers are generally used for a practical input circuit. The circuit is arranged so that proper input voltage must be applied to all inputs before the base voltage will become sufficiently positive to cut off *T1.*

If the second transistor, *T2*, had been left off of the basic transistor switch, the result would have been an AND-NOT element.

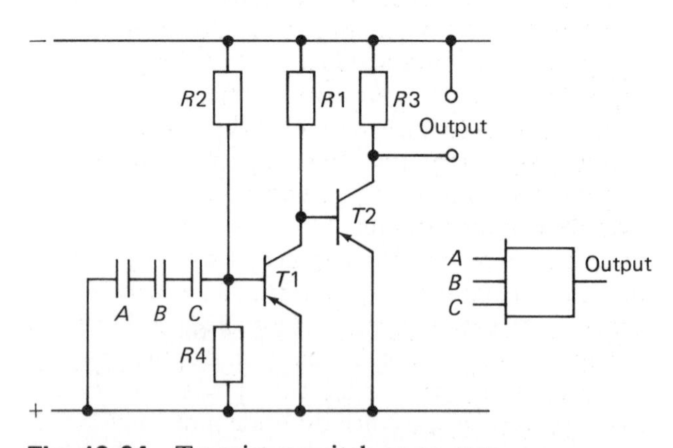

Fig. 13-64 Transistor switch as an AND.

13-23 THE TRANSISTOR SWITCH AS AN OR ELEMENT

Figure 13-65 shows the basic transistor switch connected to three contacts in parallel. If any one of these contacts—*A, B,* or *C*—were closed, the base of *T1*

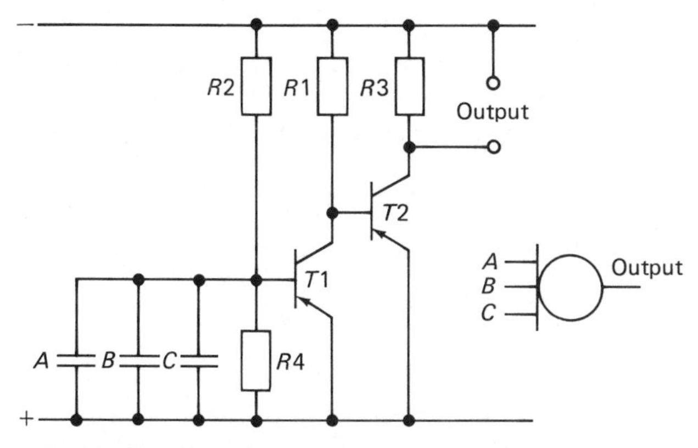

Fig. 13-65 Transistor switch as an OR.

would be connected to the positive line and *T1* would be cut off.

The actual input circuit for the transistor switch is made up of resistors or diodes or both. When they are properly connected, the OR input circuit must make the base of *T1* positive when any one or a combination of its inputs has the proper input voltage applied.

If the second transistor *T2* had not been used in the transistor switch circuit, the result would have been an OR-NOT, or NOR, element.

The transistor logic circuits we refer to as *second-generation logic circuits.* The transistor printed circuits boards with their various logic circuits are still found in many industrial control systems. They may be in operations for years to come. However, new logic systems are using ICs as the logic gates. Integrated circuits are considered as third generation or state-of-the-art devices.

The transistor logic systems with their discrete components are repairable. Anyone involved with the troubleshooting and repair of the transistorized second-generation logic modules should be familiar with the electronic circuits for the various logic gates. The internal circuitry of logic gates will not be presented in this book because the present state of the art for modern logic systems is to use ICs. Troubleshooting techniques for modern logic systems involve tracing highs and lows (1s and 0s) with a logic probe. If a faulty IC is found, nothing can be done about repairing its internal circuit. Replacing it, however, does not require a knowledge of its internal circuitry.

13-24 SHIFT REGISTERS AND COUNTERS

In many industrial control systems it is necessary to count something that affects a controlled process. When the count reaches a certain number, a control action is initiated. An automatic palletizer is a typical example. Suppose the control system called for a number of given items to be loaded on a pallet. After the desired count is reached, the pallet is moved out. The counter is reset, and the cycle is repeated with a new pallet in place.

Various counting methods have developed to handle almost any application imaginable. When the user understands the various type of counters that are available for digital and logic control systems, selecting a counter for a given process becomes an easy task.

There are basically three types of counters used in industrial control applications. The counters have such names as *decade counter, binary counter,* and *ring counter.* Pure "counting" is usually done with a decade or binary counter; the ring counter is most often applied for sequencing and is the solid-state equivalent of a stepping switch. Pure counting decade counters are usually chosen because they are easy to understand and apply. They are easy to decode, easy to read out (visually), and by definition easy to adapt to the decimal system. They also make clean subtract or add-subtract counters. Binary counters, in comparison, use fewer stepping (transfer) elements but take more logic circuitry to decode. They also are harder to read out and do not lend themselves easily to down counting.

The JK flip-flop is the building block for shift registers and counters using ICs. The truth table for a JK flip flop and its IC symbol is illustrated in Figs. 13-66 and 13-67 on the next page. The JK flip-flop also has a set and reset mode. These flip-flops programmed as counters are used in a wide variety of counting applications. Counters using flip-flops are classified either as asynchronous counters or synchronous counters. When the output of a flip-flop is used as the clock input for the next flip-flop, we call the counter an *asynchronous counter.* These counters are also called *ripple counters* in some electronic applications. A *synchronous counter* is a sequential circuit. All the flip-flops in a synchronous counter are controlled by the same clock pulse. The synchronous counter is faster in its counting operation than the *asynchronous counter.* Fast counting operations would generally use synchronous counters.

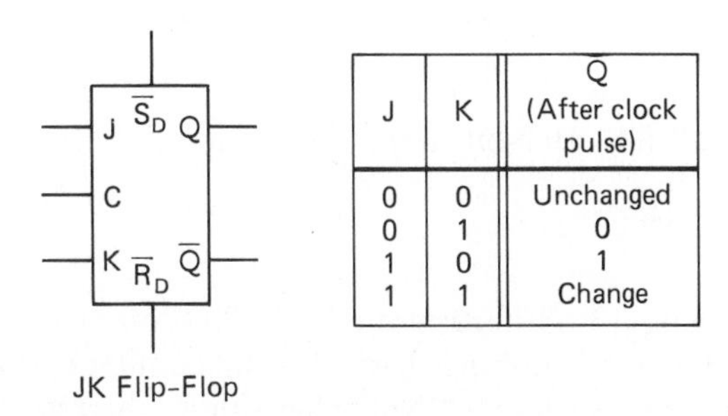

J	K	Q (After clock pulse)
0	0	Unchanged
0	1	0
1	0	1
1	1	Change

Fig. 13-66 Integrated circuit JK flip-flop with truth table.

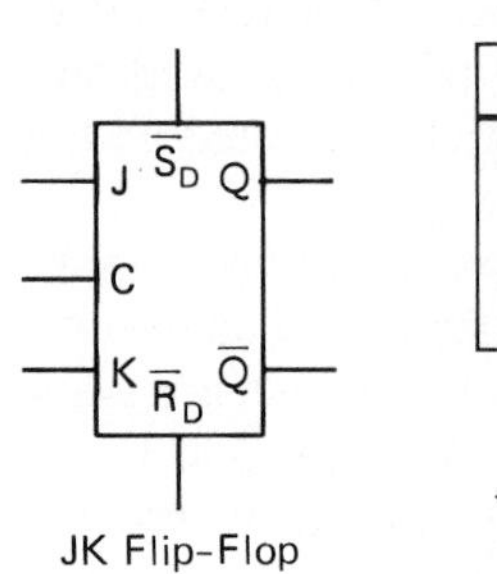

Set-Reset mode

$\overline{S}_D$	$\overline{R}_D$	Q	$\overline{Q}$
0	0	1*	1*
0	1	1	0
1	0	0	1
1	1	**	**

*Not desired in R.S. mode.
**No change from previous state.

J	K	Q_n	Q_{n+1}
0	0	0	0
0	0	1	1
0	1	0	0
0	1	1	0
1	0	0	1
1	0	1	1
1	1	0	1
1	1	1	0

Clocked Mode $\overline{S}_D = \overline{R}_D = 1$
On = State prior to clock pulse going to logic 0.
On + 1 = State after clock pulse going to logic 0.

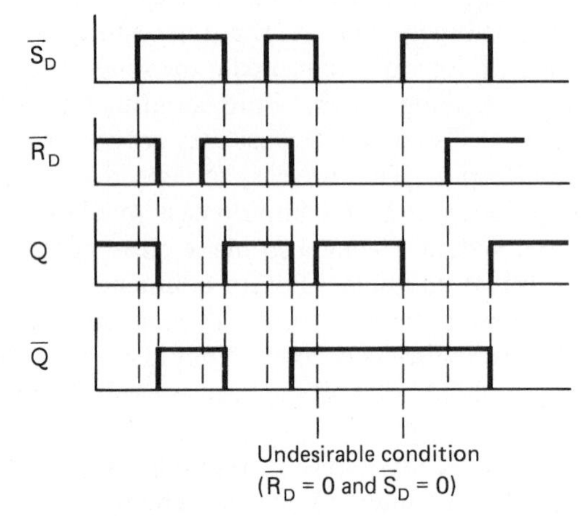

Fig. 13-67 JK flip-flop set and reset mode.

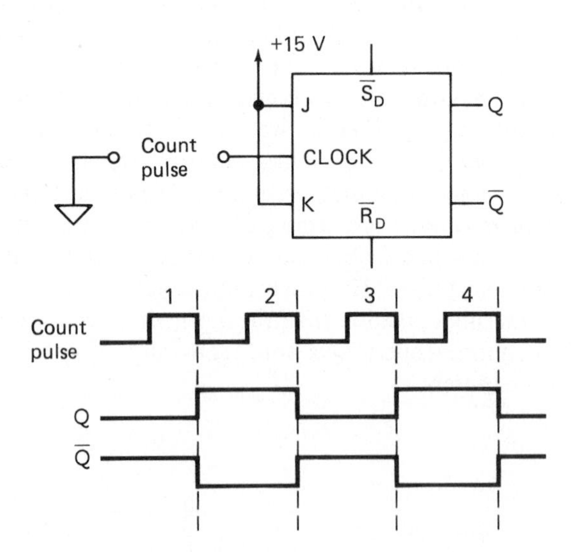

Fig. 13-68 One-bit counter with timing chart.

Figure 13-68 illustrates the JK flip-flop as a 1-bit counter and its timing chart. A 1-bit counter is also a divide-by-two circuit. Notice the timing chart verifies this. A timing chart or diagram can be used to identify the conditions required at the J and K inputs of various counter flip-flops. The information contained in the chart is no different than that present in a truth table. The timing chart provides a visual reference as to what is occurring in each stage of the counter as the count is incrementing.

Inasmuch as we will deal with binary counters, we should review binary numbers. If a counter is counting in binary numbers and using JK flip-flops, the outputs would be called Q_A, Q_B, Q_C, Q_D. Table 13-2 is for converting the output of a 4-bit binary counter to decimal numbers.

TABLE 13-2
Binary-to-decimal Conversion

Binary Number				
Q_D	Q_C	Q_B	Q_A	Decimal Number
0	0	0	0	0
0	0	0	1	1
0	0	1	0	2
0	0	1	1	3
0	1	0	0	4
0	1	0	1	5
0	1	1	0	6
0	1	1	1	7
1	0	0	0	8
1	0	0	1	9
1	0	1	0	10
1	0	1	1	11
1	1	0	0	12
1	1	0	1	13
1	1	1	0	14
1	1	1	1	15

Figure 13-69*a* shows the method of connecting JK flip-flops to give an asynchronous counter. Figure 13-69*b* illustrates the connections of flip-flops to provide a 4-bit synchronous counter with a ripple carry.

Figure 13-69*c* is a circuit for a decade down counter. Notice the timing chart indicates how the outputs on Q_A, Q_B, B_C, and Q_D change with each of count pulses. High and Low waveforms represent 1s and 0s. Notice the binary numbers that appear at the

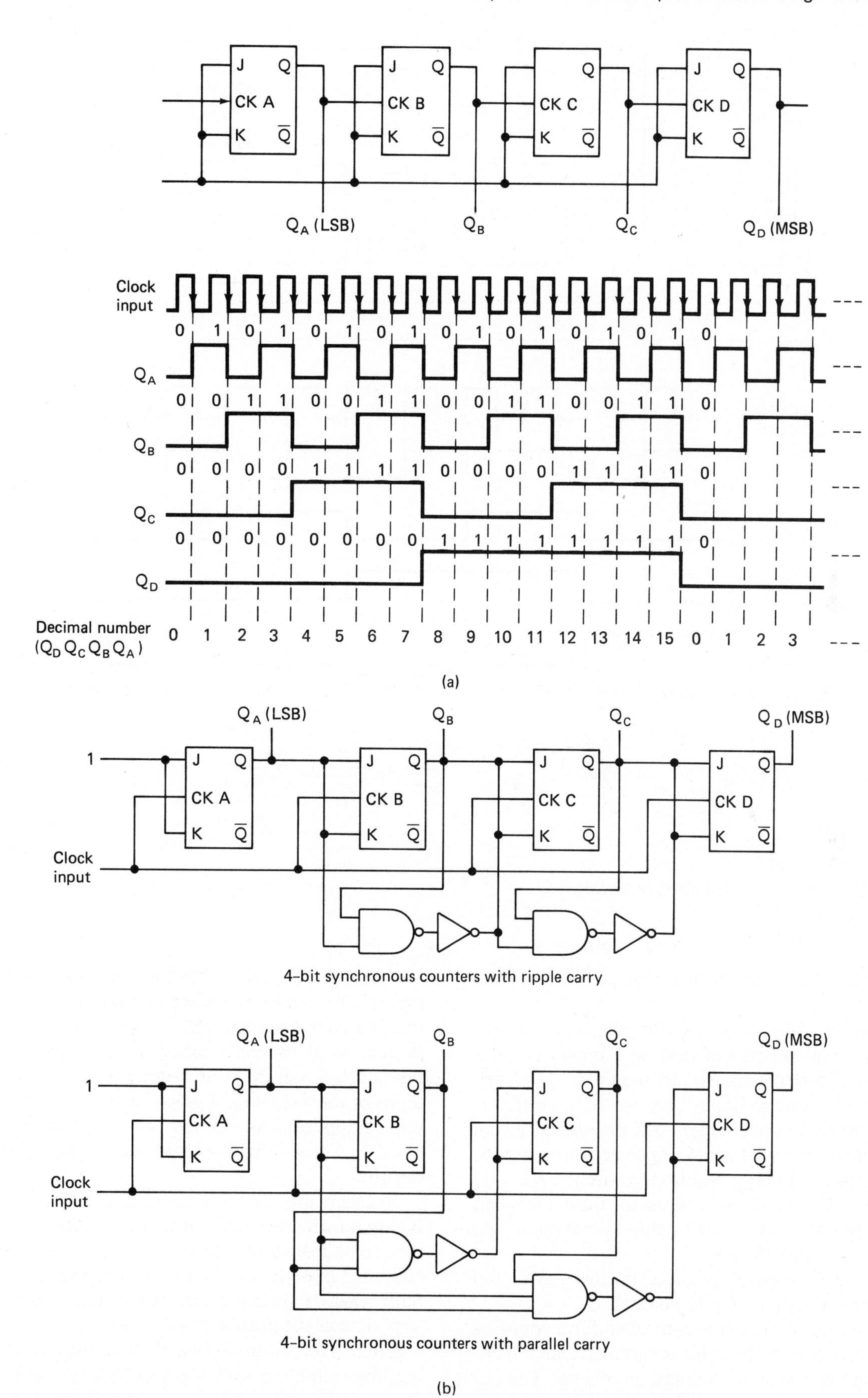

Fig. 13-69 Counters: (*a*) Four-bit (module 16) binary ripple counter. (*b*) Four-bit synchronous counters.

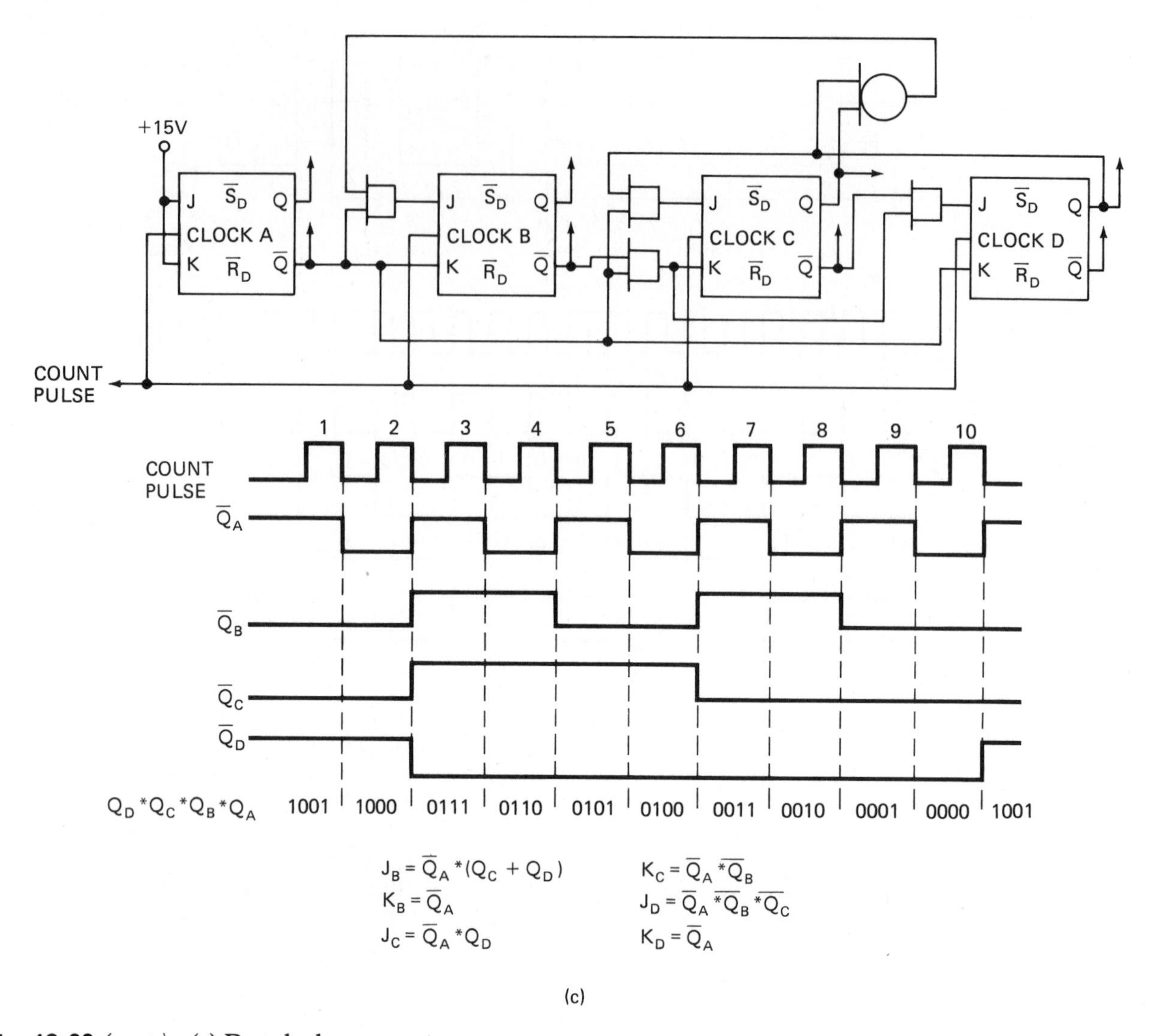

Fig. 13-69 *(cont.)* (*c*) Decade down counter.

bottom also correspond to the highs and lows of the timing chart.

Figures 13-70*a* and 13-70*b* illustrate binary counters. The old way of making a binary counter shows that five-step memories are required. The latest generation of logic indicates that one IC is all that is required to make the counter. All the circuits are in the one chip. External wiring of power source, inputs, and outputs is all that is required to wire the IC.

Figures 13-71*a* and 13-71*b* illustrate the old way and the current way of using one IC chip to make a decade counter. (See page 198.)

Figure 13-72 on page 199 illustrates how one IC chip can be used as a divide-by-12 counter.

Once a binary counter is connected for a counting operation, it may be desirable to have a decimal readout instead of a readout in binary numbers. The circuit of Fig. 13-73 on page 200 illustrates how the IC-type 9311 can be used to decode binary numbers and provide a decimal number readout.

A group of cascaded flip-flops used to store related bits of information is known as a *register.* A register that is used to assemble and store information arriving from a serial source is called a *shift* register. Each flip-flop's output in a shift register is connected to the input of the following flip-flop, and a common clock pulse is applied to all flip-flops, clocking them synchronously. The shift register is a synchronous sequential circuit.

Registers are used as data-storage devices, to store binary information. A register might accept and store input data from a keyboard and then later in a control sequence present this data to a microprocessor chip. Shift registers are often used to momentarily store binary data at the output of a decoder.

Circuits for automatic warehouses and manufacturing lines often use shift registers to keep track of the location of items as they move down conveyors.

A shift register can also be connected as a counter for some specific applications. A shift register can be con-

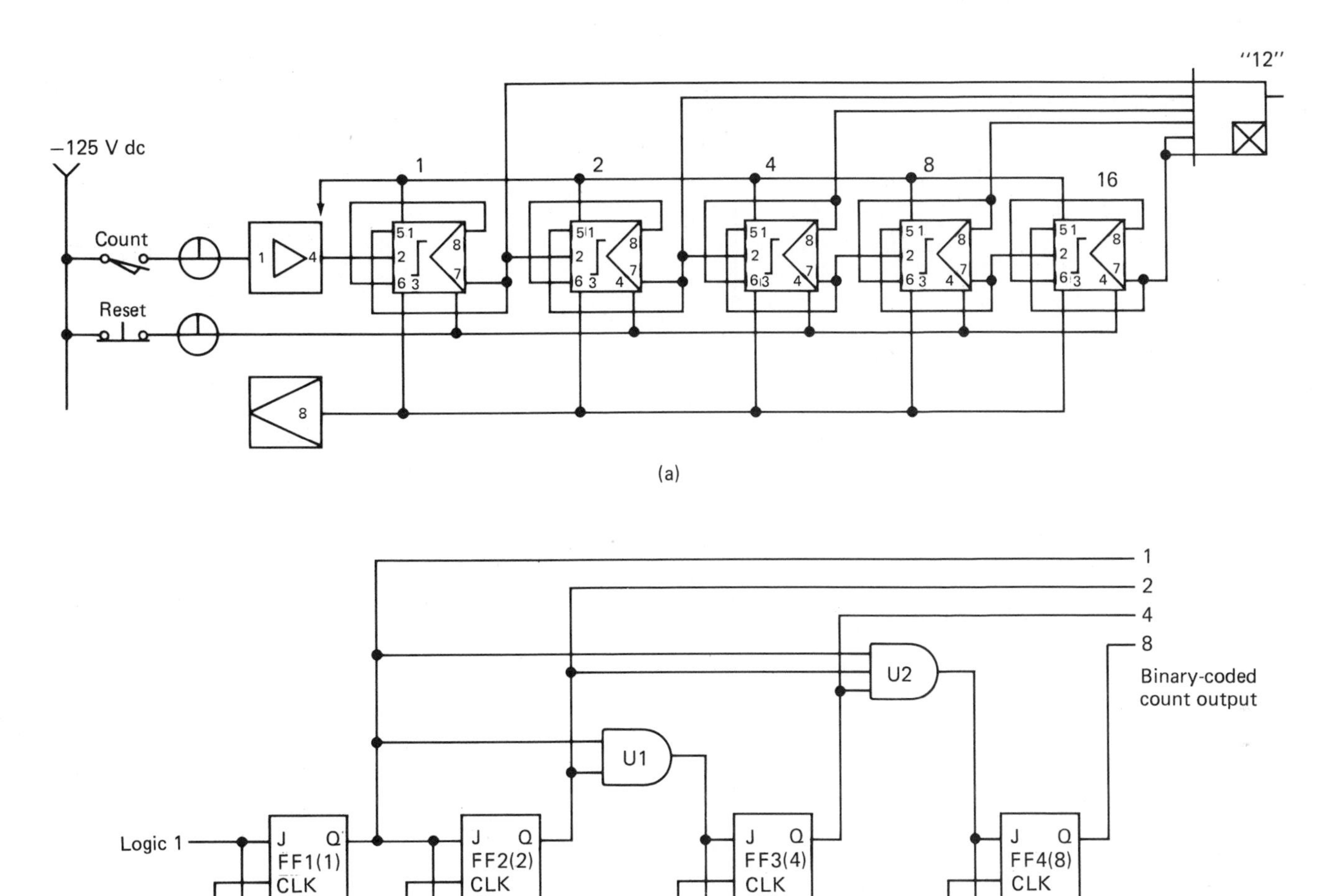

Fig. 13-70 Binary counter circuits: (*a*) Old method—made from step memory. *(Courtesy General Electric Company)* (*b*) New method—one IC.

nected as a ring counter and latches. A shift register can also be used as an up-down counter.

There are five classes of shift registers. They are

1. Serial-in, serial-out
2. Parallel-in, serial-out
3. Serial-in, parallel-out
4. Parallel-in, parallel-out
5. Parallel-in, parallel-out bidirectional

Figure 13-74 on page 201 illustrates examples of IC shift registers.

Figure 13-75 on page 201 displays all the logic that it takes to make up a 4-bit shift register. This is all found in one IC. This is certainly an easy way of obtaining a shift register.

The physical appearance of ICs on printed circuit boards used in conjunction with a microprocessor trainer is illustrated in Fig. 13-75. Integrated circuit boards contain the advance logic system used in the control of many of the automatic machines of modern industries.

SUMMARY

Static-and-logic control systems are built by properly interconnecting five basic logic building blocks: AND, OR, NOT, MEMORY, and TIME DELAY. The information section of the control circuit usually consists of conventional contact-type sensing devices. Signals of relatively high voltage from the information section of the system are modified by signal converters to the proper low-voltage direct current required by the logic elements of the decision section. The action section of the system converts the low-voltage, low-power signal from the decision section into the power required by the device to be controlled.

Earlier logic elements were built around the basic transistor switch but differ in the application of specific circuitry. Modern-day logic elements are built with ICs. Large-scale integration makes it possible to put many logic circuits into one chip.

There are three methods used to develop logic control circuits. The direct conversion method converts

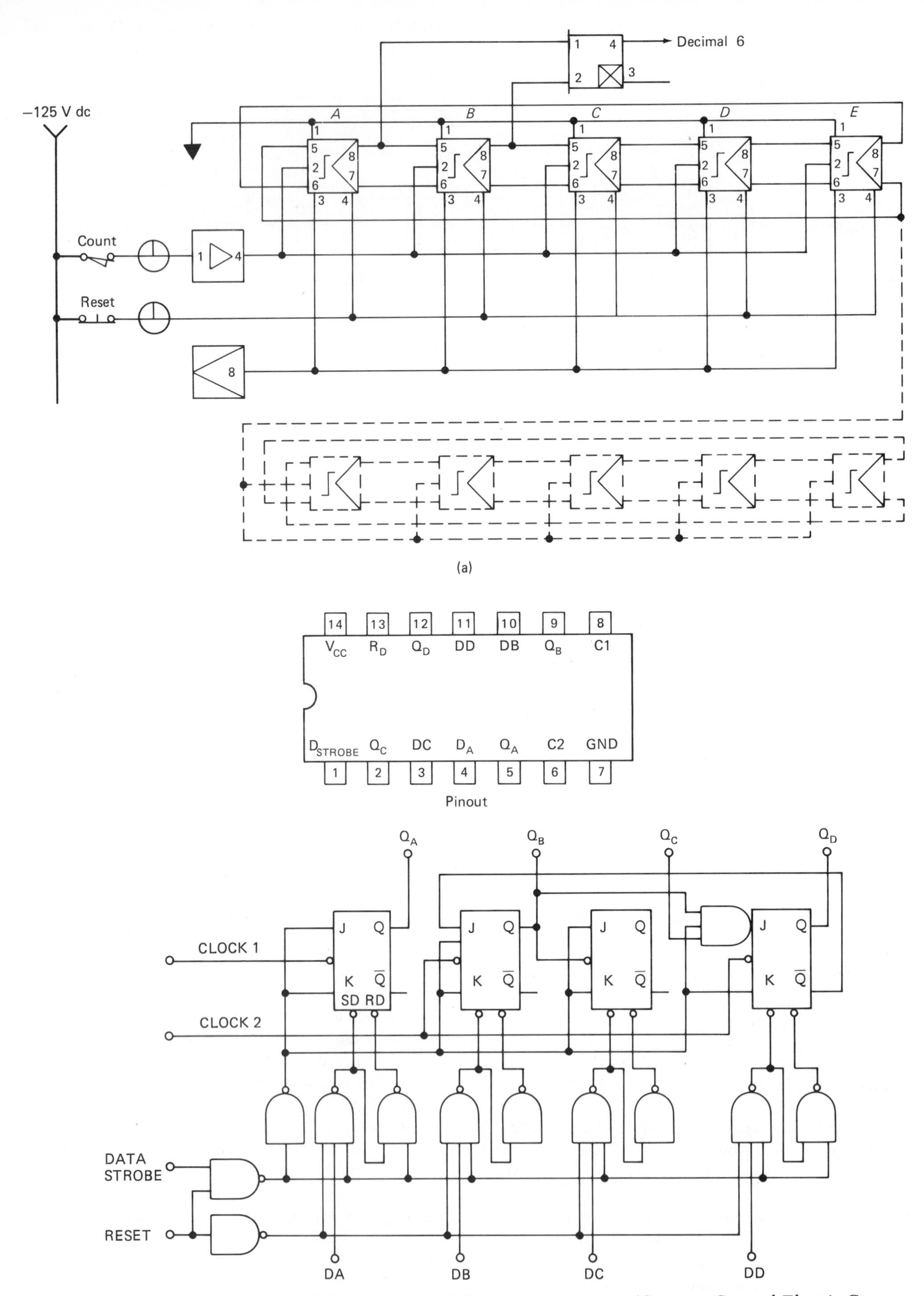

Fig. 13-71 Decade counters. (*a*) Old method—made from step memory. *(Courtesy General Electric Company)* (*b*) New method—one IC.

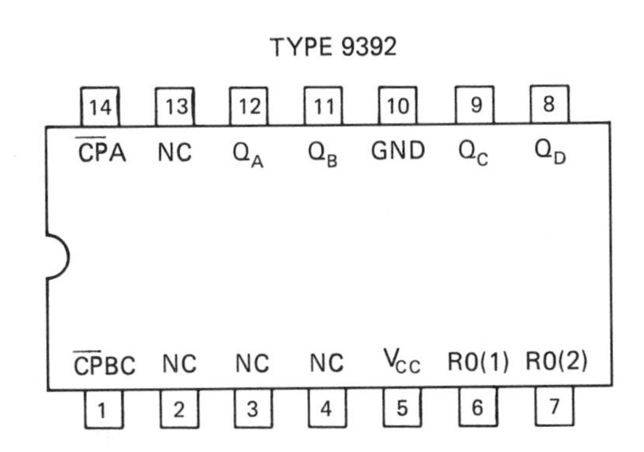

PIN NAMES

RO	Reset-Zero Inputs
$\overline{CPA}$	Clock Input
$\overline{CPBC}$	Clock Input
Q_A, Q_B, Q_C, Q_D	Count Outputs

(a) Pinout

COUNT	OUTPUT			
	Q_D	Q_C	Q_B	Q_A
0	0	0	0	0
1	0	0	0	1
2	0	0	1	0
3	0	0	1	1
4	0	1	0	0
5	0	1	0	1
6	1	0	0	0
7	1	0	0	1
8	1	0	1	0
9	1	0	1	1
10	1	1	0	0
11	1	1	0	1

NOTES:
1. Output Q_A connected to input $\overline{CPBC}$
2. To reset all outputs to LOW level both RO(1) and RO(2) inputs must be at HIGH level state.
3. Either (or both) resets inputs RO(1) and RO(2) must be at a LOW level to count.

(c) Truth table

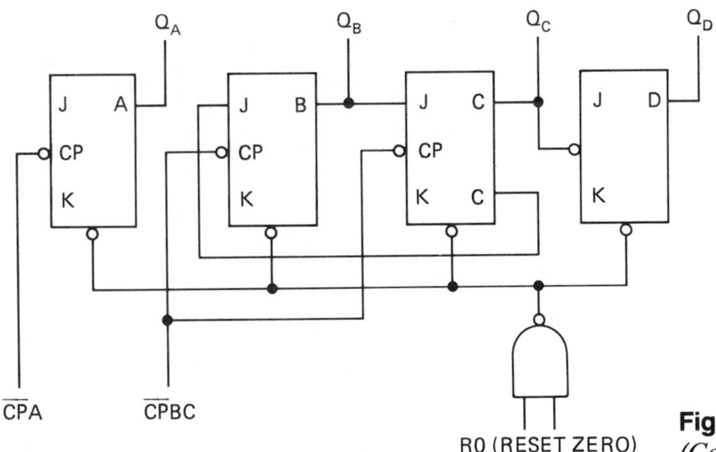

(b) Logic diagram

Fig. 13-72 Typical divided-by-12 counter. *(Courtesy Fairchild Camera and Instrument Corp.)*

each rung of a relay ladder circuit into the equivalent logic circuit. The logic circuits of each line are combined into the completed logic diagram.

The second method is to develop the logic circuits from circuit specifications. Talking your way through the circuit specifications can produce English logic expressions that can be converted to logic diagrams.

The third method of circuit conversion is the boolean algebra method. Each line of a ladder diagram can be converted to a boolean algebra expression. The boolean algebra expressions can be converted to equivalent logic circuits.

Boolean algebra can be used to give the simplest possible circuit, eliminating all redundant contacts. Mapping can also be used to eliminate redundant contacts and thus to simplify circuits. Veitch diagrams and Karnaugh maps are used in this process of simplification.

There are five steps in logic circuit development. They are

1. Convert the specifications into English logic statements.
2. Draw each statement in logic symbol form.
3. Integrate and interconnect the individual statements.
4. Combine functions where possible to simplify the circuit.
5. Review the overall circuit and check it against the specifications.

Counting circuits are used in many industrial logic control circuits. There are several basic counters used in control work. They are the decade counter, binary counter, up-down counter, and ring counter. Counters are made from JK flip-flops. Depending on

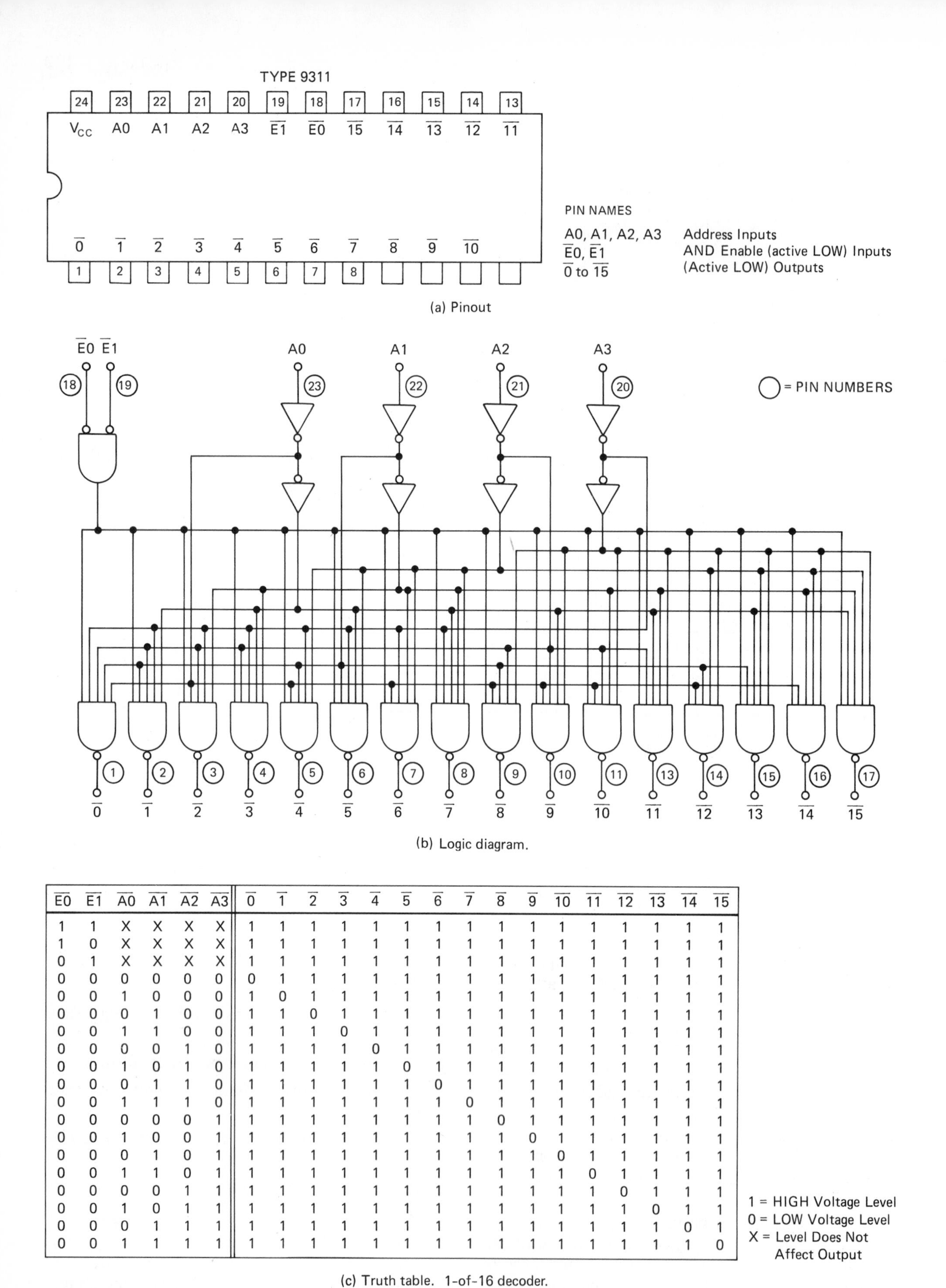

$\overline{E0}$	$\overline{E1}$	A0	A1	A2	A3	$\overline{0}$	$\overline{1}$	$\overline{2}$	$\overline{3}$	$\overline{4}$	$\overline{5}$	$\overline{6}$	$\overline{7}$	$\overline{8}$	$\overline{9}$	$\overline{10}$	$\overline{11}$	$\overline{12}$	$\overline{13}$	$\overline{14}$	$\overline{15}$
1	1	X	X	X	X	1	1	1	1	1	1	1	1	1	1	1	1	1	1	1	1
1	0	X	X	X	X	1	1	1	1	1	1	1	1	1	1	1	1	1	1	1	1
0	1	X	X	X	X	1	1	1	1	1	1	1	1	1	1	1	1	1	1	1	1
0	0	0	0	0	0	0	1	1	1	1	1	1	1	1	1	1	1	1	1	1	1
0	0	1	0	0	0	1	0	1	1	1	1	1	1	1	1	1	1	1	1	1	1
0	0	0	1	0	0	1	1	0	1	1	1	1	1	1	1	1	1	1	1	1	1
0	0	1	1	0	0	1	1	1	0	1	1	1	1	1	1	1	1	1	1	1	1
0	0	0	0	1	0	1	1	1	1	0	1	1	1	1	1	1	1	1	1	1	1
0	0	1	0	1	0	1	1	1	1	1	0	1	1	1	1	1	1	1	1	1	1
0	0	0	1	1	0	1	1	1	1	1	1	0	1	1	1	1	1	1	1	1	1
0	0	1	1	1	0	1	1	1	1	1	1	1	0	1	1	1	1	1	1	1	1
0	0	0	0	0	1	1	1	1	1	1	1	1	1	0	1	1	1	1	1	1	1
0	0	1	0	0	1	1	1	1	1	1	1	1	1	1	0	1	1	1	1	1	1
0	0	0	1	0	1	1	1	1	1	1	1	1	1	1	1	0	1	1	1	1	1
0	0	1	1	0	1	1	1	1	1	1	1	1	1	1	1	1	0	1	1	1	1
0	0	0	0	1	1	1	1	1	1	1	1	1	1	1	1	1	1	0	1	1	1
0	0	1	0	1	1	1	1	1	1	1	1	1	1	1	1	1	1	1	0	1	1
0	0	0	1	1	1	1	1	1	1	1	1	1	1	1	1	1	1	1	1	0	1
0	0	1	1	1	1	1	1	1	1	1	1	1	1	1	1	1	1	1	1	1	0

1 = HIGH Voltage Level
0 = LOW Voltage Level
X = Level Does Not Affect Output

(c) Truth table. 1-of-16 decoder.

Fig. 13-73 Decoders—these logic elements are used to convert binary numbers to decimal numbers.

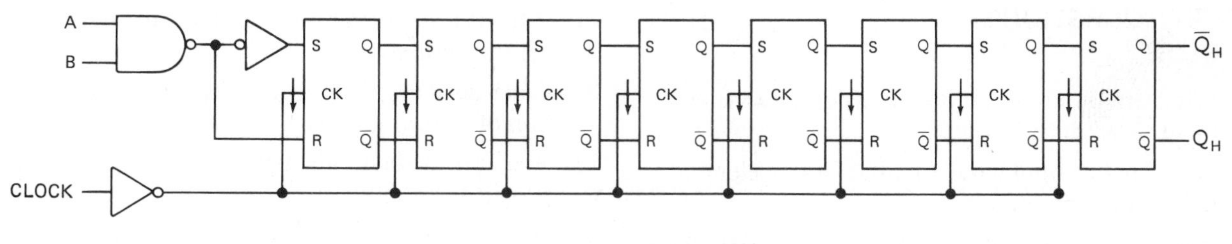

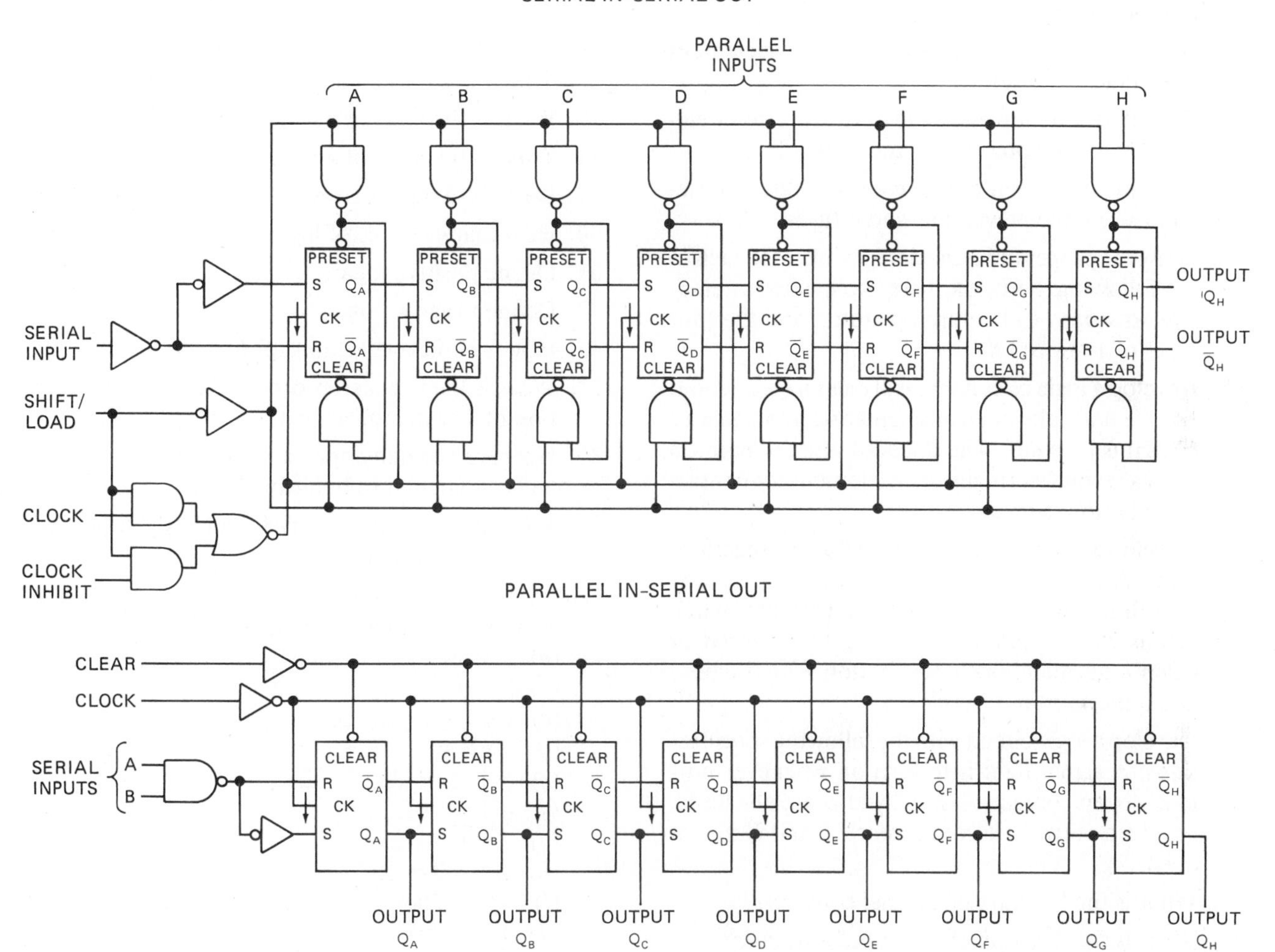

Fig. 13-74 Examples of IC shift registers.

their mode of operation, counters can be asynchronous or synchronous.

Flip-flops can be used to construct register circuits. A register is a circuit that can be used to store information. The shift register is a synchronous sequential circuit.

Registers are used as data-storage devices. Circuits for the control of automatic warehouses often use shift registers to keep track of the location of items as they move down conveyor belts.

Many complex logic circuits are constructed on one chip by what is known as large-scale integration (LSI).

Fig. 13-75 Integrated circuits on circuit boards. Integrated circuits are used to solve advanced logic control problems.

REVIEW QUESTIONS

1. What are the input requirements of an AND element to produce an output?
2. What are the input requirements of an OR element to produce an output?
3. What are the input requirements of a NOT element to produce an output?
4. What are the input requirements of a NOR element to produce an output?
5. Describe the action of the OFF-return MEMORY element when the power fails and returns.
6. Describe the action of the retentive MEMORY element when the power fails and returns.
7. Draw the proper industrial logic symbol for each of the following: AND, OR, NOT, NOR, NAND, retentive MEMORY, OFF-return MEMORY, and the four types of TIME DELAY.
8. Develop a logic circuit that will energize a solenoid when a normally open momentary-contact START button is pressed. The solenoid must remain on until a second normally open momentary-contact STOP button is pressed.
9. Develop a logic circuit with the following specifications: A coil must be energized when a normally open limit switch LS1 is closed, a pressure switch PS1 is closed, and a thermostat T1 is closed or when a normally open pushbutton PB1 is closed and a thermostat T2 is closed.
10. Develop a logic circuit with the following specifications: A signal light is to turn on whenever normally open contacts *A, B, C,* and *D* are all closed, provided normally closed contacts *E* and *F* are not open.
11. What is the function of a signal converter?
12. What is the function of an output amplifier?
13. What are the advantages of static control?
14. What are the essential logic functions used in static logic control?
15. What is the difference between retentive MEMORY and an OFF-return MEMORY?
16. Define a NAND logic element.
17. How do the symbols for a time-delay energizing circuit and a time-delay de-energizing circuit differ?
18. What is the difference between ANSI logic symbols and standard static logic symbols?
19. How is isolation of pilot devices accomplished with logic control circuits and with electromagnetic circuits?
20. What affect does electrical noise have on logic circuits?
21. What is a static amplifier used for in a static logic control system?
22. What devices make up the decision-making section of a control circuit?
23. What are the first four steps in the development of a logic diagram?
24. What is the accepted procedure for maintaining logic control systems?
25. What is a truth table?
26. What do a 1 and a 0 represent in logic work?
27. What is the logic of a NAND function?
28. What is the logic of an exclusive OR?
29. Describe a digital circuit.
30. How many classes of logic circuits are there?
31. Define positive logic.
32. Define negative logic.
33. How is boolean algebra applied to electric circuits?
34. What is the five-step procedure for the simplification of a switching arrangement?
35. How are normally open and normally closed contacts designated in boolean algebra?
36. What do symbols × and + represent in boolean algebra expressions?
37. Complete the following problems with boolean algebra postulates and theorms.
    (*a*) $1 + 1 =$ ______
    (*b*) $1 + 0 =$ ______
    (*c*) $1 \cdot 0 =$ ______
    (*d*) $A \cdot A =$ ______
    (*e*) $A + A =$ ______
    (*f*) $A + \bar{A} =$ ______
    (*g*) $A \cdot \bar{A} =$ ______
    (*h*) $A + AB =$ ______
    (*i*) $A(A + B) =$ ______
    (*j*) $\underline{A} + \bar{A}B =$ ______
    (*k*) $\bar{\bar{A}} =$ ______
    (*l*) $\bar{A} \cdot \bar{A} =$ ______
38. How are redundancies eliminated in logic circuits?
39. What is DeMorgan's theorem?
40. What is the rule concerning double negatives?
41. What is a Veitch diagram and how is it used?
42. What is a Karnaugh map and how is it used?
43. How is an electric circuit converted to a logic equation?
44. What do bars over letters represent?
45. How are the locations of 1s on a Veitch diagram determined?
46. How are the locations of 1s on a Karnaugh map determined?
47. What is the function of "I don't care" variables?

48. When circling, or looping, 1s, is the objective to get as many or as few circles or loops as possible?

49. List the steps in simplifying an electric ladder diagram and converting it to a simplified electric circuit or logic diagram.

50. What are redundant contacts?

51. What are some of the applications for counters in control work?

52. What are the three kinds of counters?

53. What is the difference between an asynchronous and a synchronous counter?

54. Define a decade counter.

55. Give the decimal equivalents of the following binary numbers:

(*a*) 01010 (*e*) 110111
(*b*) 1101 (*f*) 1011001
(*c*) 10011 (*g*) 100111
(*d*) 1001 (*h*) 1110101

56. Give the binary equivalents for the following decimal numbers:

(*a*) 13 (*e*) 67
(*b*) 19 (*f*) 80
(*c*) 26 (*g*) 93
(*d*) 36 (*h*) 156

57. Define an up counter.

58. Define a down counter.

59. What is the function of decoder?

60. What are some of the applications of shift registers?

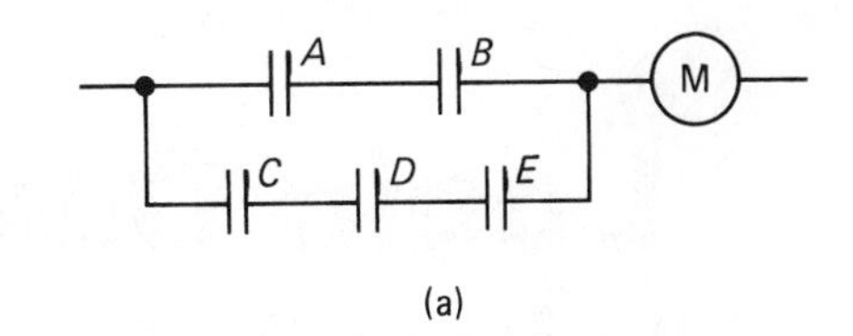

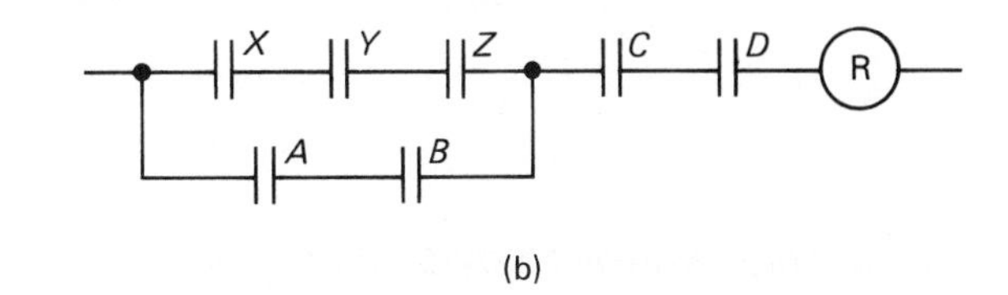

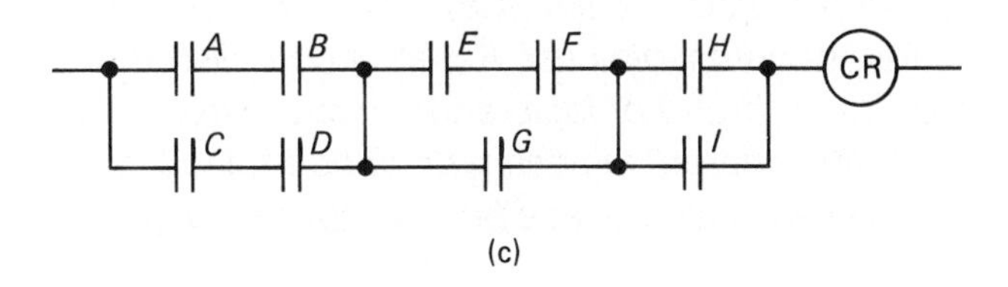

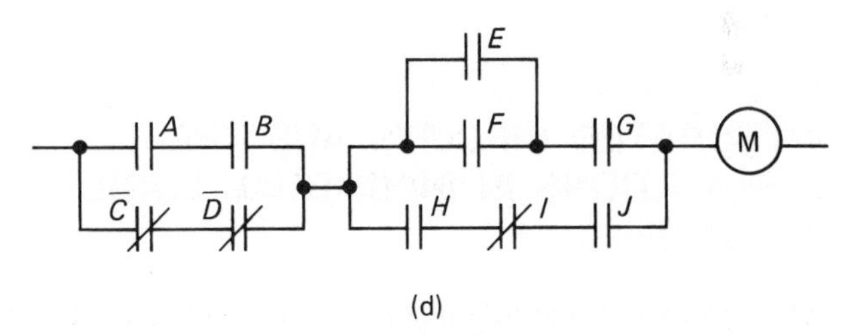

Fig. 13-76 Circuits for Problem 13-1.

REVIEW PROBLEMS

P13-1. Write the boolean algebra expression for each of the circuits in Fig. 13-76*a* through *d*.

P13-2. Simplify each of the following logic expressions using Veitchs diagram and Karnaugh map.

a. $ABC\overline{D} + ABCD$
b. $AB\overline{C}D + AB\overline{C}\,\overline{D} + A\overline{B}\,\overline{C}D + A\overline{B}\,\overline{C}\,\overline{D}$
c. $\overline{A}BC\overline{D} + \overline{A}BCD + \overline{A}\,\overline{B}CD + \overline{A}\,\overline{B}C\overline{D}$
d. $AB + A\overline{C} + \overline{A}C + \overline{A}\,\overline{B}\,\overline{D}$
e. $AB\overline{C} + ABC + \overline{A}BC + \overline{A}B\overline{C}$
f. $ABC + AB\overline{D} + AC + \overline{A}\,\overline{B}\,\overline{C}\,\overline{D} + \overline{A}C$
g. $A\overline{C}\,\overline{D} + \overline{A}\,\overline{C}\,\overline{D} + ABCD + BCD +$
$\overline{B}CD + ACD + B\overline{C}\,\overline{D}$
h. $\overline{A}\,\overline{B}\,\overline{C} + \overline{A}B\overline{D} + \overline{A}B\overline{C}\,\overline{D} + \overline{A}C\overline{D}$
i. $\overline{A}\,\overline{B}\,\overline{C} + A\overline{B}\,\overline{C} + ABC + ABC\overline{D}$
j. $AB\overline{C}\,\overline{D} + A\overline{B}\,\overline{C}\,\overline{D} + A\overline{B}\,\overline{C}D + AB\overline{C}D +$
$\overline{A}\,\overline{B}C\overline{D} + A\overline{B}C\overline{D}$
k. $\overline{A}\,\overline{B}\,\overline{C}\,\overline{D} + A\overline{B}C\overline{D} + \overline{A}\,\overline{B}C\overline{D} + \overline{A}\,\overline{B}\,\overline{C}\,\overline{D}$
l. $\overline{A}\,\overline{B}\,\overline{C}\,\overline{D} + \overline{A}\,\overline{B}\,\overline{C}D + ABC\overline{D} + AB\overline{C}D +$
$A\overline{B}CD + \overline{A}B\overline{C}D$
m. $AB\overline{C}\,\overline{D} + AB\overline{C}D + A\overline{C} + \overline{A}C$
n. $A\overline{C}D + \overline{A}\,\overline{C}\,\overline{D} + ABCD + BCD +$
$\overline{B}CD + ACD + B\overline{CD}$
o. $A\overline{B}\,\overline{C} + \overline{A}\,\overline{B}\,\overline{C} + A\overline{B}C + A\overline{B}\,\overline{C} + AB\overline{C}$

P13-3. Write a boolean algebra expression for the logic circuit of Fig. 13-77, simplify the expression, and redraw the logic circuit in its simplest form.

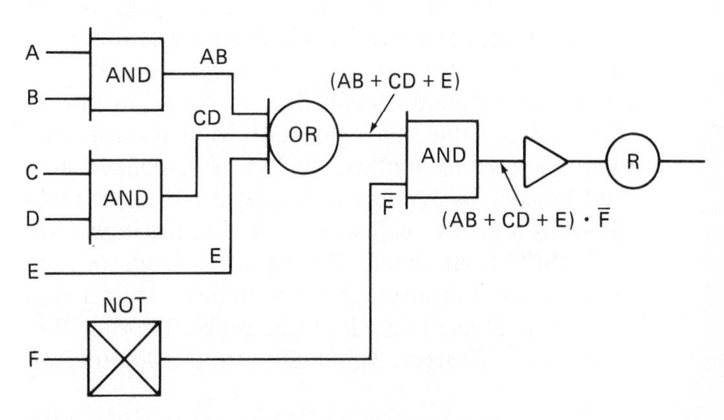

Fig. 13-77 Logic circuit for Problem 13-3.

14 Logic Control Systems Using Integrated Circuits

In this chapter we will study a number of different logic and solid-state control systems that use integrated circuits (ICs). We will see how different companies use ICs to accomplish their particular brand of logic-and-static control. The benefits inherent in each type of control will be presented. We will see how reliability, safety, machine uptime, and machine control are accomplished by different manufacturers of logic systems. Conventional electric circuits for control will be converted to equivalent logic control systems. Circuits will be analyzed with emphasis on maintenance and troubleshooting.

14-1 INTEGRATED CIRCUITS AND THEIR APPLICATIONS IN INDUSTRIAL LOGIC CONTROL

The companies that manufacture static-and-logic control devices may all take a different approach to the design of logic control switching circuits. Regardless of the design and the components used, the logic of each company's logic elements is the same as that of the others. The definition of the AND logic element is the same regardless of how the element is made. Truth tables for the same types of logic elements manufactured by different companies are the same. Although standard symbols have been established by industry organizations, symbols used by different manufacturers of logic elements are not all the same. Each company may use symbols that are very different from those of the other companies.

Troubleshooting principles are the same for all logic systems. The truth table for a solid-state logic device will generally use a 1 and 0 to indicate the condition of inputs and outputs. Logic 1 is defined as an "ON" or "high" and logic 0 is defined as an "OFF" or "low." The voltage value of a logic 1 or 0 will vary from one manufacturer to another. If logic 1 is defined as a voltage such as 5 V and logic 0 as 0.3 V, in troubleshooting a device such as an AND if all the inputs are 5 V, the output should also be 5 V. Logic probes, which have indicator lights that indicate logic 1 and logic 0, are easier to use for logic troubleshooting than voltmeters. Systems with built-in troubleshooters use devices such as small LEDs that are on when a logic device is on and are off when the logic element is off. If the logic of a circuit produces an output with a certain condition of inputs and the expected output is not produced, the logic element is probably defective and should be replaced.

Many logic systems use logic elements that are incorporated in such a way that they are not repairable. For example, if a logic element using an IC were bad, the IC would have to be replaced. Integrated circuits are sealed in plastic or ceramic material so they are not repairable. If a person were repairing a logic system using ICs, knowing the internal circuitry is nice, but not necessary. Verifying that a circuit is faulty by testing for logic 1s and 0s is all that is really needed. Replacing the faulty board will get the logic system back in operation.

When faulty logic boards, or cards, are found and replaced, what is done with the faulty logic elements? Some logic elements are designed so they are unrepairable. The first-generation logic elements, which use magnetic amplifiers as the logic-and-static switching devices, were unrepairable. Replacement of the logic elements was the only way to get the logic control system back into operation. Some of these early logic systems, such as Westinghouse's Cypak and General Electric's Static, are still found in older mills and factories. Even though static logic systems using magnetic amplifiers have not been manufactured for many years, some of these old systems are still in operation and may be in operation for years to come.

Second-generation logic systems use transistors and

discrete components. When these logic cards are replaced by field maintenance electricians, they are generally sent to a central repair shop to be repaired. A technician with proper tools, meters, parts, and equipment plus electronic knowledge is required for this type of card repair. Not all field maintenance electricians are expected to repair logic cards, but they are expected to be able to isolate and replace faulty logic circuits in a faulty control system. Some companies may return their faulty logic cards to the factory for repairs. If the number of logic elements or electronic solid-state devices that fail each year does not justify the cost of equipping a shop and hiring a technician, some plant maintenance departments will depend on the manufacturer's service engineers for repairs of logic systems. The extent of the job desired will dictate how much time is spent analyzing the internal circuits of logic elements.

In the pages that follow, industrial logic systems produced by several manufacturers will be discussed. We will study their symbols and circuits. It is essential that construction electricians who install the logic control systems as well as the maintenance technicians who troubleshoot and repair the systems understand the various static-and-logic control systems on the market.

As noted previously, first-generation logic used magnetic amplifiers. The Westinghouse Cypak and the General Electric Static Control are examples of early first-generation logic control systems. When the transistor was developed, many companies produced logic cards consisting of a printed circuit board with transistors and other discrete components. These transistor logic systems we refer to as *second-generation logic devices.* With the development of ICs, we are now in the modern generation of logic control.

Integrated circuits are manufactured with small logic circuits such as the familiar AND, OR, NAND, NORS, etc., or they may contain thousands of transistors with complex logic circuitry. The cost of IC components is so low that the older transistor logic systems are no longer competitive. The solid-state Westinghouse Numa Logic, the solid-state Allen Bradley Card Lock, the Square D NORPAK, and numerous other industrial logic systems use the IC as the heart of their logic systems. Logic functions are "built" on cards with IC chips and discrete components. Integrated chips perform the decision making in the control system.

Integrated circuits are produced at such low prices that a single IC containing several logic functions such as an AND, OR, inverters, etc., are only a few cents each. Technicians should be impressed with the technology that is required to develop and produce ICs. Our main concern, however, should be in the knowledge of how to use an IC to produce automatic control for motors. We will approach our solid-state logic study by looking at ICs, as the little black boxes that control motors automatically.

14-2 DEVELOPING INDUSTRIAL CONTROL CIRCUITS USING INTEGRATED CIRCUITS

Figure 14-1, on the next page, illustrates a small number of the ICs from which we will develop logic control circuits. Study these chips and the logic circuits or gates they contain.

The ladder diagram of Fig. 14-2, on page 207, is for two compressors. When the pressure switch PS closes the first time, CR1 energizes. This causes CR3 and M1 to energize. The start coil for the motor starter is M1. Motor M1 is now running. When the pressure switch opens, CR1 and M1 drop out, but CR3 remains energized because it is sealed in through its own contact N.O. CR3.

When the pressure switch closes the second time, CR2 energizes because of the N.O. CR3 contact that is now closed. Coil CR2 causes CR3 to de-energize and M2 to energize. Motor starter M2 starts motor M2 and it will continue to run until the pressure switch opens. When the switch opens, everything de-energizes and the sequence is set to begin again once PS closes. From this description of the relay circuit, it should be evident that each time the pressure switch calls for a motor to start, the motors take turns running.

Let's convert this circuit to logic using ICs. Let's also shoot for an IC circuit containing only three logic elements in the decision-making portion of the circuit.

The first step in converting the circuit of Fig. 14-2 to a logic circuit is to identify the input and output devices.

The pressure switch is the input device and motor starters M1 and M2 are output devices. The next step in the conversion of the circuit to logic is to write the boolean algebra expression for each rung of the ladder diagram. This would be done as follows:

$$CR1 = PS \cdot (\overline{CR3} + CR1) \cdot \overline{CR2}$$

$$CR2 = PS \cdot (CR3 + CR2) \cdot \overline{CR1}$$

$$CR3 = (CR1 + CR3) \cdot \overline{CR2}$$

$$M1 = CR1$$

$$M2 = CR2$$

To start our logic diagram, we place our two vertical lines on each side of the circuit and position our input and output devices with the inputs on the left and outputs on the right. (See Fig. 14-3 on page 208.)

To complete our logic diagram we have only to start with output devices and work toward the input of our boolean algebra expressions. Next, we will draw the logic circuit for the expression for CR1 and connect it to M1. We should also draw the logic circuit for the expression of CR2 and connect it to M2. We will complete the circuit by drawing the logic circuit for the

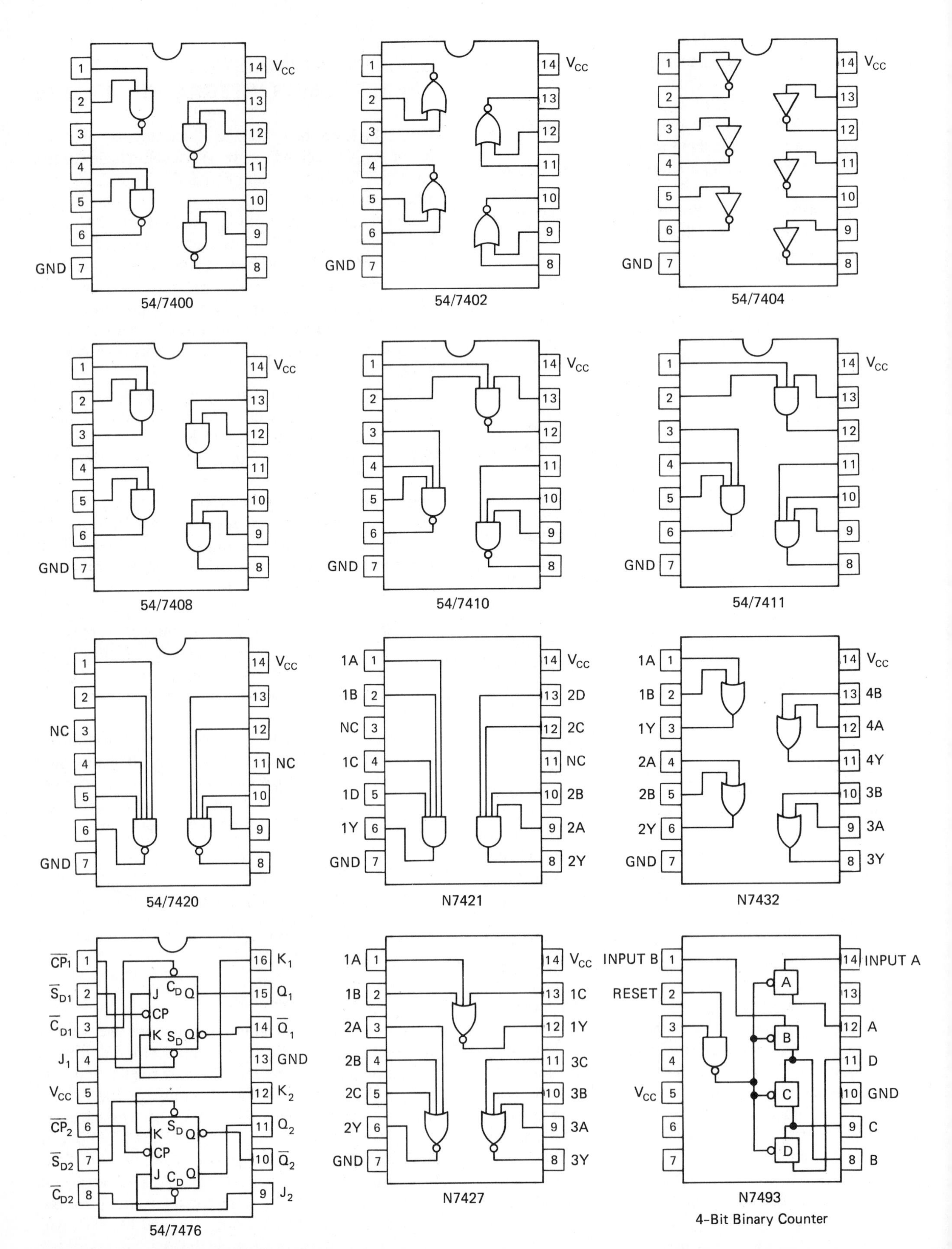

Fig. 14-1 Integrated logic circuits.

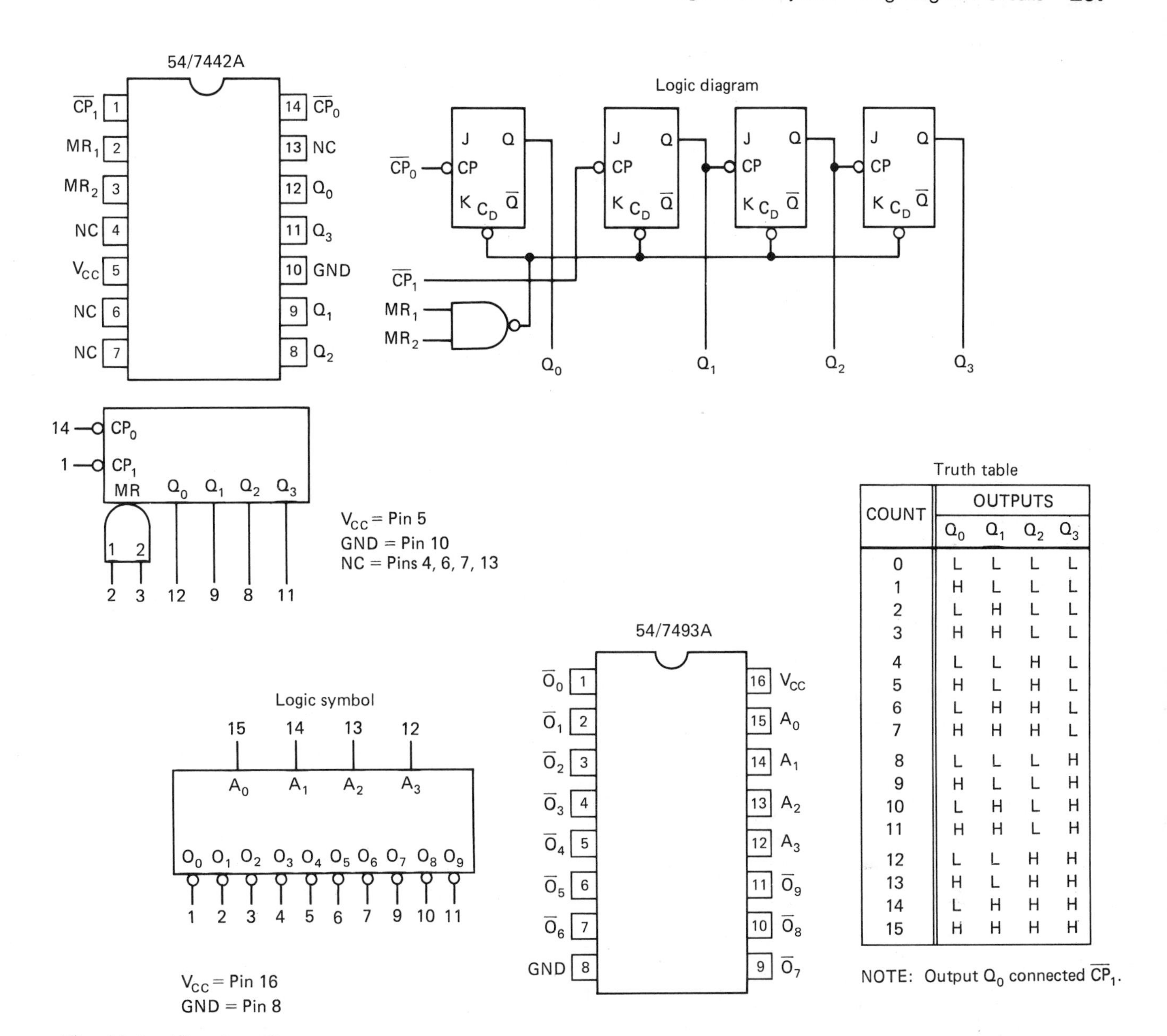

COUNT	OUTPUTS			
	Q_0	Q_1	Q_2	Q_3
0	L	L	L	L
1	H	L	L	L
2	L	H	L	L
3	H	H	L	L
4	L	L	H	L
5	H	L	H	L
6	L	H	H	L
7	H	H	H	L
8	L	L	L	H
9	H	L	L	H
10	L	H	L	H
11	H	H	L	H
12	L	L	H	H
13	H	L	H	H
14	L	H	H	H
15	H	H	H	H

Fig. 14-1 (Continued)

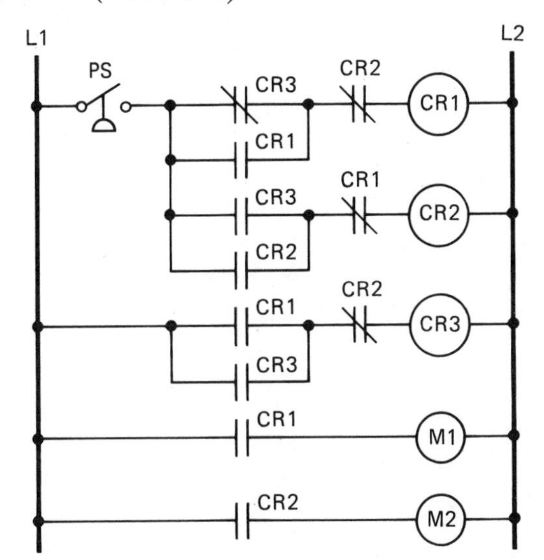

Fig. 14-2 Motor control circuit to provide sequence control to M1 and M2.

expression for CR3 and connect it to all points identified CR3. Our circuit will be complete but will use a lot more than three logic elements.

In Fig. 14-4 we have the logic circuit for each line of the electric ladder diagram in Fig. 14-2. Also, in Fig. 14-4 on the next page, we tied the logic circuits of Fig. 14-3 together to form an IC logic control circuit.

We have converted the electric circuit of Fig. 14-2 to a logic diagram using ANDs, ORs, and inverters. We developed this circuit from direct conversion of boolean algebra expressions. One of our objectives when we started our circuit conversion was to end up with a logic circuit with three logic elements. We ended up with 10 logic gates. It's obvious there must be other ways of converting or developing control circuits. Let's look at some other methods.

If we consider the operation of our original electric circuit (Fig. 14-2), we will find that an "English" description of the circuit would indicate the circuit "tog-

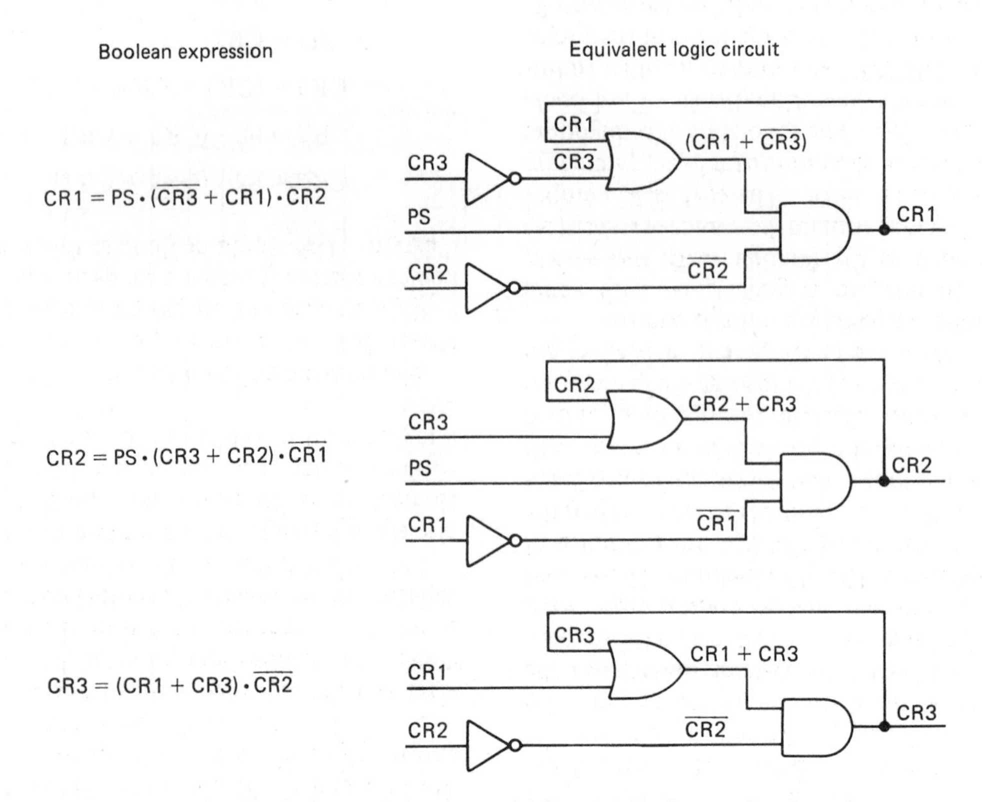

Fig. 14-3 The equivalent logic circuit for each boolean algebra expression—developed for the circuit in Fig. 14-2.

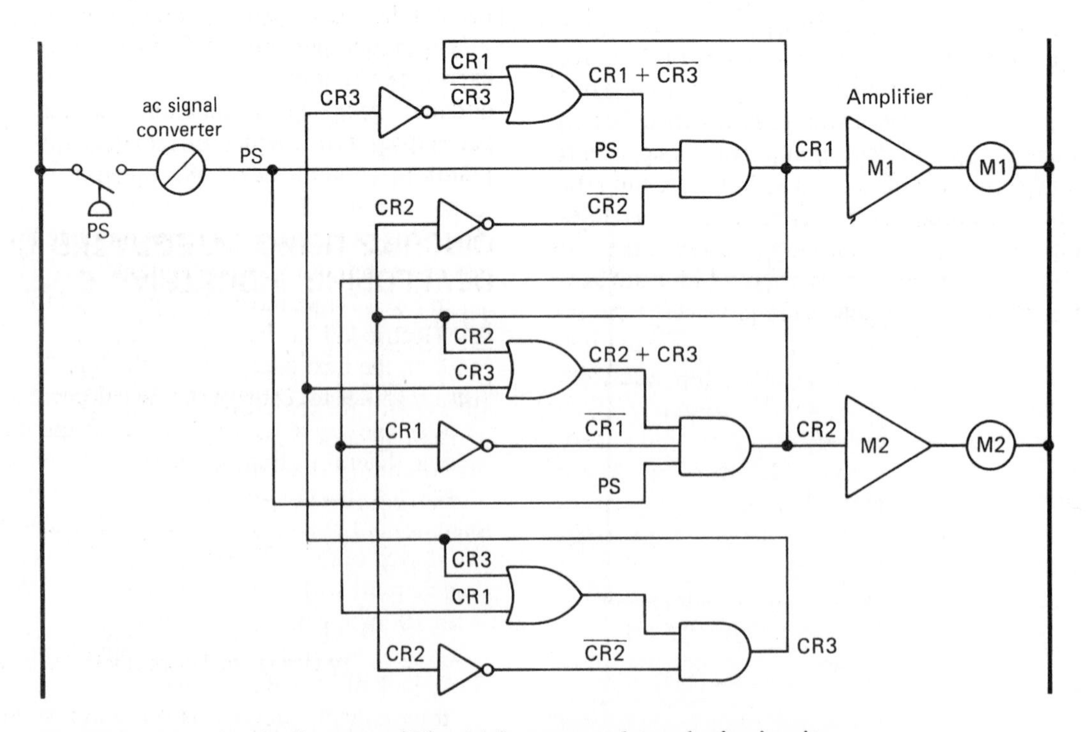

Fig. 14-4 The logic relay circuit diagram of Fig. 14-2 converted to a logic circuit.

gles" or switching points. When the pressure switch closes, motor M1 starts. The next time it closes, motor M2 starts. As the switch continues to open and close, the motors toggle, or take turns running. What kind of logic component toggles or can be made to toggle? The JK flip-flop is a single logic component that can be connected to toggle.

The JK flip-flop is illustrated in Figs. 13-66 and 13-67. The truth table for the JK flip-flop is also included. Study the truth table for the JK flip-flop. You should be able to see that when the J and K inputs are tied together and connected to a logic 1, the output will toggle on the trailing edge of the clock pulse.

Every time a clock pulse is applied, the Q and $\overline{Q}$ outputs do not change. When the clock is removed or returns to logic 0, Q and $\overline{Q}$ will reverse their output conditions. If we look at this another way, the trailing edge of the first clock pulse may turn Q on and $\overline{Q}$ off. The trailing edge of the next clock pulse will turn Q off and $\overline{Q}$ on. This toggling action will continue with continuing clock pulses. The J and K inputs must be held at logic 1 for the toggling action to continue.

Let's develop a logic circuit that toggles and apply it to the conversion of the electric control circuit of Fig. 14-2. We will develop a three-element logic circuit that will produce toggling of the two motor starters.

Figure 14-5 illustrates the start of our conversion by connecting a pressure switch to a signal converter. The two motor starters M1 and M2 that will be the outputs of logic circuit are connected to two amplifiers.

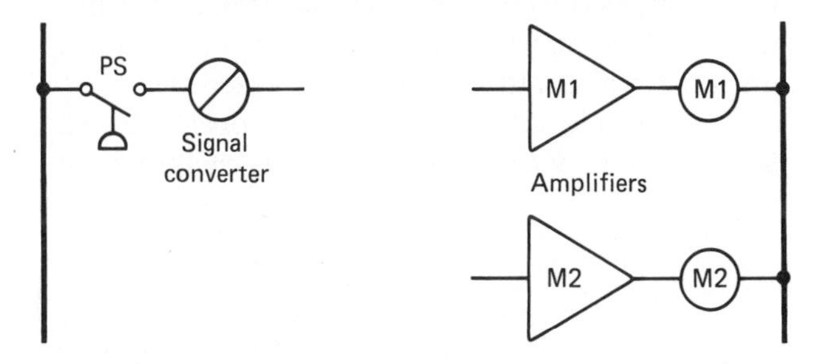

Fig. 14-5 The first step in relay-to-logic conversion is to identify inputs and outputs.

To the circuit of Fig. 14-5 we add a JK flip-flop and two AND gates. Figure 14-6 illustrates this. This figure shows the three logic elements that will complete the decision-making part of our control circuit. We connect the output of the signal converter to the C input of the JK flip-flop. The J and K inputs of the JK flip-flop are tied together and to logic 1 (+5 V can be logic 1). The outputs of the JK flip-flop, Q and $\overline{Q}$, are connected to two AND gates. One AND has its output connected to amplifier 1. The other AND is connected to amplifier 2.

If we study the circuit of Fig. 14-6, we can see that when the pressure switch closes, we will obtain an input to C of the JK flip-flop and an input to each of the two AND gates. Because a JK flip-flop always has an output, either Q or $\overline{Q}$, we will be delivering an input to one of the amplifiers. Let's say M2 is on because $\overline{Q}$ is on. Coil M2 will remain energized until the pressure switch opens.

When the pressure switch opens, the output of the signal converter is logic 0 and we will now lose our inputs to the AND gates. We will not have an input to the amplifiers and the motor starters will be off. Coil M2 will de-energize and its motor is now off. When the input from the signal converter to the C input of the JK flip-flop went to 0, or "off," the JK flip-flop toggled.

Prior to the input going off, Q was logic 0 and $\overline{Q}$ logic 1. When the input signal goes to 0, the flip-flop toggles and Q becomes a logic 1 and $\overline{Q}$ becomes a logic 0. When the pressure switch closes again, there is an input to C and the two AND gates. This time M1 energizes. When the pressure switch opens, M1 de-energizes. The motors will take turns running because of the toggling action of the logic control circuit. The toggling of the motors will continue as often as the pressure switch opens and closes.

Notice the symbols for signal converters in Fig. 14-6 are ac signal converters. A dc signal converter could also have been used. Its symbol is also shown in Fig. 14-6.

A dc signal converter may not be anything more than a voltage divider consisting of two resistors. The

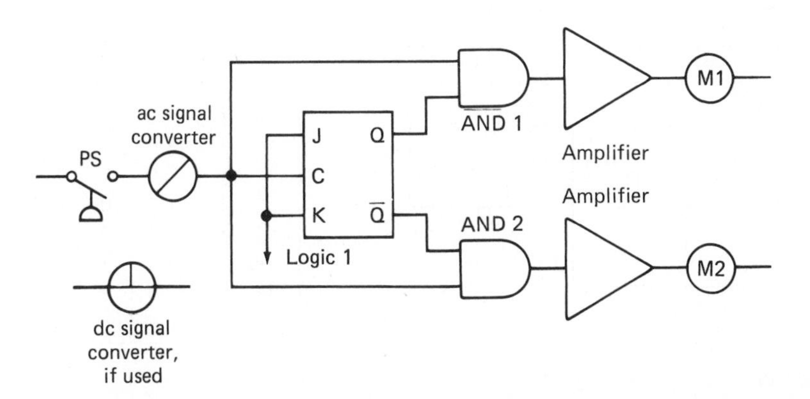

Fig. 14-6 Three logic elements (two ANDs and one JK flip-flop) will give the equivalent control as the electric relay circuit in Fig. 14-2.

requirements for a high voltage for the pilot device and a low voltage for the input to logic elements can be accomplished by using a resistor network. The ac signal converter or initial input may be a transformer with high voltage on the pilot device and transformer primary. The secondary of the transformer produces the low voltage that is then rectified and applied to the input of the low-voltage logic elements.

To get the output signals from the low-voltage logic elements to the high-voltage output devices, output amplifiers are used. The amplifier may consist of an optical coupler driving a triac. Figure 14-6 shows the circuit symbol circuit for an amplifier. The optical coupler in the input circuit of the amplifier provides an input signal from low-voltage logic circuits to the gate circuit of a triac. The triac turns on output devices such as motor starters, solenoid valves, etc.

The IC logic gates for the circuit in Fig. 14-6 would cost less than \$1. Troubleshooting can be done with a voltmeter, logic probe, or built-in LEDs. Logic 1 is 5 V; logic 0 is 0.3 V. Three voltage readings can indicate any faulty components. The advantages of solid-state ICs in this problem should be quite evident. The only way this problem could have been made simpler is if we had done it with programmable logic arrays. In a later chapter we do it with IC programmable logic arrays as well as on a programmable controller.

14-3 BINARY COUNTERS AND THEIR INDUSTRIAL APPLICATIONS

The JK flip-flop is used for numerous applications. It is widely used in counters and shift registers. Figure 14-7 represents a binary counter. The table in the figure also indicates the condition of the output, which provides different binary numbers. For example, the binary number 0101 represents the decimal number 5. The binary number 1011 represents the decimal number 11. The decimal number 13 is 8 + 4 + 1, and so the binary number for 13 would be 1101.

In Fig. 14-7 four JK flip-flops are required to make a 4-bit binary counter. By using the IC chip 7493, illustrated in Fig. 14-1, a 4-bit binary counter can be obtained with one chip and very little external wiring. All that is required is a clock pulse and connections to the A-B-C-D outputs.

A rotary switch such as a drum switch, ratchet switch, or a stepping relay is often used in conventional control circuits as scanning switches to find closed pilot devices. For example, if the four tanks in Fig. 14-8 are to be filled in sequence, we would require a scanner to look at each level switch for the tanks to determine if a tank required filling. When all the tanks are full, the scanner would continue to scan the level switches to look for a low tank. When the scanner would locate a low tank, it would stop on the low tank, initiate the control to start the filling, and wait until the tank is full to start scanning again. Let's convert the specifications of this tank-filling problem into a solid-state logic control circuit using ICs.

To start our circuit conversion, we first identify our inputs and outputs to match the needs of the problem (see Fig. 14-8).

Level switches LS1, LS2, LS3, LS4 can be any type of switch such as a float or electrode level switch. The requirements of the level switches are that they close at some predetermined low level and stay closed until they reach some predetermined high level. They do not close again until the low level is reached.

The solenoid valves are electrically operated. The solenoid valve SOL VE in Fig. 14-8 is energized only

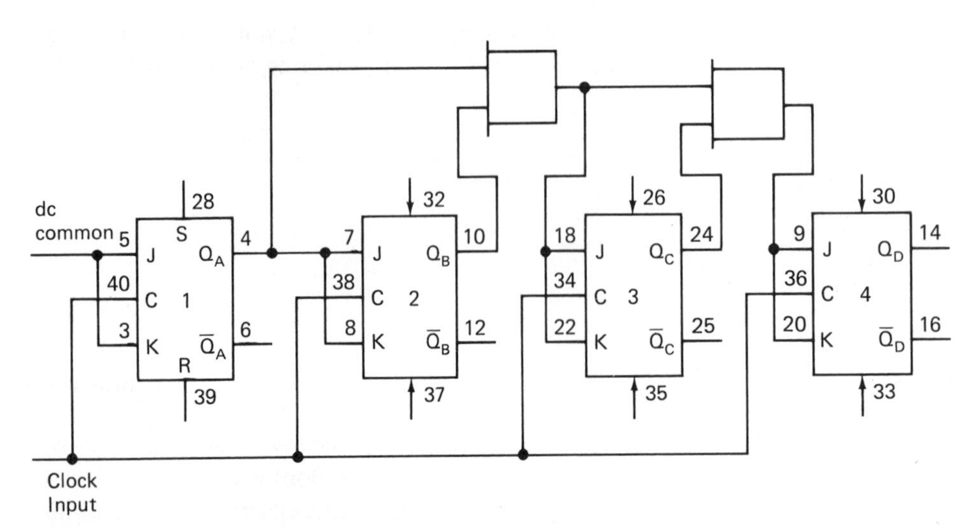

State	8 Q_D	4 Q_C	2 Q_B	1 Q_A
0	0	0	0	0
1	0	0	0	1
2	0	0	1	0
3	0	0	1	1
4	0	1	0	0
5	0	1	0	1
6	0	1	1	0
7	0	1	1	1
8	1	0	0	0
9	1	0	0	1
10	1	0	1	0
11	1	0	1	1
12	1	1	0	0
13	1	1	0	1
14	1	1	1	0
15	1	1	1	1
0	0	0	0	0

Fig. 14-7 Four-bit synchronous binary counter.

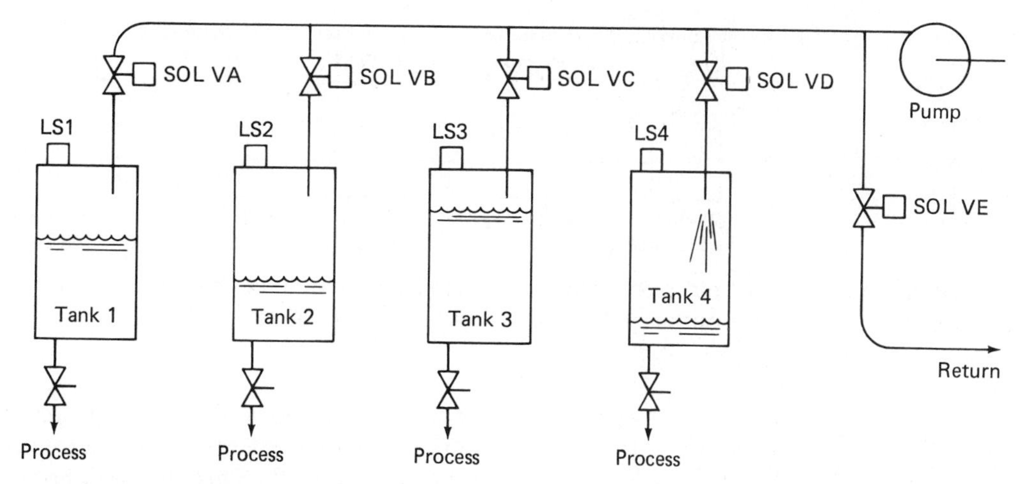

Fig. 14-8 Layout for tank-filling problem. Tanks are to be scanned for a low tank. Tanks are filled in sequence as they go low.

when all other solenoid valves are closed. The drain of the tanks is determined by the process needs. Manual valves are shown at the bottom of the tanks. Solenoid valves SOL VA, SOL VB, SOL VC, and SOL VD in Fig. 14-8 control the filling of tanks 1, 2, 3, and 4.

To develop our circuit let's start with our two vertical lines and identify the inputs and outputs that are required. Figure 14-9 illustrates this part of our circuit.

Most integrated logic systems have a device called a *clock*. The purpose of the clock is to provide a pulse at some specific time interval. For example, a clock may put out one pulse a second or up to 100,000 pulses a second depending on its design. A circuit for a clock

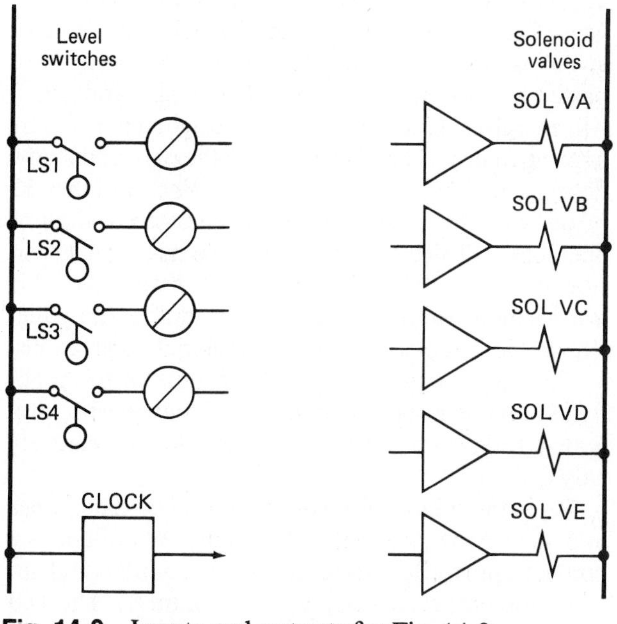

Fig. 14-9 Inputs and outputs for Fig. 14-8.

that would put out one pulse a second is illustrated in Fig. 14-10 on the next page. We need a clock with the binary counter chip to develop a scanner for our tank-filling problem. Clocks are available in ICs.

14-4 DECODERS AND THEIR USES

The output of a binary counter is a binary number which can be decoded with ANDs and inverters to come up with the decimal equivalent of a binary number. An IC chip such as a 7442 (Fig. 14-1) will decode binary numbers from the IC binary counter chip 7493 and produce the equivalent of a decimal count such as 1, 2, 3, 4, etc., or 2, 4, 6, 8, or any combination desired. All the outputs of a 7442 decoder go from a logic 1 to logic 0 one at a time as the decoder converts the binary numbers from the counter to the decimal numbers. The outputs of the decoder are connected to inverters. Figure 14-11, on the next page, is the scanner for the tank-filling problem. The clock produces one pulse a second. The binary counter counts in binary numbers from 0 to 15. The decoder, through inverters, provides us outputs on a binary count of 2, 4, 6, 8. Our scanner is looking at the level in the tanks every 15 s with a 2-s delay between each tank. Scanning the level of the tanks every 15 s is more than adequate for industrial applications. An increased clock speed of 100,000 pulses per second, would make the scan so fast it would almost be instantaneous.

Figure 14-12, on the next page, illustrates how a tank is scanned. If the tank is low, the AND will have an input from the level switch. If the count from the decoder is the correct number, the AND will have both inputs producing an output to the solenoid valve. The solenoid valve energizes, starting the filling of the tank.

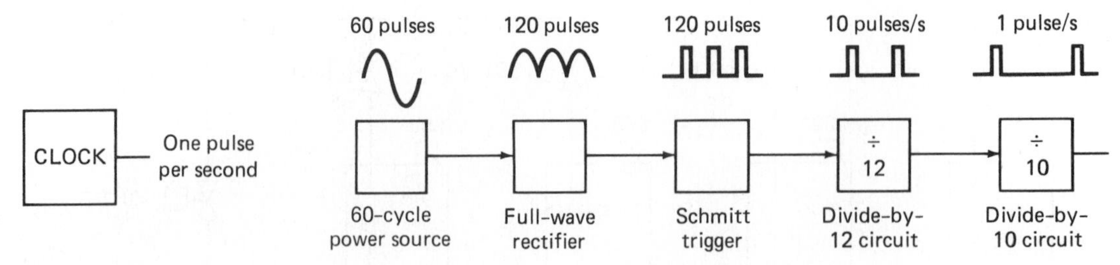

Fig. 14-10 Clock that produces one pulse per second. Divide-by-12 and divide-by-10 circuits are counters that provide one output pulse after 12 and 10 input pulses.

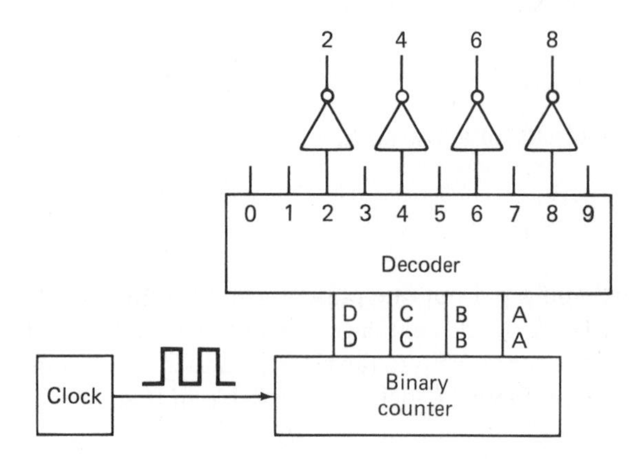

Fig. 14-11 Scanner to produce one output at a time with timed intervals.

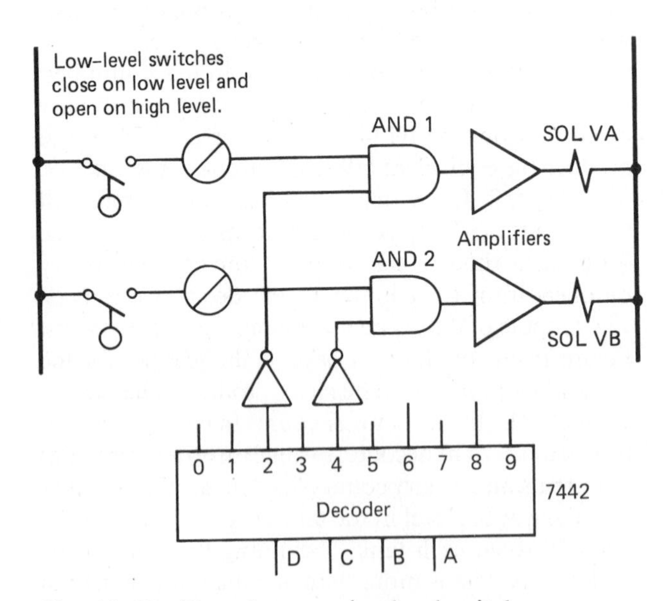

Fig. 14-12 Decoder scanning level switches.

When the tank is filling, the scanner must stop until the tank is full. This is accomplished by the circuitry in Fig. 14-13. In this circuit if any of the ANDs are providing an output to the solenoid valves, there is an input to the NOR. The NOR gives a logic 0 output anytime it has a logic 1 at any of its inputs. With no output from the NOR (due to a NOR input) the AND at the clock is shut off. This will cause the binary counter to stop counting which will maintain the decoder's input to the AND for the low tank. As soon as the tank is filled, the level switch opens, giving a logic 0 to the AND. Its output goes to logic 0. The NOR no longer has an input, and so it gives a logic 1 output. This does two things. It allows the clock pulses to reach the binary counter. It starts counting and it will also energize SOL VE. Solenoid valve E must be energized in our problem when none of the tank-filling solenoids are energized.

The complete circuit for our tank filling problem is illustrated in Fig. 14-14 on page 214. Notice how the logic circuit provides a simple solution to the problem. The only easier method is with P.C.s. We will write a program for this problem when we discuss programmable control.

Troubleshooting this circuit required first understanding the logic of the circuit and the sequence of operation. A voltmeter or logic test probe looking for logic 1s or 0s will verify if components are faulty.

Figure 14-15, on page 214, represents three conveyors A, B, and C. The boxes traveling down conveyor C are counted by breaking a light beam. When a given number of boxes (12 for our problem) have passed the counter, gate 1 will de-energize and gate 2 will energize. Gate 1 returns to its original position by a spring return. Gate 2 is moved to the right and it now deflects the boxes to conveyor B. When another 12 boxes passes the counter, gate 2 de-energizes and gate 1 energizes. Deflector gate 2 returns to its original position and gate 1 deflects the boxes to conveyor A. This arrangement would be ideal for an automatic palletizer. This problem has many industrial applications.

To design a conventional relay circuit for this problem would be rather difficult. To design it with integrated solid-state logic circuits should be relatively easy.

Begin the circuit design by drawing two vertical lines and identifying the input and outputs. We will show a contact from the light-activated relay (counter) and not show the entire circuitry for the counter. The two outputs will be the solenoid for gate 1 and the solenoid

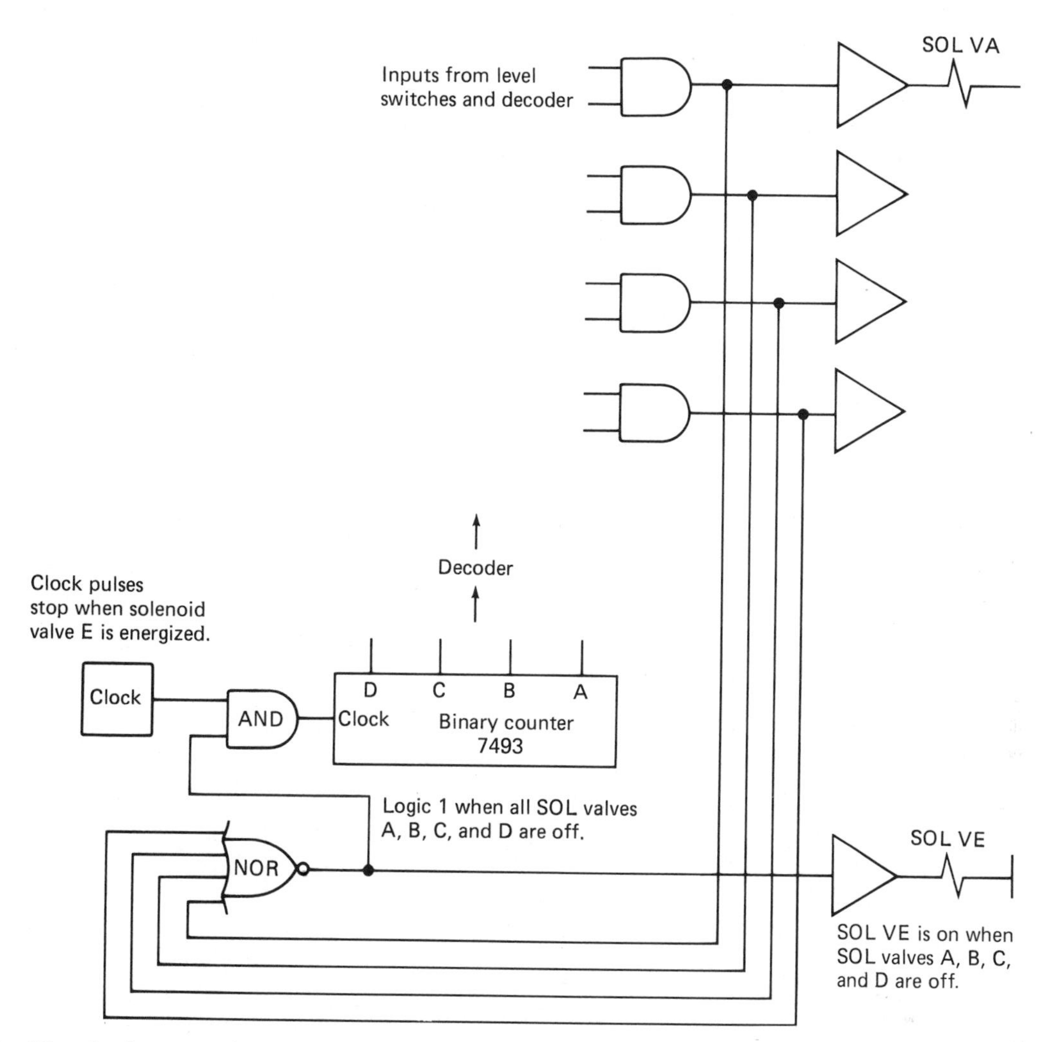

Fig. 14-13 The clock output is stopped to binary counter when a solenoid valve is energized. Solenoid VE is energized when valves A, B, C, and D are off.

for gate 2. This is shown in the logic circuit of Fig. 14-16 on page 215. In this circuit we show the gates being connected to a JK flip-flop. We are doing this because of the toggling action of the flip-flop. If gate 1 is on, gate 2 is off, and vice versa. We never will have both gates on or both off at the same time because of the toggling action of the JK flip-flop. If $\overline{Q}$ is a logic 1, then Q is a logic 0. If Q is logic 0, then $\overline{Q}$ is logic 1.

In this circuit we have the counter contact connected directly to the binary counter. Every time a box passes the counter it will advance the binary count by 1.

When the counter receives 12 inputs, the outputs of the counter will be the binary number 12. This number will be 1100. The binary number 1100 indicates that the output of the binary counter is on, the C and D outputs are on, but the B and A outputs are off. This combination, because of the two inverters, causes the AND to give an output.

When the AND gives an output, the JK flip-flop has an input. Nothing will happen because the flip-flop will switch only as the input pulse goes negative, or turns off.

Notice the AND output also goes to the reset of the binary counter. This resets the counter to the binary equivalent of 0 (0000), and the AND is now off. The input to the flip-flop is removed and it toggles, which switches its output and the deflector gates reverse their positions.

When we initially start up, if we want gate 1 to energize first, pressing the pushbutton will set Q on and gate 1 will energize. Pressing the pushbutton would only be required at start-up.

This application again should show that the logic for what would be a difficult problem in conventional relay control results in a simple, inexpensive circuit. Troubleshooting will also be simple.

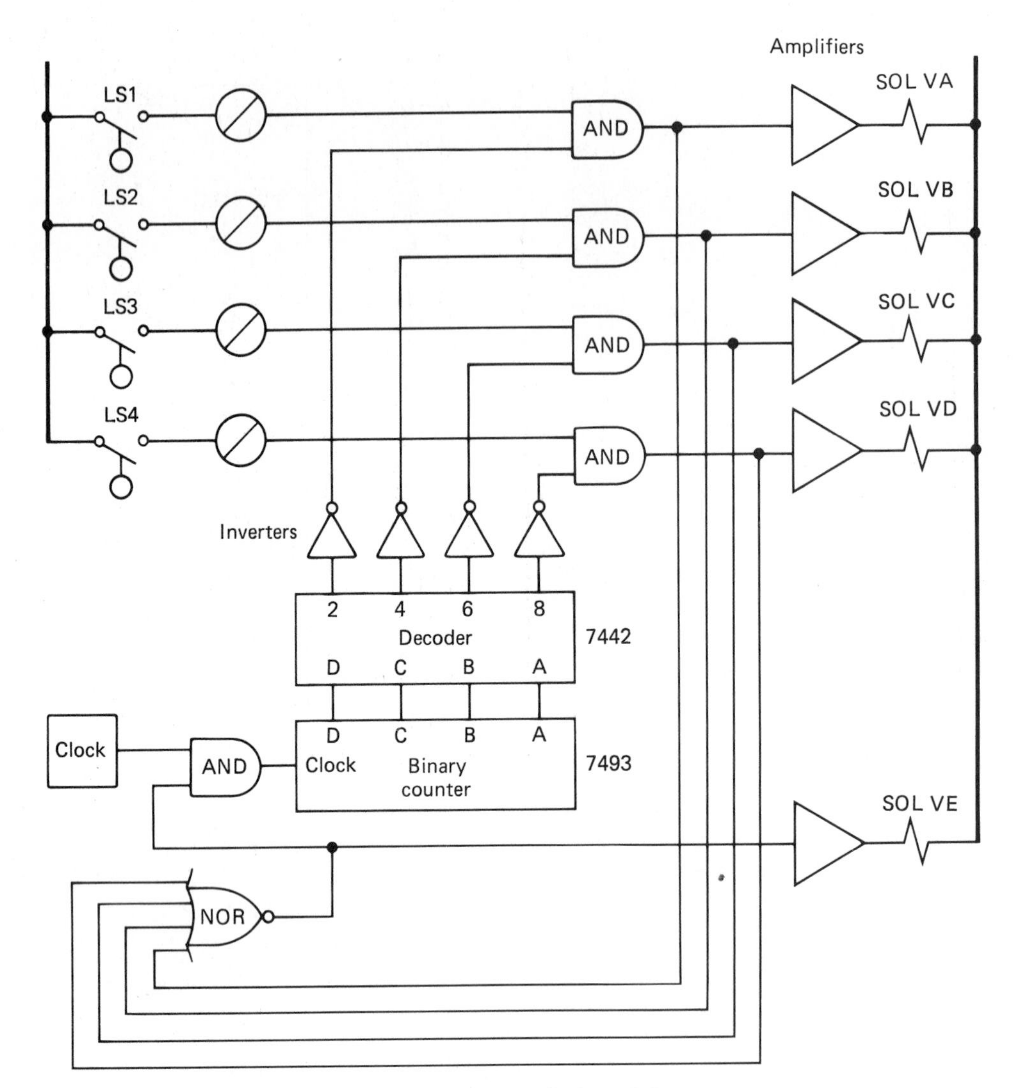

Fig. 14-14 Complete circuit for the tank-filling problem of Fig. 14-8.

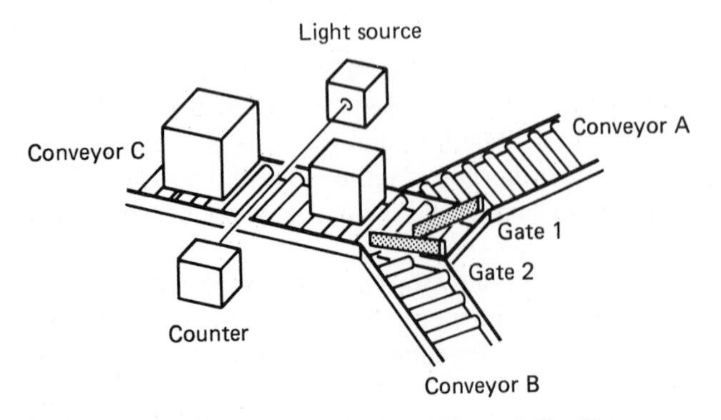

Fig. 14-15 Three conveyors A, B, and C. Boxes from some source are traveling down conveyor C. Boxes are diverted to conveyor A and B by gate 1 and gate 2.

14-5 LOW-COST CONTROL CIRCUITS USING INTEGRATED NAND LOGIC CHIPS

A 7400 IC is a quadruple two-input positive NAND gate. Figure 14-17 shows how NAND gates can be connected and converted to equivalent AND, NOT, and OR logic. One NAND with its inputs tied together will give the same logic as a NOT, or an inverter. It takes four NANDs to make a memory circuit. Figure 14-17 also shows that NAND circuits can be simplified to eliminate redundencies.

In boolean algebra we have a postulate, or theorem, that says $\overline{\overline{A}} = A$. What this means is if we NOT, or invert, a logic input twice its logic value is still the same as the input. For example, in Fig. 14-17 if a logic input is connected to two NANDs in series, the NAND is as much good as a piece of wire and can be replaced with a

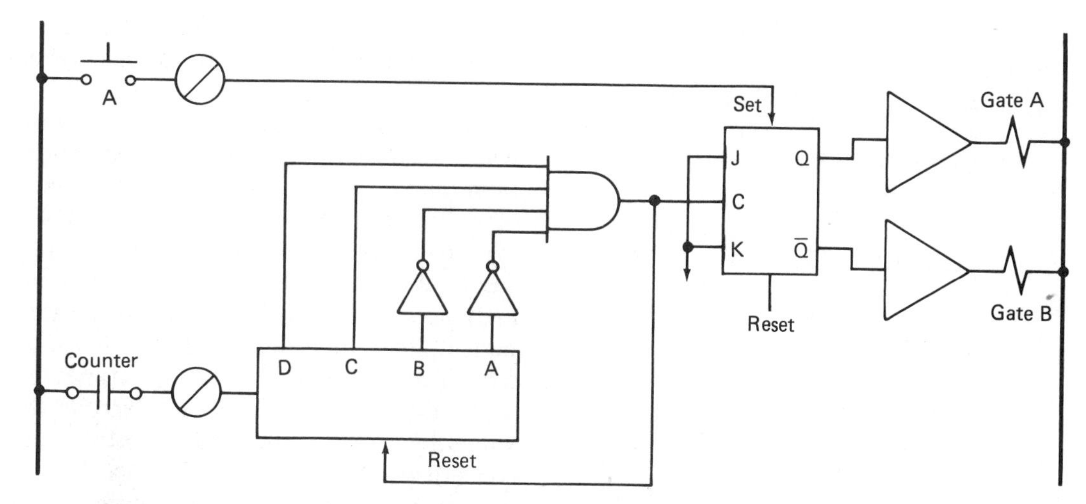

Fig. 14-16 Integrated circuit controller for the conveyor problem of Fig. 14-15.

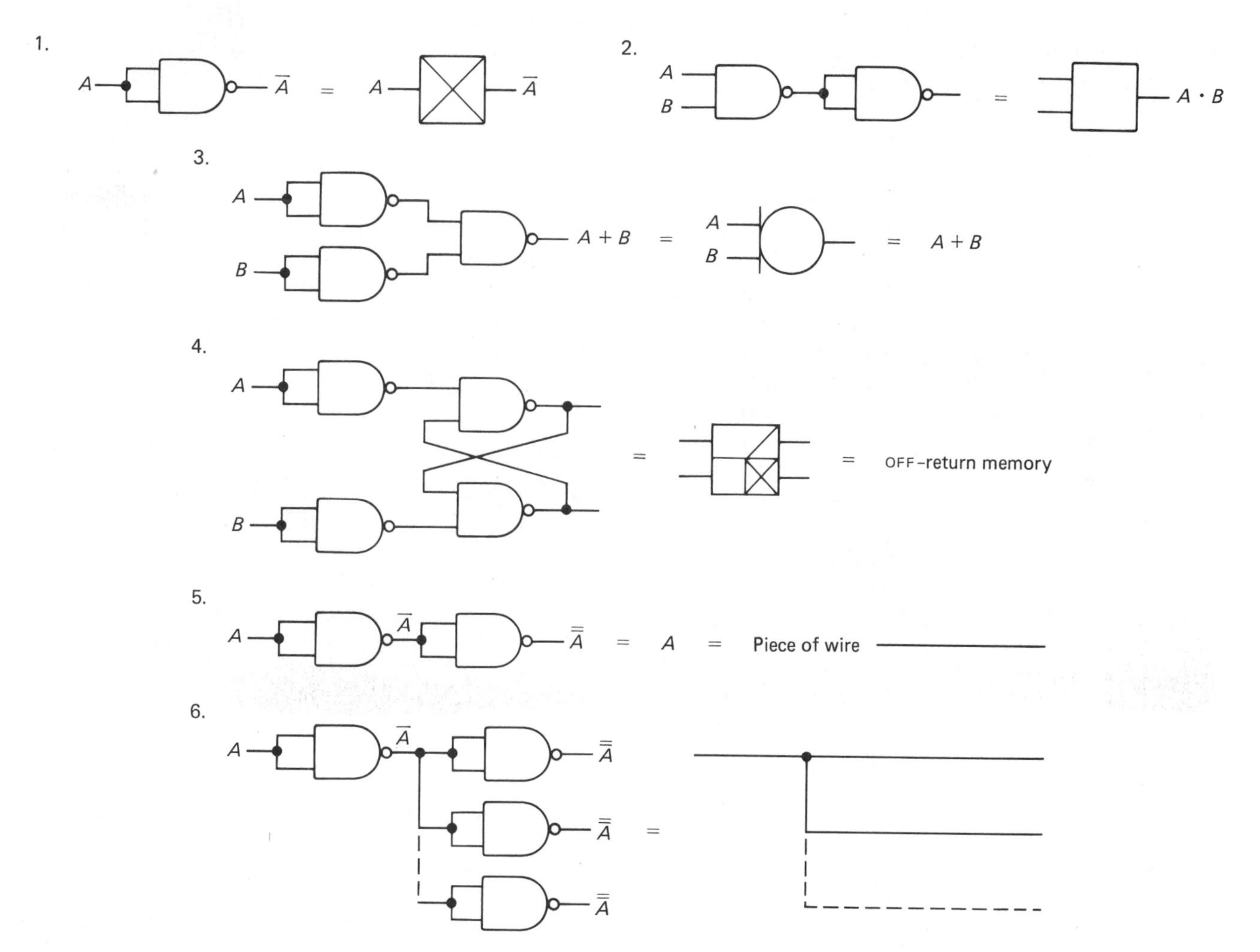

Fig. 14-17 NAND gates and their equivalent circuits.

piece of wire. Notice how this rule can be used to eliminate redundancies as well as simplify NAND circuits.

Figure 14-18, on the next page, represents a one-point annunciator. When the trouble contact closes, a bell will ring and a flashing light will come on. When trouble is acknowledged by pushing the ACKNOWLEDGE button, the bell will go off and the light will be on steady. When trouble clears up, the light will go off and the circuit waits for another trouble to occur.

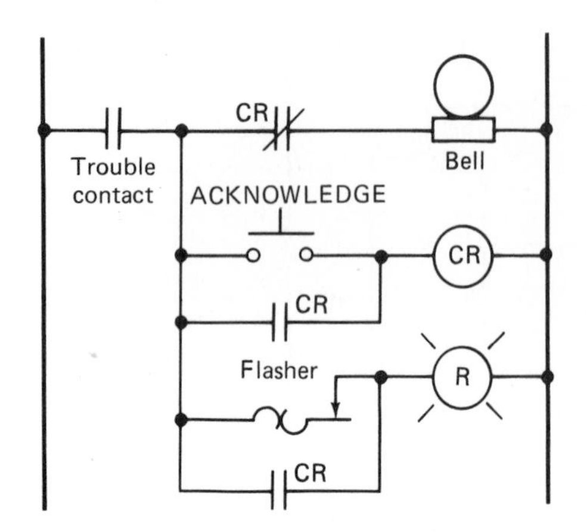

Fig. 14-18 Electric circuit for one-point annunciator.

Figure 14-19 is the logic circuit developed from the electric circuit of Fig. 14-18. The electric circuit is converted to a logic circuit using ANDs, ORs, and NOTs. Remember the objective is to do the logic of this circuit as inexpensively as possible by using NAND logic. To get a NAND logic circuit, we replace the conventional logic gates in Fig. 14-19 with equivalent NAND circuits. Figure 14-20 illustrates this. Remember two NANDs make an AND. Three NANDs are required to make an OR. One NAND can be used for a NOT.

When we converted the logic circuit for a one-point annunciator (Fig. 14-20) we ended up with 12 NAND gates. This would require four ICs. When we simplified the circuit by eliminating double negations, we ended up with 7 NAND gates. Seven NAND gates would require two ICs with one spare NAND gate. This achieved our goal of showing that just a few cents worth of ICs could solve an industrial application problem such as a one-point annunciator. Notice this is the cost for the logic only, which is the decision-making part of the circuit. Amplifiers, inputs, bells, and lights would cost extra. The final simplified circuit for our one-point annunciator using NAND logic appears in Fig. 14-21.

In our one-point annunciator NAND logic circuit we have attempted to show how a logic system can be developed from one logic element only. We have shown that with NANDs only we can have a logic circuit which in conventional logic would contain ANDs, ORs, NOTs, and memories. There are advantages to the use of NANDs only. We need only one type of logic chip. We do not require numerous different replacement chips. We also saw that the NAND logic circuits simplified into considerably simpler circuits, with less components than the original circuit. Once a circuit is simplified, it becomes difficult sometimes to recognize where the ANDs, ORs, and NOTs were in the original circuit. This is no problem as far as troubleshooting is concerned. A NAND logic is not limited to small circuits such as one-point annunciators. It can also be used for some very complicated circuits.

Later in this chapter, we will see that we can use straight NOR logic and accomplish the same results as with all NANDs.

There are thousands of applications for ICs and hundreds of ICs from which to choose. We have touched on only a few of these ICs. Next we will look at some of the industrial logic systems using ICs in the heart of their logic elements. We will look at Westinghouse Electric Corporation and its Numa-Logic® line of solid-state control next. We will start with Westinghouse's description of the advantage that Numa Logic offers for rapid troubleshooting and maintenance.

14-6 WESTINGHOUSE SOLID-STATE NUMA-LOGIC CONTROL

Westinghouse Numa-Logic is offered as the control with built-in machine troubleshooting. It is one of the easiest of the industrial logic control systems to troubleshoot.

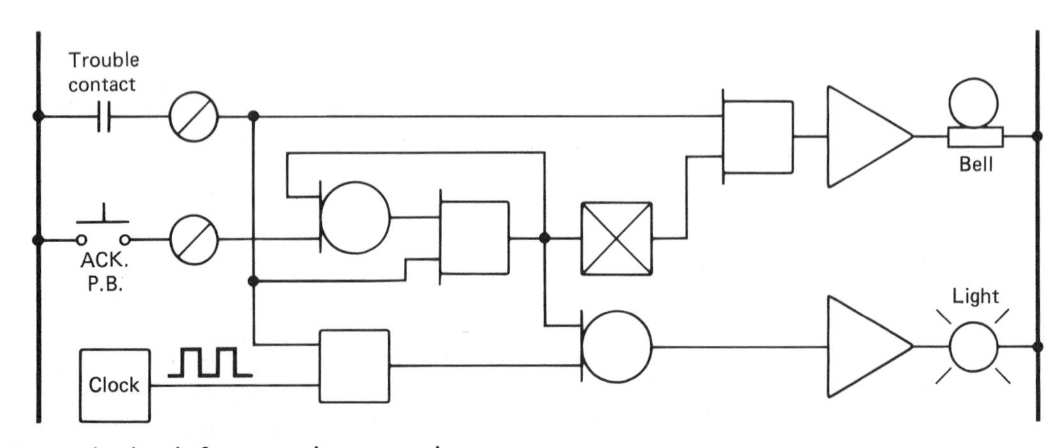

Fig. 14-19 Logic circuit for one-point annunciator.

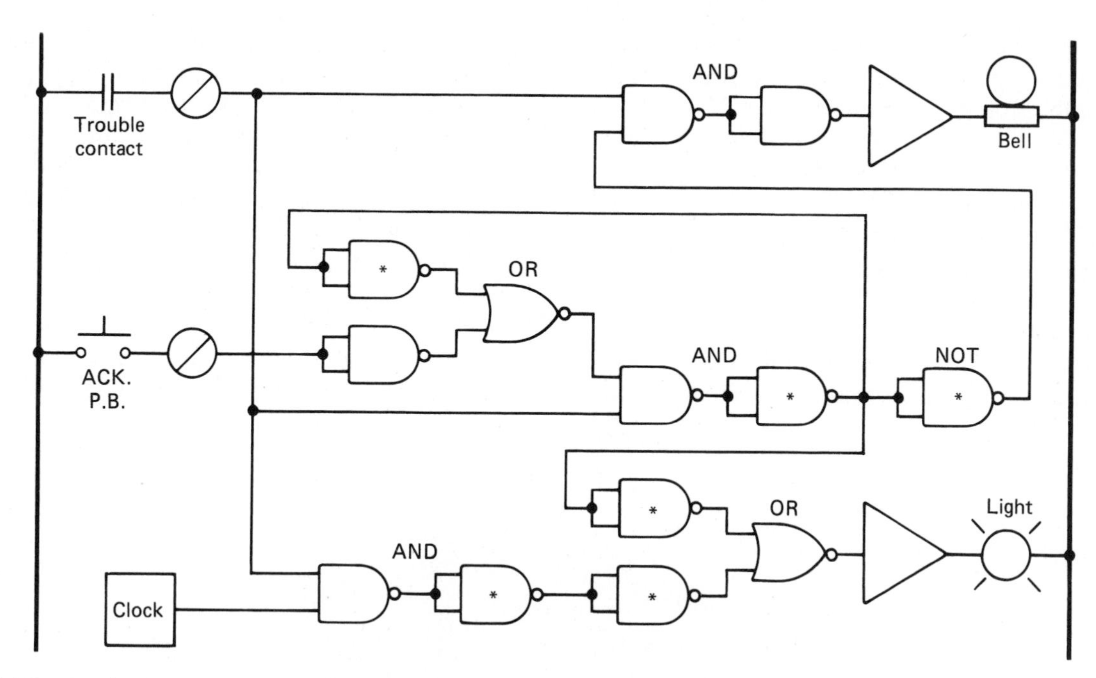

Fig. 14-20 Logic circuit of Fig. 14-19 converted to NAND-logic. The * symbol indicates NANDs that can be eliminated by the rule of double negatives—$\bar{\bar{A}} = A$.

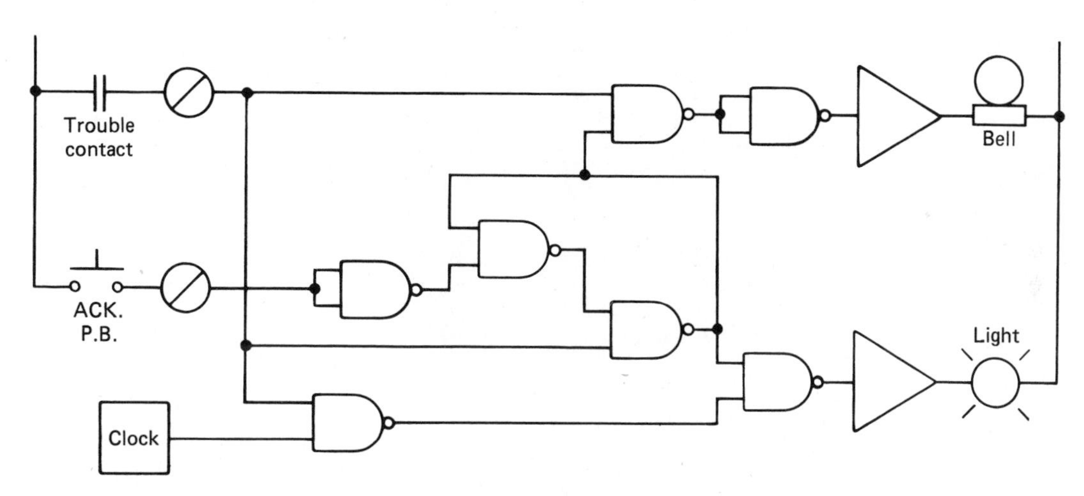

Fig. 14-21 The NAND equivalent circuit of Fig. 14-20 with redundancies eliminated by the rule $\bar{\bar{A}} = A$.

The most common cause of electrical downtime is generally failure of, or trouble with, the machine pilot devices (solenoids and limit switches). Correction of this problem is often as simple as replacing the faulty device.

Although replacing a faulty device may be a simple operation, finding it is another problem—a complicated, time-consuming one. And with today's high-speed production schedules, even a short downtime can be extremely costly. Something as simple as a faulty solenoid can cost thousands of dollars in lost production.

For example, let's say the slide unit at one of the stations on a large transfer line stops. (No one knows it, but the problem is simply a faulty limit switch.) An electrician is called, comes out, inspects the slide unit, and tries to operate it manually. When it won't work, the electrician gets the prints and a tester, goes to the correct door on the relay panel, and begins to inspect the circuit thoroughly. The electrician will spend about an hour with the prints and a test lamp searching through the maze of wires for the problem. Locating this type of problem could easily cost 85 min of production time.

Westinghouse Numa-Logic solid-state controls feature a built-in machine troubleshooter that literally pinpoints the problem. The front of each Numa-Logic control module is a pictorial display that shows the system's status at a glance. Long periods of troubleshooting are eliminated.

When that slide unit stops, the operator or supervisor can simply look through the glass window at the panel and locate the Numa-Logic modules for the particular workstation involved. The illuminated display on the front of the modules indicates that a particular limit switch is faulty and is causing the trouble. Now the operator or supervisor can call the electrician and tell him or her that a specific limit switch at a particular workstation is bad. So the electrician can bring the correct replacement. By using the Numa-Logic solid-state control the 85 min of downtime can be reduced to 8 min or less. Figure 14-22 shows the Westinghouse Numa-Logic modules.

Electrical grounds are another cause of downtime. Standard relay panels are equipped with ground lights that indicate that there is a grounding. Unfortunately, these ground lights don't indicate how many grounds there are, where the grounds are located, or that partial

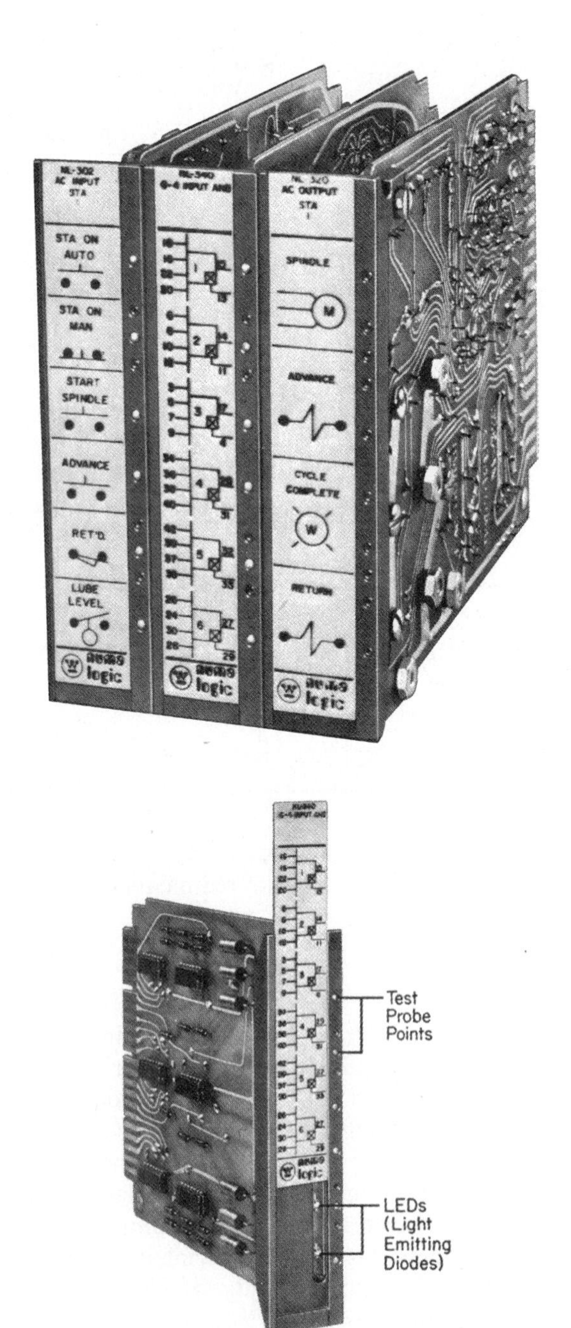

Fig. 14-22 Westinghouse Numa-Logic modules. *(Westinghouse Corp.)*

grounds are developing. It may take an electrician 2½ to 4 h of searching and testing to find the ground (if it can be found at all).

With a Numa-Logic solid-state system, this time can be reduced to 5 min. When a ground is present, a red light on top of the panel lights up. Once again, the operator or supervisor can determine the trouble instantly by looking at the pictorial displays on the front of the modules (in the case of ground detection there is a special group of modules that will show all the critical limit switches and solenoids on the machine). The operator will be able to tell at a glance exactly what limit switches or solenoids have grounded or are in the process of grounding.

The above indicates the advantages that have been engineered into Numa-Logic control. Numa Logic is the current generation of solid-state logic control manufactured by Westinghouse. The system consists of racks that contain the logic cards. The heart of the logic cards are solid-state ICs. Solid-state reliability plus the inherent long life of proven solid-state components ensure trouble-free operation, less machine downtime, and more production uptime.

Numa-Logic controls do the same job as electromechanical relays do in controlling machines and process line functions. The difference is that Numa-Logic contains no moving parts while providing instant troubleshooting.

14-7 VARIOUS NUMA-LOGIC MODULES

Westinghouse Numa-Logic solid-state controls were designed expressly for all types of automatic control systems. One of their special uses is in the control of machine tools. Numa-Logic provides the user with conveniences not associated with normal electronic technology. All the logic needed to control a complicated automatic machine such as a large transfer line has been packed into six basic motion modules. Various combinations of these basic modules replace all the relays normally associated with the logic portion of a machine-tool control panel. And the Numa-Logic panels are usually only half the size of the conventional relay panels (see Fig. 14-22 for module construction). In addition, two interface modules provide computer monitoring and computer control, and a ground detection module provides direct identification of grounded limit switches and solenoids. In other words, six basic modules and three auxiliary modules make up the complete line of Westinghouse Numa-Logic solid-state controls.

The heart of the Numa-Logic pictorial display is the faceplates on the front of the modules. Each faceplate is custom-built to depict the particular operation it serves. Should a user want to replace a module, the custom faceplate slides right off for insertion on the new module.

A further convenience of the Numa-Logic controls is the handy dwell-time adjustment on the time-delay modules. Even when the modules are plugged in the panel racks and operating, adjusting the dwell time is a simple process. A small, knurled knob is located on the top of the module right behind the faceplate. Changing the dwell time merely requires reaching over the top of the module and turning the knob. No special tools needed. No power down, no dangerous digging through the panel.

The Numa-Logic built-in machine troubleshooter is an exclusive feature of the system and pictures machine pilot devices on the faceplate lens. This feature provides visual determination of machine motions. Light-emitting diodes illuminate the faceplate as machine motions change, including each logic function. This is extremely helpful for external troubleshooting. Each function also has built-in test points. This speeds up troubleshooting by as much as 50 percent by pinpointing faulty pilot devices.

FAIL-SAFE INTERLOCKING

Exclusive electrical and mechanical interlocking of Numa-Logic controls ensures personnel and equipment safety. A short power pin electrical interlock ensures that motion will not engage while the module is being inserted or removed. Broken wires inhibit ac and dc outputs to prevent unwanted motions from occurring. Polarity interlock on each module protects against accidental dc power supply reversals. Mechanical interlocks prevent insertion of components in any except the proper position. Numa-Logic controls are designed to operate at temperatures up to 185°F (85°C). No temperature-controlled or air-conditioned rooms are required. The logic control works in all types of temperatures.

ELECTRICAL TRANSIENTS

Numa-Logic modules utilize energy rejection and circuit slow-down techniques to prevent transients from causing unwanted machine motions. Numa-Logic is immune to electrical transients.

All modules have test points on the faceplate for checking the outputs. The input and output modules can have LEDs for visual monitoring.

INPUT MODULES

Input modules convert incoming signals from ac, dc, proximity, photocell, and analog devices to dc logic levels. Each input circuit contains both "true" and NOT outputs. Faceplate lenses are available blank, or with standard logic symbols or custom symbols.

OUTPUT MODULES

Output modules convert low-power dc logic levels to ac or dc power levels to drive pilot devices. Faceplate lenses are available blank, or with standard logic symbols or custom symbols.

The logic modules contain the sequencing required to control machine motions or process functions. Modules are available for the AND, OR, timer, shift register, retentive memory, flip-flop, one shot, digital timer, digital counter, JK flip-flops, etc. Faceplate lenses are available with either standard English logic or ANSI (American National Standards Institute) symbols.

Numa-Logic solid-state controls are simply printed circuit boards (called *modules*) containing solid-state components which perform the functions of electromechanical relays. The modules plug into prewired mounting racks to complete the circuitry.

A typical Numa-Logic installation consists of a series of input and output power modules, and a sufficient number of logic modules to control the machine or process function. A mounting rack and separate power supply are required to complete the installation. Control sequences are "built-up" using the English logic system of independent AND and OR gates instead of traditional relay ladder diagram symbols. However, the end results are the same.

All Numa-Logic modules are rack mounted. Removal of two captive bolts allows the rack to drop down 90 degrees for easy wiring. (Another safety interlock immediately cuts the power when the rack is dropped. To operate the machine with the rack dropped, a defeater switch must be intentionally tripped.)

When modules are inserted in the rack, they make contact with a module connector, which must be wired to complete the circuit. The connectors are the standard industrial type, with 44 posts for solderless connections. Each post will accept up to three terminations of no. 22, 24, or 26 wire. The cantilever spring contacts of the connectors are capable of a large deflection without making a permanent set.

Wire wrapping is the technique used for making wire connections from point to point on the mounting rack to complete system circuitry. Wrapping is performed using a manual, air-driven, or electric gun. Wire-wrapping is normally already done when controls are installed in the user's plants. However, field modifications can easily be made using a hand-operated wire-wrap gun.

Block diagrams of basic control systems are shown in Fig. 14-23. Notice that each of the block diagrams has circuit inputs, logic functions, and circuit outputs.

The circuit inputs of each of the control examples (a), (b), and (c) are such items as limit switches and pushbuttons. The circuit outputs for each example are solenoids, motor starters, and lamps.

In the Example (a) of a common control application, the logic is provided by the sequencing of the machine. In Example (b), the control system uses hard-wired magnetic relays to execute the logic of the circuit. In Example (c) the input devices are in high-voltage circuits. Either ac or dc signal converters are required to convert the high-input voltages to the low voltages with which the solid-state logic devices operate. Various logic gates provide the logic control system in Example (c).

In order to work with solid-state logic control systems one needs to understand the equipment available. One needs to understand the relationships among the building blocks of control systems. A good way to accomplish this is by first examining the relay control functions and then to relate them to the solid-state functions.

We shall now take a look at how Westinghouse converts relay schematic diagrams to solid-state diagrams. Westinghouse uses what they call the *circle method* to convert a relay schematic or ladder diagram to an equivalent solid-state logic diagram. Although there are a number of methods to accomplish this, the circle method is one of the simplest and fastest. We will explain this method by means of an example.

Figure 14-24 is a pictorial representation of a drilling station on a simple machine tool. This example was chosen because such a drilling station is found on most machine tools, and the complexity of the circuit is equivalent to that of some 90 percent of the relay diagrams you may encounter.

Start-up for the unit requires setting a selector switch to AUTOMATIC or MANUAL and energizing the spindle motor with a pushbutton.

When set for automatic operation, the unit cycles upon command from the central control of the machine. First, the "advance" hydraulic control valve is energized by a solenoid, causing the unit to advance to the workpiece. Second, when the unit's full advanced position is reached, a limit switch is closed, energizing a dwell timer and a "full depth" indicator at the operation's station. Third, after a short time delay the advance hydraulic valve is de-energized and the "return" valve is energized. Fourth, when the full return position of the unit is reached, a second limit switch is closed, de-energizing the "return" valve. All cycling controls then return to normal and the unit is ready for the start of another cycle.

Figure 14-25, on page 222, shows a ladder diagram for our example as it would be drawn normally. Since the circle method requires the grouping of series and parallel contacts, we have redrawn this diagram to make these circuits easier to recognize, as shown in Fig. 14-26 on page 223. The line numbers on the left vertical bus provide the correlation between the two versions.

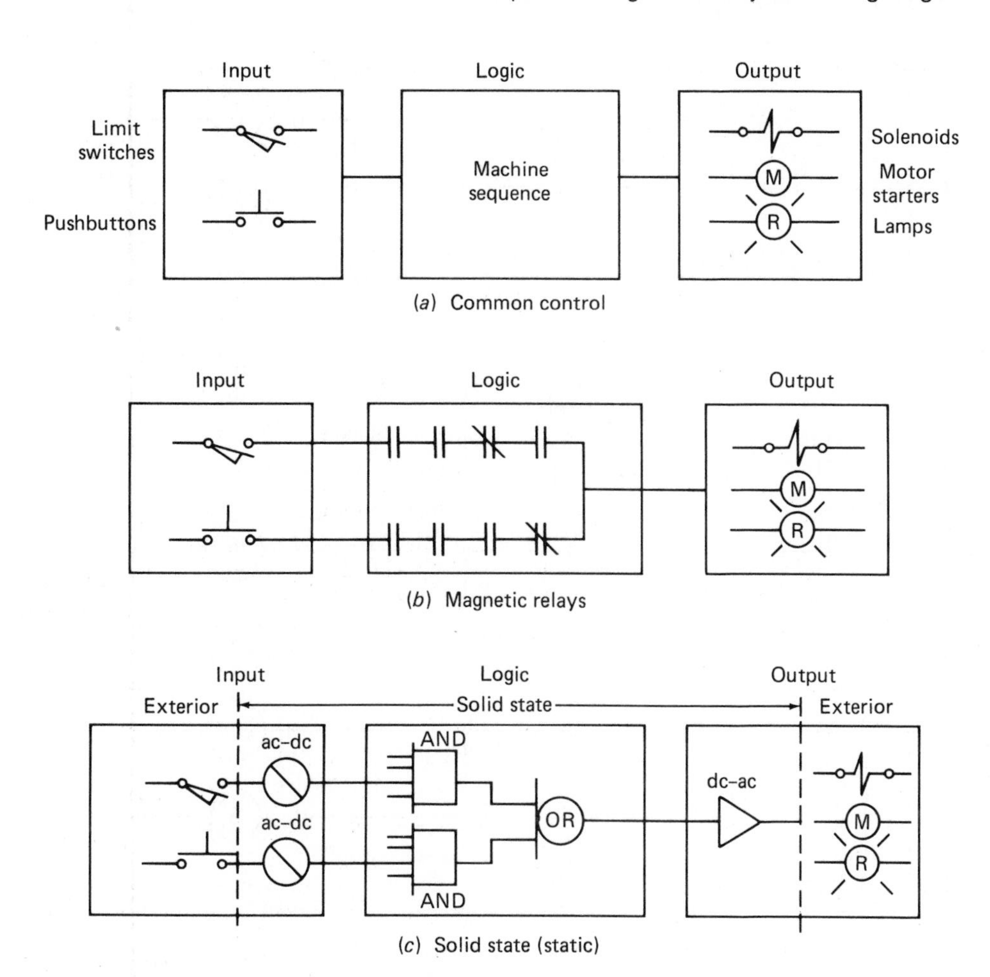

Fig. 14-23 Block diagrams of basic control systems.

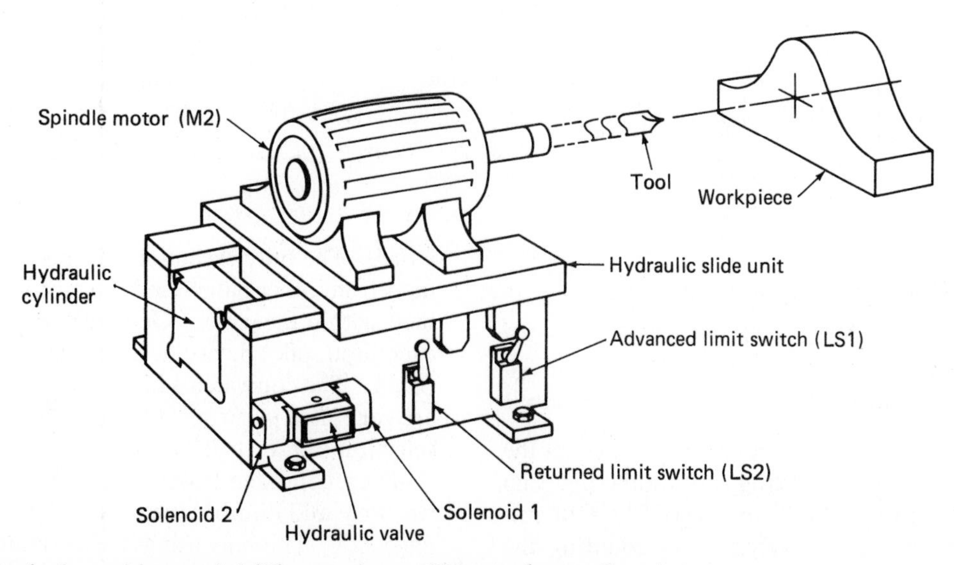

Fig. 14-24 Typical machine tool drilling station. *(Westinghouse Corp.)*

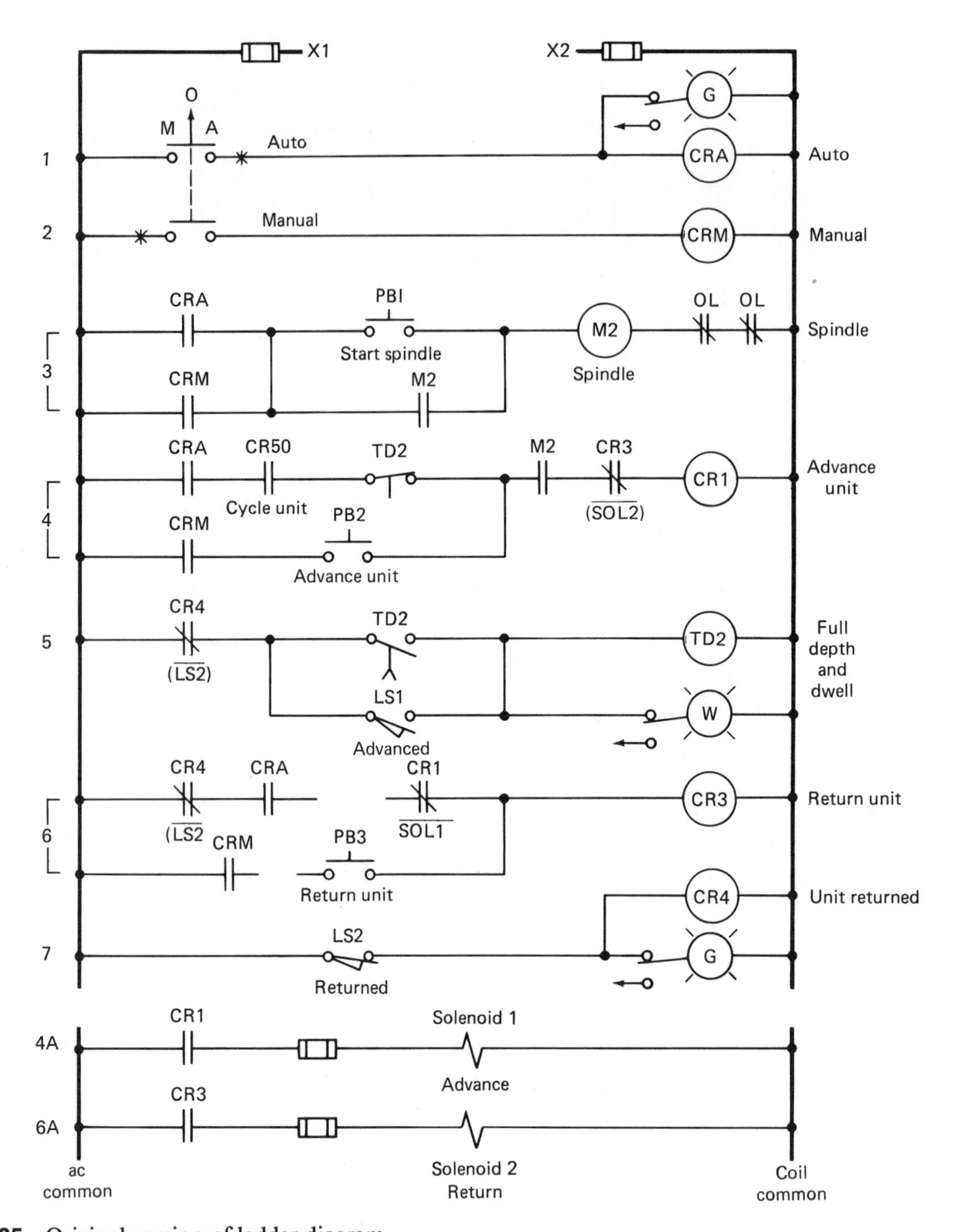

Fig. 14-25 Original version of ladder diagram.

14-8 CONVERSION OF RELAY LADDER DIAGRAM INTO WESTINGHOUSE NUMA-LOGIC

The first step in the conversion process is to select the inputs. This is done by circling and numbering each input as shown in Fig. 14-27 on page 224. Our primary concern is with identifying and counting the 120-V ac pilot devices that are located outside the panel. The only exceptions to this are interlock contacts from other panels and the motor starter seal contacts within the panel. Generally, only one contact of each input pilot device need be wired to the solid-state panel. The equivalents of the closed or open contacts are generated electronically in the ac input circuits by selection of the "true" or NOT function.

The second step is to select the outputs. Here again, we circle and number them as shown in Fig. 14-28 on page 225. The only things we are concerned with are machine outputs driven from relay contacts. Outputs

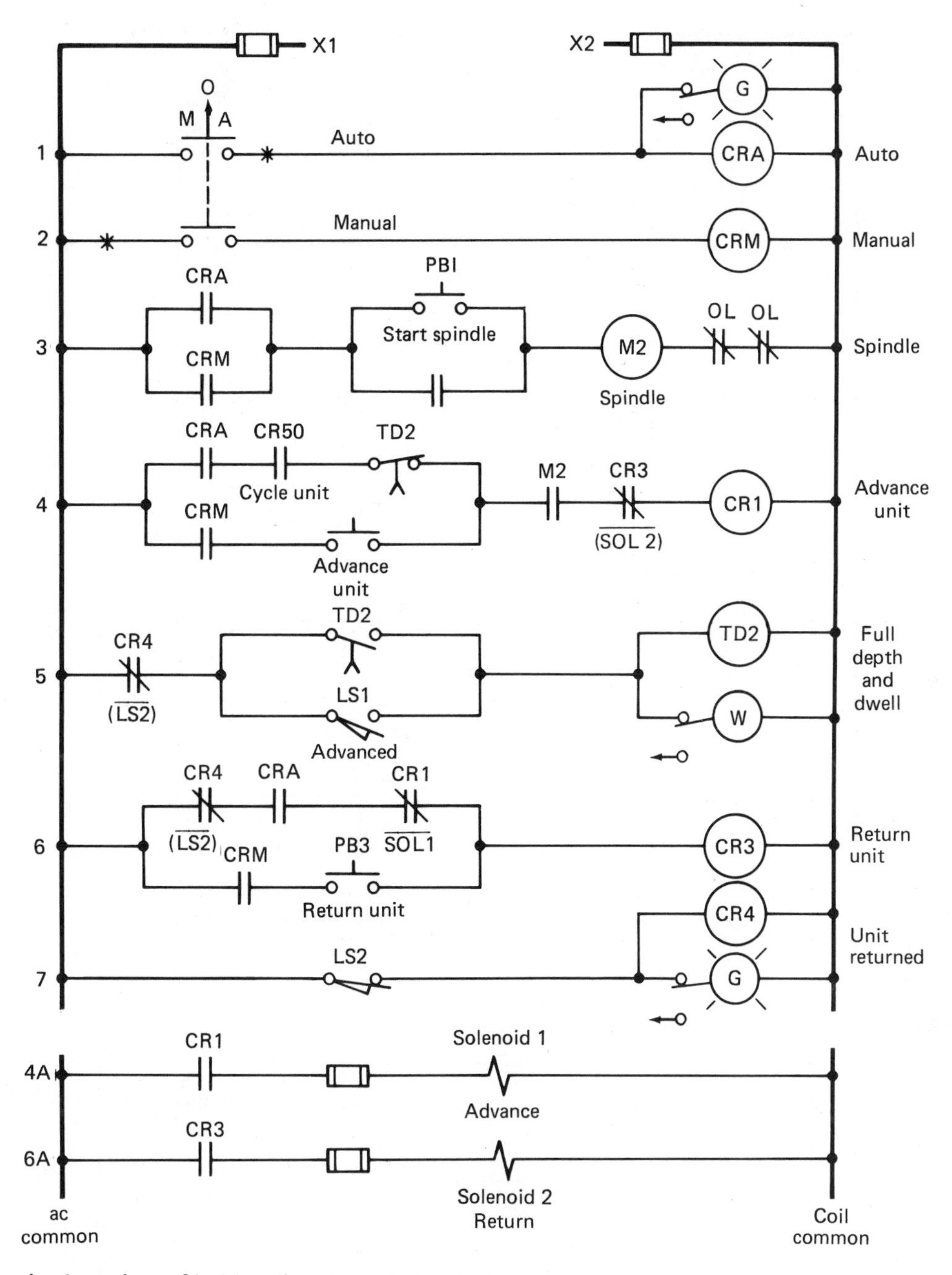

Fig. 14-26 Revised version of ladder diagram. *(Westinghouse Corp.)*

driven directly from limit switches or pushbuttons, such as lamps, remain as they are in the relay schematic and do not require solid-state ac outputs.

It is now a simple matter to convert our findings to solid-state symbols as shown in Fig. 14-29 on page 226. As shown in this example, we will need eight ac inputs and four ac outputs. The rectangles numbered TB1-1 through TB1-12 are terminal strips located on the module mounting rack.

The basic rules to follow when starting circuit conversion are to

1. Start conversion from output side of relay diagram (right vertical, X2) and work toward the input side (left vertical, X1).

2. Convert each series of individual contacts to a logic AND as shown in Fig. 14-30 on page 227. Squares in series are logic AND circuits. Place a square around each contact first. Then count the number of contacts and draw the AND symbol with the appropriate number of inputs (2 or 4). If there are more than four contacts, then more than one AND symbol must be used. Unused inputs are connected to used inputs as shown. This is read A and B for the two

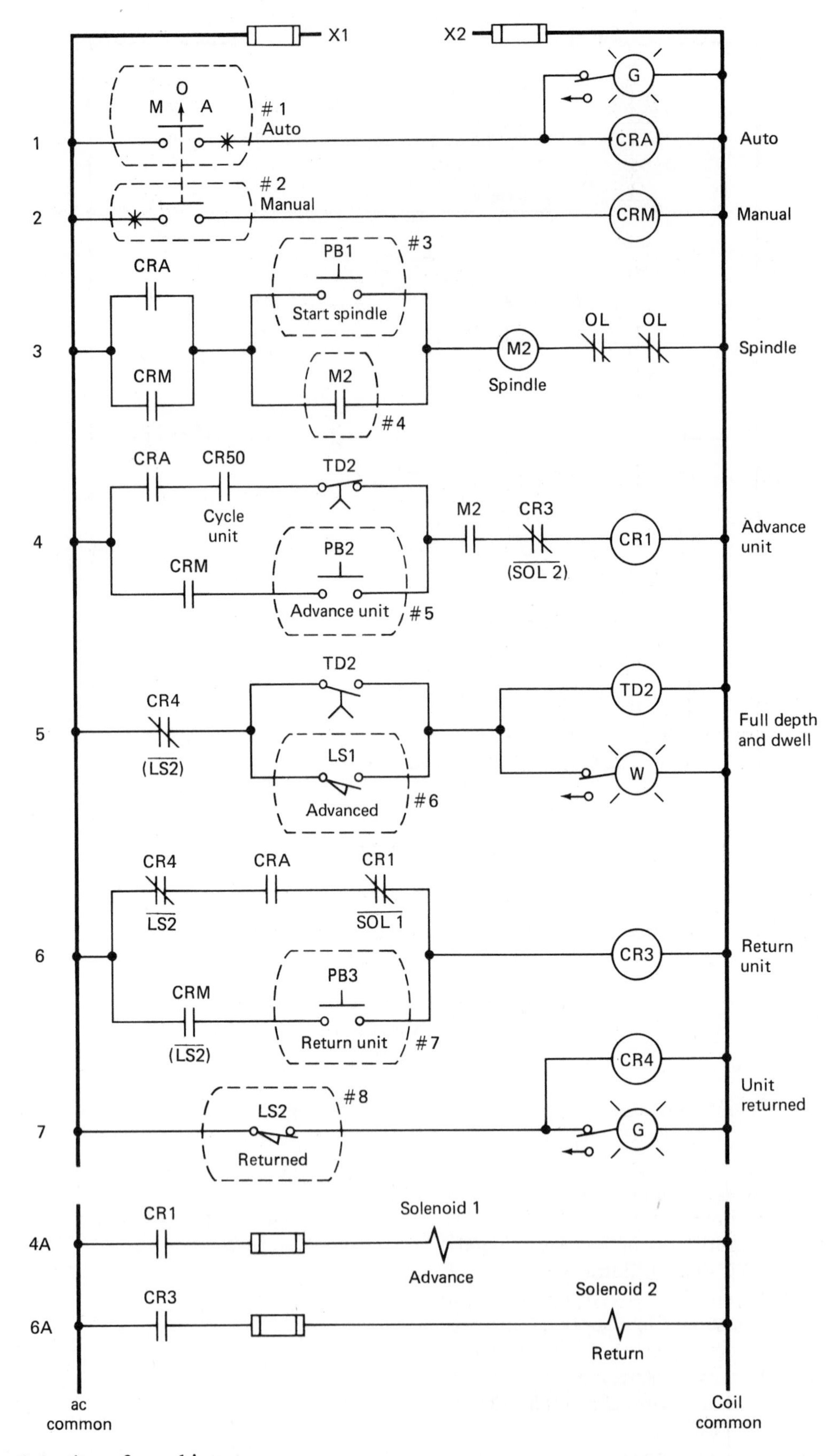

Fig. 14-27 Selection of panel inputs.

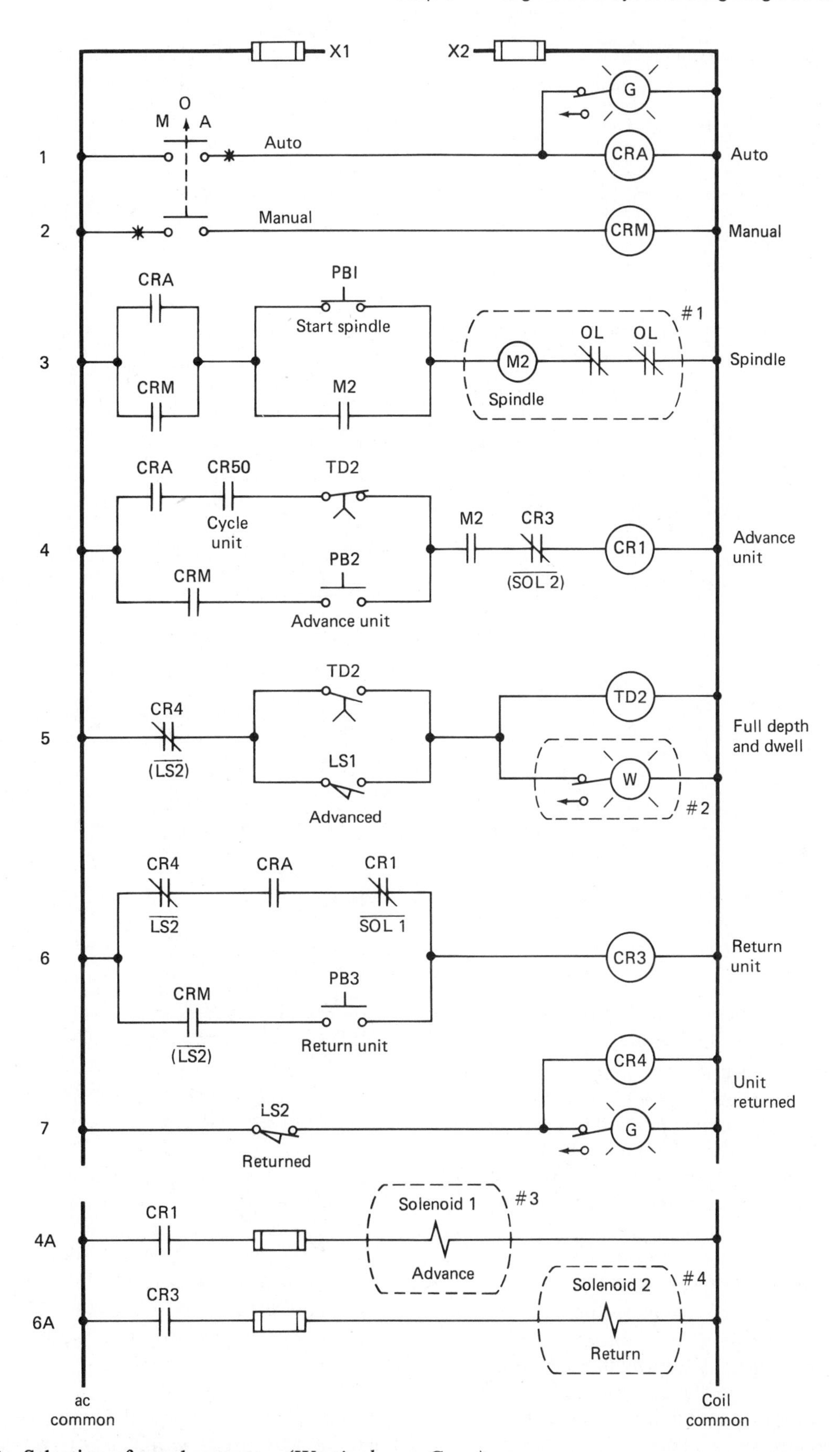

Fig. 14-28 Selection of panel outputs. *(Westinghouse Corp.)*

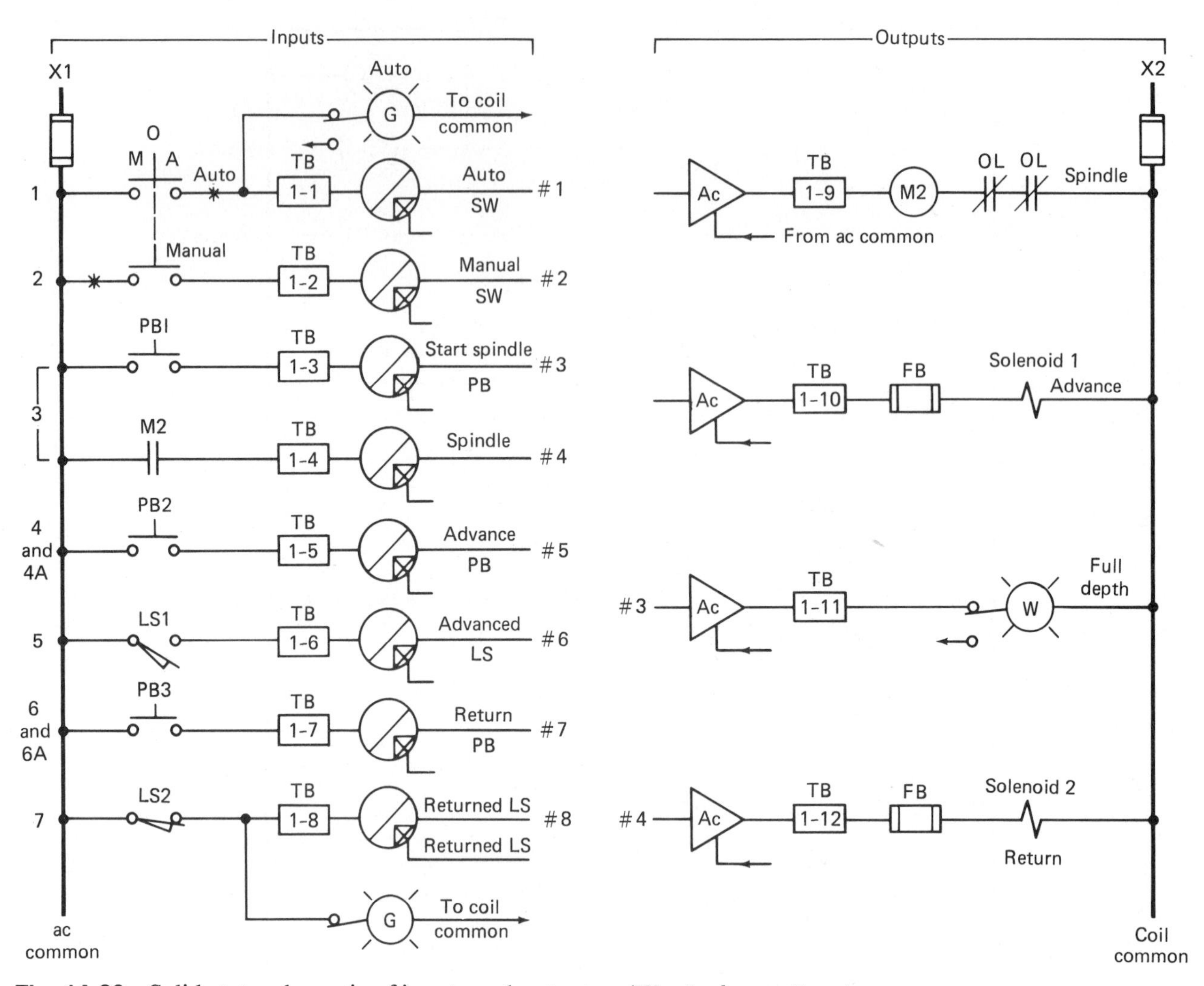

Fig. 14-29 Solid-state schematic of inputs and outputs. *(Westinghouse Corp.)*

large squares in series on line 3 of Figs. 14-30 and equals a logic AND. It is written: A · B.

3. Convert each branch (parallel) circuit to a single relay contact as shown in Figs. 14-30 by placing squares around the complete circuit from where the current divides to where it rejoins. This is represented by squares G, I, and J on line 4 and P and Q on line 5. Label the equivalent contacts and modify the AND symbol as shown in the logic equivalent drawing of Fig. 14-30. We would now have AND 1, 2, and 3 drawn on our logic diagram.

4. Convert legs of branch (parallel) circuits to OR circuits by first circling each leg of the circuit and treating it as one contact. Vertically stacked circles are logic OR circuits. In the example shown in Fig. 14-30, the upper leg containing PB1 and M2 is circled and treated as OR gate 1. The lower leg containing CRA and CRM is circled and treated as OR gate 2. The two equivalent contacts are then drawn as inputs to a two-input logic element (line 3, Fig. 14-30). This is read: PB1 or M2 equals a logic OR, or PB1 + M2. We have identified S + Spdl PB as C and Spdl ON as D. We could have written OR = C + D and OR 2 = E + F for the OR circuits in line 3 of Fig. 14-30.

5. Convert series contacts within the legs of branch circuits to logic AND circuits as is done on lines 4 and 6. Place a square around each individual contact and then convert each series of contacts (horizontal squares) to a logic AND symbol. The output of each AND symbol connects to one input of the previously drawn OR symbols (OR1 and OR3) as shown in Fig. 14-30.

Figure 14-30 combines all five rules. Note that when there are circles within circles, the outer circles are converted first (line 6). Contacts CRA, CR50, and TD2 are converted to three inputs of a four-input AND. The total branch circuit of line 5 is identified as

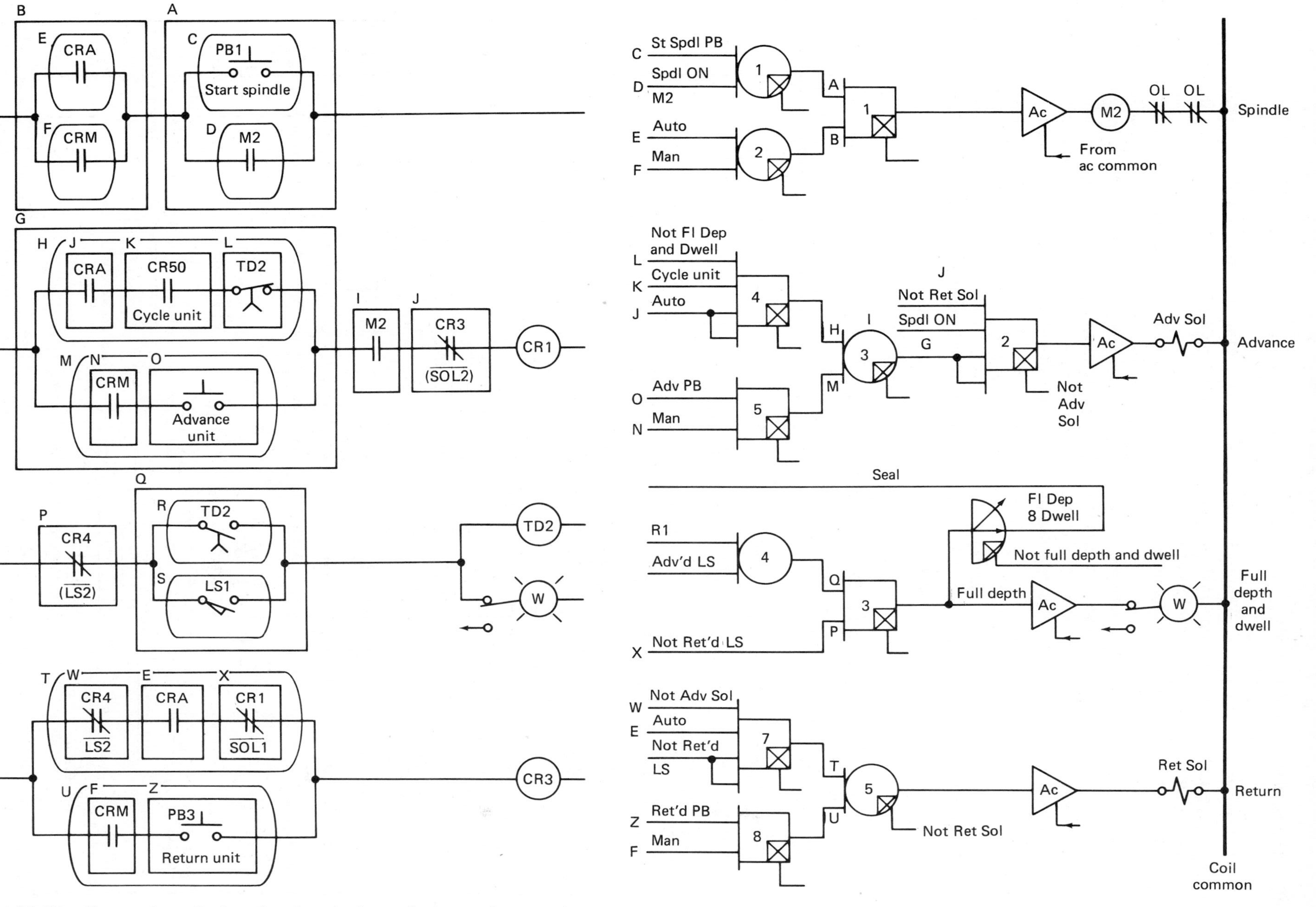

Fig. 14-30 Conversion of relay circuit to logic equivalent. *(Westinghouse Corp.)*

the fourth input to the AND gate 4. The upper and lower legs of the branch circuit are converted to a two-input OR gate (line 6) whose output is the fourth input to AND gate 2. The series of individual contacts, Ret'd PB and CR5, is converted to two-input AND 7; and Not Adv Sol, Auto, and Not Ret'd LS to three-input AND 7. The outputs of these two AND gates are connected to the inputs of OR gate 5. Using the letters we assigned as inputs, the boolean algebra expression for each line in Fig. 14-30 is

$$\text{M2} = (\text{C} + \text{D}) \cdot (\text{E} + \text{F}) \qquad \text{(line 3)}$$
$$\text{Adv Sol} = (\text{LKJ} + \text{NO}) \cdot \text{I} \cdot \text{J} \qquad \text{(line 4)}$$
$$\text{TD2 W(light)} = \text{P} \cdot (\text{R} + \text{S}) \qquad \text{(line 5)}$$
$$\text{CR3} = (\text{WEX} + \text{FZ}) \qquad \text{(line 6)}$$

Figure 14-31 shows the composite solid-state diagram for our machine tool example. The logic input designations have been changed to correspond to the correct functional input title. As you will note, the inputs are connected to the logic components. We will soon see that Westinghouse Numa-Logic circuits use *hex locators* to avoid drawing a lot of vertical lines.

The circuit of Fig. 14-31 represents the completed logic conversion of the conventional relay circuit in Fig. 14-26. This method of circuit conversion is called

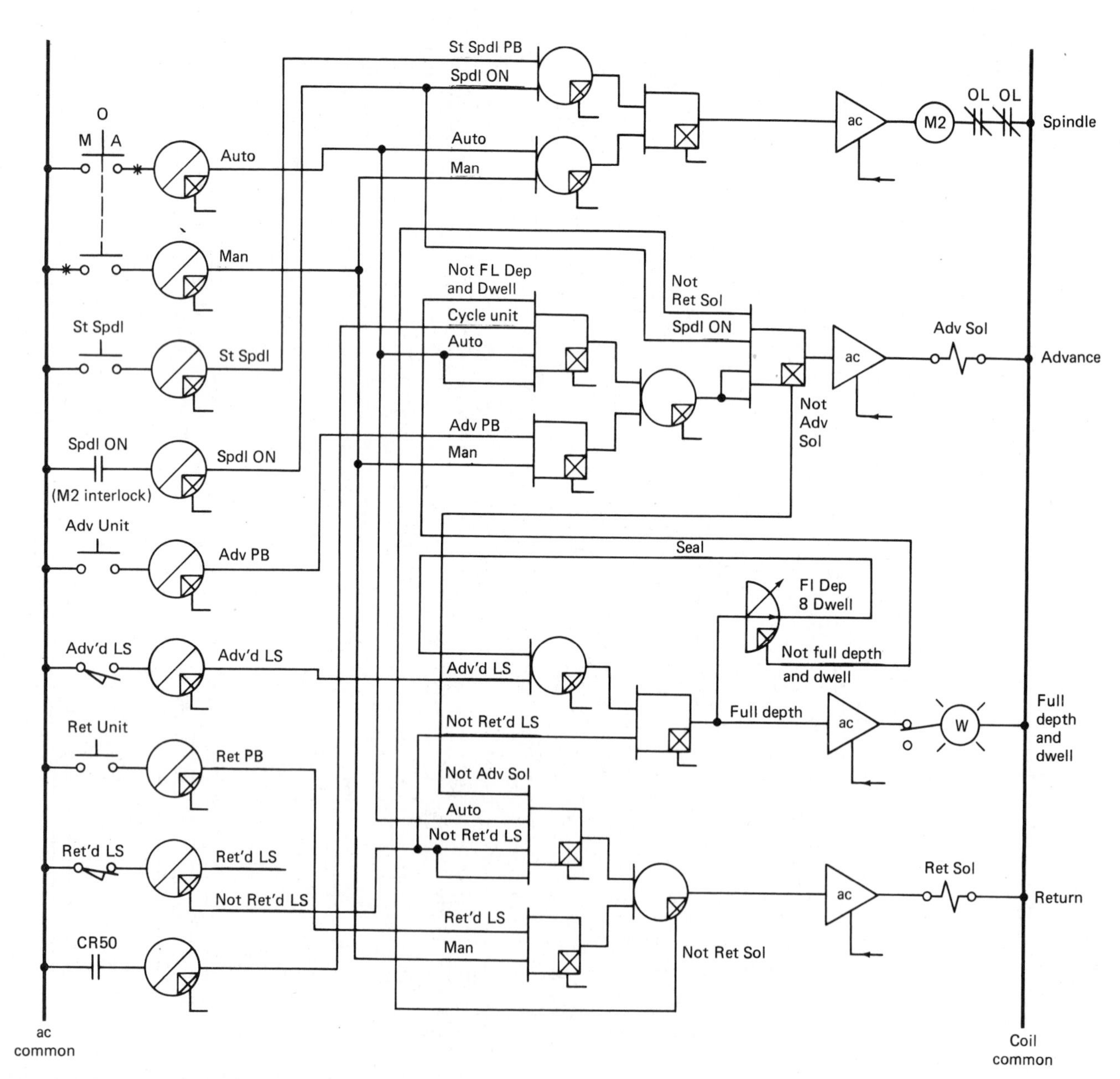

Note: CR 50 is a logic-level signal from machine's master-control calling for drill unit to cycle

Fig. 14-31 Complete solid-state schematic. *(Westinghouse Corp.)*

the circle method. This circle method is also useful in converting electric circuits to programs for programmable control.

14-9 CONVERTING WORD DESCRIPTIONS TO SOLID-STATE DIAGRAMS

Another method used to develop solid-state logic diagrams is called *word description.* A word description usually is the starting point for making a control diagram, whether it be a conventional ladder or solid-state diagram. Therefore, it often is desirable to convert directly from the word description to a solid-state schematic, thereby eliminating the effort required to develop a relay ladder diagram first. To examine this direct approach, we will use the same machine tool example covered in the relay-to-solid-state conversion.

The first step is to detail the input and output devices that are required. These are determined as functions of

The required machine motions (outputs);

The required machine sensing devices (inputs);

The operator controls (inputs).

Although these functions already may have been determined by the time the word description has been prepared, let's examine them for our example so that we better understand them.

In Fig. 14-24 the hydraulic slide unit requires ac output for the following functions:

1. To start the spindle motor
2. To energize the advance solenoid to operate the hydraulic cylinder, advancing the slide
3. To turn on the indicator light at the operator's station, indicating the drill is at full depth
4. To energize the return solenoid to operate the hydraulic cylinder, returning the slide

The drilling unit also provides ac inputs consisting of sensing the slide position by the

1. Advance limit switch
2. Return limit switch

The remaining ac inputs are operator controls consisting of

1. Selecting manual or automatic operation by a manual-off-automatic (M-O-A) selector switch
2. Starting the spindle motor by a start spindle pushbutton
3. Manually advancing or returning the slide unit by pressing either an advance pushbutton or a return pushbutton

With this background, the logic circuits can be drawn directly from word descriptions which describe the combinations of inputs required to produce each desired output.

START SPINDLE MOTOR

The word description is: "Spindle Motor will run when

1. M-O-A selector switch is set at either manual OR automatic AND
2. Start spindle pushbutton OR spindle starter holding interlock (M2) is made."

The first expression has two parts separated by an OR, indicating the need for a two-input OR solid-state device, as shown in Fig. 14-31.

The AND between the two expressions dictates that both conditions must be met. Therefore, the two OR circuits must feed a two-input AND device.

Now to complete this circuit, we simply add the input interfaces as shown in Fig. 14-29. Note that the starter interlock contact M2 operates on 120 V ac and, therefore, requires its own ac input.

ENERGIZE ADVANCE SOLENOID AUTOMATICALLY

The word description is: "Advance solenoid will be energized automatically when

1. M-O-A selector switch is set at "Auto" AND
2. Cycle unit is on (this is a logic-level signal from the machine's master control calling for the drilling unit to cycle) AND
3. Drill has NOT completed full depth and dwell AND
4. Spindle motor is on (M2) AND
5. Return solenoid is NOT energized."

For the system to operate automatically, the description calls for five requirements to be met simultaneously, necessitating a five-input AND device. To make a five-input AND circuit we use a two-input and a four-input AND device. We feed the first four requirements into the four-input AND, taking its output to feed one of the inputs of the two-input AND along with the fifth requirement. See Figs. 14-30 and 14-31.

ENERGIZE ADVANCE SOLENOID MANUALLY

The word description is: "Advance solenoid will be energized manually when

1. M-O-A selector switch is set at 'Manual' AND
2. Advance pushbutton is depressed AND
3. Spindle motor is on (M2) AND
4. Return solenoid is NOT energized."

For the system to operate manually, the description calls for four AND requirements, which, of course, can be handled nicely by a four-input AND, as shown in Figs. 14-30 and 14-31.

Assembling the pieces of our word description give us the complete schematic for controlling the energize advance solenoid function.

14-10 CONVERTING BOOLEAN EQUATIONS TO SOLID-STATE DIAGRAMS

Another method of designing a control system is to use boolean algebra. In addition to being a shorthand method that saves considerable time, boolean algebra has the unique characteristic of permitting algebraic manipulations that result in circuit simplification, thereby reducing the control hardware required.

The subject of boolean algebra and circuit simplification is covered in Chap. 13. Our concern here is only to examine the method of making a solid-state diagram based on boolean analysis. Again, we will use the machine tool example covered in the preceding pages.

Written in boolean equations, the control circuit shown in Fig. 14-25 would be

$$\text{Auto} = \text{Auto SW} \cdot \overline{\text{Man SW}} \qquad \text{(line 1)}$$

$$\text{Man} = \overline{\text{Auto SW}} \cdot \text{Man SW} \qquad \text{(line 2)}$$

$$\text{Spdl} = (\text{CRM} + \text{CRA}) \cdot (\text{PBl} + \text{M2}) \qquad \text{(line 3)}$$

$$\text{SOL1} = \text{CR1} = \overline{\text{SOL2}} \cdot \text{M2} \cdot [(\text{TD2} \cdot \text{CR50} \cdot \text{CRA}) + (\text{PB2} \cdot \text{CRM})] \qquad \text{(line 4 \& 4A)}$$

$$\text{Fl Dep \& Dwell} = \overline{\text{LS2}} \cdot (\text{TD2} + \text{LS1}) \qquad \text{(line 5)}$$

$$\text{SOL2} = \text{CR3} = (\overline{\text{LS2}} \cdot \text{CRA} \cdot \overline{\text{LS1}}) + (\text{CRM} \cdot \text{PB3}) \qquad \text{(line 6 \& 6A)}$$

$$\text{Ret'd} = \text{CR4} = \text{LS2} \qquad \text{(line 7)}$$

The conversion to a solid-state diagram is quite easy, requiring only simple examination. All we have to do is keep in mind that the symbol · equals AND, + equals OR, and a bar above equals NOT. The letter and number combinations are the same abbreviations used in standard relay circuitry.

The first boolean equation is read: "Automatic operation equals automatic switch on AND manual switch NOT on." The second equation, of course, establishes the need for a two-position selector switch which provides circuit inputs "Auto" and "Man."

In examining the third equation: SPDL = (CRM + CRA) · (PB1 + M2) we first note the two parenthetical expressions are separated by an AND, indicating a two-input AND gate. Each parenthetical expression has two units separated by an OR, indicating that the two logic inputs to the two-input AND are two 2-input OR gates. The abbreviation SPDL, obviously, is the output to the spindle motor. The equivalent solid-state circuit is shown in Figs. 14-30 and 14-31.

The fourth equation is

$$\text{SOL1} = \text{CR1} = \overline{\text{SOL2}} \cdot \text{M2} \cdot [(\overline{\text{TD2}} \cdot \text{CR50} \cdot \text{CRA}) + (\text{PB2} \cdot \text{CRM})].$$

Note that CR50 is a logic signal from the master control. This equation is a little more complicated in that it has two minor expressions within a major expression. This is no problem, however, if we remember to go from primary elements to secondary elements to tertiary elements, etc. There are three primary elements ($\overline{\text{SOL2}}$, M2, and the major expression) connected by ANDs, indicating three inputs to a four-input AND. There are two minor expressions connected by an OR, indicating a two-input OR (whose output is one of the inputs to the four-input AND). The first minor expression has three elements connected by ANDs, indicating three inputs to another four-input AND (whose output is an input to a two-input OR). The second minor expression, of course, indicates a two-input AND (whose output is the other input to the two-input OR). The equivalent solid-state circuit is shown in Figs. 14-30 and 14-31.

In the fifth equation—Fl Dep & Dwell = $\overline{\text{LS2}}$ · (TD2 + LS1)—we pick up the delay timer, but it is not too obvious. We see that the parenthetical expression calls for a two-input OR with one of the inputs being TD2—a time delay. It follows that this input must be supplied by a delay timer, which is located last in the string of logic units. The equivalent solid-state circuit is shown in Figs. 14-30 and 14-31.

The last two equations follow the same procedure.

Once the logic equations have been developed, they should be converted to logic circuits one equation at a time. When all the equations have been converted to logic circuits, the circuits should be joined together at the appropriate locations. Inputs and outputs should be connected and then the circuit is complete. Regardless of the methods used to develop a logic diagram, the logic circuits should be the same and give the same circuit operation.

Most Westinghouse Numa-Logic drawings will include a rack layout drawing. In the connection diagrams used to form the rack drawing, sometimes instead of using symbols, they will use the part number for the logic modules as well as pin numbers for the module. Figure 14-32 indicates the labeling of components and their meanings as found on Numa-Logic diagrams.

Another drawing that will be included with the logic drawings for a circuit is called a *power-up circuit.* Figure 14-33, on page 232, illustrates the power-up circuit. On this drawing are all the power circuits for the motors and other output devices. The relay circuit for

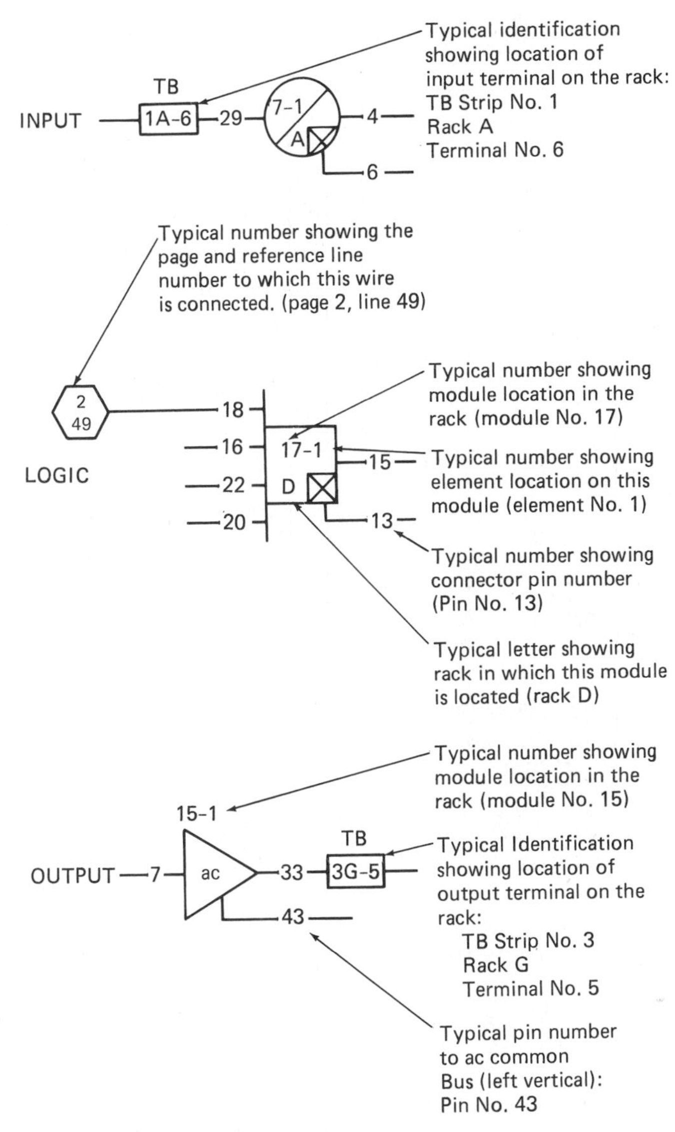

Fig. 14-32 The labeling of components and their meanings. *(Westinghouse Corp.)*

low-voltage protection is also on this circuit. The logic 5-V power supply is also on the circuit (Fig. 14-34 on page 233). Figure 14-35, on page 233, is a connection diagram for the Westinghouse Numa-Logic amplifier.

We will now convert the tank-filling circuit of Fig. 14-14 to a Numa-Logic circuit. In this figure we used a binary counter as a decoder to make a scanning circuit to look at the level switches in sequence. To introduce you to a new device we will use the Numa-Logic shift register to make a scanning circuit.

14-11 APPLICATIONS OF SHIFT REGISTERS

Figure 14-36, on page 234, illustrates the connections for two 5-bit shift registers. When a serial input (logic 1) is applied to pin 9, nothing will happen. When a clock pulse arrives at pin 5, on the trailing edge of the clock pulse, the logic 1 at pin 9 will move to pin 31. Now if the serial input is removed nothing will happen. We will still have a 1 at pin 31. At the trailing edge of the next clock pulse, the logic 1 will move to pin 33 and pin 31 will be a logic 0. Each succeeding clock pulse will move the logic 1 one more step in the shift register. When the logic 1 arrives at pin 39, it will also be the serial input for pin 9 on the lower shift register. The next clock pulse will move the logic 1 from pin 39 of the top register to pin 31 of the lower register. Each clock pulse will move the logic 1 one more step in the lower register. When it reaches pin 39 on the lower register, the next clock pulse drops it out of the circuit. Nothing happens now until we have another serial input. The register can be loaded with several logic 1s and they will all move simultaneously down the register until they drop out of the bottom at pin 39 of the lower

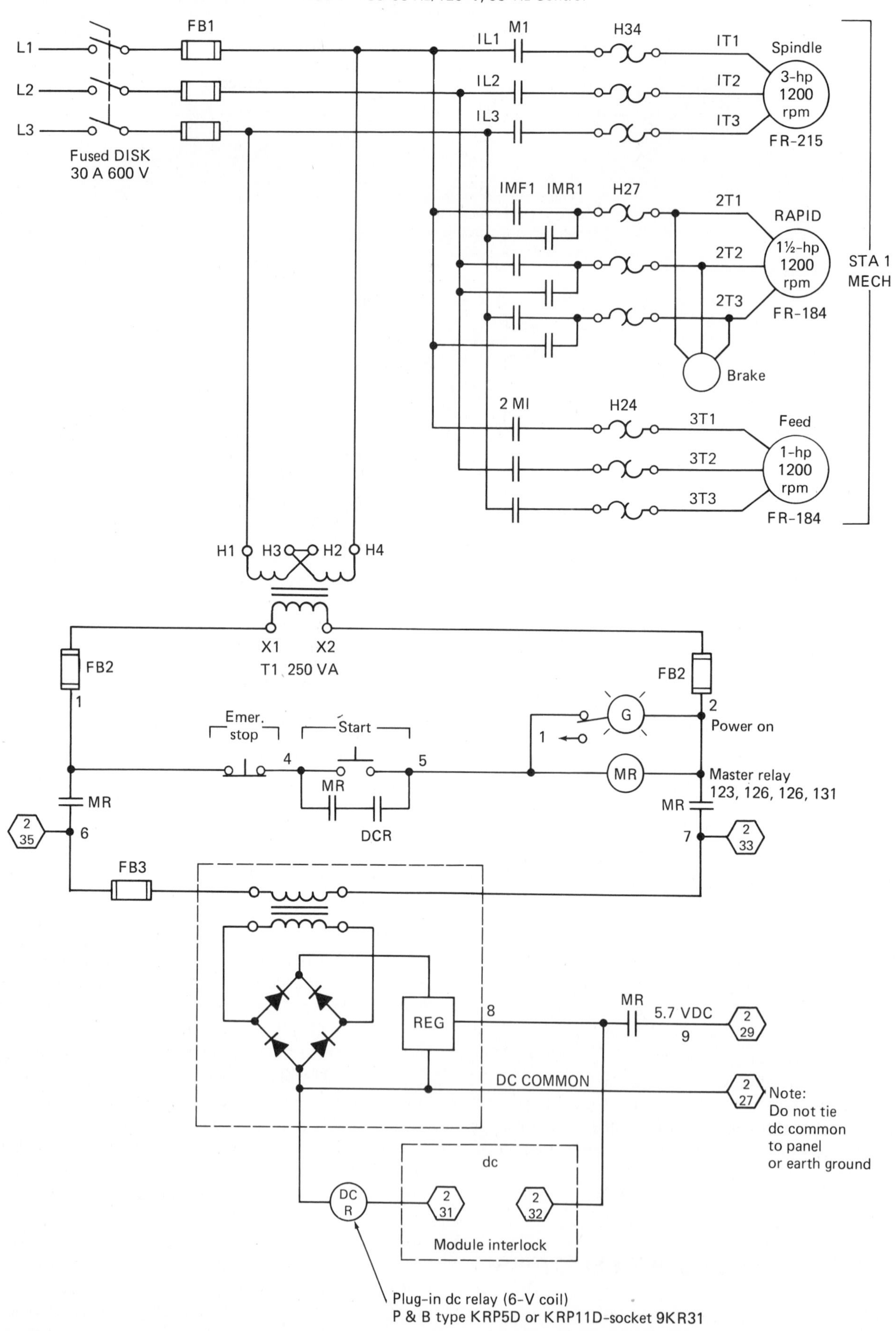

Fig. 14-33 Typical power-up drawing with dc modules interlocked. *(Westinghouse Corp.)*

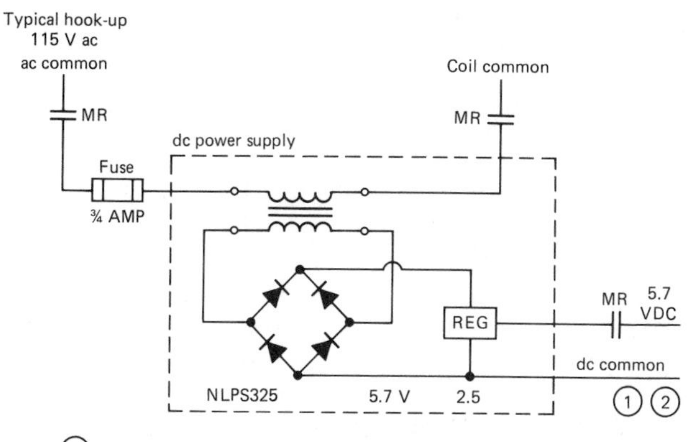

Fig. 14-34 A Numa-Logic power supply. *(Westinghouse Corp.)*

register. A momentary input at pin 13 will clear the shift registers of all logic 1s if desired.

To start our Westinghouse Numa-Logic circuit for a scanner which will look at the tank levels (Fig. 14-14) we need a clock. This can be made from two time-delay circuits. This is illustrated in Fig. 14-37 on page 235. Changing the delay adjustments will give us any clock pulse rate we desire.

14-12 USE OF HEX LOCATORS IN WESTINGHOUSE NUMA-LOGIC CIRCUIT DIAGRAMS

In Fig. 14-37 notice that instead of drawing a wire from pin 31 on the shift register to pin 3 on the memory we used hex locators. If we follow pin 31 to the left margin, we are at hex locator 20. The hex locator at pin 31 has an 18 in it. This tells us if we find a hex locator next to a device and locate that number in the left margin and move to the right, we will find the device's hex locator next to another device and that these devices connect together. If you had started on pin 3 of the memory we have hex locator 20. Pin 3 is on line 18. This is indicated by the hex locator 18 in at the left margin. We would now go to hex locator 20 in the left margin and travel right across the circuit until we find a hex locator with an 18 in it. This tells us again that pin 10 of OR 1 and pin 31 of the shift register connect together.

The operation of the circuit thus far is that when we apply power, the time delays operate by producing an on/off output at TD1 pin 15. This 101010 combination of the delays output is our clock signal. A clock signal at pin 9 does nothing until a serial input is applied. A momentary serial input is applied to pin 5 of

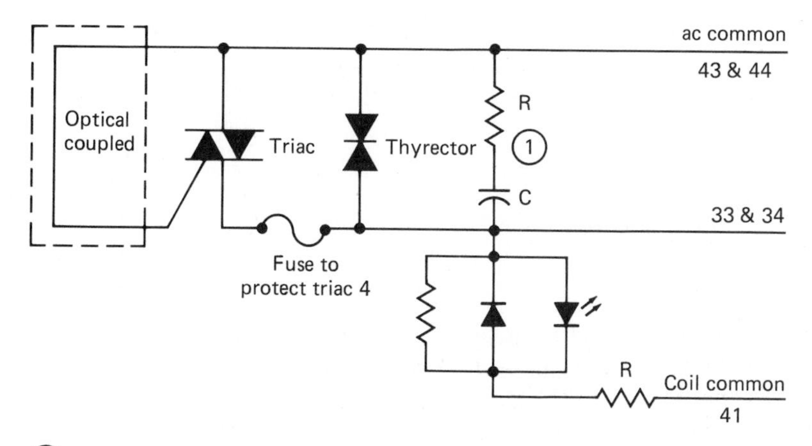

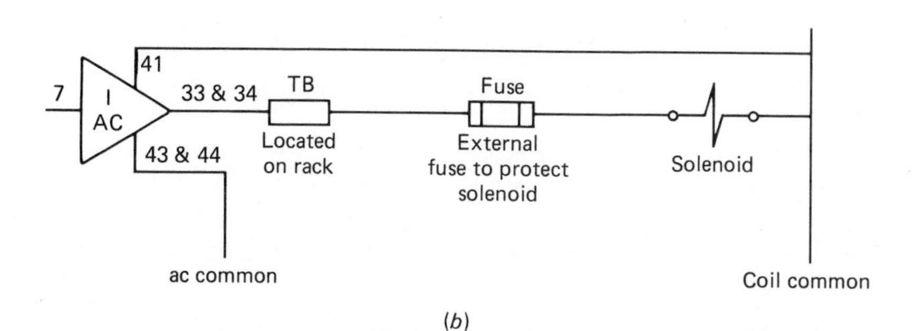

Fig. 14-35 Connection diagram for Westinghouse Numa-Logic amplifier. *(a)* output circuit *(b)* hookup. *(Westinghouse Corp.)*

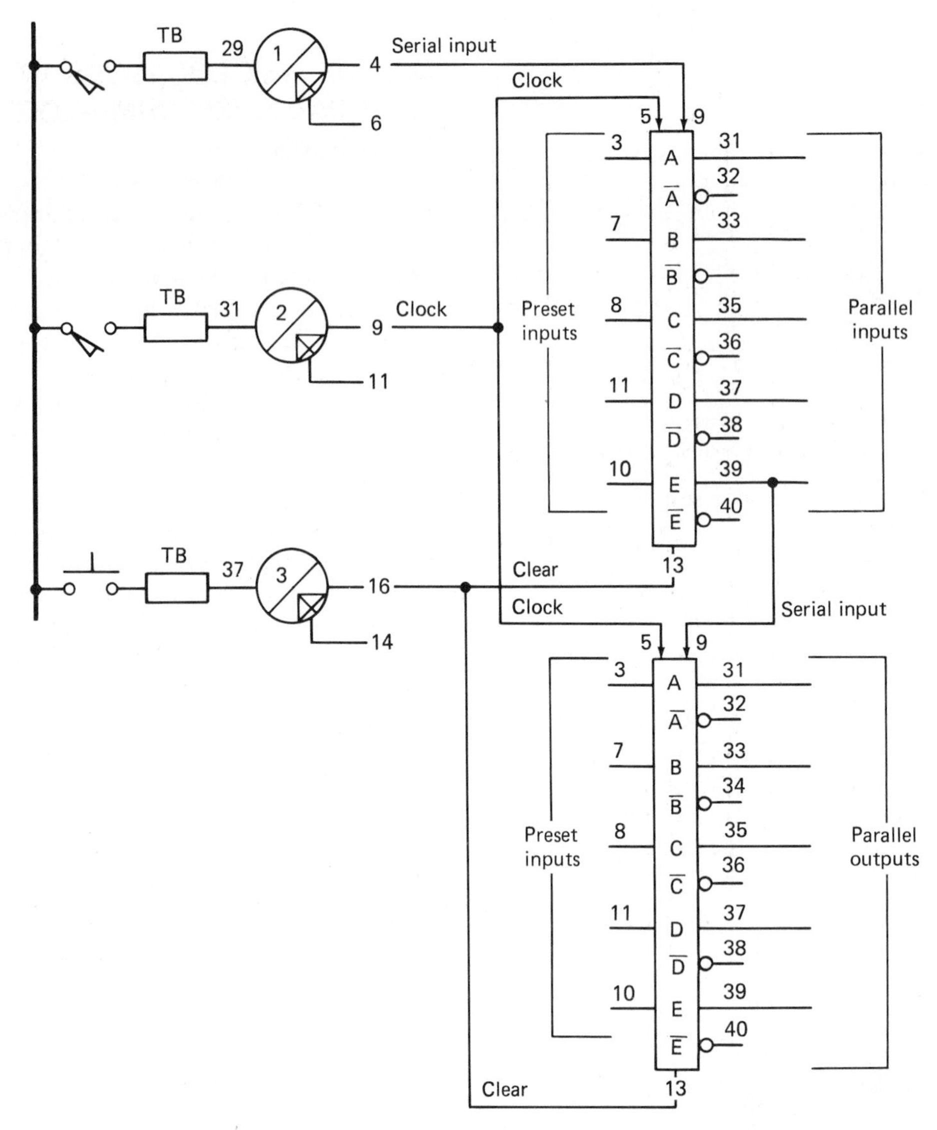

Fig. 14-36 Five-bit shift register with preset inputs. *(Westinghouse Corp.)*

the memory when the serial input pushbutton is pushed. (The purpose of the memory is to prevent the need of synchronizing the pushing of the serial input with a clock pulse.) When the clock pulse arrives, on the trailing edge of the pulse, the logic 1 from the MEMORY and OR gate will move from pin 5 of the shift register to pin 31.

Once we have a logic 1 at pin 31 of the shift register, the memory has an input at pin 3 and turns off the memory. The serial input no longer exists. The logic 1 at pin 31 will move to pin 33 on the next clock pulse. It will continue to move on each succeeding clock pulse until it reaches pin 39. Pin 39 is connected to pin 10 of OR 1 (the hex locators indicate this). If pin 39 is on, then OR 1 is on and we have another serial input. The next clock pulse will move a logic 1 back to pin 31 of the shift register when it drops out of the circuit at pin 39.

The logic again moves to each output of the register with succeeding clock pulses. If our clock gave a pulse every 5 s, our logic 1 would take 25 s to move through the register and would repeat the operation. This circuit will work very well for a scanner.

To stop the scanner an AND circuit (in dashed lines from the timer to pin 9 of the shift register) would stop the clock pulses if the lower input at pin 8 of the AND is missing. As we complete the circuit, we will see how this will be added to the circuit to stop the scanner when it finds a low tank. A closed float switch represents a low tank.

Figure 14-38, on page 236, shows the level switch and output solenoid valves have been added to complete the circuit. Pin and circuit numbers have been added. The circuit in the figure is ready to wire.

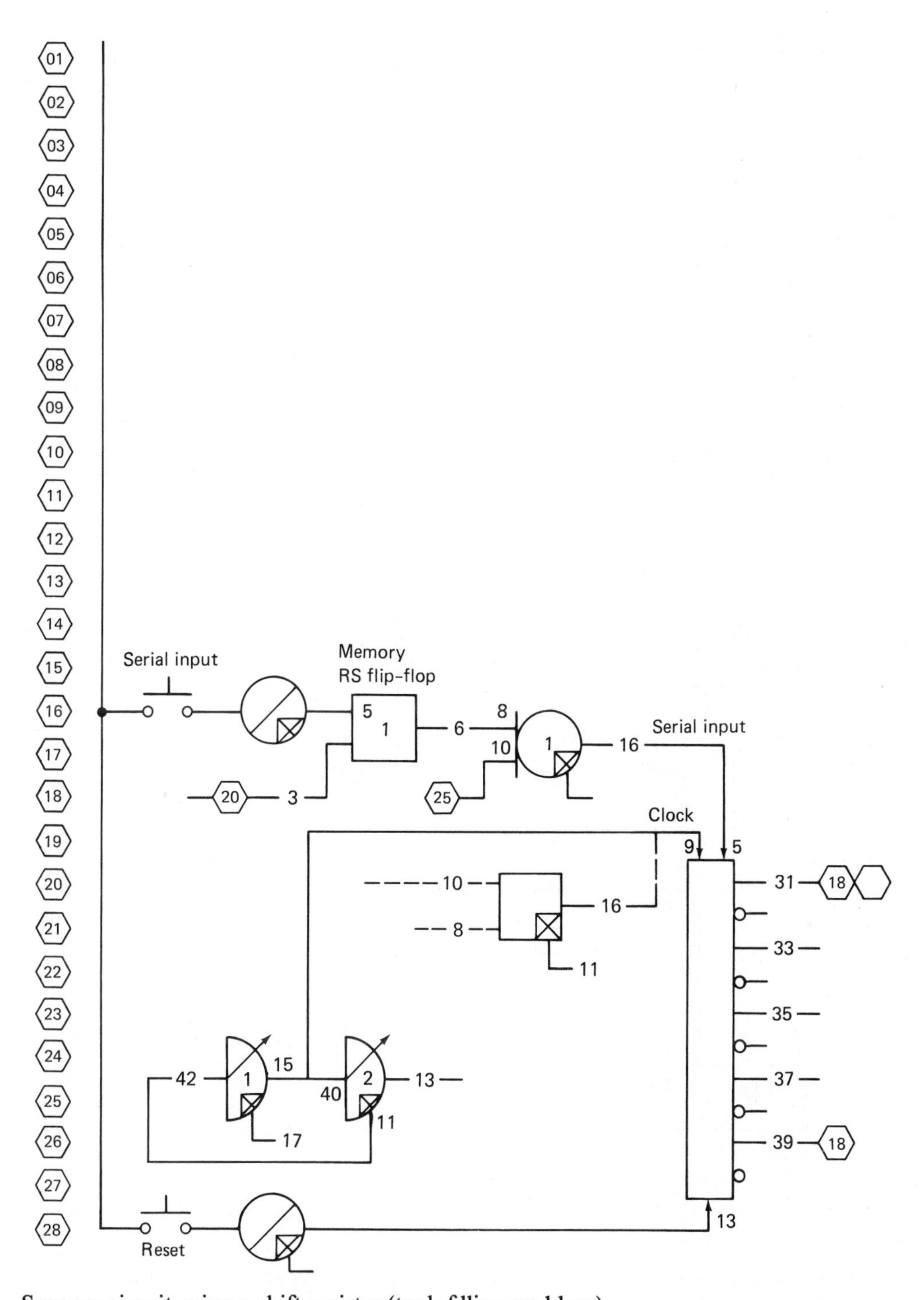

Fig. 14-37 Scanner circuit using a shift register (tank-filling problem).

14-13 DIFFERENCE BETWEEN GE STATIC AND WESTINGHOUSE NUMA-LOGIC

Figure 14-39, on page 237, illustrates the difference between the GE Static Control circuit and the Westinghouse Numa-Logic circuit for a one-point annunciator. We developed this circuit in Fig. 14-19. Notice Numa-Logic does not have a module of NOT gates. The NOT is built into all their logic. An AND module can give AND logic or NAND logic and an OR module can give OR or NOR logic. Notice how the NOT gates differ for the GE and Westinghouse circuits.

Figure 14-40, on page 238, illustrates how hex locators are used to eliminate a lot of the wiring of Fig. 14-39. Follow this example for understanding hex locators and you should have no trouble with Westinghouse Numa-Logic circuits.

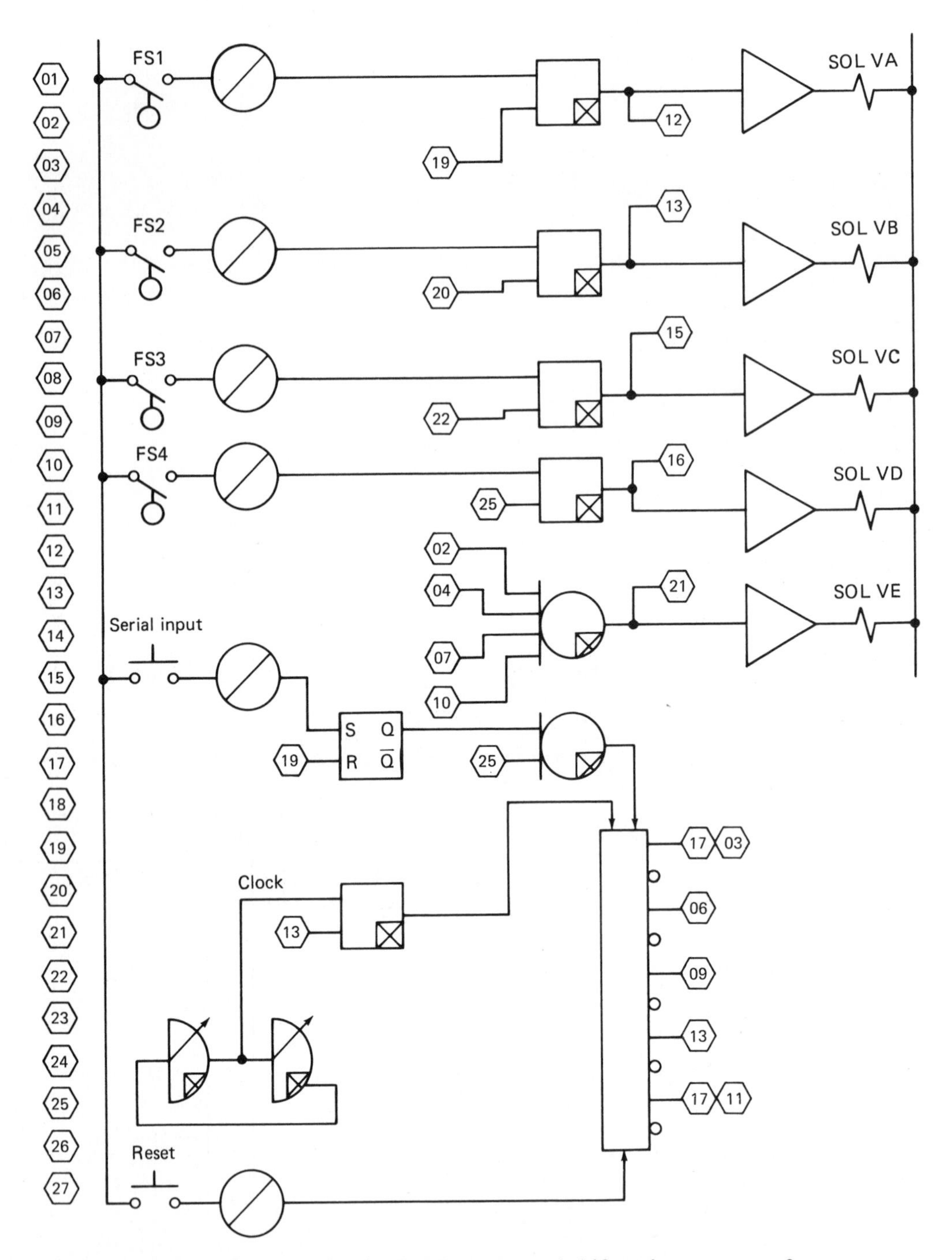

Fig. 14-38 Complete Numa-Logic circuit with hex locators and shift register scanner for tank-filling problem in Fig. 14-10.

SUMMARY

Modern industrial logic-control systems are manufactured with the integrated circuit (IC) as the heart of the logic device. Different techniques, circuits, construction methods, and components are used by the various manufacturers of control systems. The common logic gates (AND, OR, NOT, memory) are all used, but their packaging varies from company to company.

Troubleshooting principles and techniques are essentially the same for all logic-control systems. Logic probes that indicate logic 1 (ON condition) and logic 0 (OFF condition) are commonly used to troubleshoot the logic cards in a logic-control system. Some logic-control systems are manufactured with built-in trouble-shooting devices such as LED's to indicate the inputs and outputs of logic circuit. With the knowledge of truth tables for the various logic components, a technician can troubleshoot complex logic circuits with ease. The replacement of faulty logic cards is the

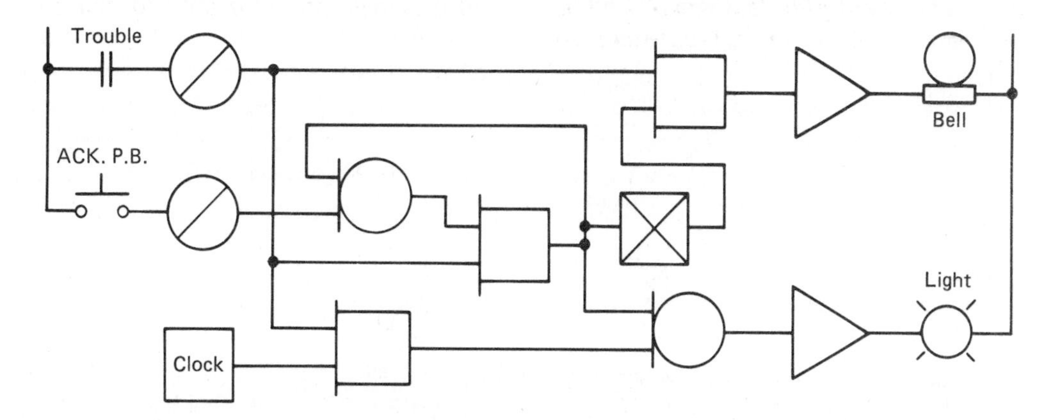

GE Static-Control logic circuit for a one-point annunciator.

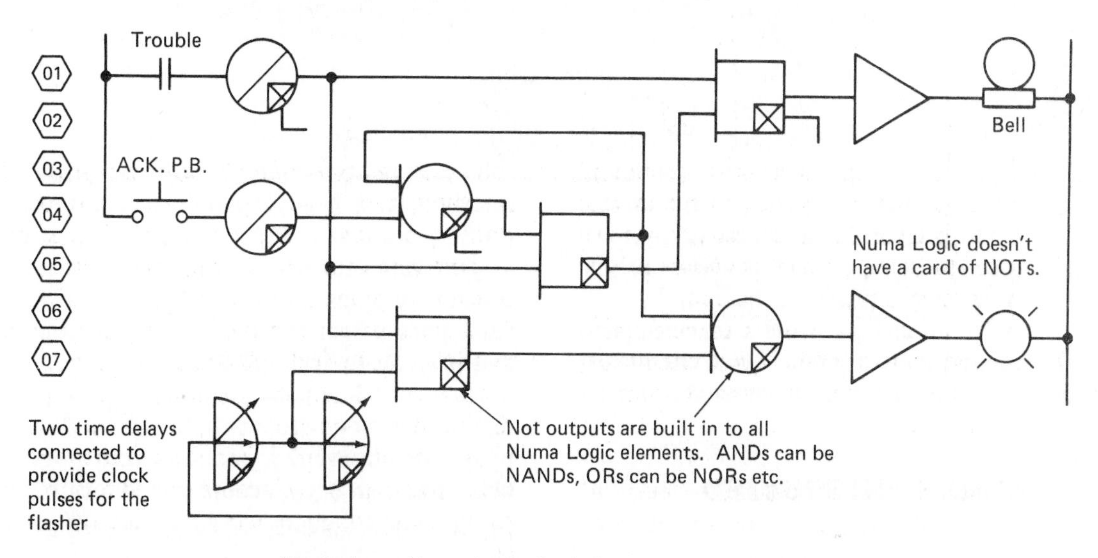

Westinghouse Numa-Logic circuit for a one-point annunciator

Fig. 14-39 These two schematics illustrate the difference between the original General Electric Static and the Westinghouse Numa-Logic drawings for a one-point annunciator.

common method for repairing logic systems in the field.

A logic card with a faulty IC is usually replaced, and is either returned to the manufacturer for repair or sent to a repair shop with the capability of making repairs down to the component level. Large companies generally have repair shops equipped with electronic test benches and staffed by electricians or technicians with knowledge of electronic repair procedures.

Logic-control circuits can be built in a number of ways. Relay circuits can be wired as logic circuits developed to perform boolean algebra operations. Boolean algebra equations can be translated into logic circuits. Another method that may be used is the English method. A story for the required control operation of a machine is developed into logic statements. The logic statements are developed into equivalent logic circuits. Once the logic circuit is obtained it can be converted into any of numerous commercial logic-control systems.

Many control systems require counting devices. Binary counters are available in the form of inexpensive ICs. These are ideal for mating counting systems with logic-control systems. Decoders can convert binary numbers into decimal numbers and control signals for many applications. Counting systems can be used with shift registers to give control to IC logic systems involved in storage, sorting, and other similar control systems.

Examples of commercial logic-control systems (using integrated circuits) are Numa-Logic (Westinghouse), Cardlock Solid-State Control (Allen Bradley), Static (General Electric), and Norpak (Square D).

First-generation logic systems used magnetic amplifiers as the heart of their control system. The transistor generation of logic-control systems is often referred to as the second generation of logic-control systems. The integrated circuit created still another generation of logic control. Large scale integrated circuits (LSI) have now put numerous logic gates into only one chip. Program-

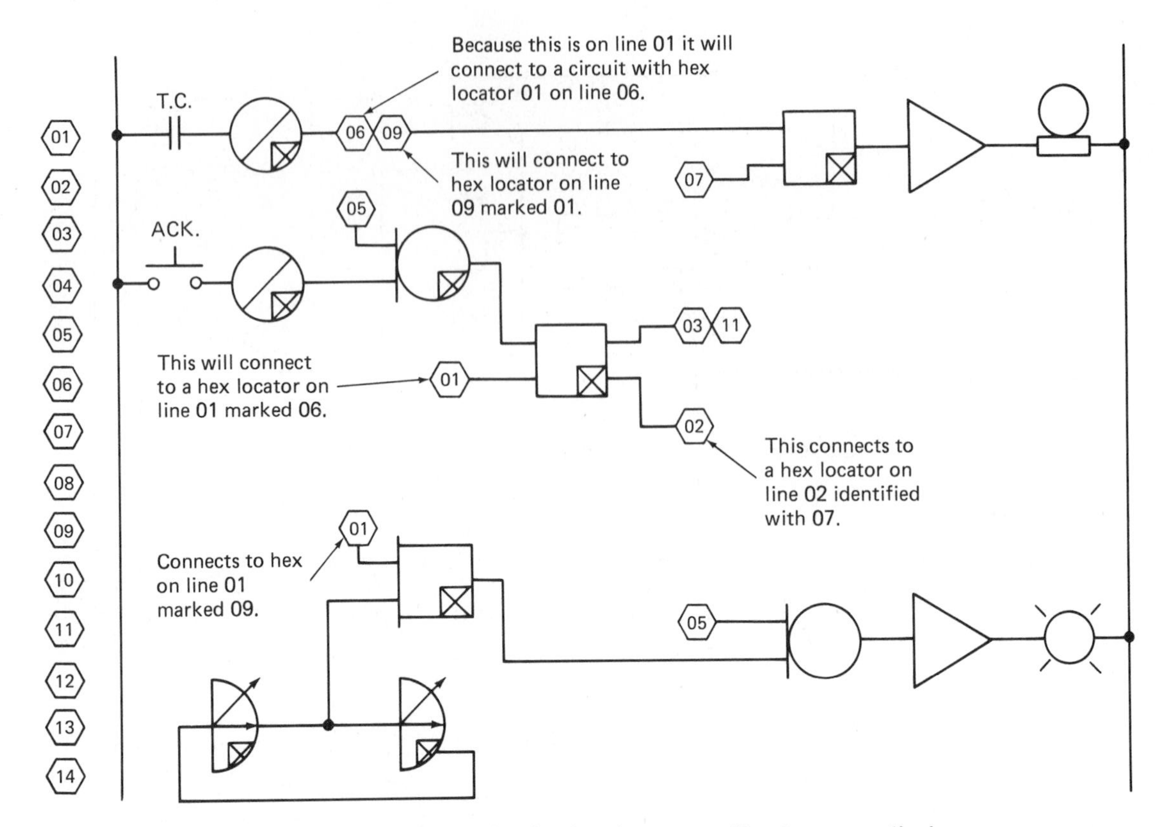

Fig. 14-40 Converting to a Numa-Logic circuit with hex locators. Hex locators eliminate long vertical lines or congested circuits.

mable-logic techniques make it possible to "burn in" numerous logic-control circuits into only one IC component.

REVIEW QUESTIONS

1. Define logic 1.
2. What is third-generation logic control?
3. Name some modern solid-state logic systems.
4. What knowledge is required by a repair technician to troubleshoot IC logic gates?
5. What are the first three steps in converting an electric relay circuit to an IC logic circuit?
6. Is Fig. 14-4 ready to wire?
7. Describe a JK flip-flop.
8. What does an "English" description reveal about a circuit?
9. How can a JK flip-flop be wired so it toggles?
10. What is a BCD counter?
11. What is the difference between the symbol for a dc and that for an ac converter?
12. How does an amplifier raise the low voltage of IC circuits to a voltage that will switch motor starters on?
13. What is an optical coupler used for?
14. What type of switches might be used as scanners?
15. How is a clock obtained from a 60 Hz power source?
16. What is a decoder used for?
17. When does solenoid valve E energize in Fig. 14-14?
18. What binary number will cause the ANDs in Fig. 14-12 to give an output?
19. What other logic circuits can be constructed with NAND gates?
20. Describe the operation of a one-point annunciator?
21. What is the most common cause of electrical downtime?
22. How does Numa-Logic minimize downtime?
23. How are grounds indicated with Numa-Logic?
24. How many "motion modules" does Numa-Logic have?
25. How does the size of a Numa-Logic panel compare to a machine panel?
26. What are the three auxiliary modules used with Numa-Logic?
27. How is dwell time adjusted?
28. How is safety interlocking accomplished?

29. What type of logic does Westinghouse Numa-Logic use?

30. How is fail-safe interlocking accomplished?

31. What can Numa-Logic replace on process line systems?

32. What boards does Numa-Logic use?

33. What does a typical Numa-Logic installation consist of to control the machine or process function?

34. How are Numa-Logic control sequences created?

35. What input and output modules are used for visual monitoring?

36. What kind of signal conversion do input modules perform?

37. What does a broken wire interlock ensure?

38. Explain the difference between an ON-delay timer and an OFF-delay timer.

39. Explain the operation of retentive MEMORY.

40. Explain the operation of a latching relay.

41. What are hex locators used for?

42. What is a power-up circuit?

43. Describe a shift register.

44. How is a shift register used as a scanner?

15
The New Generation of Solid-State Logic Control

Since the early 1800s, with the advent of telegraphy, relays have been a popular means of automatic electrical digital control. To this day, relays are unsurpassed in their ability to provide versatile, low-cost switching points for digital logic. They should be considered first in any new control system design because of their low cost and the fact that they are readily available.

As technology advanced and control speeds and equipment life increased, it was found that the relay could be replaced by the more advantageous electron tube. Electron tubes were found useful as digital switches, replacing the relay under certain conditions or for high-speed applications. In contrast to the advantage of providing higher operating speeds, electron tubes had the disadvantages of generating heat and being very fragile. These disadvantages led to the search for better ways of gaining faster and more reliable switching techniques. This search led to the magnetic amplifier-type switch. Magnetic amplifiers had the advantage of not being fragile. Their life was independent of their number of operations, but they were relatively slow and often large and heavy. The search continued for a better device.

High-speed, rugged, compact switches were made possible by the development of the transistor. In the beginning, the use of transistors as switches was found only in the computer industry. The use of transistors in an industrial control situation for the most part was ignored until the mid 1950s, when some industrial control companies began to investigate logic systems made of solid-state components. For the most part, these systems are reliable and capable of high speeds, and have an almost indefinite life since they have no moving parts. Some of the transistor's disadvantages were its size and its need for many discrete components to function in logic devices.

The development of ICs led to a reduction in the number of discrete components needed and thus the size of these logic systems. It also led to a vast increase in the operational speed of such systems; this higher speed also increased the possibility of electrical noise in the system. This was because the first IC operated on a supply voltage of approximately 5 V. Designers then looked into other ways of improving ICs and eliminating some of their disadvantages.

15-1 HIGH-THRESHOLD INTEGRATED CIRCUITS

High-threshold digital ICs exhibit characteristics attractive to industrial applications. These circuits exhibit high noise immunity by using a reverse-biased base-emitter junction as a threshold element.

The basic high-threshold IC gate is shown in Fig. 15-1. Looking at the circuit, you can see that D1 is a base-emitter junction that is operated in the reverse direction, more commonly known as zener operation. In this case, conduction occurs when the junction has approximately 6.7 V across it. The threshold for a high noise immunity IC is one forward diode drop plus one reverse diode drop, or approximately 7.5 V. The normal supply voltage for this family is 15 V ± 1 V; and in order to keep the power dissipation down, the gates have higher resistance values than comparable resistors in diode-transistor logic devices.

This gate provides basically the logic NAND function. Note that if either input A or B is below the threshold level, possible base current to transistor Q1 is routed to the low input, causing transistor Q2 to be

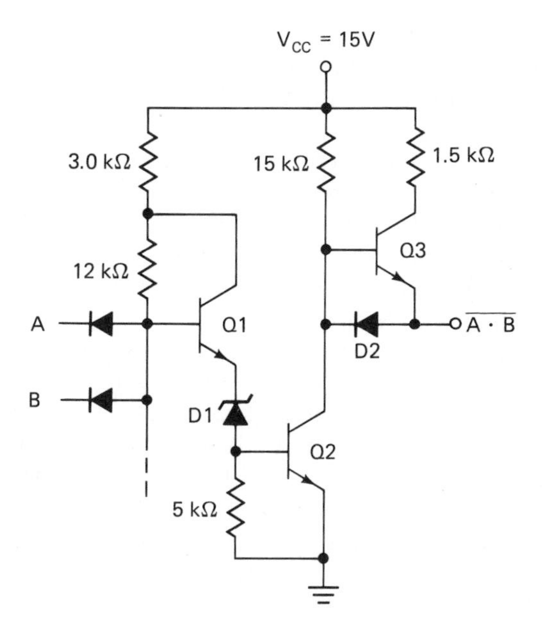

Fig. 15-1 High-threshold logic gate. *(Allen-Bradley.)*

turned off. When transistor Q2 is off, base current is supplied to transistor Q3 from the 15-kΩ resistor and load current is effectively supplied through the 1.5-kΩ resistor and the output is on. If both inputs (A and B) are above the threshold level, Q1, D1, and Q2 all turn on. When transistor Q2 is on, load current flows through diode D2 and transistor Q2. Base current is also shunted from Q3 and this transistor is off, or not conducting, in this state; therefore the output is off, or de-energized.

15-2 ADVANTAGES OF INTEGRATED SOLID-STATE DEVICES IN LOGIC CONTROL

All the different devices that perform solid-state logic control have basically the same general advantages over relay systems. These devices are designed specifically for a high duty cycle, in industrial applications where electromechanical relays are inherently limited.

Relay limiting factors might include a slow speed and cyclic rate of operation, excessive shock and vibration, or a dirty and contaminated environment. The solid-state switch, with its lack of moving parts, is not subject to these limitations to the same degree as an electromechanical relay. Operating times are measured in microseconds instead of milliseconds.

Therefore, speed, reliability, and long life are the reasons for going to solid-state control circuits. Other prerequisites must exist before justification can be given to the somewhat higher initial hardware cost of this type of system. For instance, the original equipment system designer must be convinced that the customer can understand and easily maintain the equipment on site. Hardware must be available off the shelf in the form of standardized modules. Packaging must be industrial rather than computerlike. The units must be highly immune to electrical noise (inadequate noise rejection is a common weakness of some forms of solid-state switching devices).

15-3 NOISE

A semiconductor switch is an extremely low-power, long-life, fast-acting device as compared to its counterpart, the electromechanical relay. These inherent characteristics account for its popular demand in many of today's complex or high-speed machines and processes. The same characteristics, however, make the solid-state device more susceptible to the false noise signals which are so prevalent in the industrial environment. Therefore, the degree of noise immunity is one of the most important considerations of the systems designer in the selection of solid-state hardware. No circuits are totally immune to electrical noise; however, certain precautions can be taken in hardware design to increase noise margins.

Electrical noise can be either external to the circuitry or induced within it. Examples of external sources would be the switching of inductive circuits, sparking from rotating machinery, and various signal-emitting electronic control circuits.

Internal noise may be caused by the switching of one circuit in a way that affects the state of another circuit. The amount of noise induced in the passive circuit is a function of the voltage swing, circuit change, the switching speed of the active circuit, and the inductive and capacitive couplings between the two circuits. Coupling may take place by means of a common path for the active and passive devices such as a power supply or ground leads. Noise from external sources also can be induced into the system under similar conditions.

Generally, noise is a random combination of signals from many sources, and, as such, it is extremely hard to analyze. The net result, however, is that induced positive and negative spikes relevant to the quiescent conditions of a line may cause erroneous information to be absorbed into the system. This condition must be avoided if proper operation is to be achieved by the unit.

15-4 NOISE-REDUCTION TECHNIQUES

Several schemes have commonly been employed to reduce the effect of electrical noise on a system composed of ICs. Physical shielding of the IC and its associated wiring has been utilized to reduce the effect on the circuit of external electromagnetic radiation from inducing noise. Special buffering circuits, such as signal converters, may be employed between the electronic circuits and the signal leads coming from pushbuttons and other pilot devices. These signal leads in many cases require special routing and special shielding. Extra filtering of the power supply leads is another way of reducing noise introduced in this manner. Special spacing and routing of wires, and keeping wire length as short as possible, are methods of reducing the possibility of internally generated noise.

In the past, solid-state circuits made up of discrete devices yielded a higher noise immunity because they operated with supplied voltages exceeding 10 V. Many of these circuits included special filtered inputs to reduce the susceptibility of the circuit to induced noise spikes. Standard diode-transistor logic ICs circuits do not offer this degree of noise immunity. High-threshold circuits are now available with characteristics that allow the industrial system designer to achieve with ICs the results needed and obtained with discrete-device circuits.

The switching threshold transfer characteristics exhibited by the typical high-threshold IC are shown in Fig. 15-2. For normal input low voltages, less than 1.5 V, the output exceeds 13.5 V (15 V minus 1.5 V). It will continue to do so for any input up to 6.5 V, a tested point.

OFF-RETURN MEMORY

This memory function is identical to an electrically held relay. It provides undervoltage protection. Figure 15-3 shows the basic relay and the logic of OFF-return MEMORY circuits. In the relay circuit, momentary closure of pushbutton A will energize coil E. This sets up its own holding circuit by closing the normally open E (1) contact, enabling the operator to release pushbutton A. Energy is thus maintained to coil E. Simultaneously, contact E (2) closes and $\overline{E}$ (3) opens. Pressing the normally closed pushbutton $\overline{B}$ (or the loss of power) de-energizes coil E, allowing E (1) and E (2) to open and E (3) to close.

The transition width from opening to closing is specified from 6.5 V to 8.5 V, and once the input exceeds the 8.5-V level, the output is guaranteed to be below 1.5 V. This will remain true for any further increase in the input voltage. As noted from the curve, with a 15-V supply, the worst case of noise margin in either the high or low state is 5.0 V. Normally, the low input voltage is 1 V or less, the transition region is between 7 and 8 V, and the high output voltage is better than 14 V; therefore typical noise margins of 6 V are obtained in either case. As a comparison, the transfer region for other forms of ICs (such as diode-transistor logic ICs) generally lie within the area of 0.7 to 1.5 V.

Additional noise protection can be obtained by the addition of filtering and buffering to the circuitry using high-threshold ICs. This will slow down the switching action of the IC so that a noise spike must be present for a longer period of time to cause false switching.

15-5 BASIC FUNCTION

The basic building block in the high-threshold IC line is the NAND function. If a circuit were designed with the

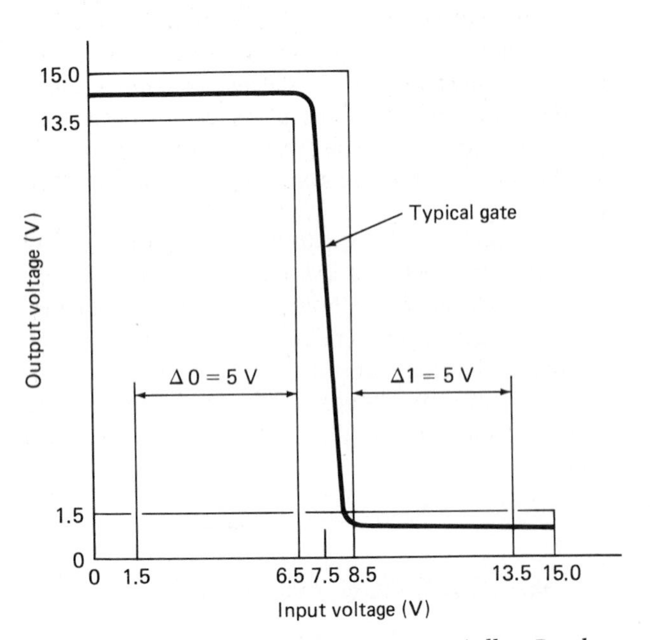

Fig. 15-2 High-threshold logic gate. *(Allen-Bradley.)*

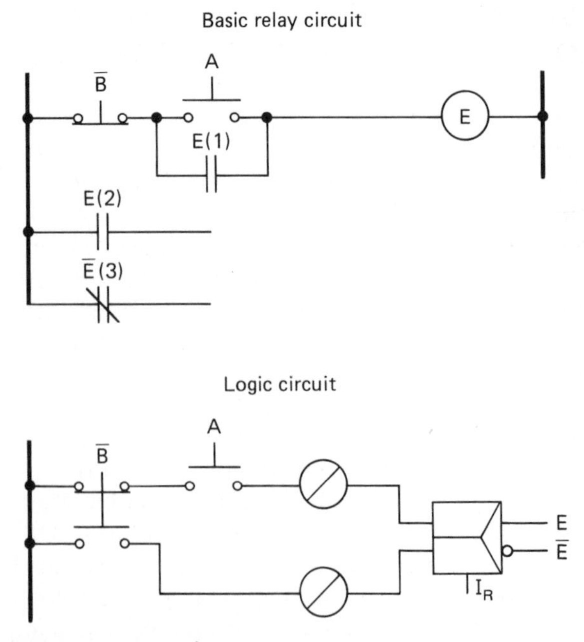

Fig. 15-3 OFF-return memory function.

internal or equivalent circuit for the NAND function drawn out every time and NAND were required, the circuit would become very complex and difficult to understand. The logic symbol shown in Fig. 15-4 has been designed to represent the NAND function and is used to simplify logic diagrams.

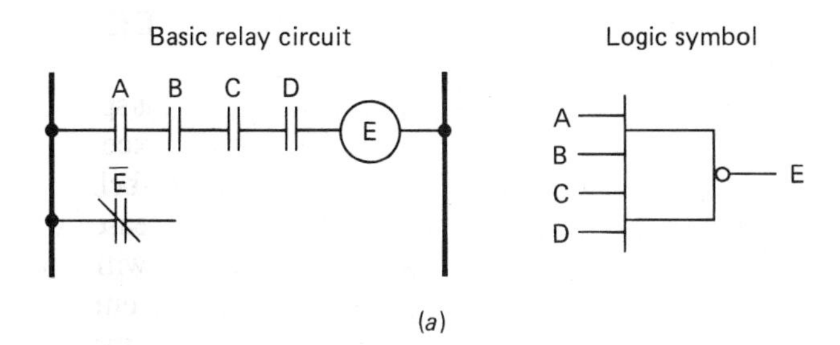

NAND Truth Table

INPUTS	OUTPUT	INPUTS	OUTPUT
A B C D	E	A B C D	E
0 0 0 0	1	0 0 0 1	1
1 0 0 0	1	1 0 0 1	1
0 1 0 0	1	0 1 0 1	1
1 1 0 0	1	1 1 0 1	1
0 0 1 0	1	0 0 1 1	1
1 0 1 0	1	1 0 1 1	1
0 1 1 0	1	0 1 1 1	1
1 1 1 0	1	1 1 1 1	0

(b)

Fig. 15-4 NAND function *(a)* basic relay circuit and logic symbol *(b)* truth table.

15-6 LOGIC FUNCTIONS

The AND, OR, and NAND elements in the Allen Bradley Cardlok Solid State control are also used in the General Electric Static and Westinghouse Numa-Logic systems. The truth tables for the logic devices are also the same. The Allen Bradley memory is a little different from that of some of the others. We will explain it next.

In the logic circuit of Fig. 15-3, momentarily closing pushbutton A provides a 1 input to the "on" side of the memory, a 1 on the "on," or E, output and a 0 on the "off," or Ē, output. Pressing pushbutton B̄ changes the memory to the opposite state by the application of a 1 to the "off" side of the memory, thus causing the "on," or E, output of the memory to go to 0 and the "off," or Ē, output to go to a 1. When power fails and is restored, an initial reset pulse is provided by the logic power supply to ensure the memory returns to the "off" state.

15-7 OTHER FUNCTIONS

The preceding discussion has generally referred to the basic high-threshold NAND gate circuit. Additional components are available, however, which exhibit the same high noise immunity characteristics obtained by reverse-biased base-emitter breakdown action. The availability of such other units as JK flip-flops, RS flip-flops, line drives, and monostable devices allows the designer or control manufacturer to design complete logic systems with a high degree of noise immunity throughout. These devices allow the build-up of the counter and shift register circuits needed in the more general and complex types of applications seen in today's technology.

15-8 APPLICATIONS

Most complex control systems are mainly multiple combinations of techniques which utilize the basic logic functions and such special components as counters and shift registers. These techniques can be modified or expanded to meet a variety of industrial control problems. If the fundamentals of each of the basic logic functions and techniques are understood, then just about any type of control system can be designed and built. Utilizing any type of solid-state digital logic and hard wiring method yields a fixed type of controller. This controller is usually designed and wired to meet a particular type of application.

15-9 COUNTERS

A *counter* can be defined simply as a device that will accept successive signals and store them, normally in coded form. In most cases, this code is in the form of either binary or binary-coded decimal, or BCD. Each of these forms has its advantages and disadvantages, and the selection of one form over the other is usually dependent upon the application. Fewer components are needed to make up a binary counter than a BCD counter; on the other hand, a more involved decoding circuit is needed to convert the stored information in the binary counter into a usable form.

Solid-state counters operate at very high speeds compared to electromechanical counters. The operating speed, while very desirable, could cause the counter to respond to any signal which appears at the counter input. Any contact-making pilot device such as a pushbutton or limit switch will produce a number of signals at each operation because of contact bounce. A counter operated by one of these devices would store the initial signal plus the signal produced by each succeeding contact bounce until the bouncing has subsided. To prevent false operation, the input to the counters should utilize a signal converter with a built-in time delay so that it will eliminate the effect of the contact bounce.

There is a unique combination of counter output signals for each different number stored within the counter. A distinct signal for each number stored can be obtained by providing an AND gate for each unique combination of counter output signals. Therefore, the output signals from a single decade BCD counter can be converted to decimal form with 10 AND gates. The output signals from each of these AND gates represent the numbers 0 through 9. The 10 AND gates in this case comprise the decoding circuit for the counter.

Every counter has a fixed storage capacity above which it will reset to 0. A single decade BCD counter can store nine signals, and the tenth signal will clear or reset the counter back to zero. A 4-bit binary counter can store 15 signals, and the sixteenth input signal will reset the counter to zero. These counters will reset to zero once their capacity has been reached. It is possible to make a counter recycle short of full capacity by decoding the point or count, stretching or storing the decoded signal for a short period of time, and using the output signal from the storage device to reset the counter. A single shot, or monostable, circuit can be used to stretch this output signal. This is needed to supply a reset time of sufficient length to completely reset the counter to zero. If the counter is to recycle at the same point every time, an AND gate can be used to decode the reset signal. If the recycle point is to be changed frequently, a decoding selector switch can be used to produce the reset signal.

Figure 15-5 illustrates a practical application of how a recycle point is frequently changed. This system consists of two palletizers supplied from a single conveyor. As the cartons pass the count limit switch, they are recorded by the counter. For maximum palletizer operating efficiency, it is desirable to alternately supply the palletizers with quantities of cartons in tier increments.

Because this manufacturer produces a number of different products, the palletizers must be versatile so that each type of product can be handled. Each product has a different-size carton and, therefore, the quantity of cartons per tier will vary from one product to the next. By means of a selector switch, the number of cartons per tier can be set. Now as the cartons pass the count limit switch, they are recorded; when the set count is reached, the diverter is activated and the cartons are then directed to the next palletizer. The system will alternate back and forth in this manner.

15-10 SHIFT REGISTER

Another basic type of control scheme is the *shift register.* A shift register is a device that accepts input signals and stores them in the sequence that they are received with respect to a time base (shift signal). Physically, the shift register consists of cascaded JK flip-flop memories coupled so that a shift signal applied simultaneously to each of the clock inputs will cause each JK memory to assume the condition of the previous JK memory just prior to the shift signal. The timebase referred to is the time duration between the recurring shift signals. This time duration is normally fixed but

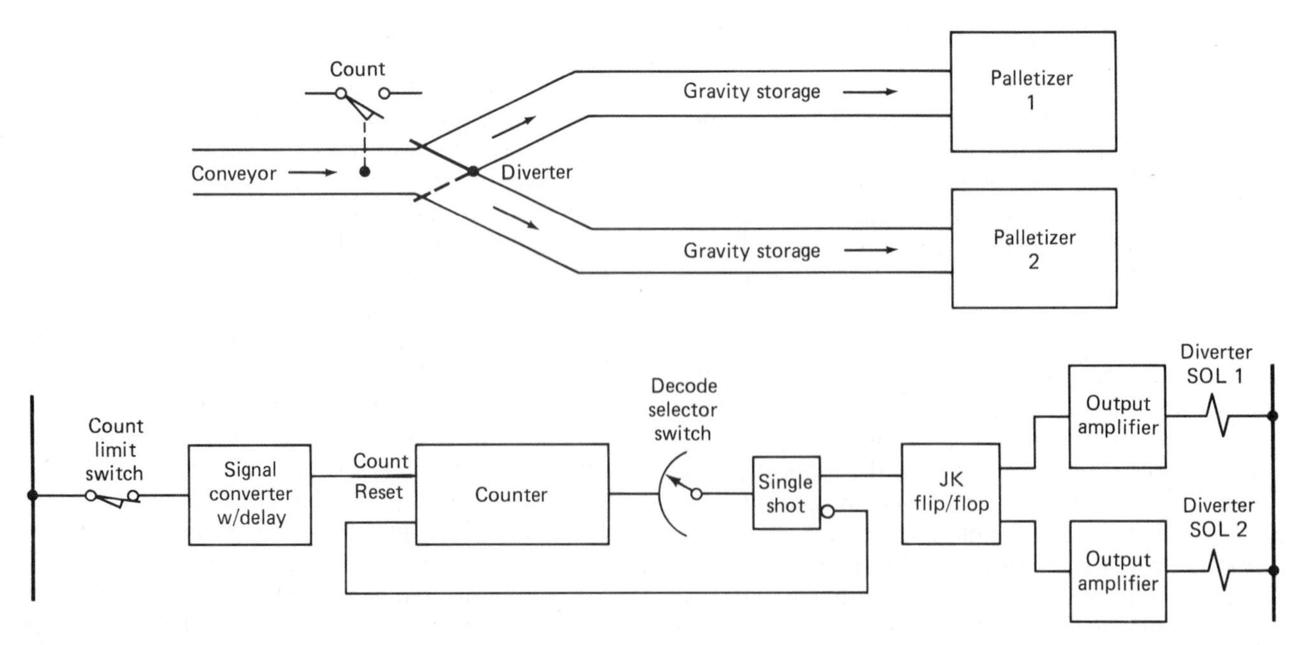

Fig. 15-5 Counter applications.

can be variable. Each of the cascaded JK flip-flops is considered one zone of the shift register, and can store one signal, or bit, of information.

The number of zones in any shift register is dependent upon the number of signals that must be stored. By paralleling cascaded JK flip-flops, each zone can store two or more signals, or bits. Therefore, a shift register of five zones that can store two signals in each zone would be a 5-zone, 2-bit shift register. The decimal numbers 0, 1, 2, and 3, when encoded to the binary form of 00, 10, 01, and 11, respectively, could be stored in each zone of this 5-zone, 2-bit shift register.

Shift registers, like counters, have a fixed storage capacity. The shift register, however, does not completely clear and reset once this capacity is exceeded. Only the final zone is affected and will be reset by each shift signal in order that it can accept the signals stored in the preceding zone. The signals in the final zone, prior to the shift signal, are cancelled and disappear.

It is possible to obtain output signals by decoding any zone of the shift register. The decoding techniques are identical to those that would be used with counters. Once decoded, signals in the shift register can be left in the shift register to be cancelled upon reaching the final zone or, by means of a direct reset, can be cancelled from the zone where they are decoded.

To cancel information in intermediate zones, signals to the direct reset must be synchronized with the shift signal. The complexity of the circuit required to synchronize the signals depends upon the speed and type of shift register used and may involve a complex storage circuit.

Figure 15-6 shows an example of a synchronous shift register for a power-and-free dispatch storage system. This system is a tire inspection line and must accept destination commands from a dispatcher or inspector. For each carrier that passes the dispatch station, it must store the commands and energize the correct storage line diverter as a carrier approaches that storage line. There are three storage lines. The dispatcher has a button for the first two lines, which is to be pressed as a carrier and tire pass the dispatch or inspection station. No action is required of the dispatcher to send tires to the reject line. The system is divided into four equal zones. The first two zones are strictly storage zones made necessary because of the distance between the dispatch station and the first storage line. The last two zones each contain diverter switches leading to storage lines. The shift register also contains four zones. The last two zones must have decoding circuits to produce the diverter operating signals.

As a pusher with a tire for line 1, which stores whitewall tires, approaches the dispatch station, the dispatcher will momentarily press the "line 1" pushbutton if it is a good tire. This produces a signal which turns on the OFF-return memory through a signal con-

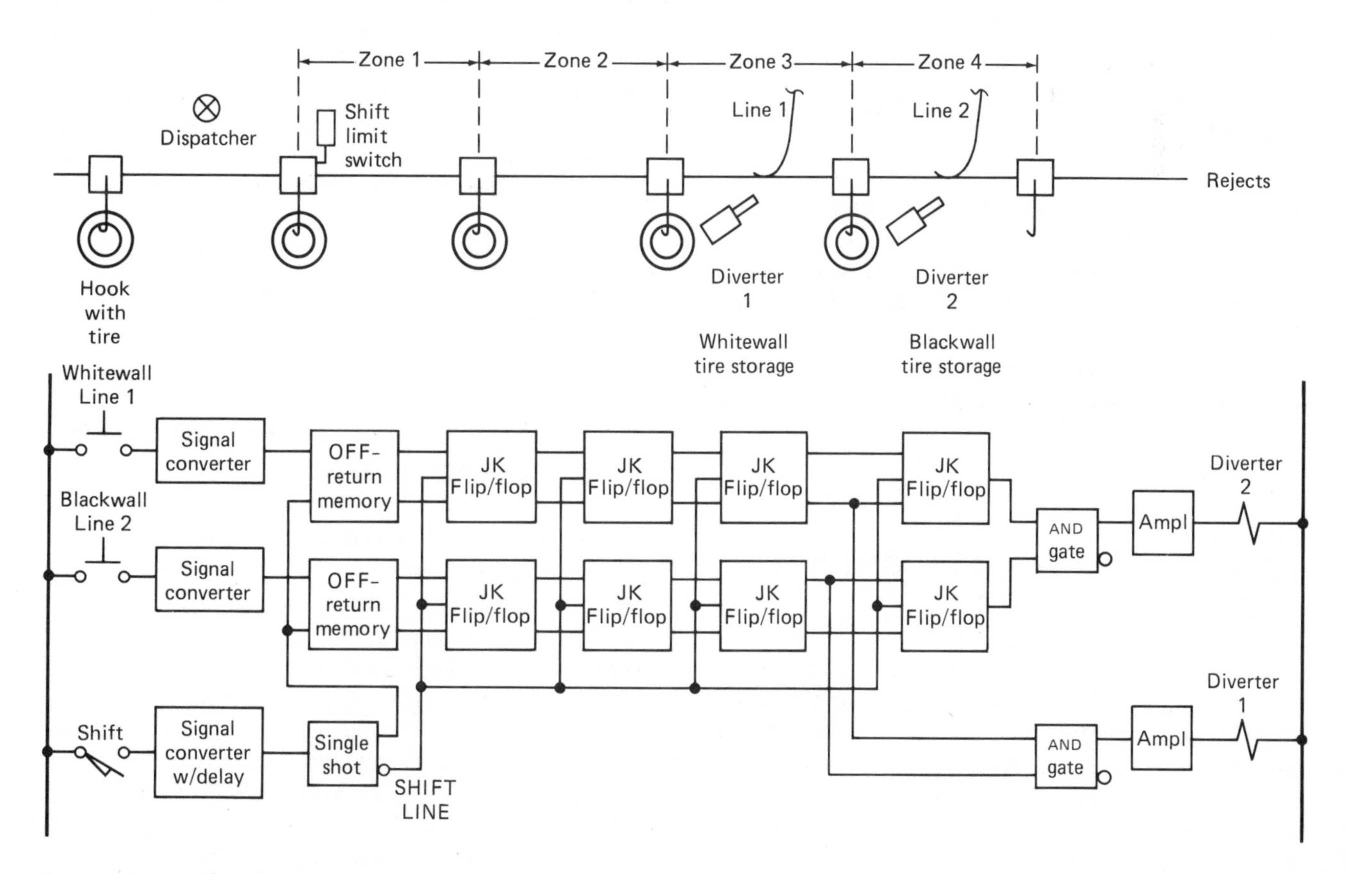

Fig. 15-6 Shift register application (tire sorter).

verter. As the pusher enters zone 1, the "shift" limit switch will close and produce the "shift" signal through the single shot, which transfers the signal stored in the two memories into the first JK flip-flop (zone 1) of the shift register. As the pusher moves into zone 1, the "shift" limit switch is released. This resets the single shot and allows the dispatcher to set in a new command for the next tire coming down the line. The dispatcher will code only good whitewall or blackwall tires. All other tires will go through the reject line. The OFF-return MEMORYS perform a simple encoding and storage function to ensure that the dispatcher signal is present and in encoded form when the "shift" signal appears. The signals are transferred on the leading edge of the "shift" signal.

As each pusher passes the entrance of its particular transfer line, the diverter is activated, which causes it to divert the tire from the main line to the storage line. If the diverter is not activated, the pusher will continue downstream until it is diverted or runs off on the reject line. In every case, the diverter is energized for one zone length and, as the next signal is transferred, will either retract or remain energized if the following tire is to go into the same storage line.

Pushers with no tires, or defective tires, are assigned the code 0, which is not decoded and, therefore, produces no diverter switch operating output signals.

15-11 LOGIC FUNCTIONS AND AUXILIARY DEVICES

The family of Allen Bradley Cardlok Solid State logic devices is illustrated in Fig. 15-7. The connection diagrams for many of the cards are given in Fig. 15-8*a* through *f* on pages 248 and 249. Application notes from the user's manual are found in these figures as well as in the Allen Bradley user manual. The application notes explain Cardlok logic applications well.

15-12 DEVELOPING SCHEMATIC DIAGRAMS AND WIRING CHARTS

To develop the schematic diagram and wiring chart for an industrial electric control circuit requires that a rack diagram be developed first. Figure 15-9, on page 249, illustrates the physical appearance of a rack layout for the logic cards used for an industrial application. Once it is decided how many cards are needed, then any number of racks and the layout of the cards can be arranged any way desired. Pins on the card plugs are wired with a wire-wrapping gun, so consideration should be given to a card location that will facilitate easy wiring.

The logic cards can be arranged in any sequence desired. Once the number of cards needed is determined, the circuit numbers for the logic elements are selected. If a card has six circuits, for example, the six circuits can be used anywhere in the logic diagram. Pin numbers are determined from the card connection diagrams in Fig. 15-8.

Look at the circuit in Fig. 15-10 on page 250. It can be seen that the wire numbers have three digits. This is so wire numbers will not be confused with the pin numbers for the logic gates. Notice in Fig. 15-10 numbers for the wires of input circuits range from 600 to 699, numbers for logic circuit wires range from 200 to 299, numbers for 24-V light driver amplifiers range from 400 to 499, and numbers for output circuits range from 800 to 899. You can use any other combination of numbers you desire as long as you are consistent in your diagram.

The wiring chart starts with a wire number and indicates what logic element the wire came from and to what logic gate inputs it is to be connected. Once a wiring chart is developed, the wiring of the racks is done by using the chart and not by using the schematic diagram. The schematic diagram finds its greatest use in troubleshooting once the circuit is in operation. A logic probe with lights indicating logic 1s and logic 0s is very effective in troubleshooting. If a bad logic circuit is found, the logic card can be replaced. If spare cards are not available, there are generally spare logic gates on the existing cards that can be wired into the logic system to place it back into operation. Faulty cards can be replaced with new cards very rapidly. This is often desirable to eliminate the downtime required for card circuit repair. Where downtime is costly, it is suggested that spare cards be kept in stock for rapid card replacement.

15-13 SUMMARY OF ALLEN BRADLEY SOLID-STATE LOGIC CONTROL

The Allen Bradley Cardlok logic system is one type of control system. However, with the use of programmable control systems in control work, many logic circuits are becoming programs for P.C.s. Industrial systems with numerous logic functions can be programmed into P.C.s more easily than they can be hard-wired into logic systems. Small control situations may continue to be wired with solid-state logic systems such as Cardlok, but the big jobs with many input/output circuits are applications for P.C.s because of the ease with which they can be programmed, monitored, and checked.

GENERAL SPECIFICATIONS—Most Functions

Logic Convention: Positive logic, using English functions
Logic Levels: Logic 1 = +13.5 to + 15V; Logic 0 = 0 to + 1.5 V
Transition Voltage: +6.5 to 8.5V (7.5V nominal)
Noise Margin: 6V typical, 5V worst case
Temperature Rating: −20°C to +60°C
Frequency Response: Approx. 25K Hz
Fan-Out: 10 load units Fan-In: 1 load unit

Type of Card				Catalog Number	Type of Card				Catalog Number
Name	Symbol	Functions per Card ↓	Remarks		Name	Symbol	Functions per Card ↓	Remarks	
AND		4	Standard and inverted outputs. Also expander input.	1720-L004	RETENTIVE MEMORY		2	Retains output status on resumption of power.	1720-L540
SEALED AND		4	1, 2, or 3 inputs may be sealed.	1720-L014	JK FLIP/FLOP		6	Operates in both synchronous and asynchronous mode.	1720-L610
NAND		6	Expander input available.	1720-L104	DECADE COUNTERS		2	BCD outputs 2 decades per card.	1720-L710
OR		8	Two inputs.	1720-L202	PRESET DOWN COUNTER		1	Can be preset to count from a predetermined number.	1720-L720
OR		4	Four inputs.	1720-L204	UP-DOWN COUNTER		1	Counts either up or down	1720-L730
NOT		12		1720-L310	COINCIDENT INPUT		1	Prevents simultaneous inputs to Up-Down counter.	1720-L731
TIMERS 0.1–0.3s		2	Convertible TDE or TDD. Range can be extended by adding capacitors.	1720-L410	4-ZONE SHIFT REGISTER		2	Serial and parallel inputs and outputs available.	1720-L810
TIMING CAPACITORS		10	Extends timing range of L410, L450, and L460.	1720-L420	REVERSIBLE SHIFT REGISTER		1	Shifts information right or left	1720-L830
SINGLE SHOTS 0.1–0.3s	SS	2	Timing range can be extended by adding capacitors.	1720-L450	INITIAL RESET		2	Upon applied power the output is logic 0 for 40 ms, then logic 1	1720-L905
SINGLE SHOTS 0.1–0.3s	SS	2	Timing range can be extended by adding capacitors.	1720-L460	INITIAL RESET		4	Upon applied power the output is logic 0 for 40 ms, then logic 1.	1720-L906
OFF-RETURN MEMORY		5	R-S Flip/flop with initial reset input.	1720-L520	LOGIC DRIVER	LD	4	Increases fan-out of any standard logic device.	1720-L1010

Fig. 15-7 Family of Allen-Bradley solid-state devices.

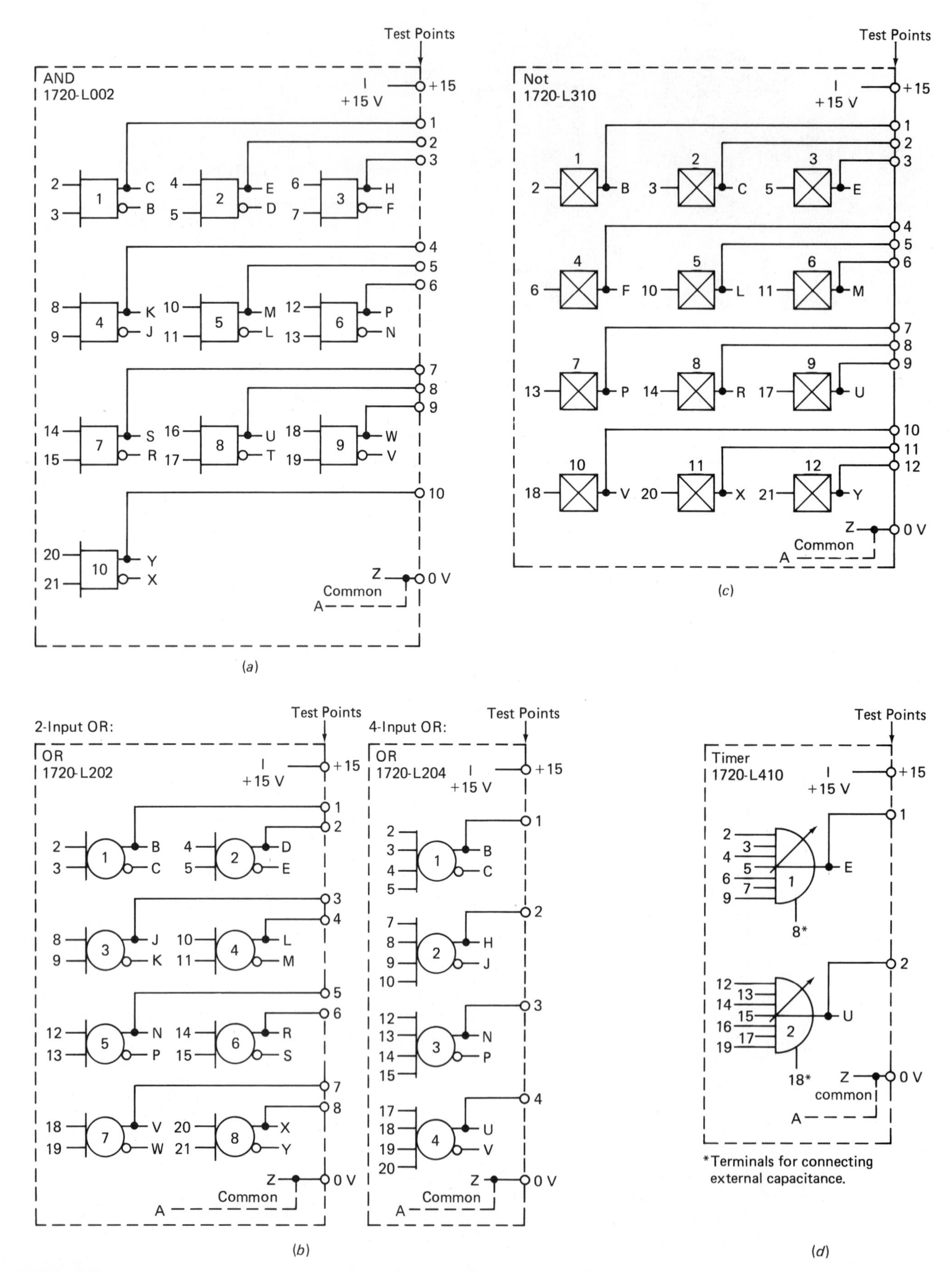

Fig. 15-8 Connection diagrams for cards (Allen-Bradley Cardlok solid-state logic control). *(a)* AND *(b)* OR *(c)* NOT *(d)* timer *(e)* JK flip-flop *(f)* decade counter.

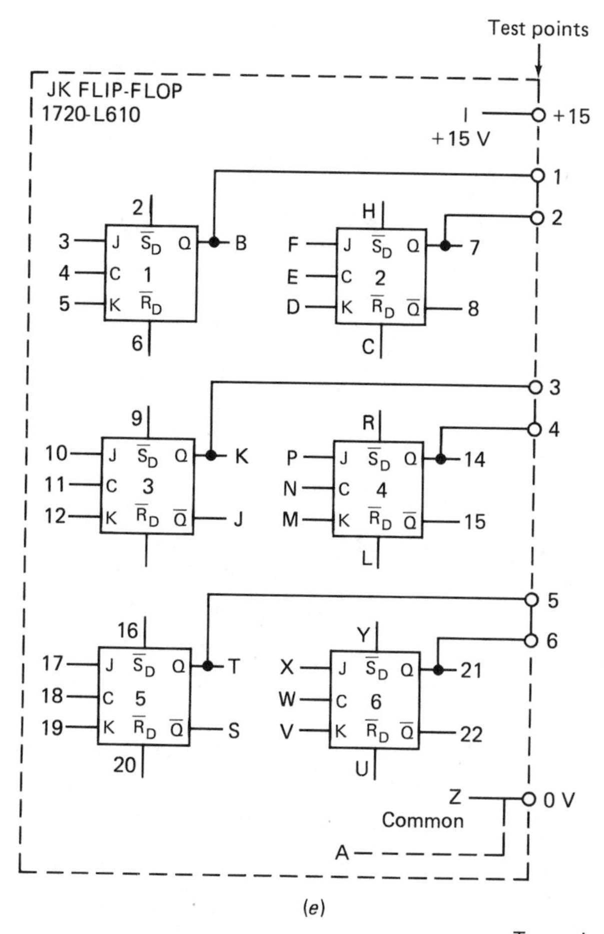

(e)

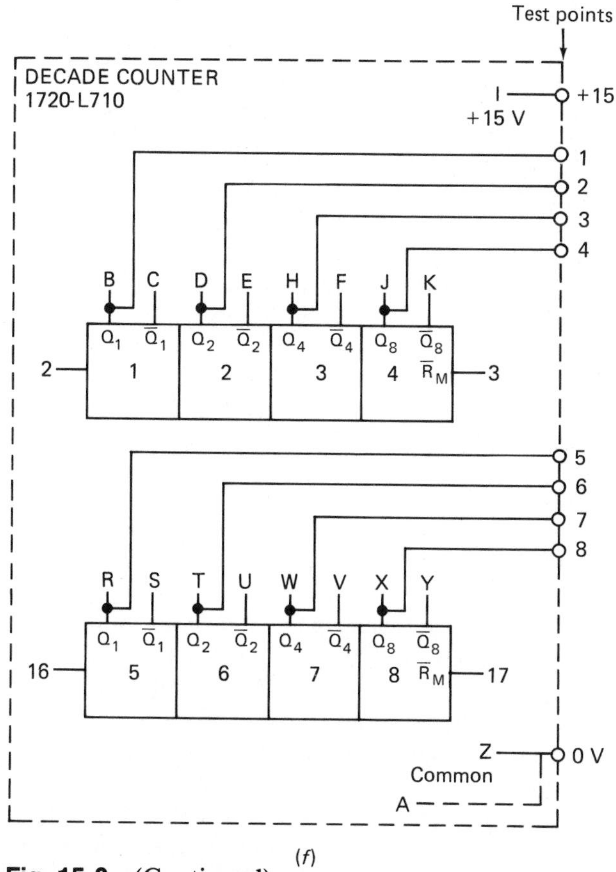

(f)

Fig. 15-8 (Continued)

(a)

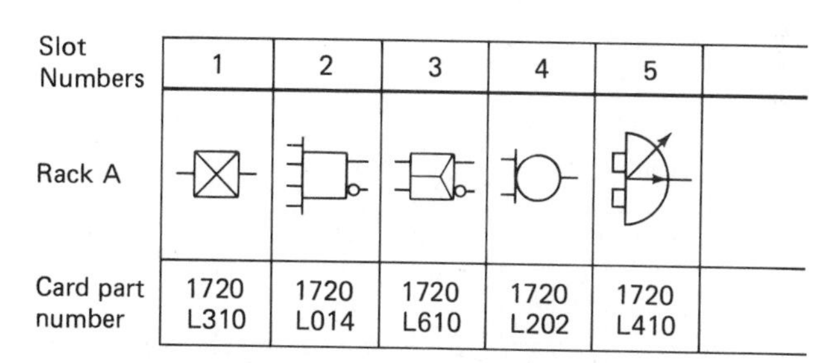

Slot Numbers	1	2	3	4	5	
Rack A						
Card part number	1720 L310	1720 L014	1720 L610	1720 L202	1720 L410	

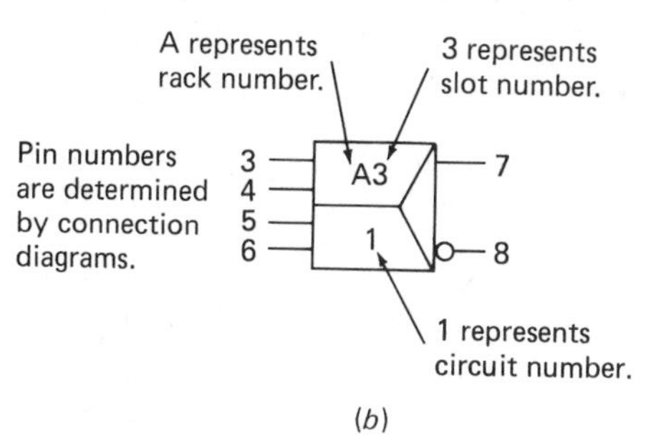

(b)

Fig. 15-9 Rack layout for Allen-Bradley Cardlok circuits. *(a)* Card rack and frame *(b)* rack layout in schematic form.

15-14 CUTLER-HAMMER LSL LOGIC

Cutler-Hammer also manufactures solid-state logic. Their system is called *LSL* or *Ladder Static Logic.*

The *LSL User's Application Manual* describes the Cutler-Hammer industrial control system, which enables a standard relay ladder design to be directly implemented with solid-state IC logic elements.

The LSL control system is based on the interconnection of elements which behave exactly like relay elements. The translation from a ladder diagram to LSL control is, accordingly, direct and requires no spe-

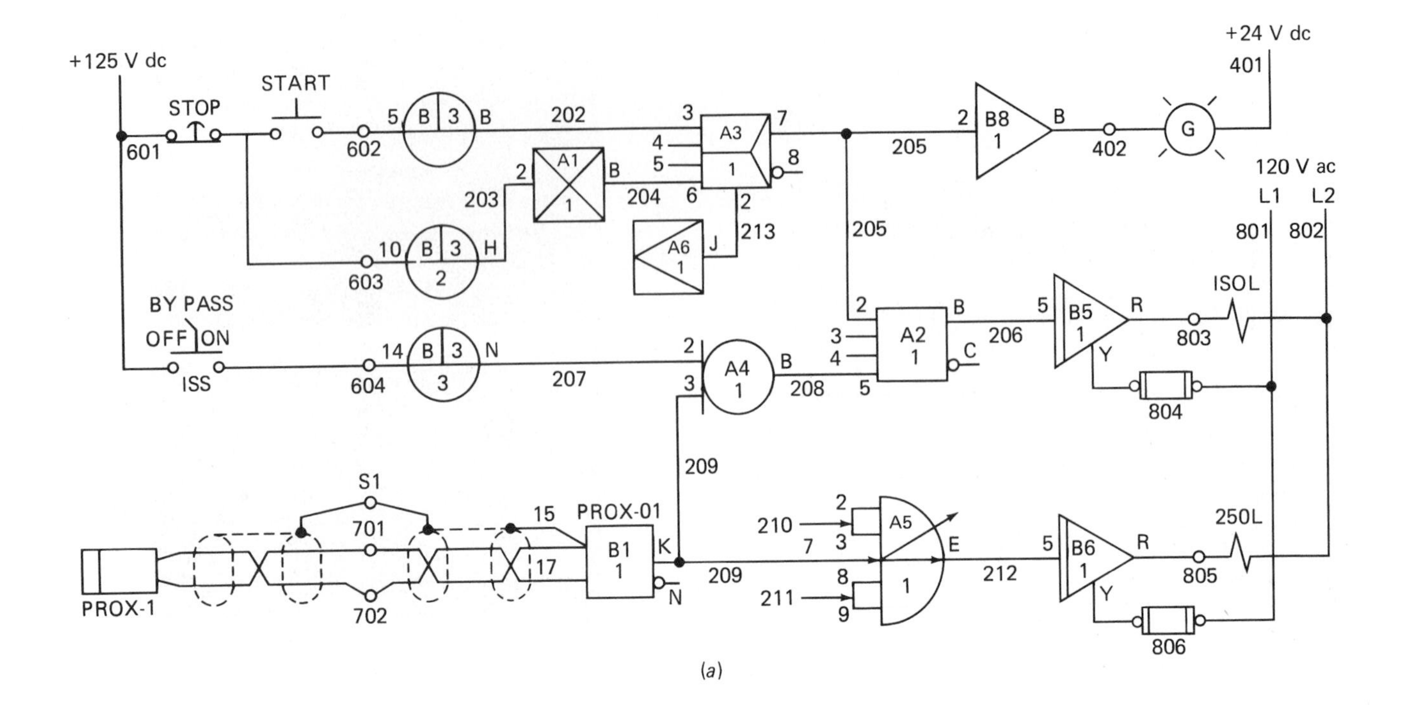

(a)

Wire No.	From	To
202	A3-3	B3-B
203	A1-2	B3-H
204	A1-B	A3-6
205	A2-2	A3-7
205	A3-7	B8-2
206	A2-B	B5-5
207	A4-2	B3-N
208	A2-5	A4-B
209	A4-3	A5-7
209	A5-7	B1-K
210	A5-2	A5-3
211	A5-6	A5-9

Wire No.	From	To
212	A5-E	B6-5
213	A3-2	A6-J
402	B8-B	Ter. Blk. 402
602	B3-5	Ter. Blk. 602
603	B3-10	Ter. Blk. 603
604	B3-14	Ter. Blk. 604
701	B1-15	Ter. Blk. 701
702	B1-17	Ter. Blk. 702
S1	B1-15	Ter. Blk. S1
804	B5-Y	Fuse T.B. 804
806	B6-Y	Fuse T.B. 806
803	B5-R	Ter. Blk. 803
805	B6-R	Ter. Blk. 805

Wire No.	From	To
201	15 V dc Power Supply (+)	Top Bus Rack A
	Top Bus Rack A	A1-1
		A2-1
		A3-1
		A4-1
		A5-1
		A6-1
	Top Bus Rack A	Top Bus Rack B
	Top Bus Rack B	B1-1
		B3-1
		B5-1
		B6-1
201	Top Bus Rack B	B8-1

Wire No.	From	To
500	15 V dc Power Supply (Com)	Bottom Bus Rack A
	Bottom Bus Rack A	A1-Z
		A2-Z
		A3-Z
		A4-Z
		A5-Z
		A6-Z
	Bottom Bus Rack A	Bottom Bus Rack B
	Bottom Bus Rack B	B1-Z
		B3-Z
		B5-10
		B6-10
500	Bottom Bus Rack B	B8-Z

(b)

Fig. 15-10 Sample schematic diagram and wiring chart. (Allen-Bradley Cardlok solid state Logic). *(a)* Schematic diagram *(b)* wire charts.

cial apparatus. Further, LSL control has the stability of a hard-wired system and is immune to the transient malfunctions of volatile programming memories. It requires no special skills or equipment to maintain LSL; any electrician experienced in relay control can readily do the job without special training or equipment.

Although LSL control retains the language of the ladder diagram, it has many advantages over relays. Among these are the long life and reliability of its IC solid-state elements, elimination of the timing races inherent in relays, selectable maintenance monitoring lights, and easily implemented computer remote monitoring. The result is a system which combines the best of both worlds: the simplicity and universality of relay logic, and the high performance and flexibility of solid-state logic.

Static control is a sophisticated technology. Simply stated, it is an electronic technique by means of which circuits are turned on and off without mechanical devices. Higher-speed operation, more complex integrated machines, and higher production requirements demand the long life and reliability that can be achieved only through the application of solid-state devices.

The computer commands and monitors the completely automated plant. For the smaller machine application that cannot justify a computer, and yet requires the advantages of solid-state technology, static logic controllers provide the answer. With, of course, one major drawback: the intensive, prolonged training required to understand (much less design and maintain) the static control system. There is no new technology to learn if you are already familiar with conventional ladder control diagrams. Ladder Static Logic employs the same transistor integrated circuitry that gives static logic control its long life—high speed, small size, light weight, and reduced maintenance. The big difference is LSL's direct application from ladder diagrams.

Cutler-Hammer's LSL approach is achieved with standard relay coil, contact, timer, and limit switch symbolism, and without concern for the electronics or the logic involved.

There is a similarity between the conventional ladder diagram and LSL. The only difference is the addition of monitoring and terminal numbers as an aid to faster comprehension and quicker troubleshooting. One drawing provides schematic connection and interconnection information and serves as the only required plan to properly wire and maintain the control panel.

15-15 DESIGNING AN LSL LOGIC SYSTEM

The design of an LSL system begins with a ladder diagram drawn as for a relay system. There are no restrictions on the manner in which the contacts and coils are interconnected other than that good relay design practice be followed. However, the following features of LSL control can be used to simplify the design:

- The designer need not be concerned with the number of contacts for any coil; any number can be associated with an LSL coil element without difficulty.
- The designer can assume that all relays energize and de-energize simultaneously and that, specifically, interconnected contacts controlled by different coils will open and close simultaneously without any "overlap."
- The designer can rely on all relays responding not only simultaneously but also in a uniform time of 4 milliseconds (ms).

After the ladder diagram is finished, wire numbers can be assigned as for a relay design. It is suggested, for reasons which will be apparent later, that these numbers be taken from the range 1 to 4999.

The next step in the LSL control design is to convert the logic portion of the relay ladder diagram to LSL logic. This consists simply of substituting LSL elements for those relay elements performing logic operations (coils, contacts, and switches). The procedure is as follows:

1. LSL coil elements are substituted for relay coils. This includes the coils of control relays, latching relays, and timing relays.
2. Normally open (N.O.) and normally closed (N.C.) contact elements are substituted for the contacts on the diagram.
3. Switch contact elements are substituted for the switches—limit switches, pressure switches, pushbuttons, etc.—shown on the diagram.

The substitution is done symbolically by drawing a diamond around the relay diagram symbol to indicate that the coil contact or switch has been converted to a static logic device.

Such nonlogic elements as solenoids, lights, or, in general, power circuits are not converted. These will be handled as output circuits, which will be discussed later.

Figure 15-11, on the next page, shows a ladder diagram for an electric relay circuit. Figure 15-12, on page 253, is the same circuit converted to LSL. The symbols used by Cutler-Hammer in their LSL logic system are in Fig. 15-13 on page 254.

Figure 15-14*a* on page 254 is a START-STOP control circuit for an electric motor. It consists of the primary power to the motor controlled by the motor starter M, a control transformer supplying power to the control, the coil of the motor starter, the motor starter overload, a START button, STOP button, and the interlock contact M from the motor starter.

Of these, only the STOP and START pushbuttons and interlock contact constitute logic; the remainder are power circuits. In particular, the motor starter coil must be classified as a power circuit. Further, all the logic elements are switch contacts, including the M contact of the motor starter. Relay contacts are those controlled by elements in the logic system, mostly control, latching, and timer coils.

Figure 15-14*b* illustrates the following points:

- The LSL elements are connected as counterparts to the switches on the relay diagram. However, the left line is not a power source but what might be regarded as a logic signal source. It is called the *logic common* and is obtained from a terminal on the master control board.
- The control inputs of the switch contact elements go to the left toward the input circuits, which will be discussed in the next section.

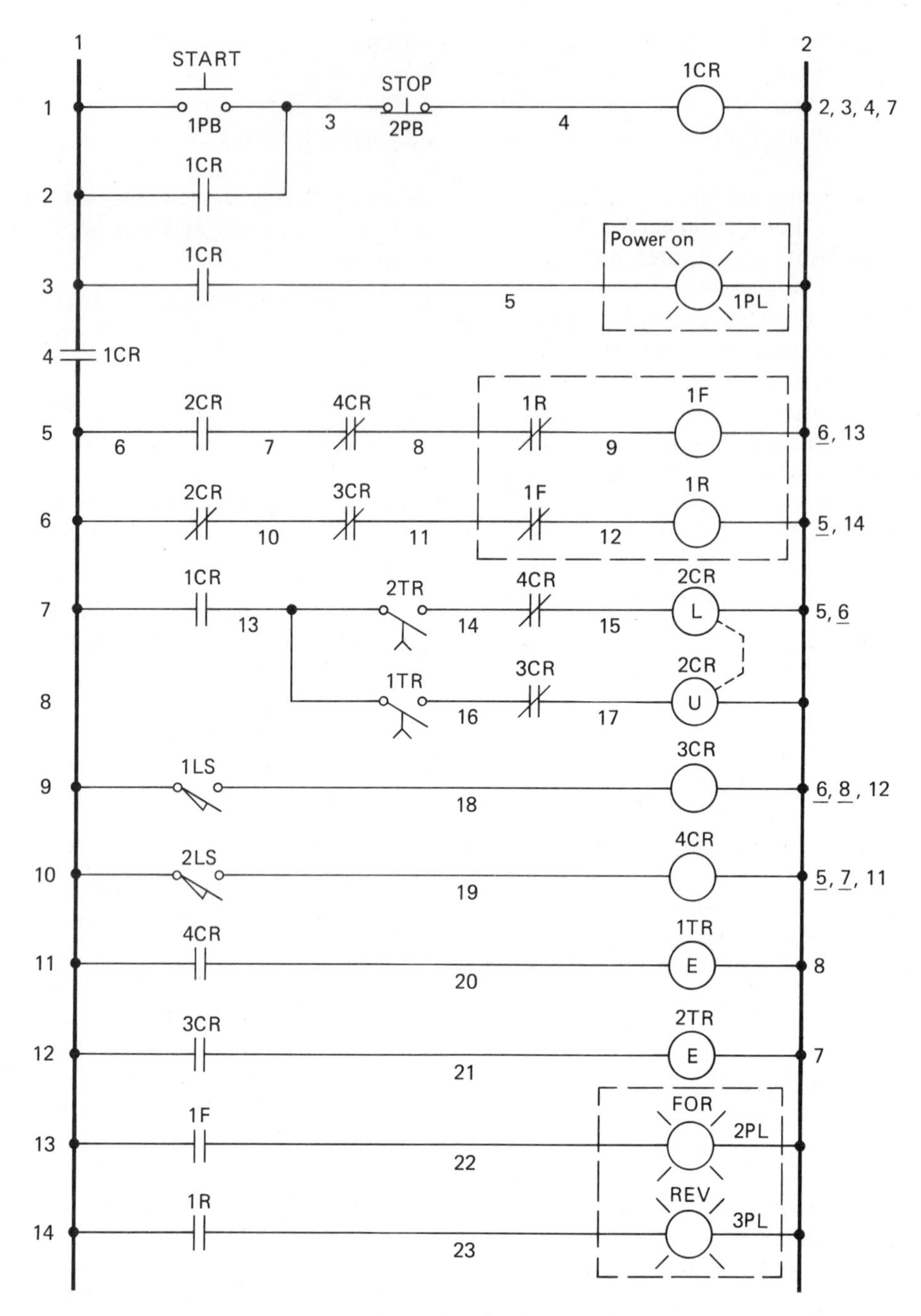

Fig. 15-11 Ladder diagram for LSL circuit design. *(Cutler-Hammer Inc.)*

- The logic controls a power device: the motor starter coil. This is done through an output circuit, which will also be discussed later.
- The diagram shows that more space is used to show LSL logic elements than to show relay elements. This space will become apparent later, when the implementation of the design is considered, for it is highly desirable in the final drawing to enter terminal and monitoring information around the symbol to facilitate the use of the diagram for maintenance. It is recommended that the symbols be spaced 1½ in. apart and the lines of the diagram be spaced 1 in., as shown in Fig. 15-15 on page 254.

A second illustration is shown in relay diagram form in Fig. 15-16 on page 254. Here, a control relay has been added to the circuit of Fig. 15-14*a* to give a jog control. The control relay and its associated contacts are a part of the logic and can be implemented with LSL control coil and N.O. contact elements; the motor starter and overload contact are power circuits and are moved to a separate drawing.

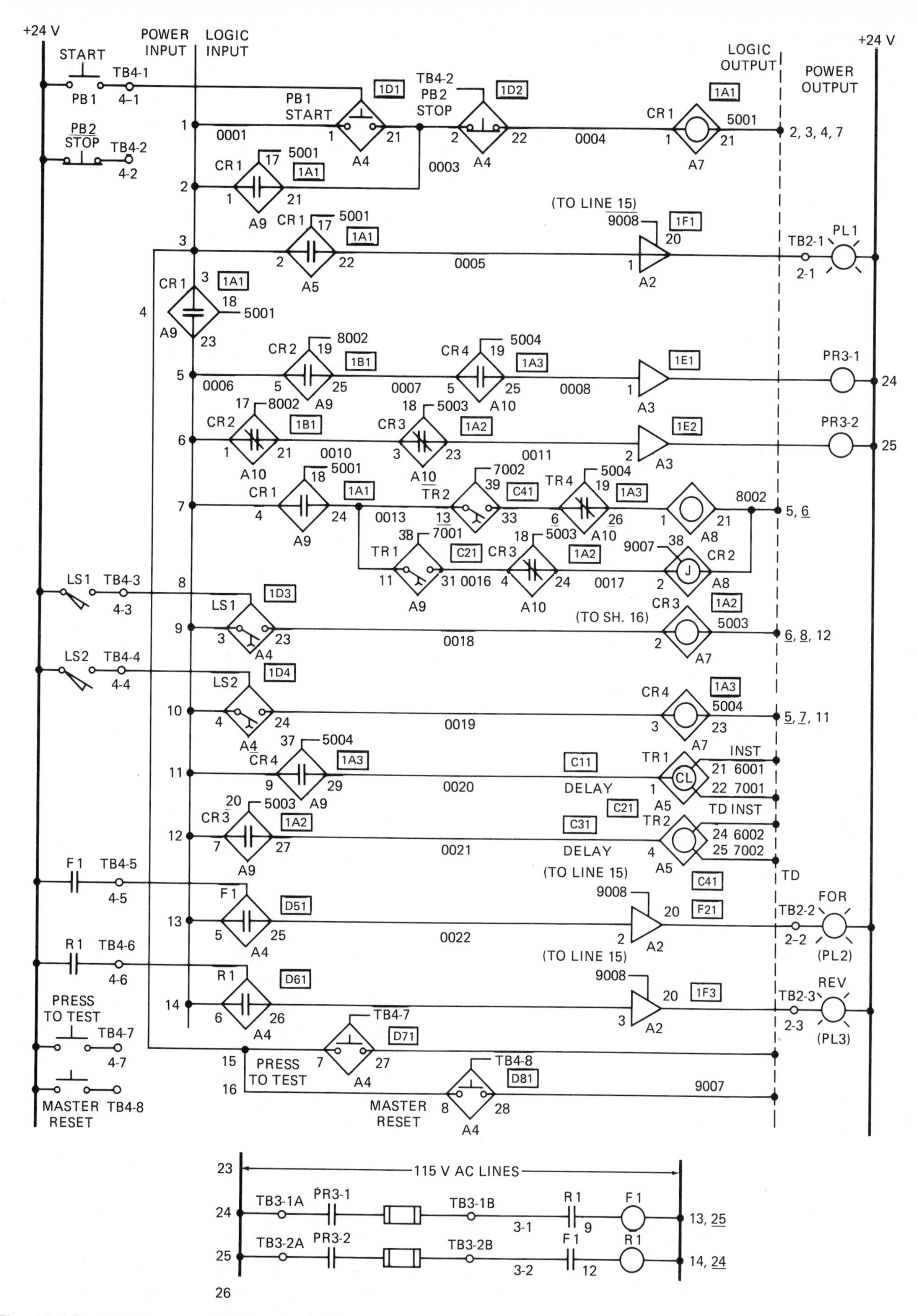

Fig. 15-12 LSL diagram for electric circuit.

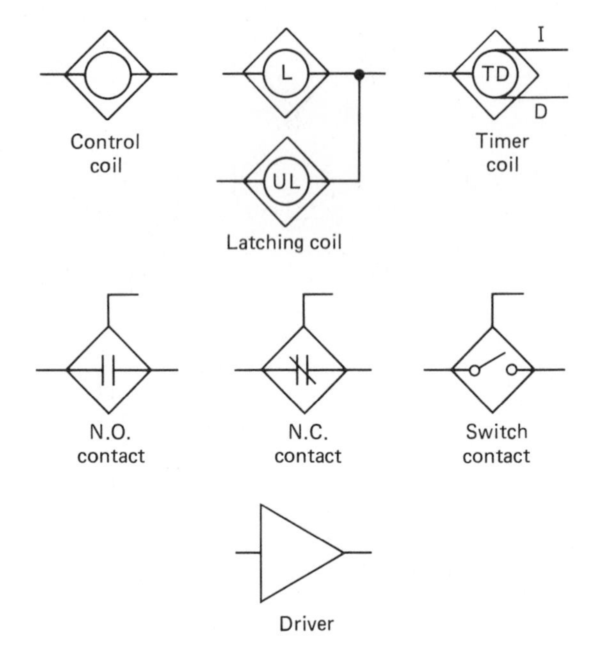

Fig. 15-13 Symbols used to represent LSL elements.

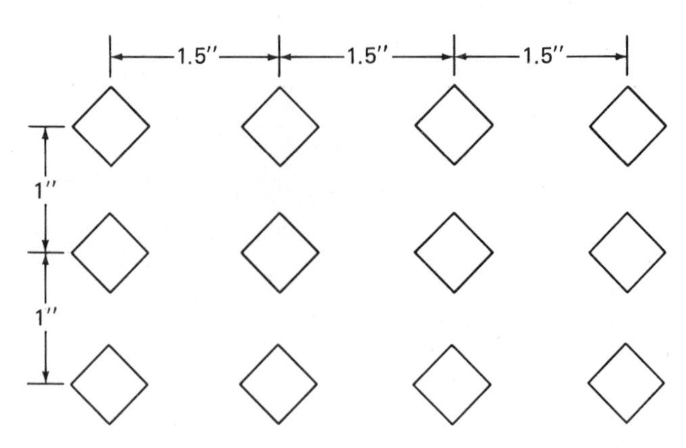

Fig. 15-15 Recommended spacing for LSL symbols.

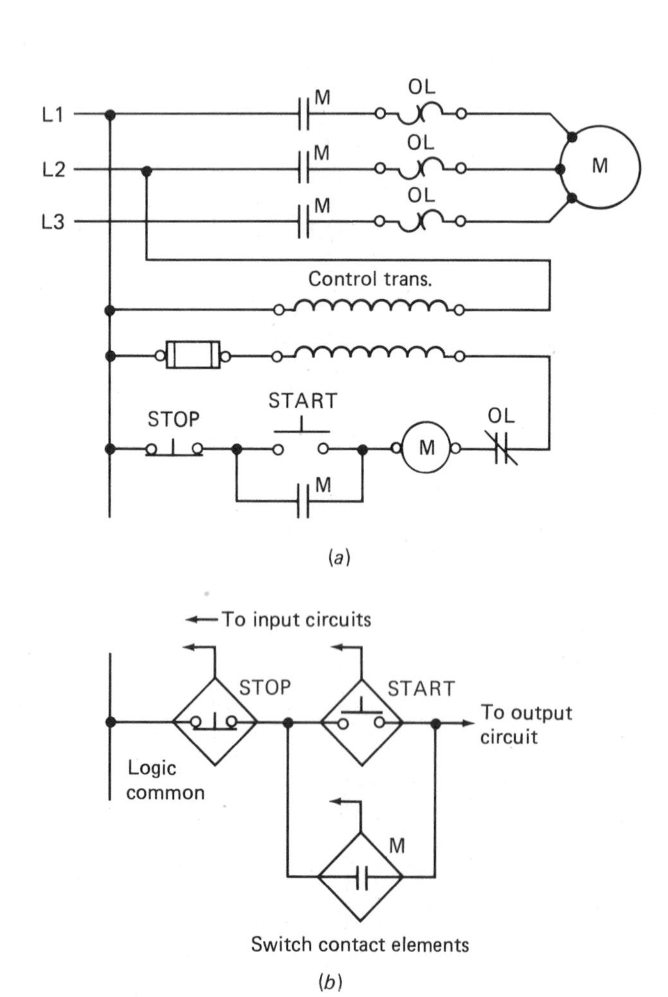

Fig. 15-14 Start-stop motor control *(a)* relay diagram *(b)* logic diagram.

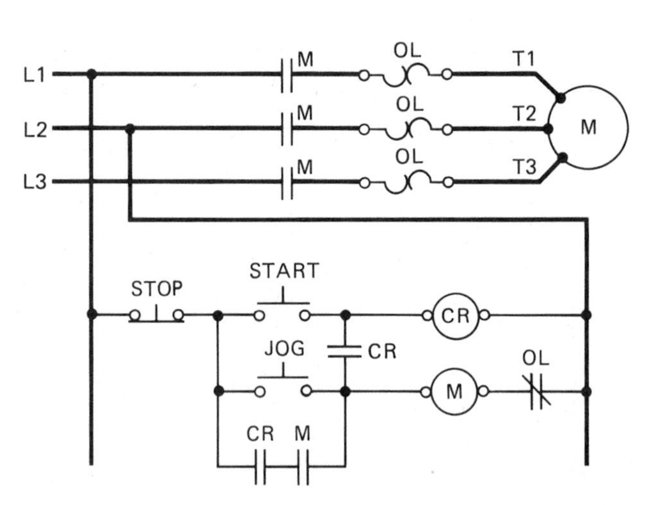

Fig. 15-16 A relay diagram to be implemented with LSL elements.

"Vertical" logic as represented by the CR contact between the START and JOG contacts must be treated with caution. Specifically, if the logic of the contact closing can be bidirectional, that is, if the logic signal can pass in either direction, the contact cannot be implemented by a signal LSL contact element. However, an examination of the relay circuit of Fig. 15-16 shows the signal through the contact actually is only required to flow "up" from the M contact to the CR coil since the START pushbutton does not control M, only CR. The logic can then be implemented with LSL elements as shown in Fig. 15-17.

The diagram also illustrates the following additional points concerning the LSL logic diagram:

- The N.O. contact elements have control inputs. These inputs are drawn to the right in the direction of the coil elements.
- The right line of the diagram is not a power connection as on a relay diagram but can be regarded as the

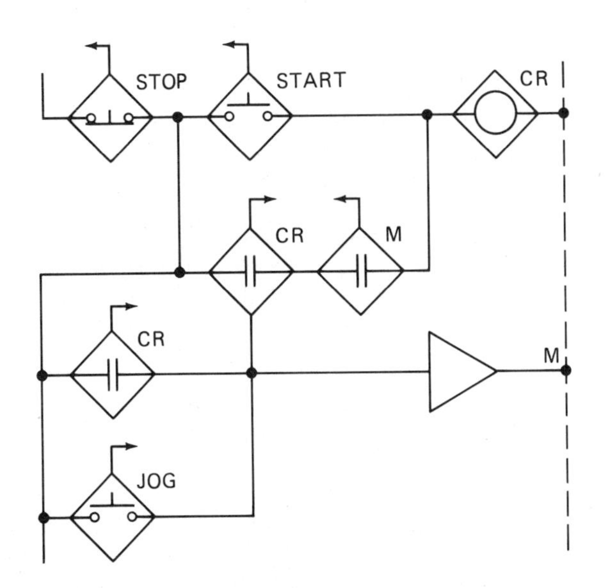

Fig. 15-17 Logic portion of diagram of Fig. 15-16 implemented with LSL elements. *(Cutler-Hammer Inc.)*

logic output of all the coil elements and other elements which terminate at the line. Accordingly, it is not a single wire but rather is a bundle representing all the output wires of the coil elements. The control inputs of the contact elements are then, in effect, connected to this bus.

Occasionally, an actual case of bidirectional logic might be encountered in a relay design. Figure 15-18 is an example. Here, with 3CR open, 4CR and 5CR are independently controlled. However, with 3CR closed, both are simultaneously energized by either 1CR or 2CR.

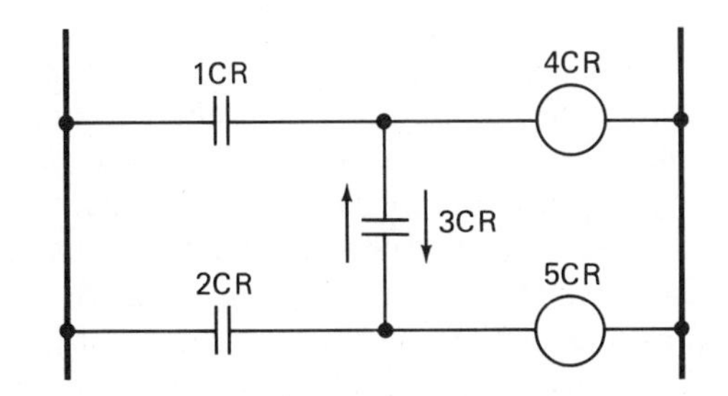

Fig. 15-18 Example of vertical and bidirectional logic.

As stated previously, the 3CR contact cannot be implemented with a single LSL element, since the logic signal must flow in two directions. The implementation requires that the logic be modified to give unidirectional flow. This is done and implemented with LSL elements as shown in Fig. 15-19.

It might be said that the bidirectional logic of Fig. 15-18 occurs rarely in practice because of the limited

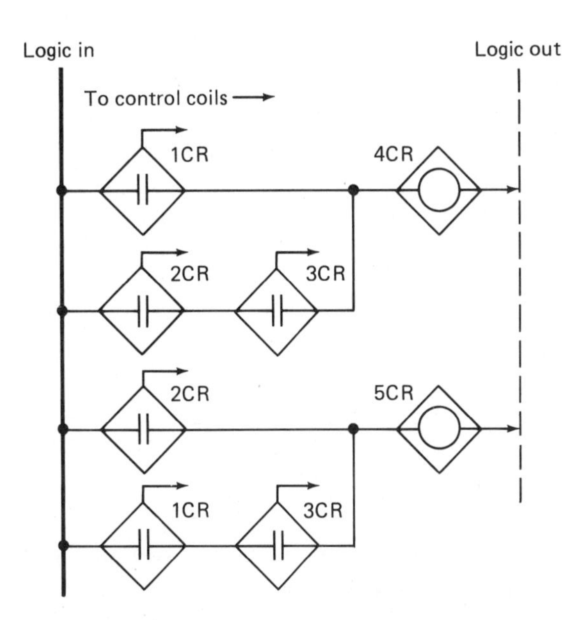

Fig. 15-19 LSL diagram showing how logic of Fig. 15-16 is modified to give unidirectional logic flow. *(Cutler-Hammer Inc.)*

situations where it can be applied. And even in those cases, the cost of the extra contacts required for unidirectional logic is trivial and the new diagram is usually easier to read and understand.

Figure 15-20 gives an example of a common relay design: a master relay is incorporated which controls through double-break contacts the entire source of power to the remaining portion of the relay control. The least expensive manner of implementing this with LSL is simply to control the primary power (110 V ac) to the two LSL power supplies, as shown in Fig. 15-21 on the next page, for when these power supplies are not energized, the LSL system is "dead" not only in a logic sense but also in a power sense.

15-16 MASTER CONTROL FOR LSL LOGIC SYSTEMS

For some reason it may be desired to obtain the master control through a power disconnect. This can be done

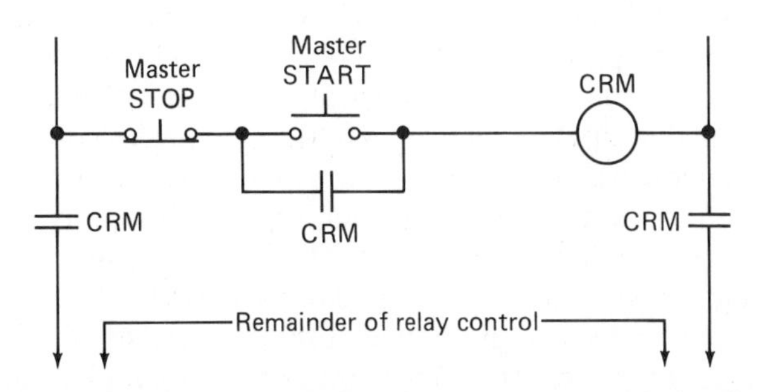

Fig. 15-20 Example of a master relay control circuit.

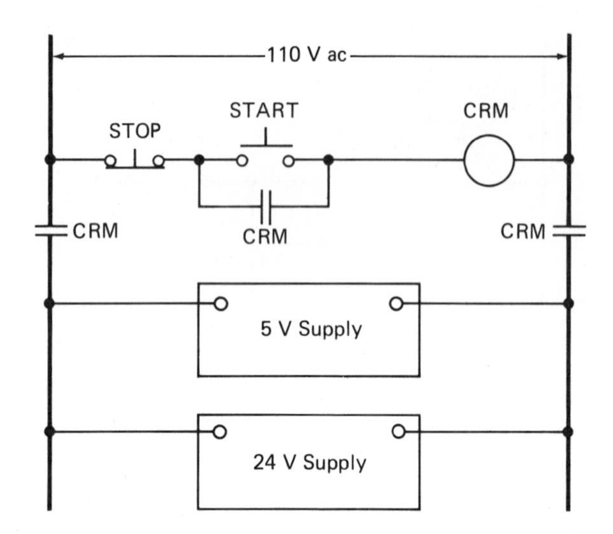

Fig. 15-21 Alternate method of controlling an LSL control from a master relay.

by incorporating a single switch contact element, controlled by a CRM contact, and energizing in common all those elements which would otherwise be connected to the left line, or logic common. However, this will usually pose a problem: The number of elements required to be "driven" will usually be far beyond the capacity of a single switch contact element. If it is controlling the LSL power supplies, the master relay circuit must be left as a power circuit, of course. Even if the master control is through logic, the master circuit is best left on a power basis because it usually controls motor starters and other power devices as well as the master relay.

The solution of the problem is through the use of logic repeaters, which are nothing but amplifiers inserted periodically into the common line driving the element inputs. There are several ways of obtaining repeaters from LSL elements; the least expensive and least complicated is to use an N.C. contact element left unenergized (the contact is, in effect, "permanently" closed).

Figure 15-22 shows the procedure. The input of the switch contact element is connected to logic common; the output of the element then drives the inputs of 11 elements plus the input of an N.C. contact "repeater." The output of the repeater drives 11 more elements plus the input of another repeater, etc. The last repeater drives 12 elements.

As seen, the contact symbol is modified slightly by drawing a coil symbol (a circle) within the diamond. This is done to mark the symbol as representing a repeater and also to indicate that the action is similar to that of repeater relays, relays introduced not for logical reasons but simply to increase contact capacity.

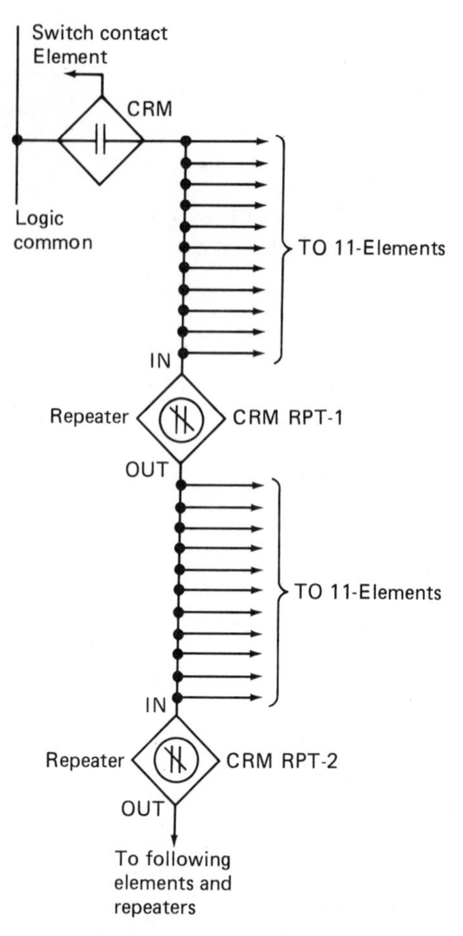

Fig. 15-22 LSL diagram showing how repeaters are used to extend capability of a switch contact element for supplying a large number of inputs to elements. *(Cutler-Hammer Inc.)*

It should be kept in mind that since contact elements are energized in pairs, N.C. contacts must be committed in pairs as repeaters.

In the previous example, the inputs of a large number of contact elements were in parallel. More rarely occurring is the case when the outputs of a large number of contact elements are paralleled. Ordinarily, this is limited to 12 paralleled contacts, but repeaters can be used here as well to obtain additional capacity. However, it must be done in such a way as not to invalidate the logic implied in the parallel connection, which is a logic OR operation.

Figure 15-23 shows the recommended method. The paralleled elements are divided into groups of a maximum of 11 each, the output of each group is connected to the input of a repeater, and the outputs of the repeaters are then paralleled to give the final logic output.

Occasionally, it is desired to energize a large number of contacts from a single relay. A single LSL coil element can control up to 12 contacts; if more capacity is required, repeaters can be inserted as in the previous examples.

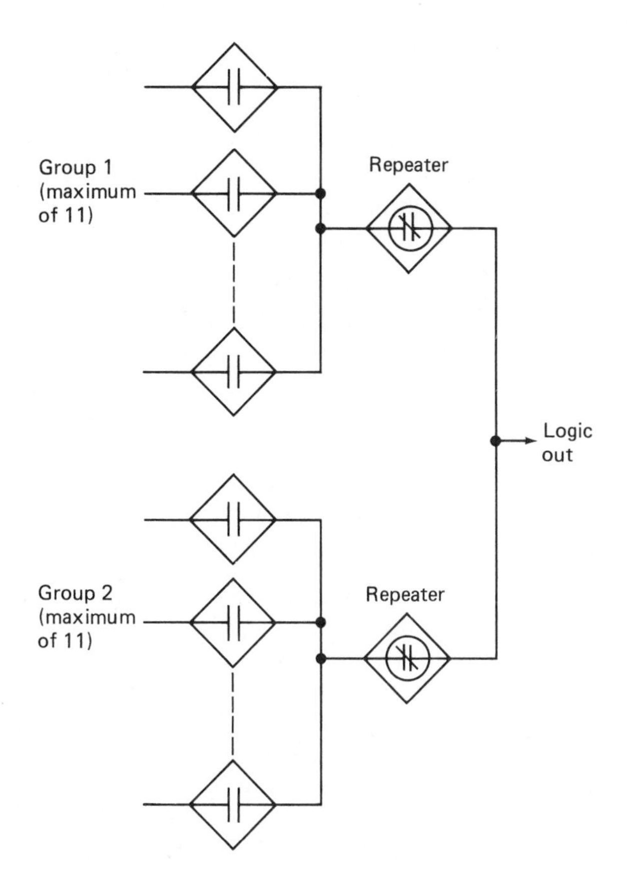

Fig. 15-23 Outputs of a large number of contact elements are paralleled through the use of repeaters.

Figure 15-24 shows how the method is presented on the diagram. The coil element can energize 12 contacts, the second repeater an additional 12, etc.

It should be emphasized that each connection from a coil output to a contact board controls two contacts. Therefore, there should be a maximum of six connections from a coil or repeater to contact control terminals.

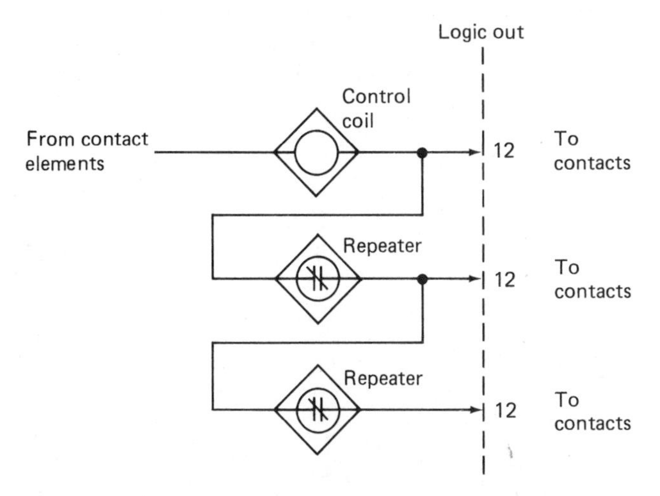

Fig. 15-24 Example of the use of N.C. contact repeaters to expand the capacity of a control coil element.

15-17 IMPLEMENTING INTERNAL SEQUENCING WITH LSL

One of the most formidable problems in relay design is to establish an internal sequencing action—one controlled by internal control relays rather than external devices—without causing logic "races." An example is given in Fig. 15-25, showing a circuit intended to be a three-station ring circuit which, once started, oscillates around the circuit with a single relay energized at a given time.

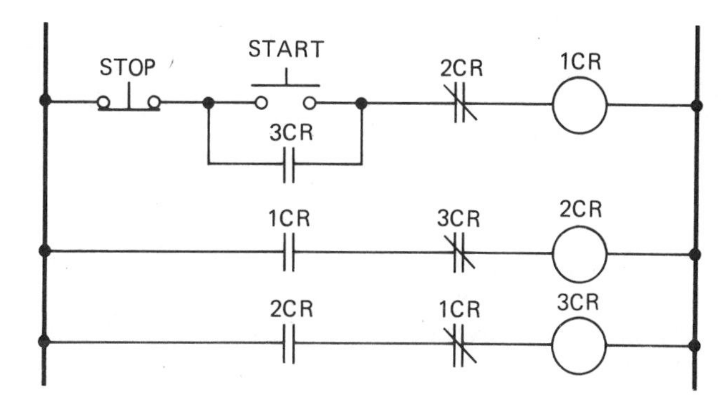

Fig. 15-25 Example of a three-station ring-sequencing circuit. The operation of the circuit with relays is highly uncertain.

Not only are the operating times of this circuit undeterminable with the relays, but also it is highly uncertain whether the circuit will operate at all, because correct action is completely dependent upon the relative times required for the contacts to open and close.

In contrast, the system implemented with LSL elements as shown in Fig. 15-26 on the next page not only will provide the desired ring oscillator action but the action can be precisely predicted as a sequence of steps, with each step controlled by an OPERATE pulse from the master control board.

The steps defining the sequence might be regarded as a succession of snapshots, where each snapshot is the logic input to all the coil elements at the instant of the OPERATE pulse. The pulse forces the elements to respond to the logic and the changed outputs of the elements create a new snapshot to define the next stop in the sequence, etc.

Now, if the START pushbutton is released at any time during the sequence, the action of step 2 above, where both 1CR and 2CR become energized simultaneously, will be eliminated and the circuit will oscillate around the ring, with a single coil energized at a given step.

Not only is the exact sequence predictable but also the frequency of sequencing is set by the master control

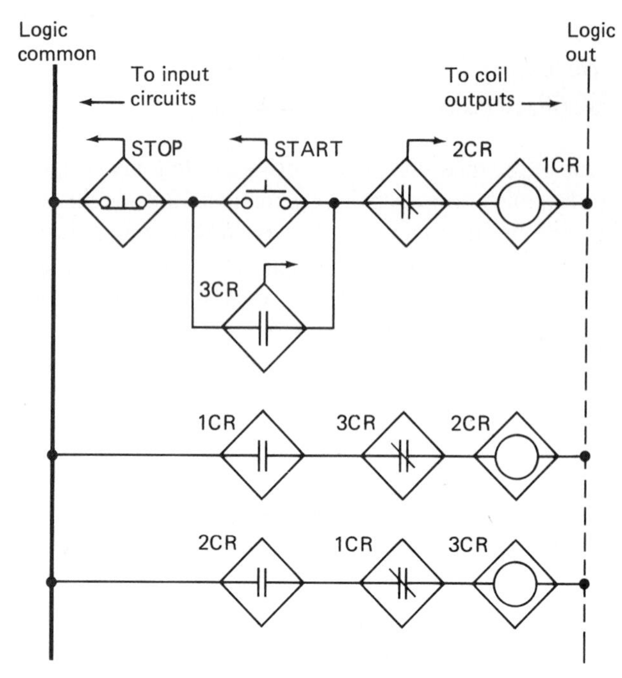

Fig. 15-26 Implementation of the diagram of Fig. 15-25 with LSL elements. The operation of the control is precisely predictable. *(Cutler-Hammer Inc.)*

board at 250 Hz (cycles per second). This means that there will be a uniform 4 ms between steps.

15-18 INPUT CIRCUITS FOR LSL LOGIC SYSTEMS

The input circuits consist first of such external switch devices as pushbuttons, limit switches, and pressure switches. These are connected in common to a source of power, either 24 V dc or 110 V ac. If 24 V dc is used, the switches are then connected directly to their associated switch contact elements in the LSL logic through 24-V input panel boards; if 110 V ac is used, the connection is through signal converters mounted on a 110-V ac input panel board.

Figure 15-27 shows the input circuits for the motor start-stop control—assuming that 24 V is used to energize the three switch devices. The example illustrates the following points:

- The input circuits are placed on the diagram at the left of the logic. Each switch device is located as near as possible at a position directly to the left of its associated switch contact element in the logic.
- The switch is shown connected to a terminal of the input panel board. This terminal is given a number and the same number is placed on the control input of the switch contact symbol. (The assignment of terminal numbers will be considered later.)

Figure 15-28 gives another example showing the input circuits for the circuit of Fig. 15-17. In this case 110 V ac has been chosen to power the pushbuttons and contact. The diagram is much the same as those for the 24-V dc input circuits of Fig. 15-27, except that signal converters are indicated between the input terminals and the connection to the logic portion of the system.

In choosing between 24 V dc and 110 V ac operation of the external switch devices, it should be kept in mind that 24-V operation is considerably less expensive (since no signal converters are required) and one-half the panel space is required for the input panel boards. As for the relative reliability of operation, extensive tests have been run in the Cutler-Hammer Development Department to assess the relative reliability of the voltages applied to high-quality limit switches in the presence of such contaminants as cutting oils, coolants, and paper dust. It has been found that 24-V operation is as reliable as 110-V operation, and in some cases more reliable, for cutting oils and coolants. Somewhat better operation is obtained at 110 V for the case of paper dust contaminants, although in either case the paper dust had to be deliberately introduced

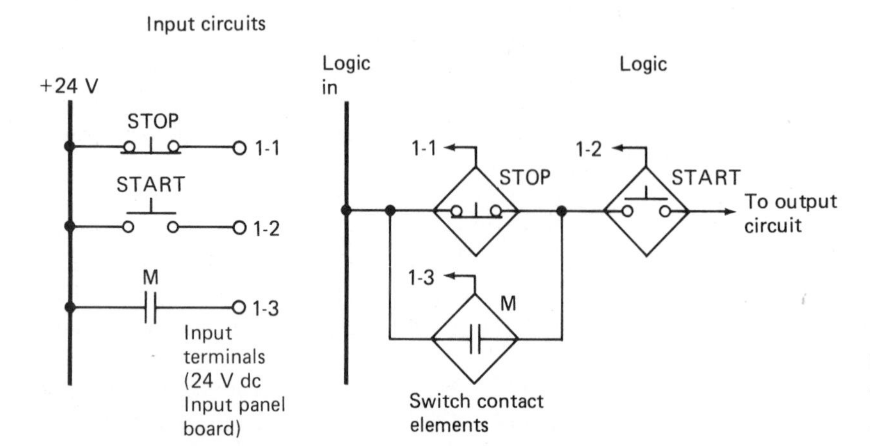

Fig. 15-27 Example showing how the input circuits are placed on the left of the LSL diagram.

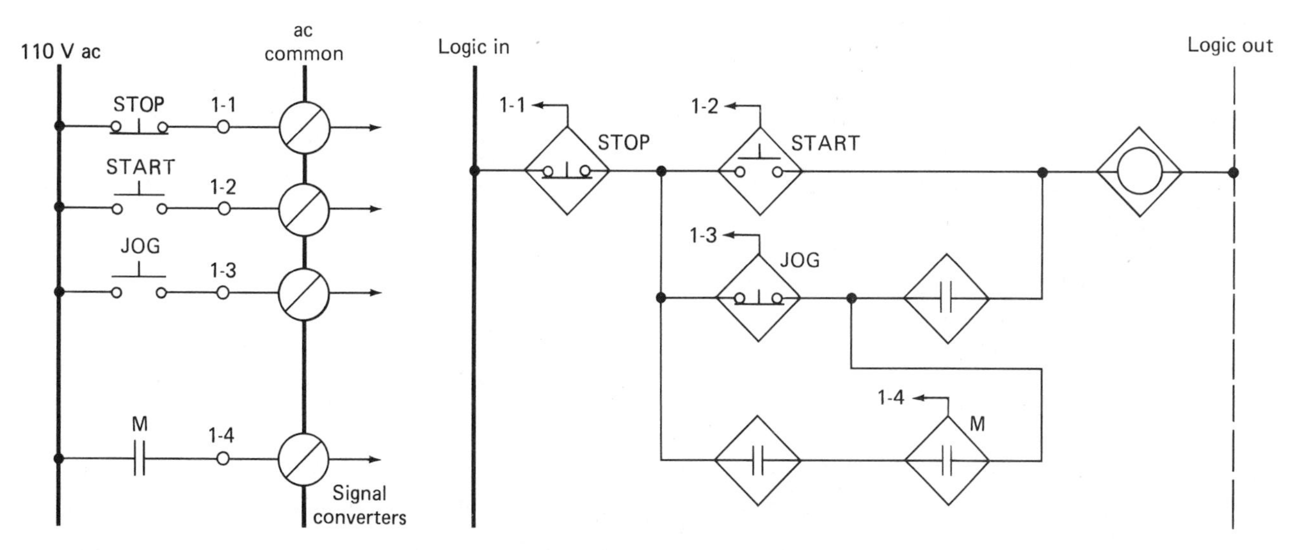

Fig. 15-28 An example of input circuits required for the LSL diagram. *(Cutler-Hammer Inc.)*

into the switch to obtain faulty operation. The results of tests indicate that, in general, highly reliable operation at 24 V dc can be obtained if high-quality sealed-switch devices are used.

15-19 OUTPUT CIRCUITS FOR LSL LOGIC SYSTEMS

An LSL output circuit is required for each external device or contact closure to be controlled from the LSL logic. The circuit consists primarily of an LSL driver element, which accepts a logic signal from an LSL element and generates a 24-V control signal. The remainder of the circuit consists of the device controlled by the 24-V signal. This might be indicator lamps operating at 24 V or powered relay coils to obtain ultimately 110-V ac control of such devices as solenoids through contacts from the relay.

If the control application requires a large number of indicator lamps, direct control of the lamps at 24-V operation should be considered. The primary advantage is that since no output signal converter (from 24 V dc to 110 V ac) is required, there is a significant savings in cost. Further, approximately one-third the panel space is required for a 24-V output panel board compared to a 110-V output panel board.

The following points concern the placing of the output circuit on the LSL control diagram:

- The output circuit is put at the right of the logic portion of the diagram and a direction-connection shown from the energizing LSL element to the driver element.
- The driver element is connected to a terminal of the 24-V output board. (The driver element is connected to a terminal only for direct control. For 110-V ac operation, to be discussed next, a signal converter in the form of a powered relay is placed between the driver element and the output terminal.)
- A field wire number may be placed on the diagram between the terminal and an indicator lamp. The lamp (and all other lamps) are connected in common to 24 V dc.

Figure 15-29 shows the output circuit required to control the motor starter M of Fig. 15-14. The output circuit now includes the coil of a powered relay (part of the 110-V output board). No terminal symbol is shown since the output terminals are connected to a contact of the powered relay and this is a power circuit placed on a separate drawing.

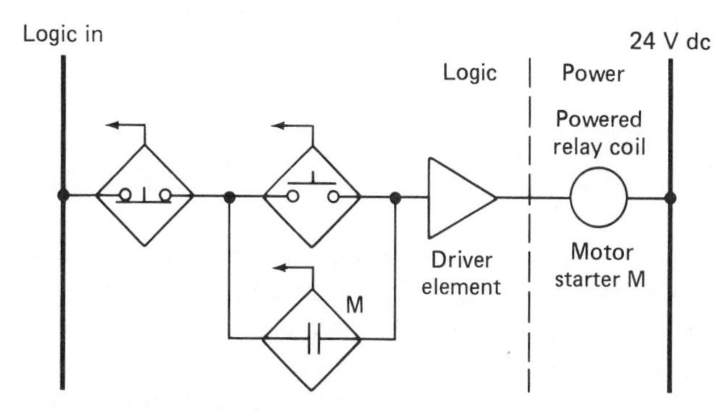

Fig. 15-29 Example showing output circuit for the LSL diagram.

Figure 15-30 is an example of a common relay circuit; the control relay CR controls two contacts to give a "double-break" solenoid control.

Figure 15-31 shows the output circuits required to implement the control with LSL elements. Note that two driver elements and powered relay coils are required, one for each contact controlled. The contacts supplied by the control relay on the relay control are now replaced by contacts of the powered relay. Both coil and contact of the powered relay are mounted on the 110-V output panel board.

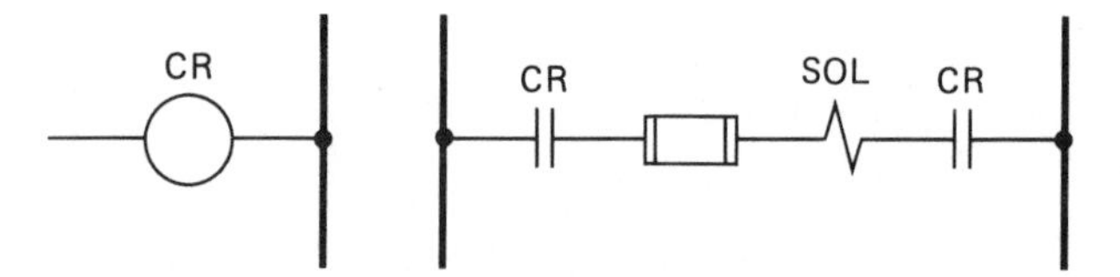

Fig. 15-30 Example of double-break control of a solenoid.

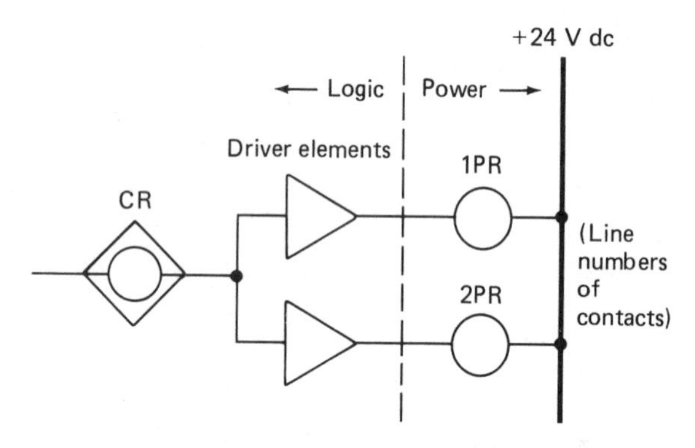

Fig. 15-31 Diagram showing how two output circuits are used to control a double-break power circuit. The designation PR indicates a powered relay.

The contact portion of the powered relay controls 110-V ac power and is accordingly placed on the power portion of the diagrams. As is standard relay diagram practice, a line number identifying the location of the contact can be placed beside the powered coil symbol.

15-20 POWER CIRCUITS IN LSL LOGIC SYSTEMS

The power circuits are designed and placed on the system diagram in a conventional manner. Figure 15-32 shows two power circuits controlled by powered relay contacts. In Fig. 15-32*a* single contact controls a motor starter coil. In Fig. 15-32*b* two contacts control a solenoid. Each output circuit is enclosed in dashed lines. Note that each circuit includes a fuse.

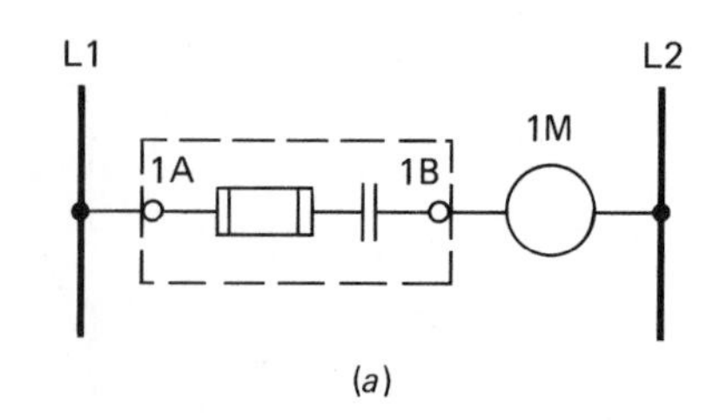

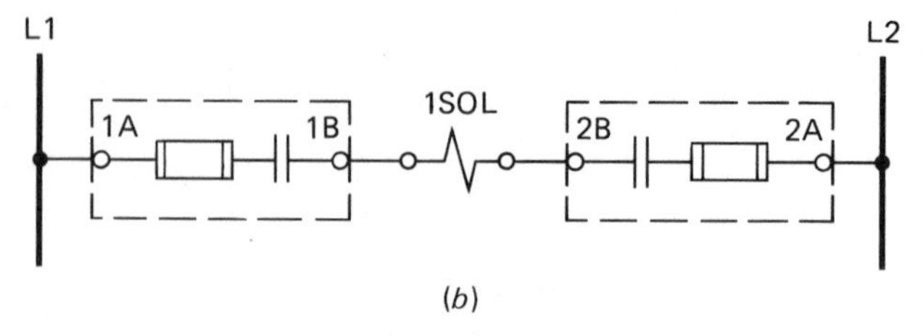

Fig. 15-32 Diagram showing how an output circuit of 110 V ac powered relay panel board controls. *(a)* A motor starter coil through a single contact and *(b)* a solenoid through two contacts.

The LSL monitoring system requires that certain connections be made from the monitored boards (control coil, latching coil, timer coil, switch contact, and driver boards) to the monitor driver board. Logic cards, referred to as *buckets,* are illustrated in Fig. 15-33. The rear wiring of the buckets is shown in this figure.

Examples of bucket pin numbers are also illustrated in Fig. 15-34 on page 262.

15-21 SUMMARY OF LSL LOGIC APPLICATIONS

The Cutler-Hammer static system is an English logic system which is applied by means of plug-in logic elements. The logic elements are packaged four to a board and have individual indication lights mounted on the front of the board. Standard mounting provides for 20 boards in each bucket. Connections are made by means of taper pins which must be installed with the use of a special insertion tool.

The heart of this system is the basic transistor switch, operating from a 10-V dc power supply. The ON signal for this system is +10 V dc and the OFF signal is 0 V dc. Recommended voltages of contact-type sensing devices are 48 V dc or 115 V ac.

When the company makes the control circuit drawing for a factory-wired control system, the symbols are coded with numbers to aid in servicing the equipment.

Fanout limitations must be calculated on a basis of a maximum of an 80-mA load and a minimum output voltage of 5 V.

Normal installations will require at least two power supplies, one 10-V supply for logic elements and one

Fig. 15-33 Interconnection and bucket wiring.

48-V supply for signal converters. A third supply of 24 V will be needed if any 24-V power ANDs are used. Loading of these power supplies should be carefully calculated as part of the reliance and safety factors in this system.

State-indication lights for normal troubles are a standard feature of every input and logic element in this system and provide visual checking of any part of the circuit without other equipment.

15-22 SQUARE D NORPAK LOGIC CONTROL SYSTEMS

Earlier in this chapter we saw how NAND logic elements could be used to design logic circuits that used logic gates such as ANDs, ORs, NOTs, and MEMORYs. We made these logic gates with NAND logic circuits. We saw how we could simplify the circuits by using the theorem, or postulate, $\overline{\overline{A}} = A$. Many redundant circuits could be eliminated by the simplification rules. The simplified circuit did not always resemble the original circuit but accomplished the same logic. The same thing we did with NANDs we can also do with NORs. The NOR circuit equivalents for ANDs, ORs, NOTs, and MEMORY" are somewhat different from those in NAND circuits. Let's take a look and see how they differ.

Square D Electric Company manufactures a logic system called Norpak. The older NOR system used transistors and discrete electrical components to make up the NOR circuit. The circuits came in "packages." The newer NOR system, manufactured with as many as 20 circuits per package, uses racks and cards. The new Norpak logic system uses solid-state ICs as the decision-making elements of the NOR logic cards.

The following description of Norpak control systems is adapted from Square D literature.

The Square D Norpak system is built around NOR logic. The NOR logic utilizes one circuit only to make all other logic elements. The NOR circuits originally used in the Square D Norpak logic control systems used the transistor as the central device. Figure 15-35, on page 263, illustrates the three-input NOR symbols with their truth table.

The NOR logic symbol used in Fig. 15-35 is for the transistorized NOR logic circuit. The truth table for the NOR indicates that when all the inputs are a logic 0, the output is a logic 1. When any input or combination of inputs is a logic 1, the output is logic 0. The logic of the NOR with all the inputs tied together or using one input only would be the equivalent logic of the NOT logic element. Figure 15-36 on page 263 illustrates the NOT logic function and the relay equivalent.

A logic function such as the AND function can be developed by using the NORs, as illustrated in Fig. 15-37 on page 263. Notice that it requires a logic 1 on each input to give a logic 1 output. A three-input AND would require four logic NOR circuits.

A logic circuit giving a two-input OR logic function is illustrated in Fig. 15-38 on page 263. Any input to the circuit will give an output. If the first NOR gate had

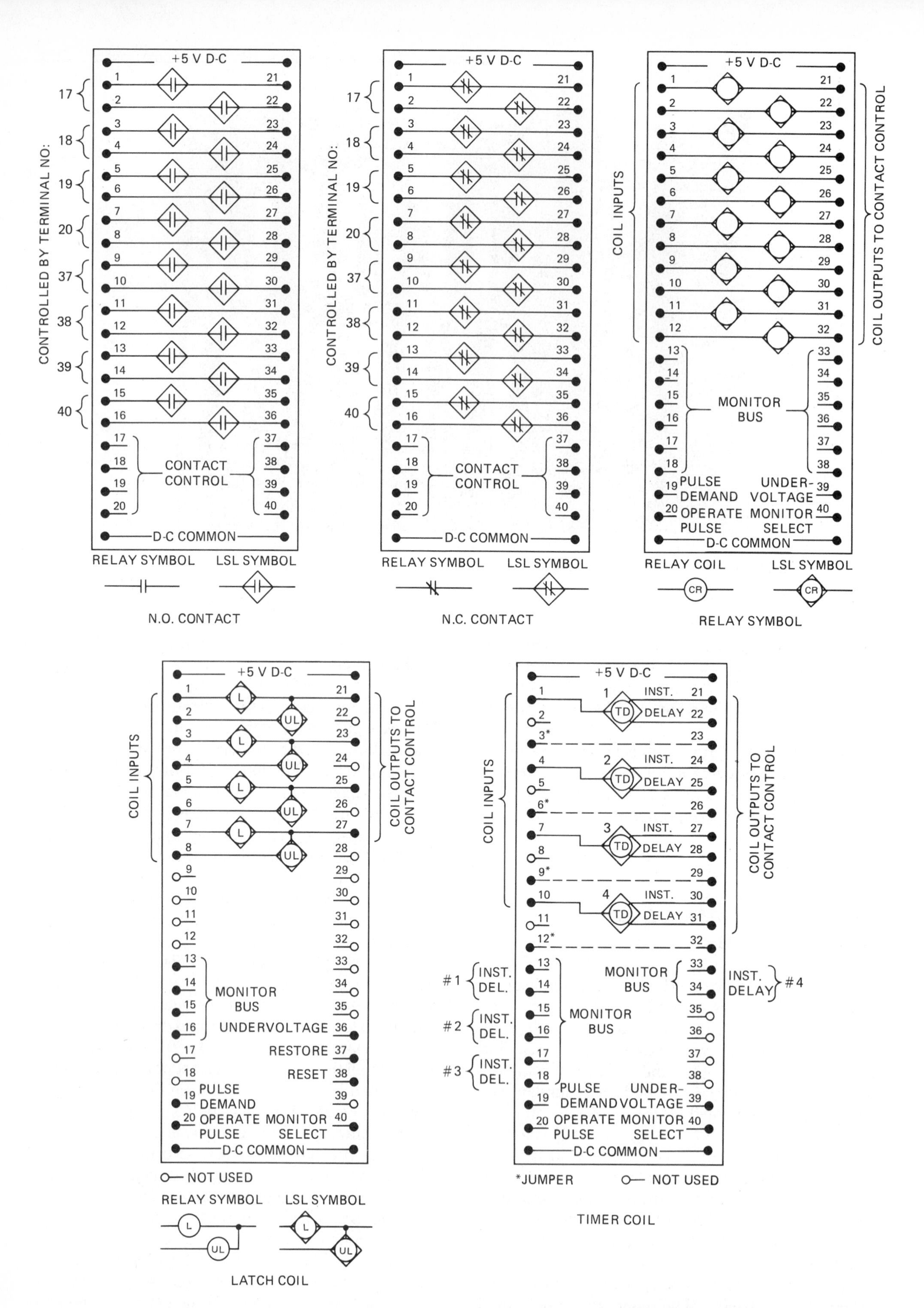

Fig. 15-34 Bucket pin numbers.

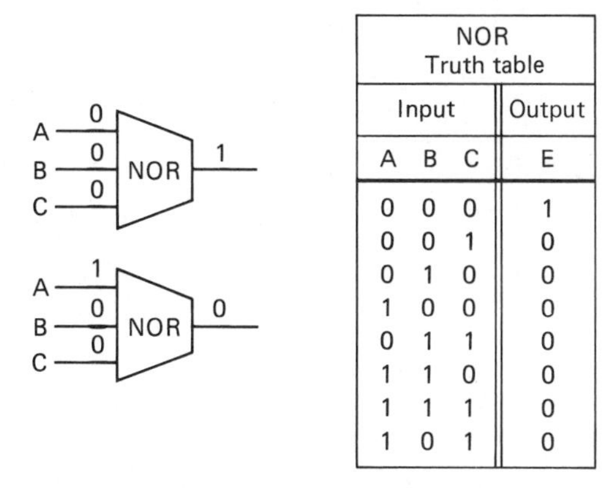

NOR Truth table			
Input			Output
A	B	C	E
0	0	0	1
0	0	1	0
0	1	0	0
1	0	0	0
0	1	1	0
1	1	0	0
1	1	1	0
1	0	1	0

Fig. 15-35 NOR logic symbol. *(Square D Co.)*

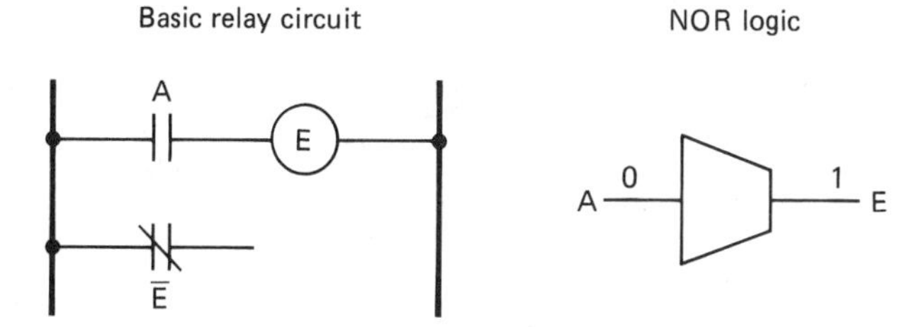

Fig. 15-36 NOR-logic NOT function. *(Square D Co.)*

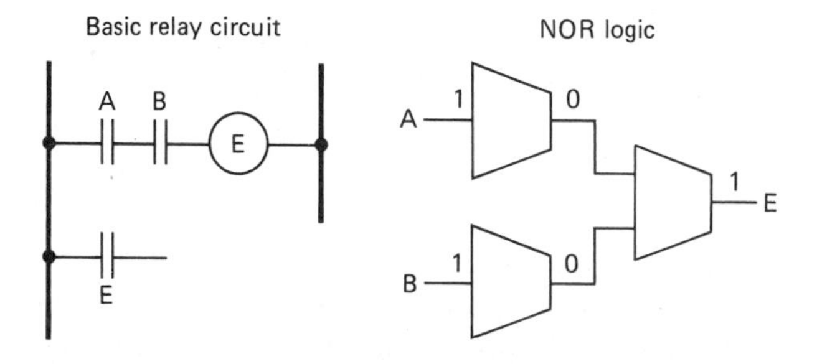

Fig. 15-37 NOR-logic AND function. *(Square D Co.)*

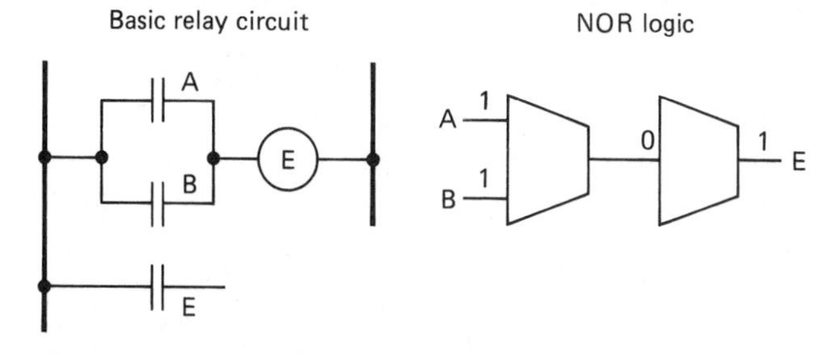

Fig. 15-38 NOR-logic OR function. *(Square D Co.)*

three inputs, the circuit would become a three-input OR. Figure 15-39 illustrates OR logic obtained by diodes in parallel.

The circuit in Fig. 15-40*a* illustrates a relay that provides low voltage protection as well as logic 1 and logic 0 output. The equivalent logic circuit (Fig. 15-40*b*) requires only two NOR logic elements. A momentary input at pushbutton A will provide an "on" output (logic 1) at E and a logic 0 at $\overline{E}$. A momentary input at pushbutton B or a logic 1 signal at the off signal line will cause the off-return memory to reverse its outputs. Output E will deliver an "off" output (logic 0) and the output $\overline{E}$ will give a logic 1 output. The "off" signal comes from the Norpak power supply. When power fails, a reset signal would be applied at NOR circuit Y, causing the off-return memory to return to its "off" conditions.

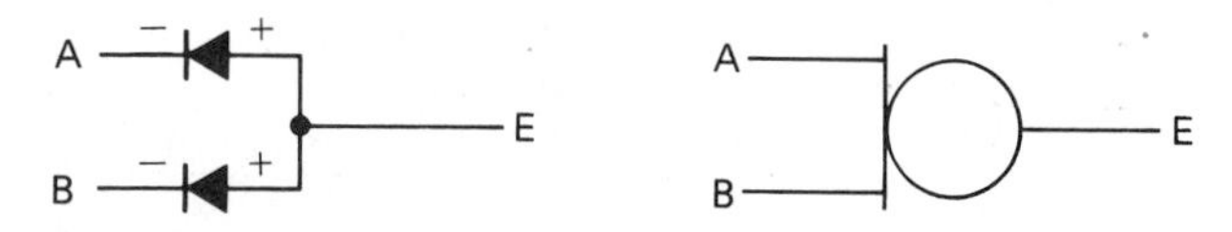

Fig. 15-39 Diode OR function. *(Square D Co.)*

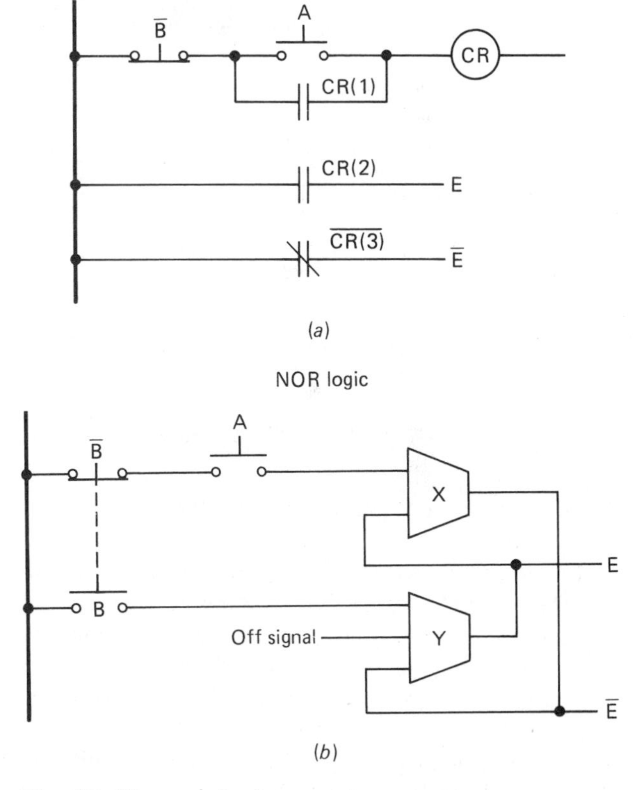

Fig. 15-40 NOR-logic OFF-return MEMORY. *(Square D Co.)* *(a)* Basic relay circuit *(b)* NOR logic.

When a memory must remember and return to its condition (on or off) before a power failure, a retentive-memory logic circuit is used. As indicated in Fig. 15-41 on the next page, this circuit uses the basic latching relay to provide the logic function of a permanent memory. If both inputs are present to the retentive memory at the same time, it will provide an overriding "off" signal. Output E will be logic 0 and output $\overline{E}$ will be a logic 1.

Square D Norpak has two types of time delays. They have the on delay or time delay after energizing and the off delay or time delay after de-energizing.

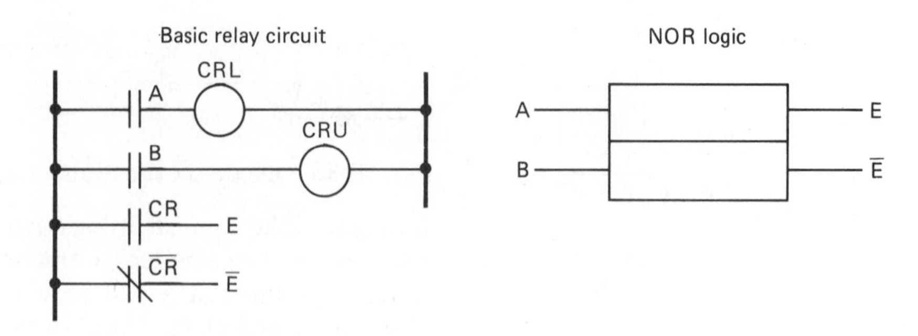

Fig. 15-41 NOR-logic retentive MEMORY. *(Square D Co.)*

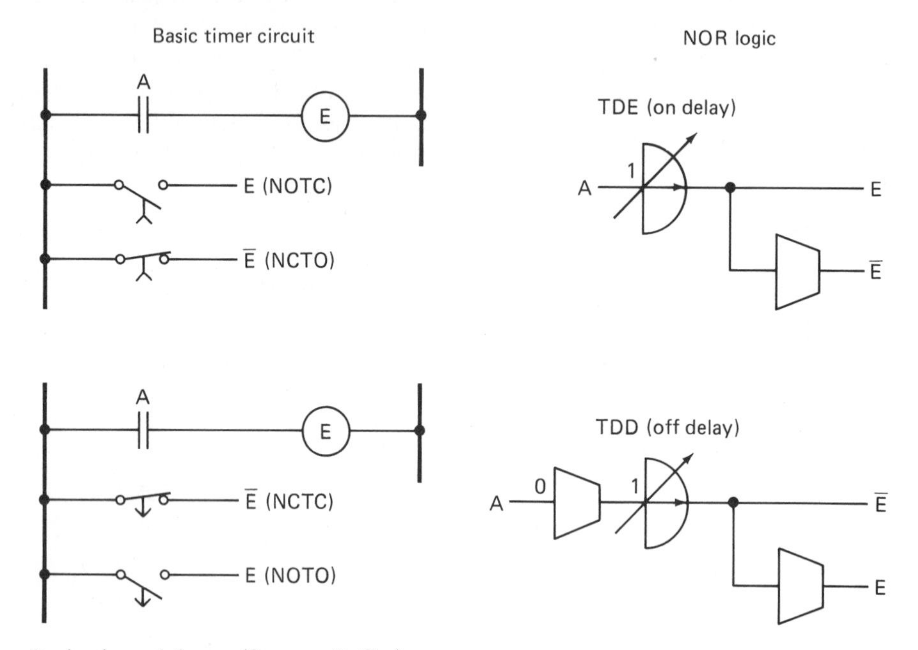

Fig. 15-42 NOR-logic time delay. *(Square D Co.)*

Notice Fig. 15-42 has the equivalent relay delay circuits and the Norpak logic circuits.

Figure 15-43 is a table of relay-equivalent NOR circuits and their NEMA symbols. This table is very useful when designing logic circuits from relay ladder diagrams or when converting logic circuits using the NEMA symbols to NOR logic or vice versa. Figure 15-44, on page 266, indicates NOR logic input expansion.

Figure 15-45*a*, on page 266, is a relay circuit that provides control of two solenoid valves. Figures 15-45*b*, *c*, *d*, and *e*, illustrate how the CRl relay circuit is converted and Fig. 15-45*f* illustrates how the CR1 relay circuit is converted to NOR logic. Figures 15-46*a*, *b*, *c*, and *d* on page 267 show the conversion of CR2, CR3, SV1, and SV2 to NOR-equivalent circuits. NOR conversion circuits, when combined, form redundancies that can be eliminated. Two NOTS or two one-input NORS in series are equivalent to NOTing an input twice. When an input is "NOTed" twice, the NOTS (2 one-input NORS in series) can be dropped and replaced with a piece of wire. This is because of the boolean algebra rule $\overline{\overline{X}} = X$.

Figure 15-47, on page 267, illustrates the combining of the NOR circuits of Fig. 15-46. This is the NOR circuit conversion of the electric relay circuit of Fig. 15-45. The NOR complement symbols are shown in Fig. 15-48 on page 267. Figure 15-49, on page 268, shows the final NOR logic development. Notice the simplification of the circuit. The circuit of Fig. 15-47 has 11 NOR gates and Fig. 15-49 has only 7. The simplification occurred by using NOR complementing.

Figure 15-50, on page 268, is the NOR equivalents of English logic symbols. Figure 15-51*a*, on page 269, is a logic circuit using English logic symbols. This logic circuit is an English logic conversion for the relay circuit of Fig. 15-45*a*. Figure 15-51*b* shows the conversion of the English logic circuit to NOR logic. Figure 15-52, on page 269, is the completed NOR logic conversion of the English logic circuit of Fig. 15-51*a*. When all redundancies are eliminated in a simplified NOR logic circuit, it becomes difficult to see where the ANDS, ORS, NOTS, and MEMORYS are. Converting a NOR logic circuit back to English is difficult.

If a conversion from a NOR logic circuit back to an English logic circuit is required, sometimes the simplest way to do it is by analyzing the physical operation of the circuit and by developing a word description of the

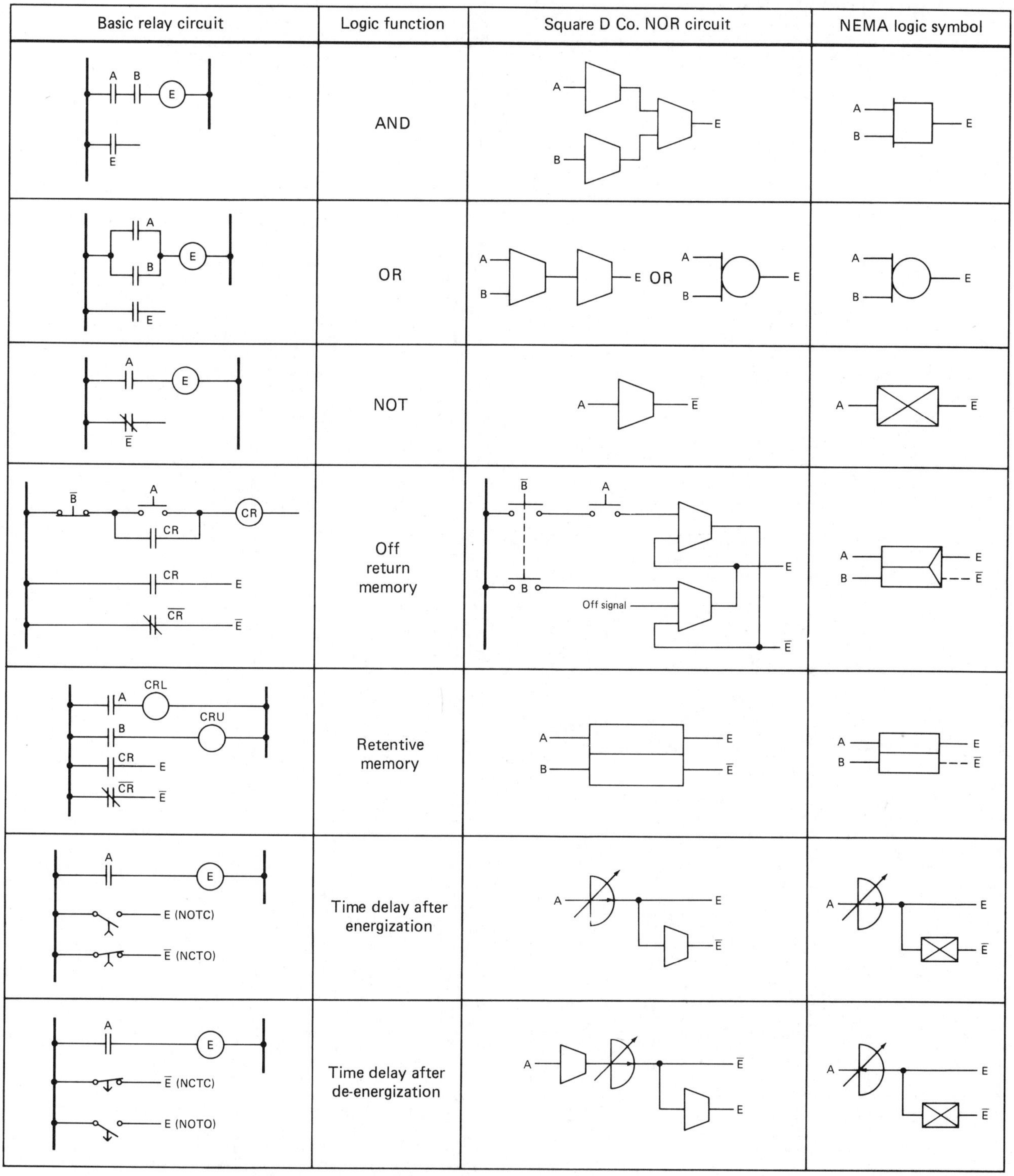

Basic relay circuit	Logic function	Square D Co. NOR circuit	NEMA logic symbol
A B E E	AND	A B E	A B E
A B E E	OR	A B E OR A B E	A B E
A E $\overline{E}$	NOT	A $\overline{E}$	A $\overline{E}$
$\overline{B}$ A CR CR CR E $\overline{CR}$ $\overline{E}$	Off return memory	$\overline{B}$ A B Off signal E $\overline{E}$	A B E $\overline{E}$
CRL A CRU B CR E $\overline{CR}$ $\overline{E}$	Retentive memory	A B E $\overline{E}$	A B E $\overline{E}$
A E E (NOTC) $\overline{E}$ (NCTO)	Time delay after energization	A E $\overline{E}$	A E $\overline{E}$
A E $\overline{E}$ (NCTC) E (NOTO)	Time delay after de-energization	A $\overline{E}$ E	A E $\overline{E}$

Fig. 15-43 Table of relay equivalents. *(Square D Co.)*

machine operation. The word description can be developed to an English logic diagram, which can then be developed into any circuit desired.

Figure 15-53, on page 270, illustrates some steps in the conversion of a word description of a planer table to a logic circuit. Each requirement of the machine is analyzed to develop the word description. For example, the planer is a machine that consists of a table which moves back and forth by means of a reversing motor. The momentary action of limit switches located at each end of the bed initiates the reversing contactor, which causes the table to oscillate back and forth. The protective controls desired are extreme operating limit switches on each end of the bed, motor overload protection, interlocking with the coolant supply, and minimum lubrication pressure.

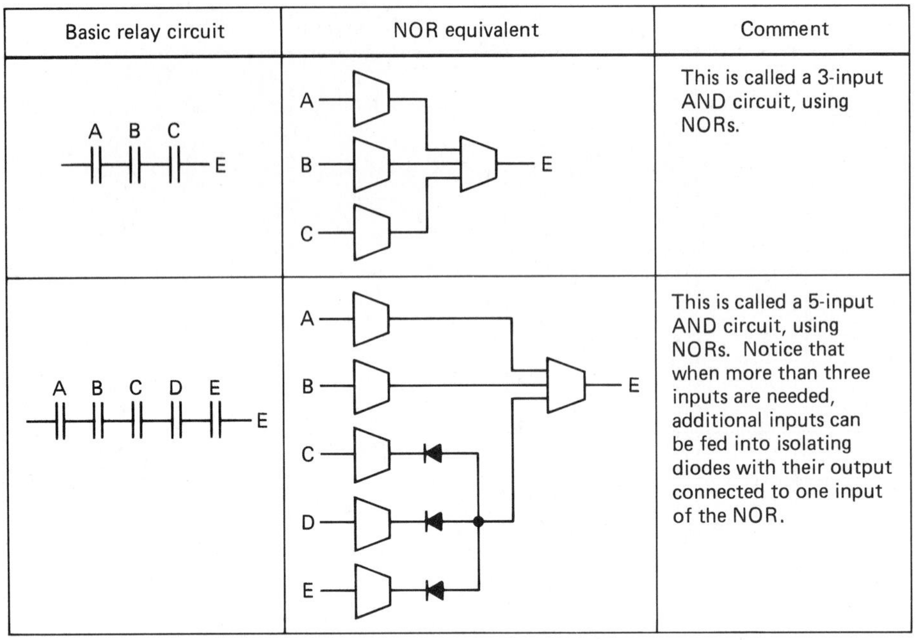

Basic relay circuit	NOR equivalent	Comment
A B C E	A B C E	This is called a 3-input AND circuit, using NORs.
A B C D E E	A B C D E E	This is called a 5-input AND circuit, using NORs. Notice that when more than three inputs are needed, additional inputs can be fed into isolating diodes with their output connected to one input of the NOR.

Fig. 15-44 NOR-logic input expansion. *(Square D Co.)*

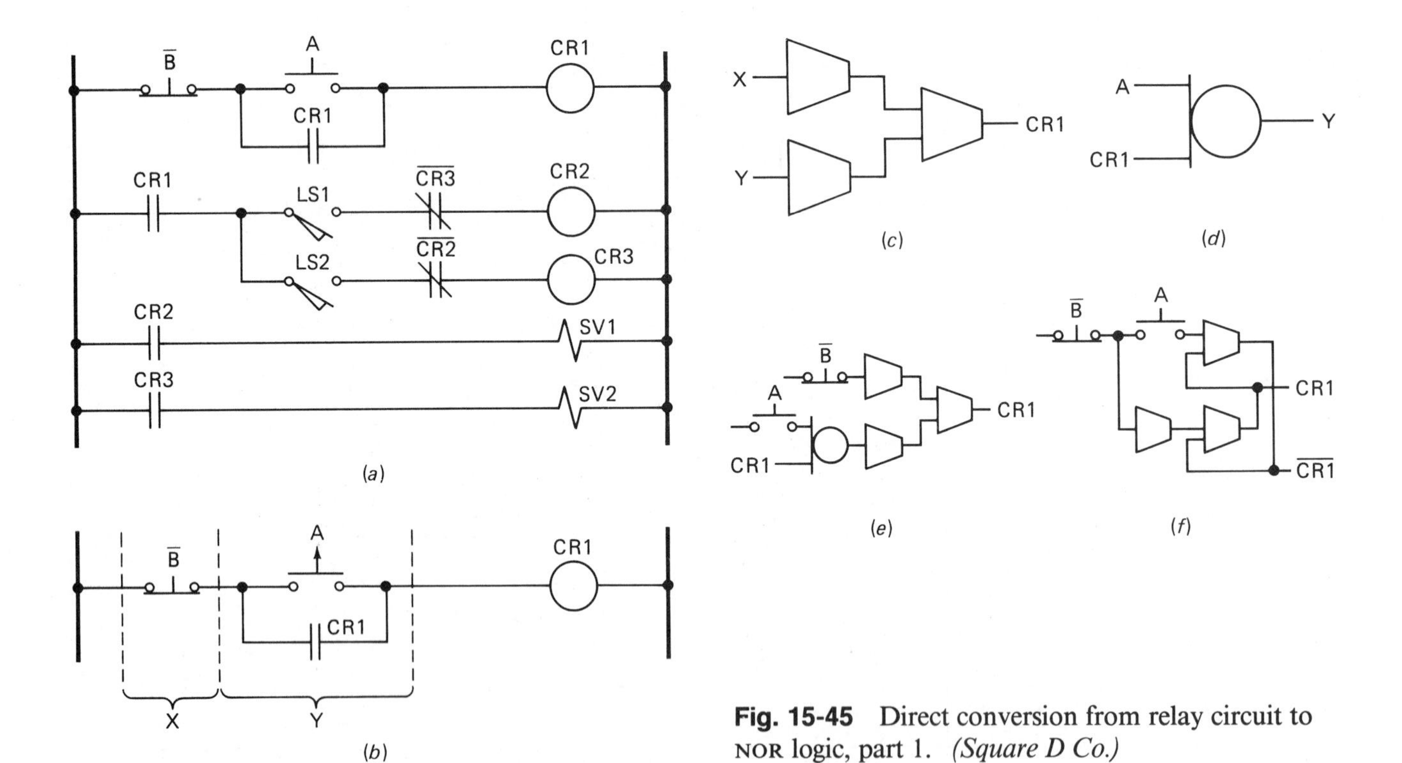

Fig. 15-45 Direct conversion from relay circuit to NOR logic, part 1. *(Square D Co.)*

What is required to provide a ready-to-run position? With minimum lubrication pressure present, the coolant motor operating, the drive-motor overload relays incorporated in the circuit, and the emergency switches normally closed, it is now desired that the machine either run continuously or jog.

We should pause at this point to visualize a table that oscillates back and forth; its movement is controlled by a limit switch at each end of the table, and emergency limit switches are mounted near the operating limit switch to prevent the table from traveling too far in the event the initial limit switch fails. Of equal importance is the fact that some type of logic must be performed to put the table into operation and also to stop the machine by either manual or automatic operation if predetermined conditions are not met.

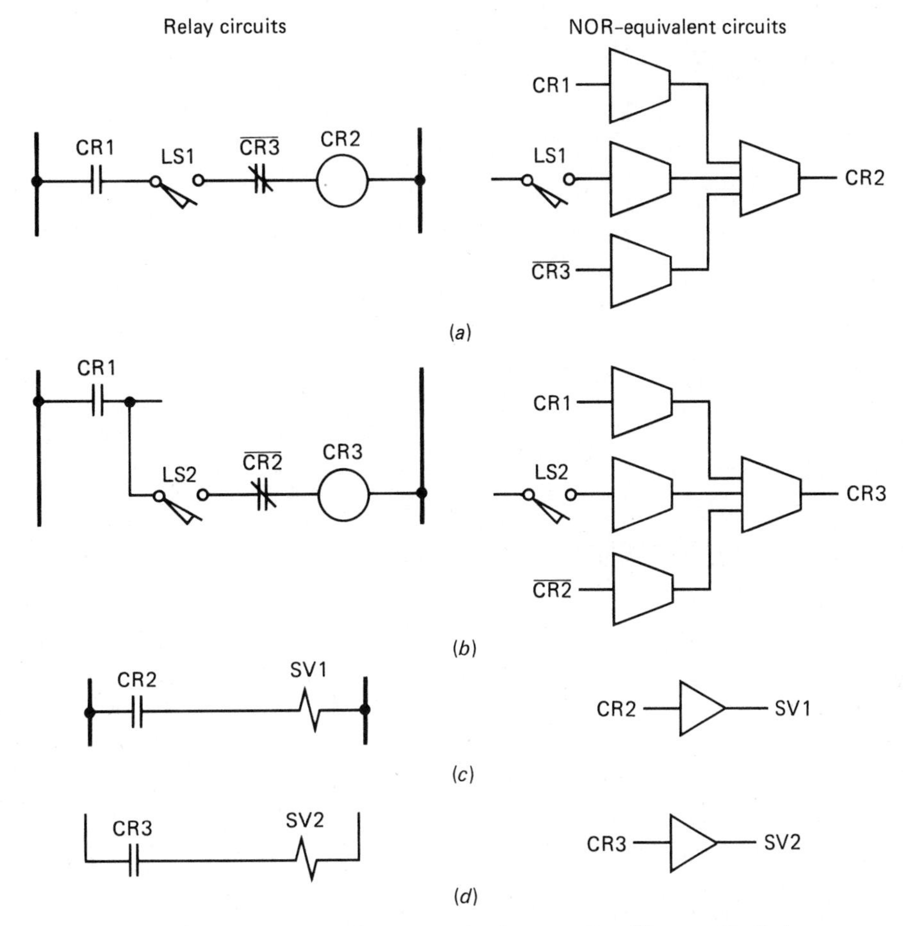

Fig. 15-46 Direct conversion from relay circuit to NOR logic, part 2. *(Square D Co.)*

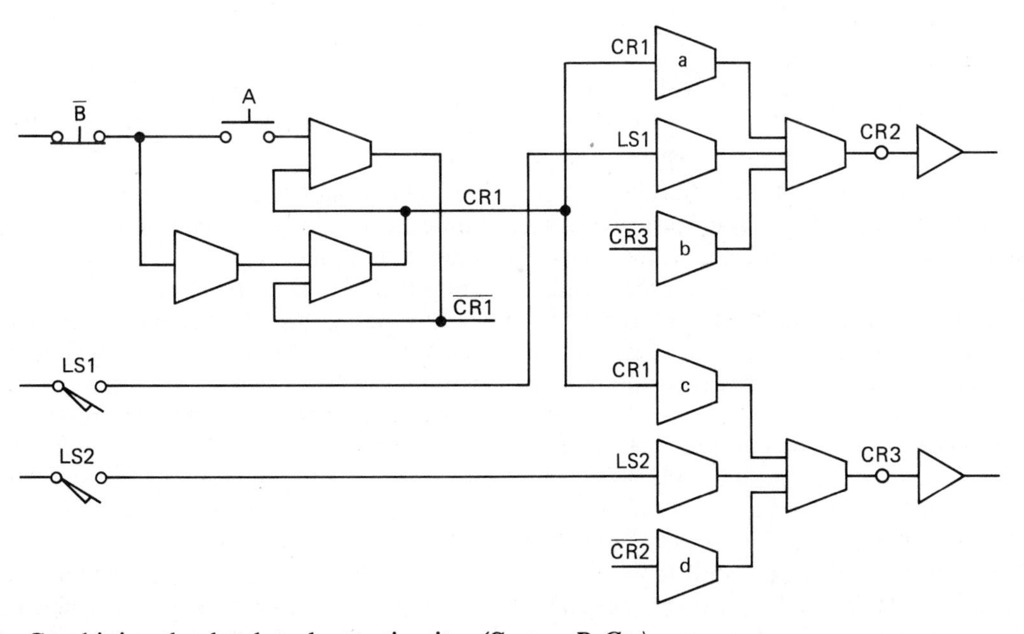

Fig. 15-47 Combining the developed NOR circuit. *(Square D Co.)*

0 — 1 $\overline{X}$ — X $\overline{CR3}$ — CR3

Fig. 15-48 NOR complements.

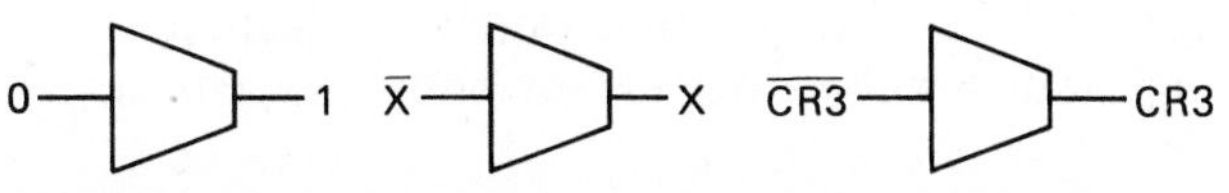

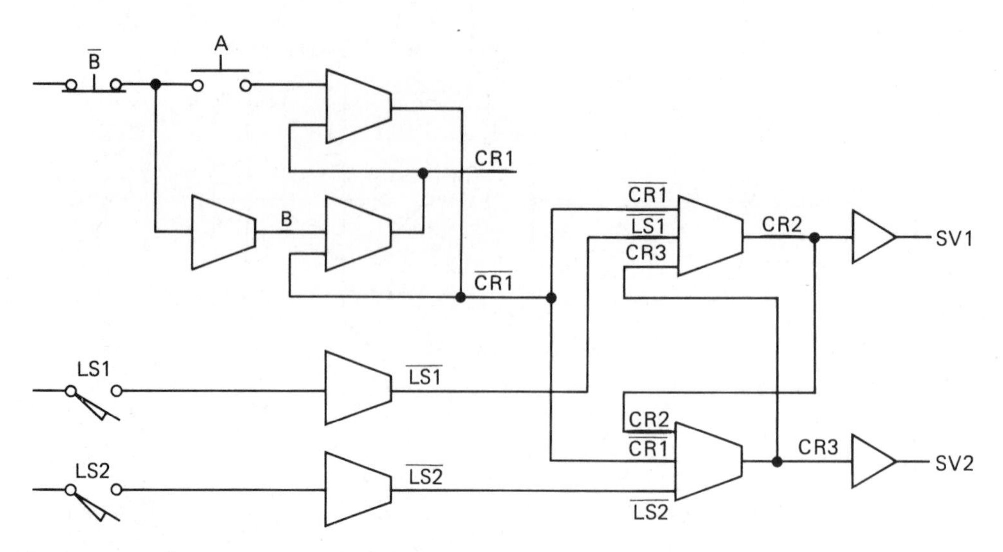

Fig. 15-49 Final NOR logic developments. *(Square D Co.)*

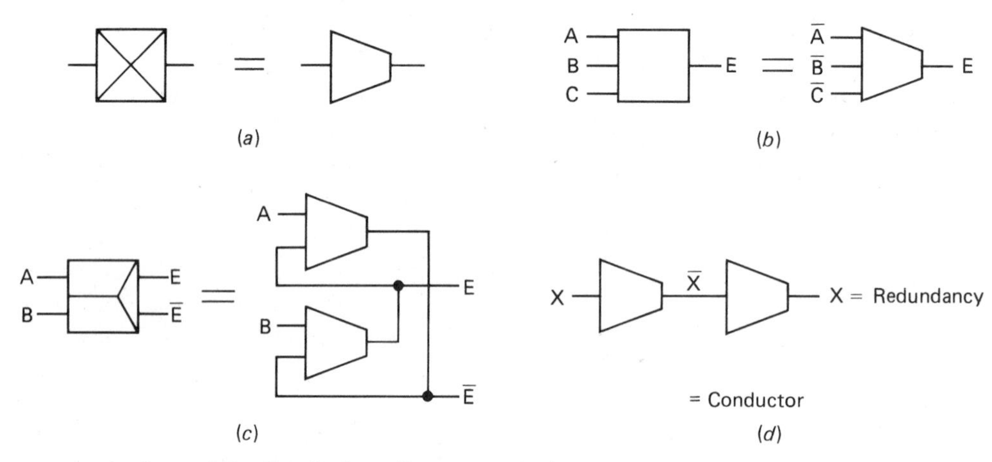

Fig. 15-50 NOR equivalent of English logic. *(Square D Co.)*

The ready-to-run circuit could be controlled by a six-input AND. In general, all forms of start-stop circuits use OFF-return MEMORY. In this case, the inputs to the AND circuit are coming from pilot devices and, therefore, the AND circuitry can be accomplished by merely connecting these signals in series. The complete ready-to-run circuit would appear as shown in Fig. 15-53*a*.

Consider the work cycle of the machine. When the machine is in the ready-to-run position, the operator presses either a right traverse or a left traverse button. For example, if the right traverse button were pressed, the table would move to the right until momentary engagement of the right limit switch. When the right operating limit switch is actuated, the right traverse contactor is de-energized and the left traverse contactor energized so that the table is driven to the left until it hits the left operating limit switch. Actuation of the left limit switch causes the reversing contactor to reverse the motor and drive the table to the right. In some applications it would be desirable to have the table stop at one end of its travel in order to blow off chips (i.e., a time delay is involved).

Examining this information reveals that the ready-to-run circuit and either the limit switches or the traverse buttons cause the machine to operate. It is a momentary action of the limit switch or momentary closing of the traverse buttons that causes action. Therefore, a memory circuit is involved. A decision must be made on whether an OFF-return MEMORY or retentive MEMORY should be used. The machine builder in this instance might insist on having the equipment start up where it left off after power failure, in which case a retentive MEMORY should be used. The work cycle part of our circuit is shown in Fig. 15-53*b*. The combined circuit is shown in Fig. 15-54 on page 270.

The Norpak circuit elements used in the past were made from transistor circuits, as illustrated in Fig.

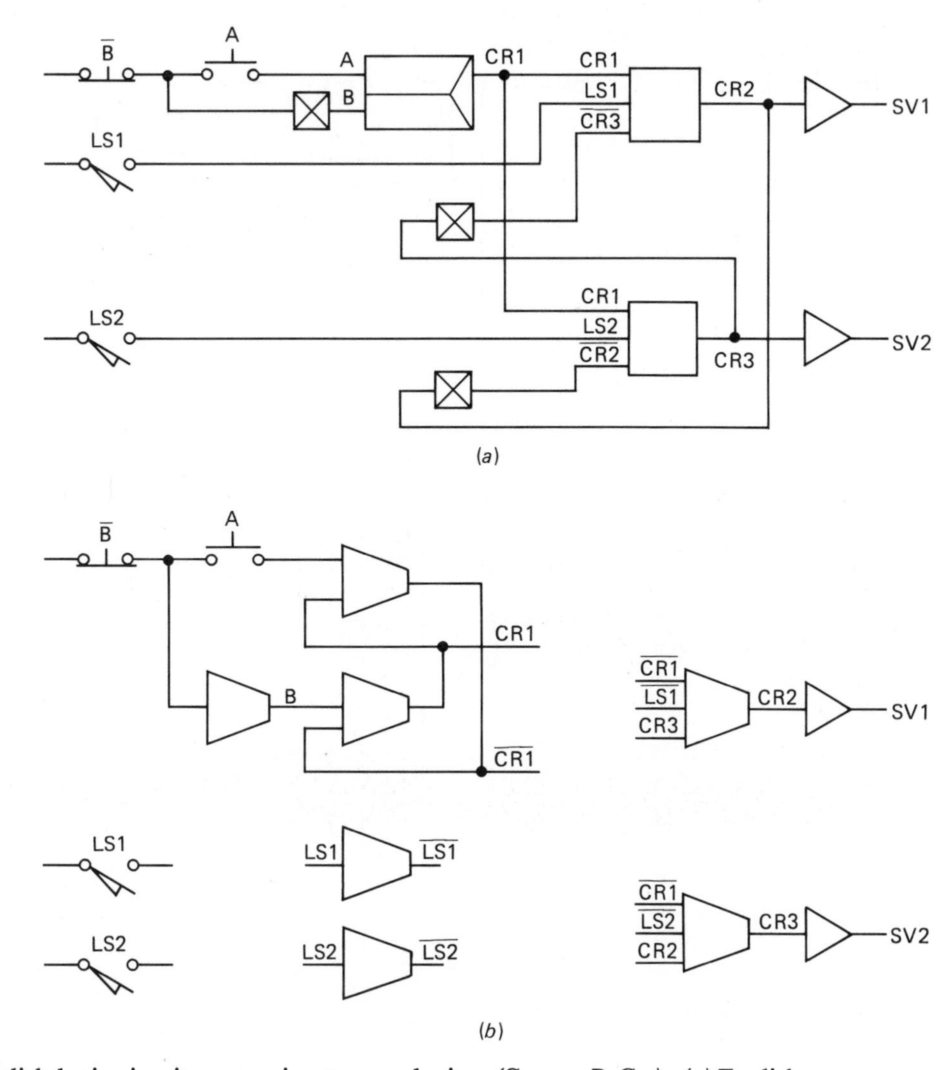

Fig. 15-51 English logic circuit conversion to NOR logic. *(Square D Co.)* *(a)* English logic circuit *(b)* NOR logic.

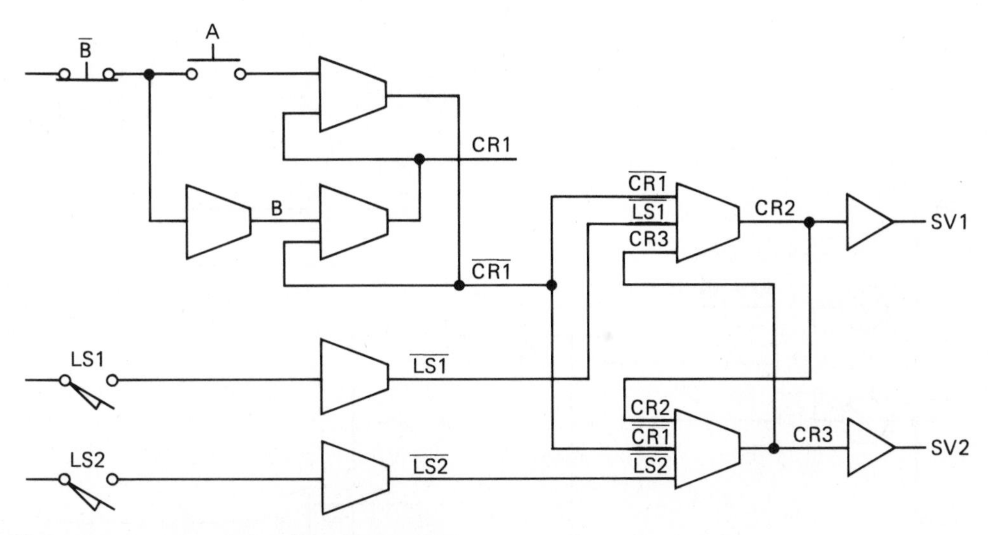

Fig. 15-52 Completed conversion of English logic to NOR logic. *(Square D Co.)*

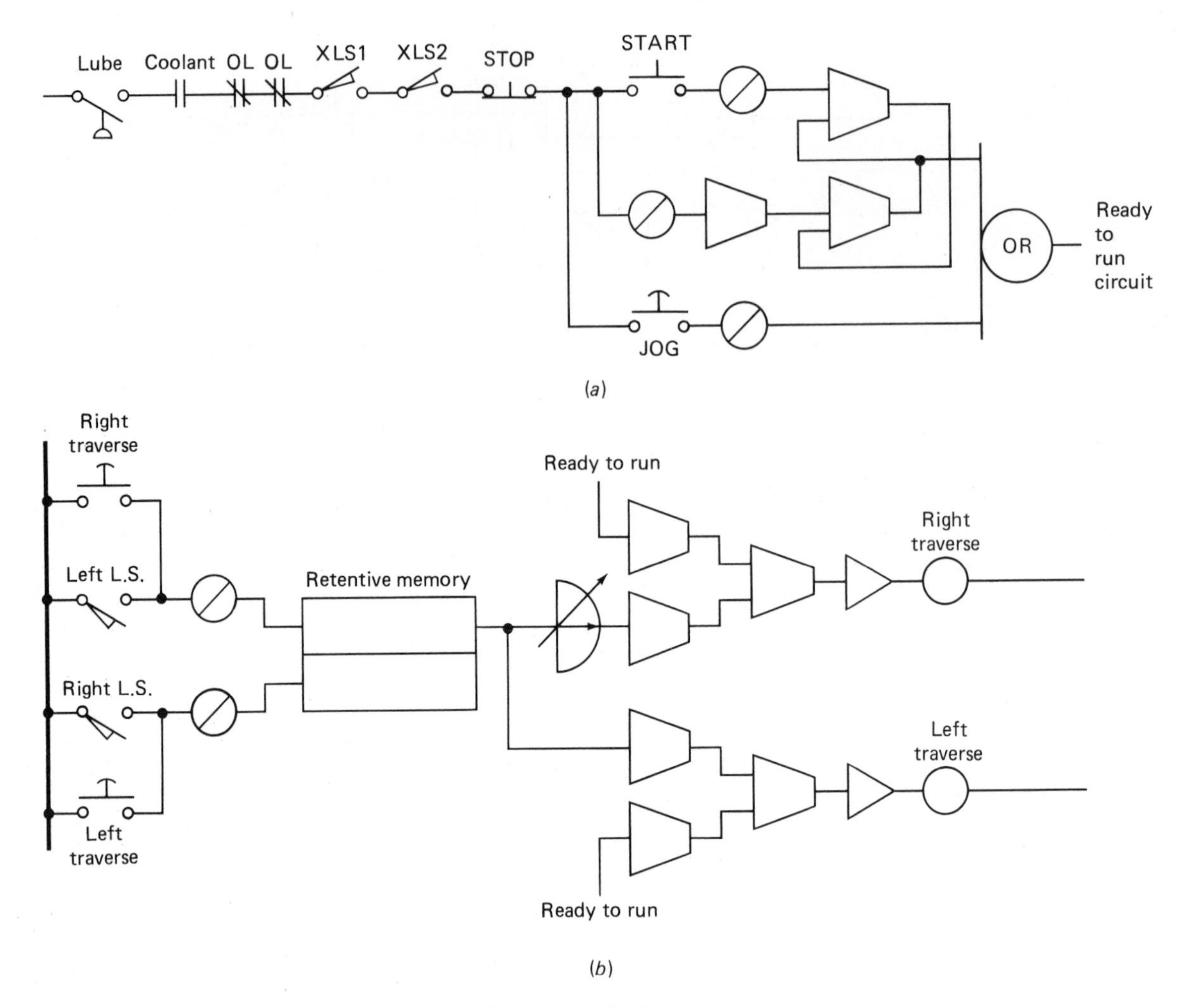

Fig. 15-53 Direct NOR logic development. *(Square D Co.)*

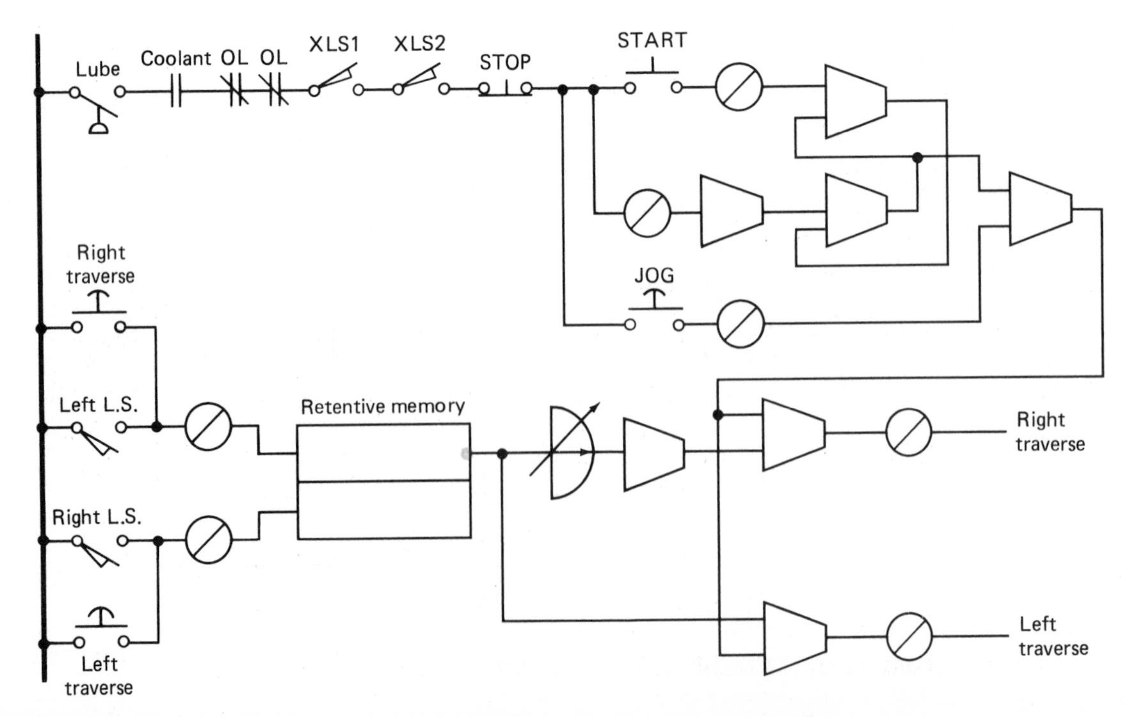

Fig. 15-54 Direct NOR logic development—complete circuit. *(Square D Co.)*

15-55. The circuit shows that with no input the transistor is off because of the +20 V connected to the base. The output at pins 9 and 10 of the circuit with no current flowing in the collector and load resistor would then be −20 V. Minus 20 V becomes the output voltage. A NOR with a logic 0 at the inputs provides a logic 1 at the outputs. The output of this circuit for a logic 1 would be −20 V. Minus 20 V is considered logic 1 and ground, or 0 V, is considered logic 0 in the Norpak system.

Figure 15-56 illustrates how the NOR transistor circuit was packaged in the early NOR logic systems. Figure 15-57 illustrates the universal NOR circuit. Inputs are terminals 1-2, 3-4, 5-6, and the output is the terminal 9-10. The input terminal 7-8 has a special direct base input to be used for a circuit input expander. It is used only where instructions so state.

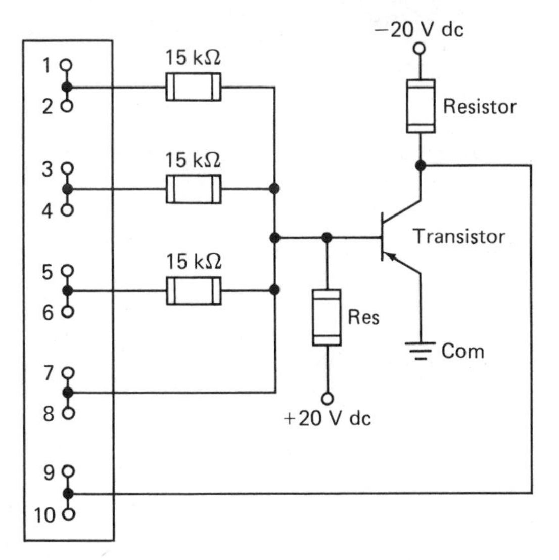

Fig. 15-55 NOR transistor circuit. *(Square D Co.)*

Figure 15-58, on the next page, illustrates the type L3 OR pack. Each wafer contains a one-input OR and a two-input OR. An OR is capable of handling 50 units of logic load. Its purpose is to provide isolation between two or more signals at a common point.

The retentive memory illustrated in Fig. 15-59 and its circuit in Fig. 15-60 consists of a memory controlled by a *saturable core.* The saturable core provides the memory feature by gating or directing a pulse signal to the "on" or "off" portion of the standard NOR memory depending on the state of the core. The state of the core, saturated or unsaturated, is set by the last input signal prior to the power failure. The retentive memory cannot be driven by an OR. Because of the off-override feature, if both inputs are present at the same time, the on output will be logic 0 and the off output will be a logic 1. Terminals G1-2 are designated as the on input to provide an on override. Terminal G1-2 would be designated as the off input and terminals G3-4 as the on input to make the retentive memory circuit provide an on override.

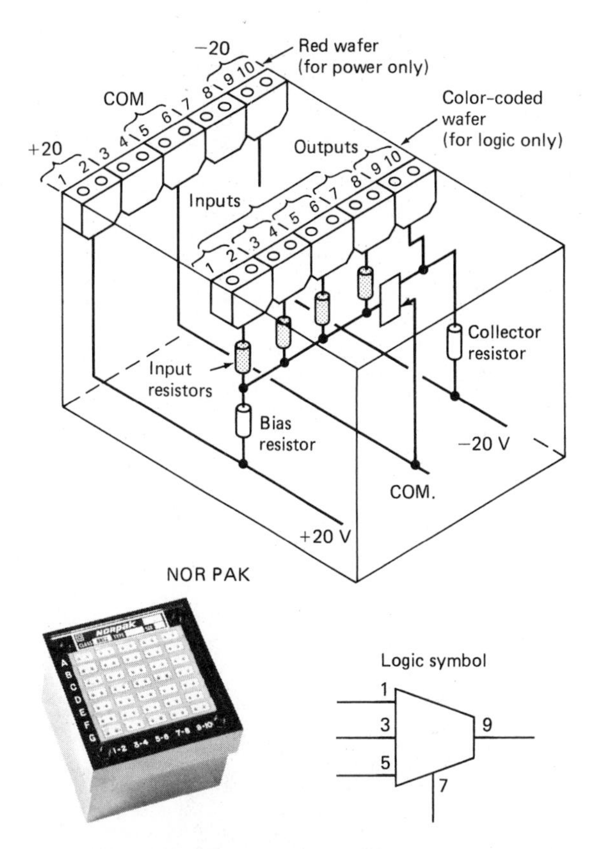

Fig. 15-56 The construction of a NOR package and logic symbol with pin numbers.

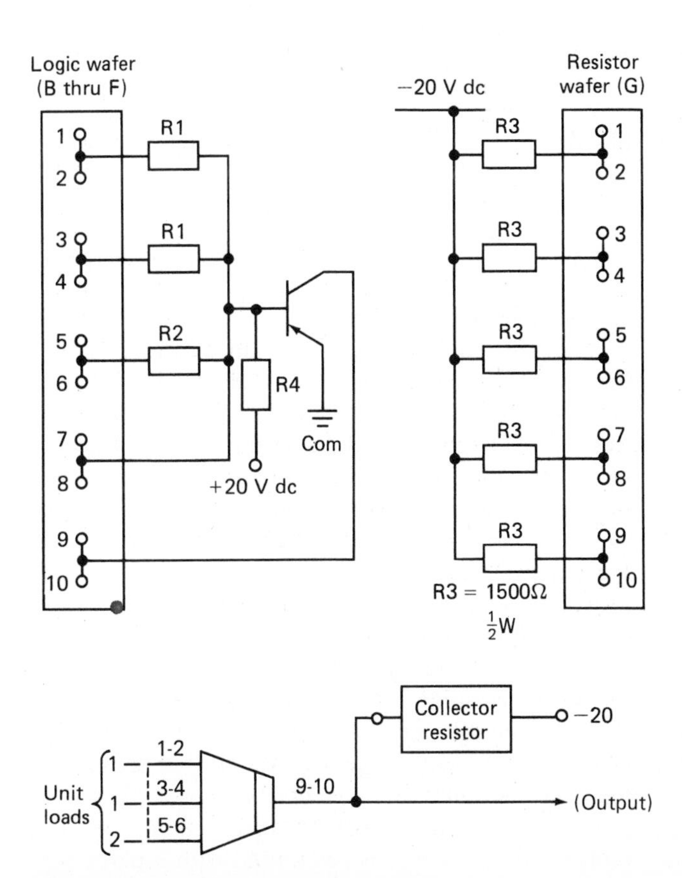

Fig. 15-57 Universal NOR circuit.

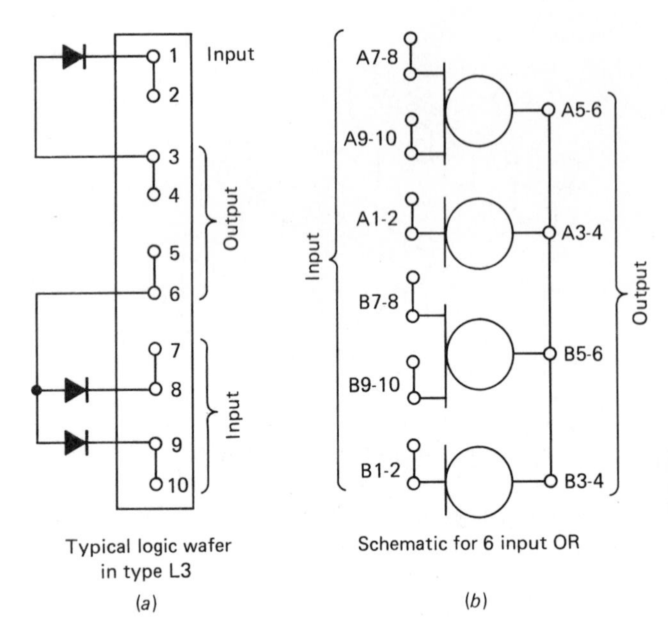

Fig. 15-58 Diode OR circuit. *(Square D Co.)*

Fig. 15-59 Retentive MEMORY package. *(Square D Co.)*

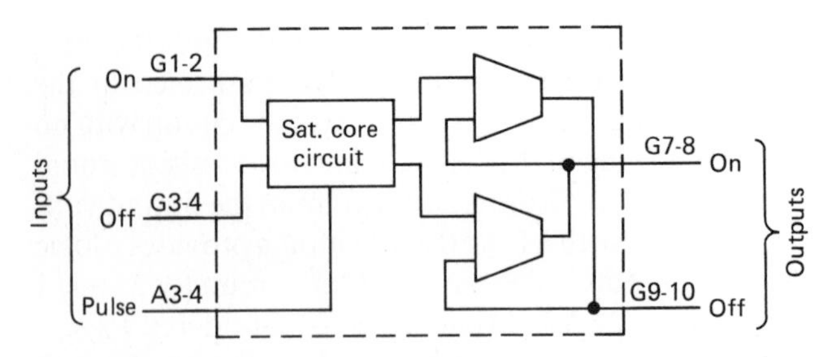

Fig. 15-60 Retentive MEMORY circuit. *(Square D Co.)*

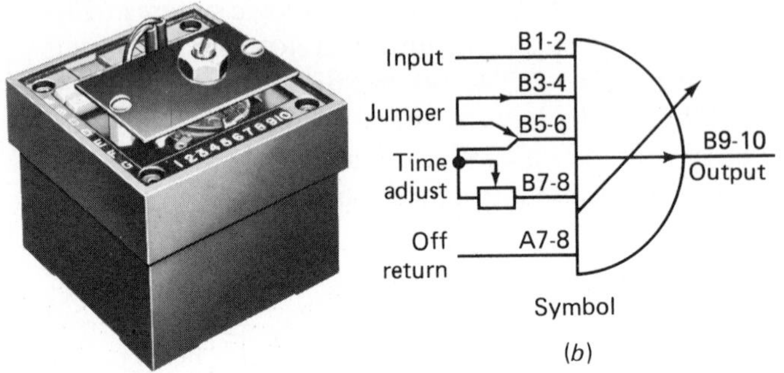

Fig. 15-61 Time-delay package *(a)* and *(b)* symbol with pin numbers. *(Square D Co.)*

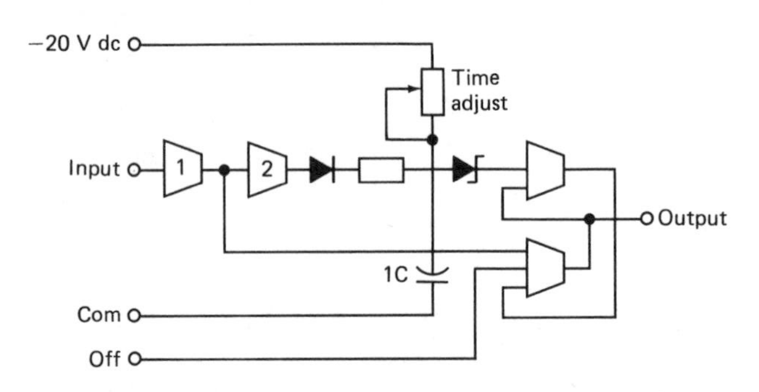

Fig. 15-62 Time-delay circuit. *(Square D Co.)*

Figure 15-61 illustrates the physical appearance and symbol of the time-delay element. The time-delay circuit is shown in Fig. 15-62.

The time-delay function is provided by an *RC* timing network and a zener diode. A logic (−20 V) signal at the input causes the blocking signal to the MEMORY from NOR 1 to go to 0 and the output of NOR 2 to go to 1(−20 V dc). This voltage changes capacitor 1*C* through the time-adjust rheostat. The charging voltage on 1*C* continues to rise to the breakdown voltage of the zener diode; at this value, the MEMORY turns on, providing a signal at the output. The timing period depends upon the setting of the time-adjust rheostat.

Since the timing network employs an *RC* time delay, the delay is subject to changes of supply voltage (−20 V) and the ambient temperature. These parameters will vary from system to system but should be relatively constant for any one system. Thus, the time-delay adjustment potentiometer cannot be calibrated until the timer is being used with a particular system.

Standard pilot devices, such as limit switches, pushbuttons, selector switches, etc., are usually used as a source of input signals to the static system. To ensure the reliability of these contact-making devices, they should be used at voltages higher than those of the 10- to 20-V signals that are found within the NOR circuitry. The higher voltages are, of course, easily obtained, but then a problem is to convert them to a usable level for the logic portion of the system.

Figure 15-63*a*, *b*, *c*, and *d* illustrates the physical appearance and the circuits for a dc-dc–type signal converter as well as the ac-dc type. A built-in neon lamp indicates the presence of an input signal. Signal converters are used to convert the high voltages from the pilot devices to the lower voltages used to drive the NOR logic circuits.

Figure 15-64, on page 274, illustrates input circuits that are used to eliminate the effect of contact bounce. Contact bounce gives false input signals that would cause counters or shift register circuits to operate improperly. The circuit and applications for a shift regis-

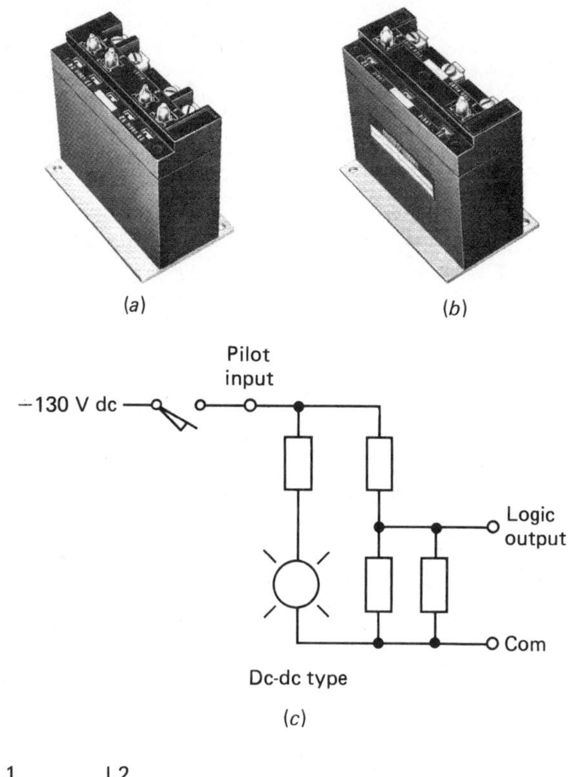

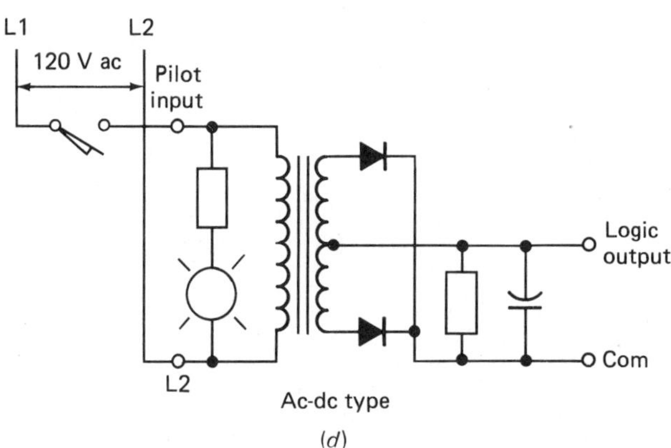

Fig. 15-63 Signal converters. *(Square D Co.)* *(a)–(b)* Physical packages *(c)–(d)* circuits.

ter and for bit binary counters will be presented in later sections.

The filter circuit of Fig. 15-65, on the next page, is used to provide a buffer against induced or unwanted input signals that could enter the logic circuit through wires coming from points outside the Norpak panel. Logic-level signals could originate from another Norpak panel, a computer, an electrical instrument, or from contact-making devices that are used at low voltages. A separate filter input should be used for each wire entering the logic panel. A logic-level signal leaving a Norpak control should always be isolated by a separate NOR having no other connections to its output.

The universal NOR can be made to function as either a NOR element or an output amplifier. Figure 15-66, on page 275, illustrates the universal NOR circuit used as an amplifier. When used as an output amplifier, the NOR has a maximum rating of 1.5 W. The maximum output rating of 1.5 W is obtained when the load resistance is 270 Ω and the inputs of all three of the universal NORs are connected together. When a universal NOR is powering a relay or other inductive load, a discharge path of an OR element or diode must be provided, as shown in Fig. 15-66*b*. Incandescent lamps have a low resistance when cold; this can result an inrush current 10 times the normal value. A current-limiting resistor should be placed in series with the lamp, as shown in Fig. 15-66*b*.

The elementary diagram for a Norpak dc output amplifier is shown in Fig. 15-67 on page 276. This is basically a two-stage transistor amplifier.

The elementary diagram for an ac output amplifier is shown in Fig. 15-68 on page 276. A logic 1 input causes the NOR to give a logic 0 output. The capacitor 1*C* charges and causes the unijunction transistor to fire. With capacitor 1*C* continually charging, the UJT continues to fire and produce the trigger pulses for the SCR. The trigger pulses are coupled through transformer T1 to the gate of the SCR. The bridge rectifier provides an ac output to the load. When the input to the NOR becomes a logic 0, the UJT turns off and no trigger pulses are applied to the SCR.

The Norpak power-supply circuit is illustrated in Fig. 15-69 on page 276. The power supply provides a −20-V dc source, a +20-V dc source in reference to common, −130-V dc source, and a pulse signal. The NOR power supplies should not be paralleled. The common terminals are tied together. The logic load should then be divided proportionately among the applicable outputs of the system power supplies.

Good wiring practice not only means ease of assembly but also is the basis of trouble-free operation. Some circuits are sensitive to external interference, but if the practices discussed here are followed, this problem can be eliminated. The layout of the components cannot be fully planned until the system design has been completed, so that every element has been taken into account. The schematic must then be studied to determine the best layout that will minimize wire runs. The various logic components should therefore be grouped to conform with the circuit functions. One should consider not only interlogic wiring but also wiring from signal-converter outputs to logic inputs, from logic outputs to monitor lights, and from logic outputs to the user's terminals and output devices. Grouping the components according to circuit function will frequently result in a few spare logic elements, but this may prove to be an advantage if circuit changes are made or an element must be replaced. If possible, one or two spare NORs should be allowed for in every 20 packs to facilitate any future changes. It is best to locate such spare NORs at the middle and end of the 20 packs. Each unused NOR of any type must have one of its inputs connected to common.

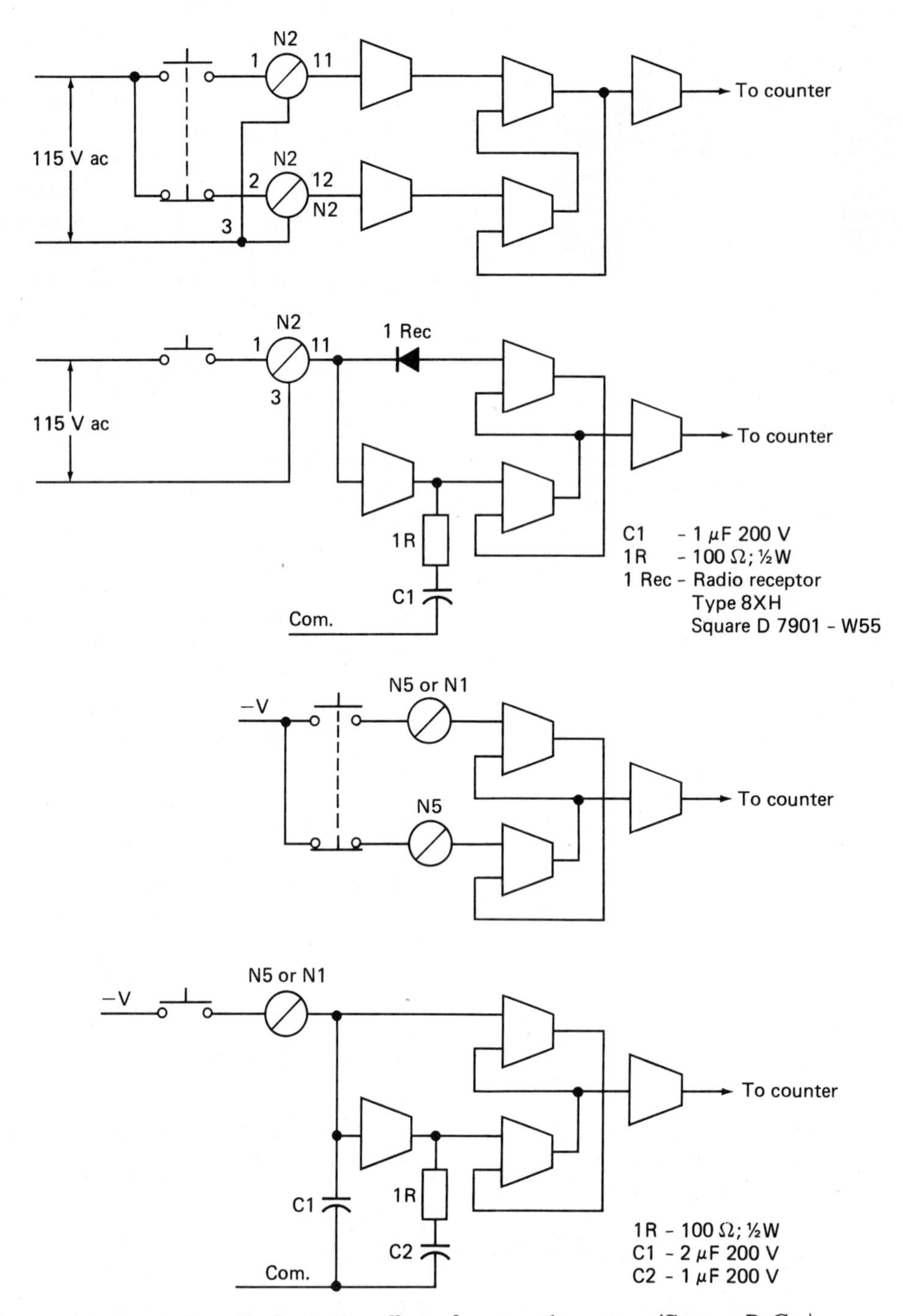

Fig. 15-64 Input circuits used to eliminate the effect of contact bounce. *(Square D Co.)*

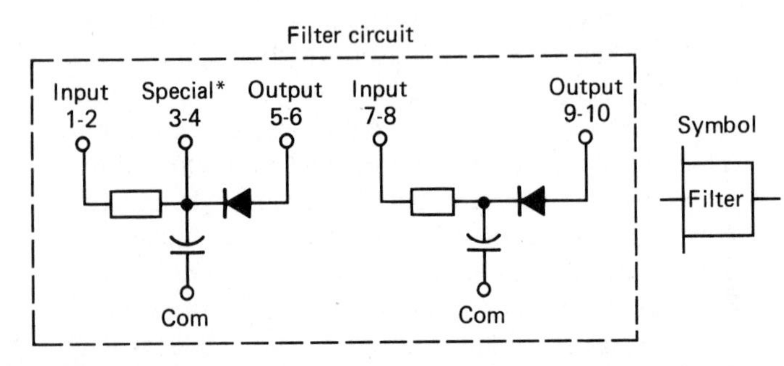

Fig. 15-65 Type-F1 filter-pak circuit. *(Square D Co.)*

Logic wiring requires care and thought; because the NOR has a high-impedance input and is switched very rapidly, it is susceptible to induced high-frequency transients. These transients, or stray pickup signals, are especially disruptive in circuits using transfer elements in one form or another or where a flip-flop MEMORY happens to be in a position subject to these stray input signals. The effect of stray signals can be eliminated if the layout of the logic components described above is followed. In addition, the following wiring practices should be adhered to. A flip-flop MEMORY circuit should be made up only of NORs of one type

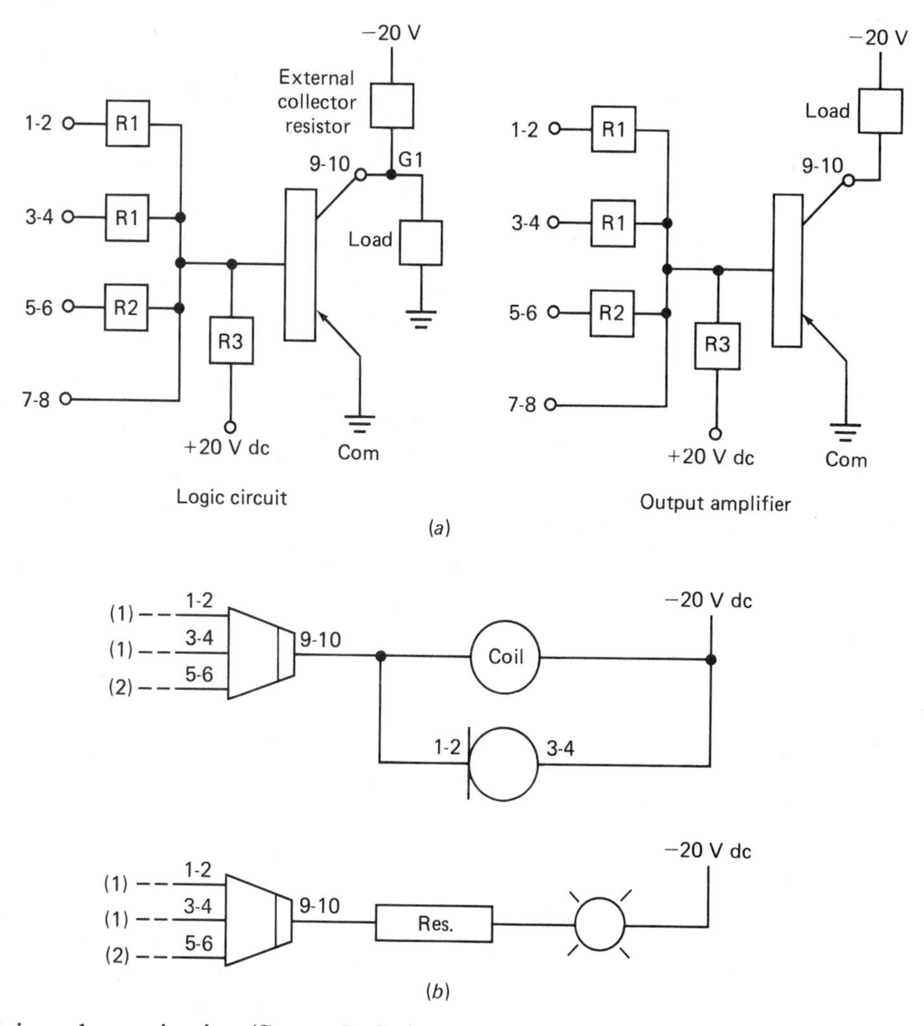

Fig. 15-66 Universal NOR circuit. *(Square D Co.)*

(standard NORS, power NORS, universal NORS) and should be contained in one pack. Neither of the NORS making up the MEMORY should be used as an amplifier to drive relays or lights. Diodes are not to be included as part of the feedback loop of the MEMORY circuits.

The output of a transfer element is a diode; this allows direct paralleling with other outputs. Each output is to be connected directly to pin 7 or 8 or a NOR, but no more than two transfer elements should be paralleled in such a manner. To accommodate more than two transfer elements, only two outputs are connected to pin 7 and 8 of a NOR, with one of its standard inputs connected to −20 V. When either transfer element pulses, the result will be a −20-V pulse at the output of the NOR. The outputs of several such NORS are then connected together by means of the usual OR circuit.

When logic-level signals are to interconnect to another remotely located Norpak panel, these signals should not be taken directly from OR circuits or from MEMORYS but should instead be isolated with NORS (preferably power or universal NORS).

When neon monitor lights are to be used, they should be restricted to use only on or within the Norpak control cabinet. If the metal case of the neon light is not in firm contact with its mounting surface, the lights may tend to flicker because of slight voltage differences that develop. Incandescent lights should be used whenever a remote indicator is needed.

The +20- and −20-V power connections to the logic packs can be made with no. 20 standard wire in series strings "jumpering" from one pack to the next. The common run of wire, however, should be kept as direct and short as possible. A no. 14 wire should be connected to common on the power supply and then directed down the center of the group of logic components to a series of terminal posts with standoff insulators that are mounted on the corners of the logic packs. Each of these points then serves as a local common point with a direct return to the power supply. The common connections of no more than three logic packs should be "jumpered" in series and brought to the local common terminal via the usual no. 20 logic wire. The common terminal at the power supply should then be connected to the chassis with a no. 14 wire under a power-supply mounting screw. The common terminal of a dc-to-dc signal converter should

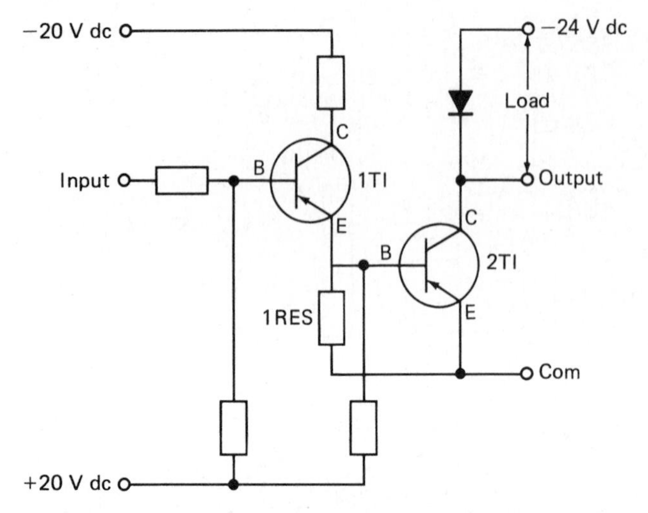

Fig. 15-67 Dc output-amplifier circuit. *(Square D Co.)*

also be connected to the chassis at the mounting screw. Wherever common is brought out of the logic circuit as a user's terminal, it should be connected to the chassis.

Stray interference can be fairly well controlled by following the practices discussed above, but to help in eliminating the paths of this interference, all logic input wires should be separated from the same channel duct conduit or harnesses that carry any power or load wiring. Stray pickup signals are usually the result of transients that are generated by the dropout of a magnet coil. When the coil is being driven by some types of output amplifiers, there is a diode built into the device that shunts out any circulating currents resulting from the transient and in so doing provides protection for the transistor and damps out the transient. This protec-

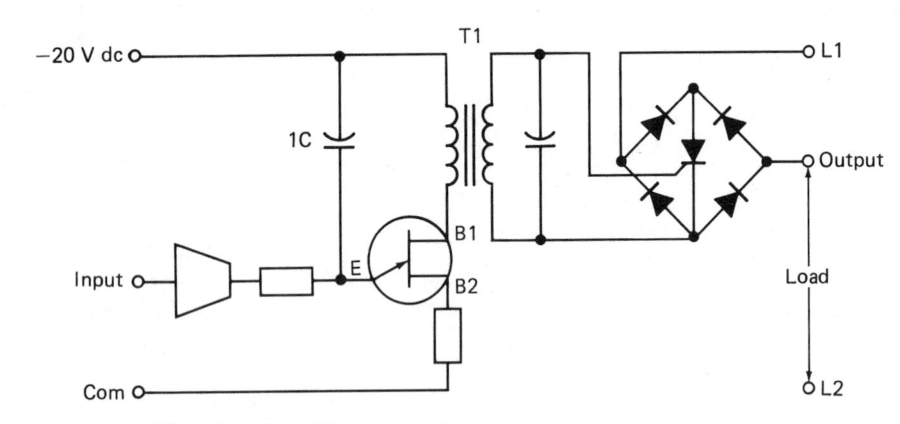

Fig. 15-68 Ac output-amplifier circuit. *(Square D Co.)*

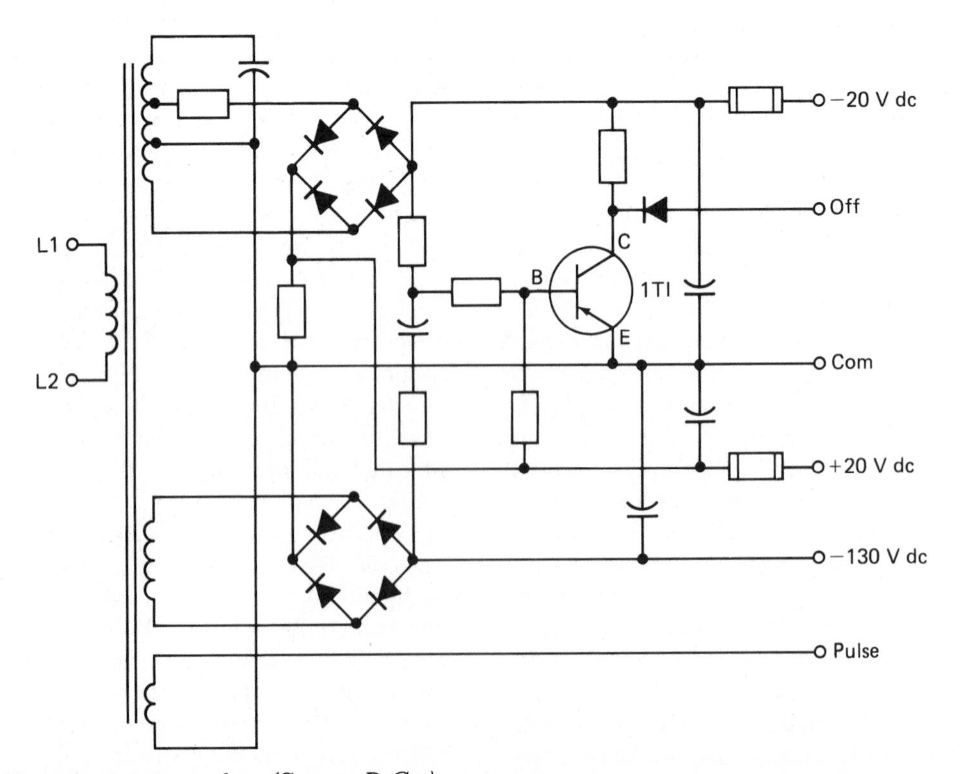

Fig. 15-69 Norpak power supply. *(Square D Co.)*

tion is lost, however, if any contacts are interposed between the load and the −24-V terminal; if the contact is opened, the shunt path around the load is broken and a very high transient could develop. To maintain this protection, provision is made on the static switch for a connection directly behind the load instead of at the −24-V terminal.

Whenever the application includes any type of counting circuit, alternator, or shift register that is driven by contact-making devices, the problem of contact bounce must be considered. Circuits designed to eliminate bounce are shown in Fig. 15-64.

NORPAK CLASS 8854

The solid-state Norpak logic system explained in the first part of this section represents an early version. Transistors and resistors provided the maximum in reliability and ruggedness. The encapsulated construction and tapered-pin termination withstood the most severe industrial applications and environments. While field modifications have always been simple to perform, maintenance personnel often prefer systems that allow the quick substitution of the logic components as a service technique. By using logic cards that can be replaced with a substitute card by plugging a new card into a socket with no wiring required, downtime in troubleshooting can be kept to a minimum. The Norpak Class 8854 is a plug-in solid-state control that offers this feature. It provides the quick change feature and with its unique construction, using a shock resistant thermoplastic case, it still retains a high degree of industrial ruggedness.

The logic circuits are mounted on printed circuit boards that are placed in what is called a *Logic Pak assembly.* The Logic Pak has contacts on the back side of the pack that plugs into sockets in the frame. The sockets are wired at the rear using wire wrap.

To identify each individual frame, a letter is assigned to the frame, the top frame being frame A, the next lower B, etc. Figure 15-70 illustrates the frame assembly with its plug-in Logic Pak. A socket address code appears on the Norpak symbol (Figure 15-71). The type number is made up from section number (1), frame letter (A) and socket number (5).

A complete Norpak solid-state control system is more than a collection of plug-in logic packs and their sockets positioned on logic frames. A complete system also includes auxiliary or related components such as power supplies, signal converters, output amplifiers, terminal blocks, a disconnect switch with circuit protector, and, frequently, electromechanical devices such as relays, starters, etc. After the need for each of these components is recognized and established, the system design is then developed sufficiently to plan the panel layout for the most effective component arrangement.

Certain rules must be followed in component arrangement and in wiring a Norpak system. Following

Fig. 15-70 Norpak socket view. *(Square D Co.)*

the rules will make wiring easier and minimize the possibility of stray pickup and electrical interference affecting the entire operation of the Norpak system. Deviations from these rules will not necessarily result in erratic performance of the system, but the possibility must be considered.

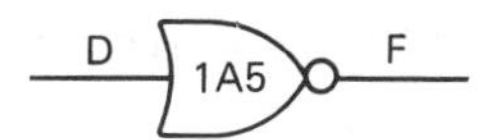

Fig. 15-71 Frame and socket location codes.

TYPICAL COMPONENT LAYOUT

Figure 15-72 shows a typical panel layout of a system. The shaded area is used to locate auxiliary power supplies, transformers, etc. Starters and relays are mounted below the frames. All ac wiring must be

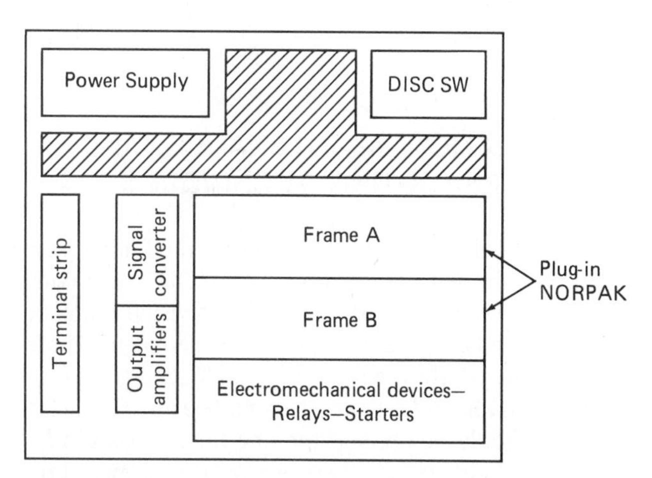

Fig. 15-72 Typical Norpack component layout. *(Square D Co.)*

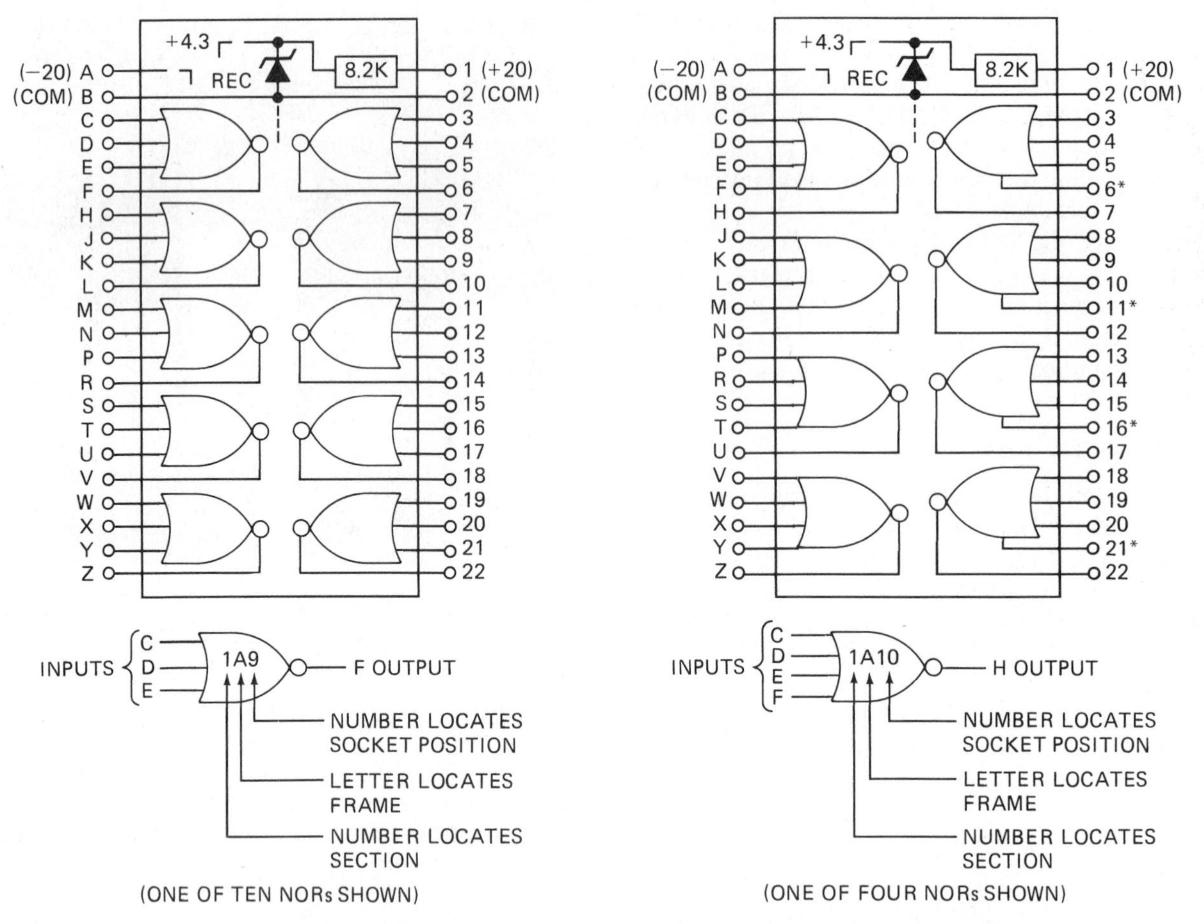

Fig. 15-73 Typical NORs and connection diagrams. *(a)* Three-input *(b)* four-input. *(Square D Co.)*

routed separately from dc logic wiring. Figure 15-73 indicates input and output pins plus frame and socket location codes.

Figure 15-74 illustrates the OR diode 20 pack. This pack provides 20 single-input isolated diodes. The diodes can be grouped to perform OR logic functions having several isolated inputs and a common output.

Figure 15-75 illustrates time-delay connections and indicates the circuit connections and circuit variations for TDE and TDD operations.

15-23 INSTALLATION, MAINTENANCE, AND TROUBLESHOOTING TECHNIQUES

Most cases of trouble will be found in the final action device or the sensing devices. When the signal converters are equipped with state lights, a visual check of these will indicate if a sensing device is responsible for the trouble. In the circuit of Fig. 15-76 the lights associated with LS1 or LS2 or both should be on. The light for PS1 should be on, and the one for PS2 should be off. Any light which is not giving the proper state indication will lead you to the actual cause of the trouble.

Suppose the system is not equipped with state lights. This in itself will make servicing the system a slower process but does not materially alter the steps to be used. The panel should be equipped with a tester designed by the manufacturer to be used to test for on and off signals at the input and output of a logic element. It becomes necessary to use the test lead and check the input and output of each logic element from the amplifier to the signal converter until the faulty one is located.

There is no need to check the logic elements until all original inputs are determined to be correct and present at the output of the signal converters. A check of the input to the amplifier will then tell you whether there is any trouble in the logic section. Whenever the input to the amplifier responds properly to changes in inputs to the logic section, the logic elements are operating properly.

Many times it is necessary to check part of the system under input conditions not represented by the condition of the sensing-device contacts. For instance, the NOT in Fig. 15-76 can be checked only when PS2 is closed. This condition can be simulated by applying the proper on voltage to the input of the NOT. This is best done by using the testing device available from the manufacturer as part of the component line for the system.

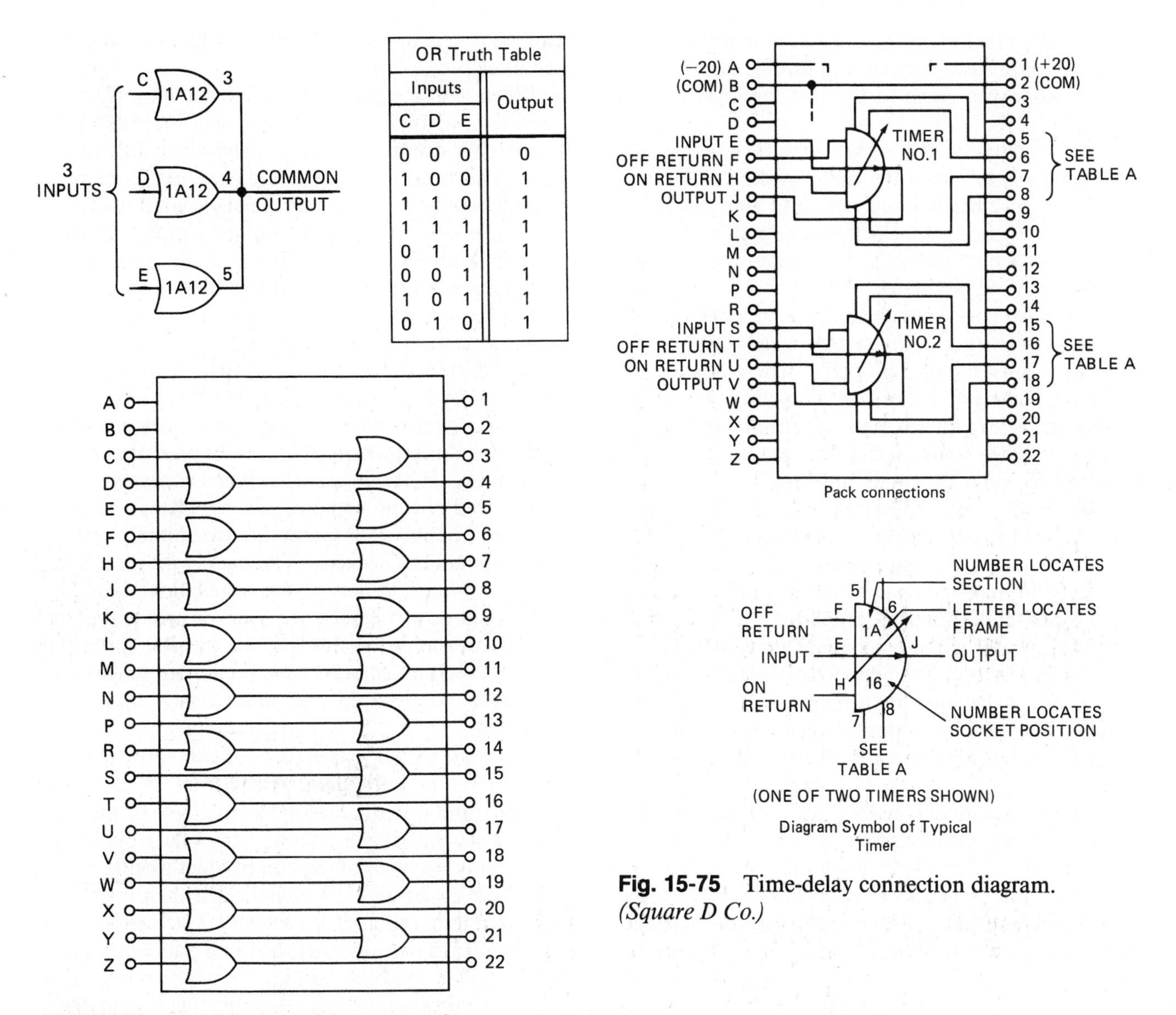

OR Truth Table			
Inputs			Output
C	D	E	
0	0	0	0
1	0	0	1
1	1	0	1
1	1	1	1
0	1	1	1
0	0	1	1
1	0	1	1
0	1	0	1

Fig. 15-74 Norpak OR package connection diagram. *(Square D Co.)*

Fig. 15-75 Time-delay connection diagram. *(Square D Co.)*

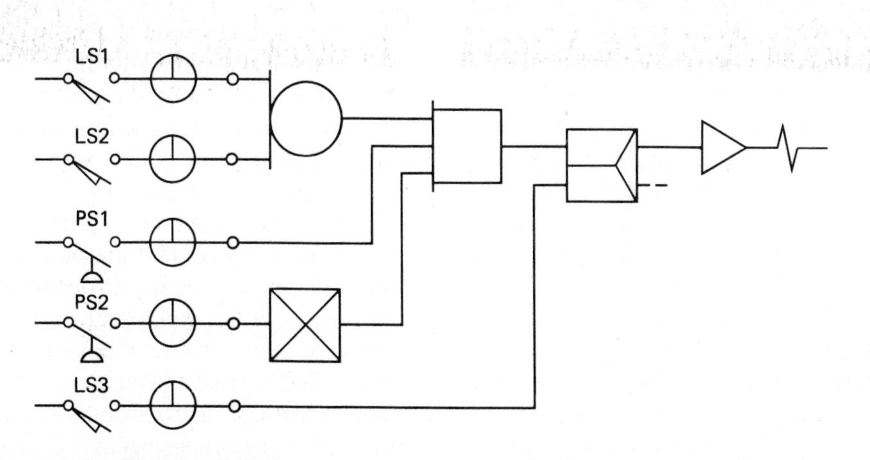

Fig. 15-76 Static control with state lights.

One of the real advantages of a static control system is that in most cases the source of trouble can be determined while you stand in front of the logic control panel.

The function of the person who is servicing the equipment is to locate the cause of the trouble and repair it in an efficient manner so as to limit the downtime of the system. An efficient system should be followed in order to locate trouble and eliminate it with a minimum of wasted effort.

Failure to operate on initial start-up might be caused by errors in wiring. Any system which has been in operation must be wired properly. Therefore, there is no need to check it out wire by wire. Many servicers with years of experience seem never to learn this fact and waste many hours in circuit tracing.

Consider again the circuit of Fig. 15-76. The solenoid does not operate when it should. We will assume that all logic elements are equipped with state-indication lights, as are the signal converters. The first step would be to make a voltage check at the input of the solenoid or the output of the amplifier, whichever location is accessible without opening connections or disturbing the wiring. The presence of proper voltage at the output of the amplifier would indicate a bad solenoid or broken connections between the solenoid and amplifier. A further check must then be made at the solenoid.

Assume that there is no output from the amplifier and that all logic elements are equipped with state-indication lights. In this case, it will not be necessary to make any voltage checks unless the MEMORY element indicates an output. With an output from the MEMORY, the trouble must be in the amplifier if it has no output. A check of the amplifier's input voltage will eliminate any possibility of trouble in the logic wiring between the output of the MEMORY and the input of the amplifier.

The above process can be quickly followed through the logic section by merely checking the state lights on each element. In this case, the AND should be on, the OR should be on, and the NOT should be on.

The approach up to this point has been to use circuit testers designed especially for the system to be serviced. Signal tracing and service can be performed by taking voltage readings at the input and output of a logic element if good voltmeters are used. The minimum acceptable voltmeter would have a 20,000-Ω/V sensitivity. Do not use neon testers or 1000-Ω/V voltmeters for this purpose. Digital voltmeters are especially good because of their high input resistance and accuracy.

If voltage testing is to be used, the servicer must know what voltages represent on and off in the particular system in question. Of those we have studied, the General Electric Company system uses 0 V for on and −4 V for off. The Cutler-Hammer Company system uses +10 V for on and 0 V for off. The Square D Company system uses −10 to −20 V for an on and 0 V for an off.

When a 0-V signal is indicated for the off or on condition, a good voltmeter may well indicate up to about 0.5 V due to current leakage through the transistors. This is normal and should cause no concern.

When more than one part of the system is malfunctioning or when the output amplifier seems to be bad, check the power supply. Power-supply troubles account for a large percentage of trouble in most electronic devices and systems and therefore should always be suspected until checked out.

Intermittent troubles are always hard to find and can be caused by bad contacts or poor connections. One factor which must always be considered when one seeks the cause of intermittent trouble is the presence of electrical noise. When a system has been in operation for a long period of time without noise problems and then develops trouble which is suspected to be noise, certain things should be checked. If any new wiring has been installed, check it out. New installations of other equipment or wiring near the panel or field wiring for the system should be checked as a possible source of noise. When no changes have been made, check out grounding connections and filters which may have gone bad.

The static elements are probably the most reliable parts of the entire system. Be sure to check fuses, switches, and other common electric parts of the system. They probably are the source of the trouble.

In many ways installing and troubleshooting static systems is easier and less complicated than installing and troubleshooting relay systems. The logic part of the circuit is generally packaged in a panel or panels where it offers ready access to most of the circuit. The modular construction makes unit replacement relatively simple, and locating trouble is much more simple than with a comparable relay system.

There is one idea which cannot be overstressed in this area of static control: follow the manufacturer's instructions. Each component line offers reliable, efficient components that the company has engineered to work together as a system. Exhaustive tests have been run on all phases of the system, including installation. Field personnel can hardly improve upon the proven methods developed by the company.

Some general statements can be made which will apply regardless of which system is being installed. First, select good sensing devices. Limit switches and other devices which employ contacts should be selected to provide good contact pressure and good wiping action. Wire each sensing device to an individual signal converter. Never run signal wiring in the same enclosure with power leads. Generally speaking, iron conduit should be used as a raceway for signal wires, but the manufacturer's instructions should be consulted on this. Some systems employ shielded cable under specific grounding requirements.

Should the power supplies be grounded? Which wires should be or can be grounded? Consult the manufacturer's instructions, as there is no general approach to these questions.

If you are building up a panel from components, there are some general principles which should be followed. Group the signal converters so that the wiring from the sensing device does not mix with logic wiring. Place the signal converters so that the wire from the logic side of the logic element is as short as is practical. It is bad practice to feed a logic element in one panel directly from a logic output in another panel.

Group the logic elements within the panel so that a lead between elements can be as short as is practical. Use the size of wire recommended and terminate it in the specified manner.

Amplifiers, relays, and other ac devices should be mounted to give good isolation from the logic elements and logic wiring. A great deal of electrical noise can be generated in these devices, particularly in the relays. Shielding helps prevent interference from electrical noise, but isolation is much more effective in most cases.

Good electrical connections are a must in static control. When relays are used for control, the voltage and current in the circuit will tolerate a great deal of poor quality work in the joints and connections. Static circuits operate at low voltage levels and with practically no current; therefore, they cannot tolerate loose or dirty connections. Bad connections are very hard to locate after installation, and, therefore, it behooves the installer to be sure each and every connection is made properly.

The logic element, whether it be an AND, OR, or NOT, is a very high-speed switch which will react to signals of such short duration that they are merely a flash on the screen of an oscilloscope. Any loose contact or electrical noise which provides even a very short pulse to the logic input can cause a malfunction of the system. This type of trouble is very hard to locate unless the signal is applied to a MEMORY element, which will indicate that it has had a momentary input applied.

Trouble in any control circuit is first indicated by some part of the machine or process not functioning properly. The cause of the malfunction could be in any part of the system—from the sensing devices, which initiate the control, to the logic elements and amplifier, which drive the final-action device.

15-24 ELECTRICAL NOISE AND ITS ELIMINATION

The term *electrical noise* has been used many times in previous discussions and deserves some clarification and explanation. This section will be devoted to the subject of noise and its elimination.

Electrical noise signals can be generated by almost any power equipment. The opening or closing of contacts under load almost always produces electrical noise. The transmission system which feeds power to the building carries electrical noise, and every piece of operating equipment connected to the wiring system of the building is a potential source of noise. Electrical noise, then, cannot be prevented, but must be kept out of the signals fed into the logic control components.

Electrical noise exists in many forms, but for the purpose of this discussion, consider it to be essentially a sudden change in current flow or voltage in a conductor. When changes in value of current or voltage are uniform or regular, we refer to this current as *alternating current* or *ac.* We then might describe electrical noise as *nonuniform* alternating current.

An acceptance of the above oversimplified description of electrical noise provides us with some rules for keeping it out of the control circuit. Signals of alternating current and, therefore, electrical noise, are transferred or coupled from one circuit to another by one or more of three ways.

The first way, and probably the most common, is through mutual electromagnetic coupling. The second is through the capacitance between conductors, or electrostatic coupling. The third is by means of common impedances between the circuits.

Most cases of electrical noise will involve a combination of the three forms of coupling, since every circuit has inductance, capacitance, and impedance in varying proportions. We will consider each form of coupling separately for simplification.

Electromagnetic coupling occurs whenever the circuit containing the noise is run parallel and close enough to the signal circuit to provide coupling through their mutual electromagnetic fields. The most effective means of reducing this form of coupling is by separation. The degree of coupling decreases with the square of the distance between the conductors. Electromagnetic fields tend to cancel each other in circuits with twisted-pair wires; therefore, if either or both the noise circuit and the signal circuit employ twisted-pair wiring, the coupling will be reduced. The third method of reducing electromagnetic coupling is by means of shielding. The noise which is the source of trouble is generally low frequency and therefore requires a ferrous material as an effective shield. Rigid conduit of the common electrical variety is ideal for this purpose, provided only that signal wires are run in the pipe. The amount of electromagnetic coupling between two circuits is proportional to the distance for which they are parallel; therefore, reducing the distance will reduce the coupling and the noise problem. Each case of noise coupling is different and may require analysis if it is severe, but most can be eliminated by using good installation practice.

Electrostatic coupling is caused by two conductors run parallel and at different voltages. The conductors form the plates of a capacitor. The air or other material between the conductors forms a dielectric. The most effective means of reducing this coupling is by shielding the signal wires. Separation of the conductors and reduction of the distance where the two circuits are parallel will also help.

Common impedance coupling occurs only in circuits which share some impedance. The most common source of this type of circuit is the temptation to use a common wire to feed one side of many sensing devices. The most effective way to eliminate this form of coupling is to use two wires from the panel to each sensing device.

The use of common arc-suppressing circuits on relay contacts and inductive loads can prevent or reduce the generation of noise signals at their source.

A summary of noise suppression would indicate that static systems should be wired with care. Separate circuits should be used for each sensing device. The use of twisted-pair leads for sensing or signal circuits might be indicated in some cases. Rigid conduit for field wiring seems to provide the best likelihood for trouble-free operation. Remember, consult the manufacturer of the system in question for specific details.

The installation of static control systems requires only the normal skills of the electrician or technician plus a little care. Follow these rules: Make every joint carefully. Never run signal wires in conduit with power wiring. Keep signal wires and logic wires separated in the panel. Follow the manufacturer's recommendations. Avoid noise sources.

Whenever possible, use the testing devices available from the manufacturer of the system in question. If it is necessary to service it with a voltmeter, use one with at least a 20,000-Ω/V rating.

15-25 RELAY CONVERSIONS

Many problems have to be converted from relay circuits to logic circuits. Figure 15-77 represents a table of relay circuits and various logic equivalents. This table is very useful in converting relay to logic circuits. The table also is useful for converting from logic to relay circuits. If a logic circuit were to be replaced with a P.C. having the same inputs and outputs, the first step would be to change the logic circuit to an electric circuit or relay ladder diagram. If it were to be programmed in a P.C. using a cathode-ray tube (CRT) programmer, the electric circuit would be numbered for programming. When programming, you would see on the CRT the electric circuit as it was being programmed into the P.C.

If the P.C. were programmed with boolean algebra expressions, we would still have to convert a logic circuit to an electric circuit to develop a ladder logic diagram for programming into the P.C.

You will have a lot of use for circuit conversion when updating control systems in industry.

SUMMARY

The Norpak system uses NOR logic and is built around a basic NOR element, which is connected in the control panel to make up AND, MEMORYS, and ORS.

The logic diagrams used with this system are different from those used with English logic systems but can be developed from English logic diagrams.

An ON signal for Norpak is −10 to −20 V dc, while an OFF signal is 0 V. The basic power supply provides +20 V d-c from the base-to-collector circuit of the transistor. In the event that there are some unused NORS, the −20-V collector voltage and the +20-V bias voltage would create a collector-to-base voltage of 40 V, which could damage the transistor. Unused NORS should have one input connected to common.

The NORS in Norpak are encapsulated in standard modules, which are then connected to the circuit by jumpers.

The system also includes a test unit which should always be used in servicing the logic section of the panel.

REVIEW QUESTIONS

1. According to Allen Bradley, what is the advantage of using ICs in their Cardlok system?
2. What is the advantage of high-threshold ICs?
3. What are the advantages of semiconductor switches?
4. How is electrical noise produced?
5. What causes internal noise?
6. What are the different noise-reduction techniques?
7. What must be developed first before a schematic diagram and a wiring chart is developed for a Cardlok system?
8. What is the advantage of a hard-wired system?
9. What is the advantage of LSL control of relays?
10. How does Cutler-Hammer define static control?
11. What type of circuitry is used in LSL?
12. What voltages are used in LSL circuits?
13. How is master control accomplished in LSL logic control circuits?
14. What is a repeater in LSL logic?
15. What is the purpose of monitor boards in the LSL logic system?
16. How is undervoltage provided in the Cutler-Hammer logic system?
17. Define fanout?

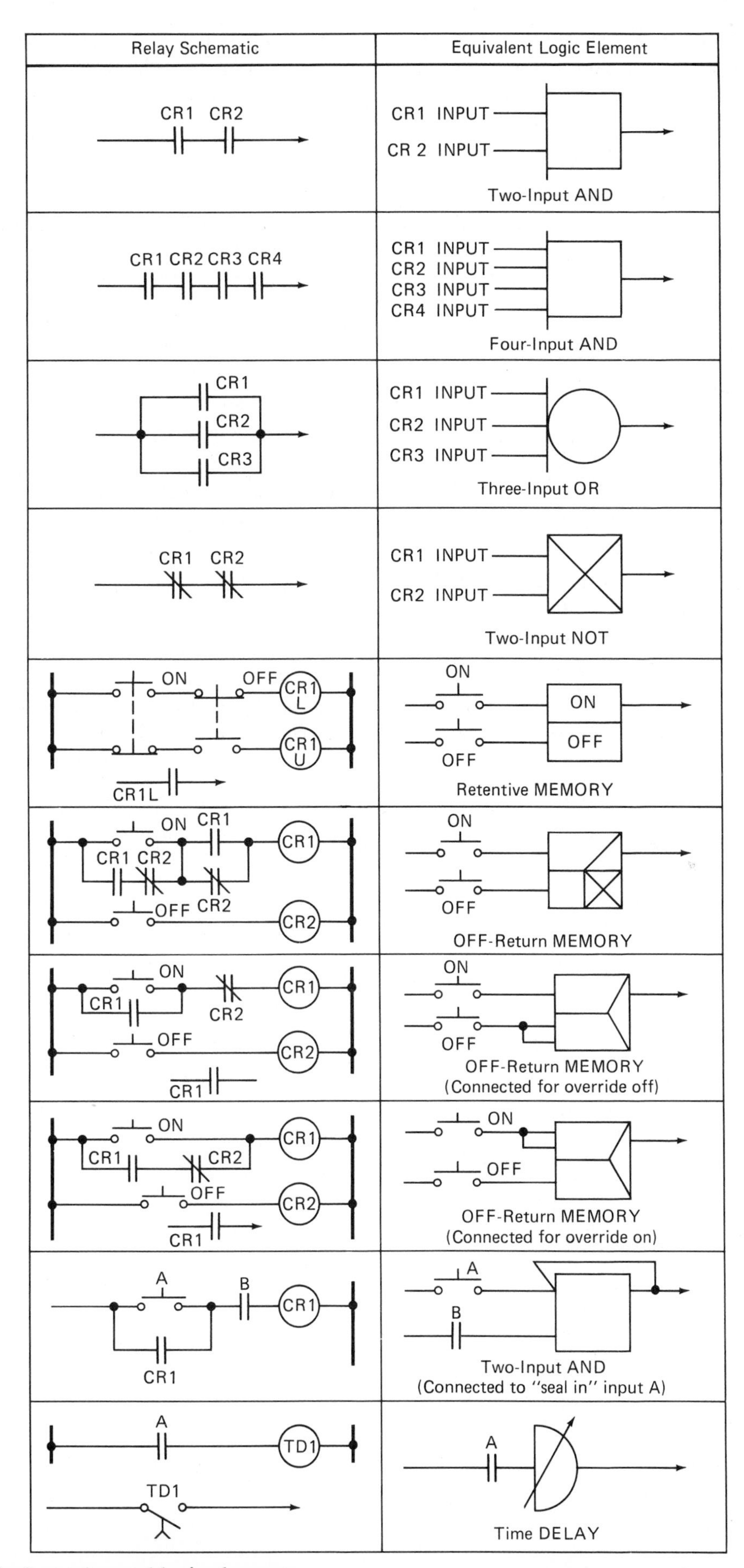

Fig. 15-77 Equivalent relay and logic elements.

18. What is a bucket in LSL logic?
19. What type of control has replaced the LSL logic control for modern industrial control applications?
20. What is meant by English logic?
21. How are fanout limitations calculated?
22. What is used to check the condition of each part of the logic system?
23. Draw the transistor circuit for a NOR.
24. Why is a NOR called a universal logic element?
25. What is meant by the term *logic concept?*
26. What is Norpak?
27. What is the heart of the NOR circuit?
28. What is the output voltage of the basic three-input NOR circuit if there is no input to A, B, or C?
29. Draw the NOR circuit for a three-input AND.
30. Draw the NOR circuit for a three-input OR.
31. What are the two types of time delays used in Norpak? How do their symbols differ?
32. What is the meaning of a NOR complement?
33. Where are most troubles found in logic control systems?
34. If logic system fails to operate on initial start-up, what might be the cause of trouble?
35. What is the advantage of using state indicating lights on logic elements?
36. How is signal tracing accomplished in a logic system?
37. How are logic systems checked for intermittent troubles?
38. What is generally the most reliable part of a logic system?
39. What are some of the general principles regarding logic system installation?
40. What is the rule for grounding logic systems?
41. What methods are used to eliminate electrical noise generated in amplifiers?
42. How is electrical noise transmitted in logic circuits?
43. Why is twisted-pair wiring employed in the wiring of many logic systems?
44. What causes electrostatic coupling?
45. When does impedance coupling occur in logic systems?
46. What circuits in logic control systems require arc suppression?
47. What are the requirements of voltmeters used in testing logic circuits?

16
Programmable Controllers

The programmable controller is fast becoming the workhorse of industry in providing control and decision making for industrial control systems. Using simple programming techniques, it is capable of providing for motor control needs with the most advanced capabilities imaginable. Its main role in control work is in decision making and task solutions.

Programmable controllers are easy to program and reprogram. All programs are entered in a P.C. by pressing program letters and numbers on a keyboard of the programmer. Changing the control functions of a relay panel requires hours of rewiring. A few minutes of pressing keys on a programmer or P.C. loader will produce the same logic operations that hours of wiring relays will produce. New logic sequences programmed into the P.C. are stored in the P.C.'s memory. Control logic can be changed as often as is desired with ease.

16-1 HISTORY OF PROGRAMMABLE CONTROLLERS

The programmable controller (P.C.) was introduced as an electrical control device for automatic control systems in 1969. The programmable controller was originally intended to be an electronic replacement for the relay panel. Initially, P.C.s did little more than on-off sequencing of motors, solenoids, and actuators. The original P.C. was capable of making a few more logic decisions than relay panels. As the years passed, the P.C. continued to evolve until it became the intelligent workhorse that it is today. The modern P.C. has many advanced capabilities not found on the early P.C.s. For example, new P.C.s have provisions for data acquisition and storage, report generation, execution of complex mathematical algorithms, servomotor control, stepping control, axis control, self-diagnosis, system troubleshooting, and talking to other P.C.s and mainframe computers.

16-2 FEATURES OF PROGRAMMABLE CONTROLLERS

Features of P.C.s that have contributed to their proliferation in industrial control are their relatively low cost, reliability, ease of maintenance, and ruggedness. Programmable controllers can withstand temperatures from 0 to 60°C and relative humidities from about 0 to 95 percent. Is a P.C. a computer? It is in the sense that any stored-program digital device is a computer. Is a P.C. a microprocessor? It isn't, but it often contains microprocessors. The main elements of a P.C. are its CPU, or central processing unit, its input/output (I/O) modules, and its power supply.

A control logic program in the form of instructions is programmed into the P.C. and stored in memory. The memory tells the CPU what events should take place and the logical sequence in which they should occur. Depending upon the P.C., the CPU scans the I/O in from 3 to 200 ms. This keeps the P.C. up to date on the status of any problem the P.C. is controlling and indicates the control action it is providing. The status of each input and output is stored in memory. Necessary control actions are initiated through execution of program instructions and the scanning of inputs and outputs. Logic determination of problem solution is determined with the logic programmed into the P.C. Inputs are signals from sensors or switches in the control loop. Outputs are signals that actuate the machine control or initiate the industrial process that is being controlled.

The P.C. differs from commercial-grade mainframes and minicomputers in that it is designed specifically for the manufacturing environment and for controlling the operation of machines and industrial processes. The P.C. would fail at many tasks that are asked of computers. The P.C. would fail miserably at crunching numbers and solving mathematical problems. But a single P.C. with 32 I/Os and selling for a few hundred dollars can handle the complex control of industrial machines and be cheaper than the relay panel it replaces. Larger P.C.s selling for hundreds of thousands of dollars may have 8000 or more I/Os and may be able to control extremely complex machines.

16-3 DIFFERENCE BETWEEN PROGRAMMABLE CONTROLLERS AND COMPUTERS

The important distinction between the basic operation of a P.C. and that of other computers is that the P.C. goes through a series of tasks in a sequential fashion. This is known as *scanning.* A typical task of the P.C. that illustrates its scanning operation is *input/output updating.* The inputs and outputs are scanned by the P.C. to check their status. Figure 16-1*a* shows that a program structure for a P.C. requires sequential execution with a scan, starting with task 1 and proceeding through task 4. Task 1 may be, for example, updating the I/O status. The task for the P.C. would be to check the status of all I/Os and determine any changes since the previous check. Task 2 for the P.C. may be solving problems by using the logic program in the P.C.'s memory. Task 3 may be internal executions that ensure a secure control system—self-diagnostics, for example. Task 4 for the P.C. may be to communicate with programming panels, with one or more of them positioned remotely from the P.C. The above task sequence could have also been the sequence of operation of an actual relay panel. The P.C. then is capable of providing the same industrial control that relays do.

Figure 16-1*b* illustrates the operations of a computer. The computer can execute tasks in any order. It does not require sequential execution, as does the P.C. If two or more events that require different tasks were to occur simultaneously, the computer would have to be provided with a means to determine which task should be accomplished first. Because of its sequential structure, the P.C. is slower than a general-purpose computer. This limitation of the P.C. scanning system can be minimized by using "smart" I/O modules. (Modules capable of making decisions are called "smart.") Smart I/O modules can make the P.C. capable of competing with general-purpose computers and minicomputers for task solution.

16-4 PROGRAMMING METHODS FOR PROGRAMMABLE CONTROLLERS

The programming languages for P.C.s and computers are very different. Those for P.C.s are designed to deal with the many I/O points that are part of the control loop. Those for general-purpose computers are usually oriented toward scientific or business data processing.

Four major languages can be used in programming the modern P.C. These languages may use boolean algebra equations, mnemonic commands, logic diagrams, and ladder diagrams. Ladder diagram logic, which is the same as that for relay control circuits, is a

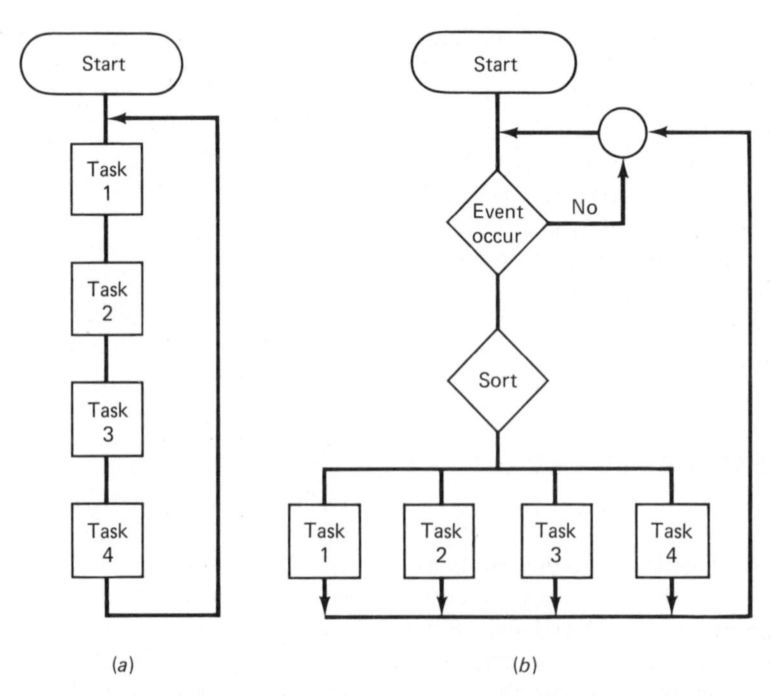

Fig. 16-1 Examples of flowcharts. (a) The P.C. requires sequential execution with a scan starting at Task 1 and proceeding through Task 4. (b) A computer permits task execution in any order.

popular choice for programming most P.C.s. It is easy to understand, and most plant or industrial electricians and technicians are accustomed to working with elementary relay ladder diagrams. Since electricians are already familiar with relay panel logic, the transition to programming P.C.s with ladder logic diagrams is relatively easy. In the discussion of various P.C.s, programming methods will be presented.

Input/output modules for P.C.s have a variety of applications. One of their functions is to reduce the high voltages on the pilot devices to the low-level logic voltages that can be handled by the P.C. Another main function is to provide a method of taking the low-level logic voltages from the CPU and converting it to the higher voltages that can operate the various motor starters, solenoids, and actuators that are used as the output of the P.C. Input/output modules also isolate the P.C. from electrical noise, which would cause problems if these signals reached the circuits of the P.C. Input/output modules must handle inputs from a variety of devices, such as limit switches, pushbuttons, selector switches, thumb wheel switches, light-emitting diodes, strain gauges, and thermocouples. They must also provide interfacing with a broad range of other control devices.

Process control is often provided by P.C.s in modern industrial control systems. Process control refers to controlling such variables as temperature, pressure, and flow on a continuous basis. Closed-loop control is used to maintain some process variable at a set point. Figure 16-2 illustrates the positioning of a steam valve used to maintain the temperature of some liquid within a tank. Earlier P.C.s were capable of providing on-off control to a process but could not provide proportional

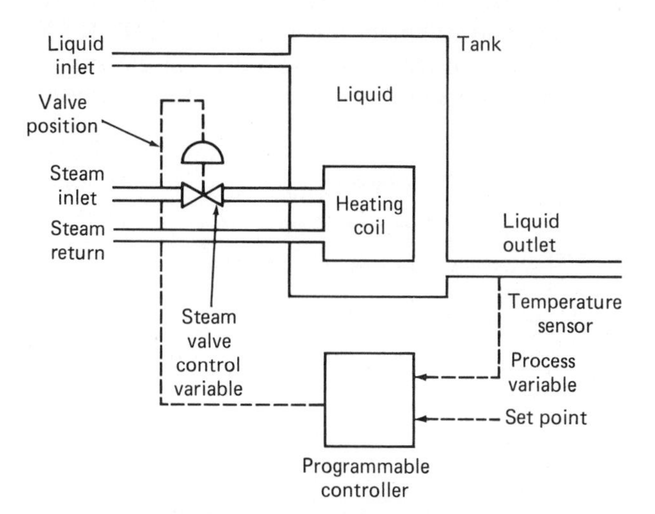

Fig. 16-2 Process control. In this example of process control, a P.C. maintains a set temperature for the liquid in the tank. The PID module makes it possible for the P.C. to maintain the steam valve at a position between on and full off.

control action, integral control action, and derivative control action. These controls, referred to as *PID* (for *p*roportional, *i*ntegral, and *d*erivative) *modes of control,* are essential in providing good process control.

16-5 SMART I/Os AND THEIR APPLICATIONS IN PROCESS CONTROL

A smart I/O was developed for the P.C. that provided PID modes of control. The P.C. now can be used when intermediate positioning of a valve is required. The process control problem of Fig. 16-2 is now within the range of the P.C. Proportional, integral, and derivative control, which for years was offered only by electronic and pneumatic instrumentation systems, can now be offered also by P.C.s.

Until the development of intelligent I/O modules, the P.C. provided only on-off commands. With only on-off control, the P.C. could not control motor speed or machine position. With smart I/O motion control, machine speed, and machine position are all in the realm of P.C. control. By using smart, or intelligent, I/Os interfaced with digital position sensors and actuators, the P.C. is capable of unlimited control action.

16-6 THEORY AND APPLICATIONS OF PROGRAMMABLE CONTROLLERS

Allen Bradley bulletin 1772 states that the memory of a P.C. can be readily changed without the extensive rewiring that is necessary to make the same changes in a hard-wired relay control system.

The controller continuously monitors the status of devices connected as inputs. Based on the user's program, the controller controls the devices connected as outputs. These input and output devices may be of different types with various voltage and current ranges. They may include

- Limit, float, selector, and/or pressure switches
- Pushbutton switches
- Thumb wheel switches
- Alarms, indicators, and/or annunciator panels
- Solenoids
- Motors and/or motor starters
- Transducers
- Various solid-state devices, including TTL and analog instrumentation

A P.C. is an electronic device that offers a means of controlling the operations of a system of electromechanical devices. A NEMA standard defines a P.C. as "a digitally operating electronic apparatus which uses a programmable memory for the internal storage of instructions for implementing specific functions such as logic, sequencing, timing, counting, and arithmetic to control, through digital or analog input/output modules, various types of machines or processes."

Programmable controllers can replace relay logic control (including timers and counters, electromechanical timing and sequencing devices, and solid-state control cards).

Programmable controllers are available in all shapes and sizes. Programmable controllers like the TI510,® the Modicon Micro 84,® and the Eagle Signal EPTAK 200® are considered to be small P.C.s. The TI510 has 12 inputs and 8 outputs. The Micro 84 has 64 discrete inputs and 64 discrete outputs. The EPTAK 200 has 128 inputs or outputs. These small P.C.s would be used for practical applications such as small transfer lines, machine tools, small assembly machines, batch sequencing, material handling, motor control devices, packaging machines, energy management systems, and numerous other applications.

Large P.C.s such as the Giddings and Lewis PC400® have 2032 inputs and outputs. The Gould Modicom 584® and the Allen Bradley PIC/PLC3® as well as the EPTAK 700® have 4096 inputs and outputs. All major companies manufacture large P.C.s. Applications for the larger P.C.s would be for such large industrial applications as power generation, energy management, manufacturing, and assembly lines.

In most industrial applications we generally find the P.C. controlling motor starters. The P.C. is the decision-making part of the control system. The P.C. makes all the logic control decisions of relays, electromechanical digital controls, or static switching systems. It makes these decisions with built-in logic and user-programmed instructions. The P.C. is fast replacing all the other types of decision-making control systems.

We will now take a look at some of the smaller P.C.s starting with the Eagle Signal EPTAK 200 Programmable Controller.

16-7 EAGLE SIGNAL EPTAK 200 PROGRAMMABLE CONTROLLER

The EPTAK 200 P.C. is a solid-state microprocessor-based industrial control system. Through monitoring of input signals, the system detects changes from such sources as pushbuttons, sensors, and limit switches. Based upon the status of these input signals, the controller system reacts, through user-programmed internal logic, to produce output signals to drive external loads such as control relays, motor controls, and annunciators or alarms.

The EPTAK 200 P.C.'s capabilities include a 420-program statement capacity, 32 timers or counters, two 32-stage bit shift registers, 128 I/O switching points, 128 internal control relays, data handling, and add-subtract arithmetic functions.

The basic operating system consists of two units, the processor unit and I/O track. Peripheral devices include the portable programmer, program recorder/loader, auxiliary interface unit, data access and display module (DAD/M), and program printer. The peripheral units offer programming, display, and diagnostic functions, and add to the convenience and flexibility of the total system. See Fig. 16-3 for a block diagram of the EPTAK 200.

PROCESSOR UNIT

The processor unit performs all of the logic, timing/counting, and arithmetic functions of the EPTAK 200 system. The power supply and processing circuits are contained in a single extruded aluminum enclosure

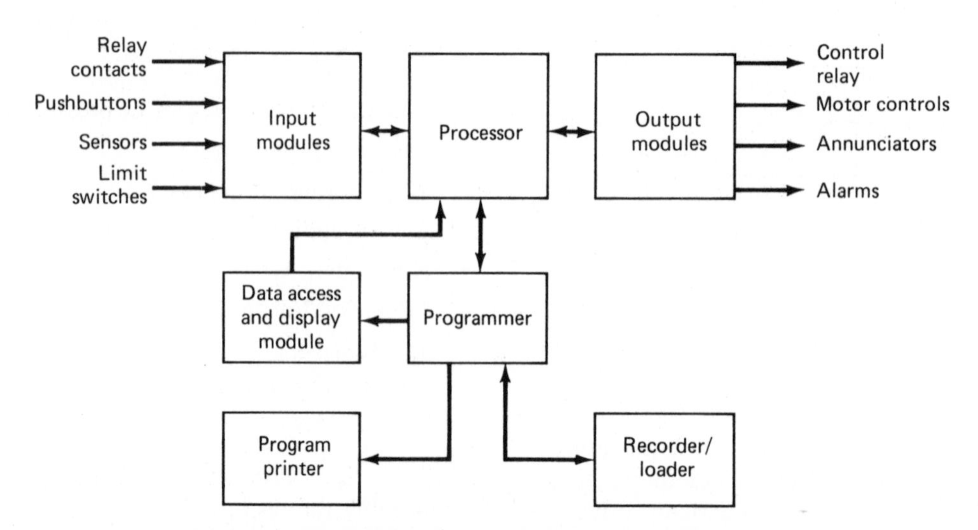

Fig. 16-3 Block diagram for EPTAK 200 programmable control.

mounted directly on the I/O track. The processor unit measures approximately 4 by 8¾ by 4 in. and weighs approximately 10 lb (Fig. 16-4). The processor unit can be removed from the track mounting by loosening four securing thumbscrews and sliding the unit free of the I/O track connector. The processor-I/O track connector is a P.C. edge connector and no cable connections are required from the processor to the I/O track. Power connections to the processor unit are to a three-terminal screw connector block on the end of the processor unit. The unit is protected with an easily accessible and resettable circuit breaker. The processor unit also houses connections for the remote programmer and an auxiliary port connector for the auxiliary interface module or the add-on I/O expander unit.

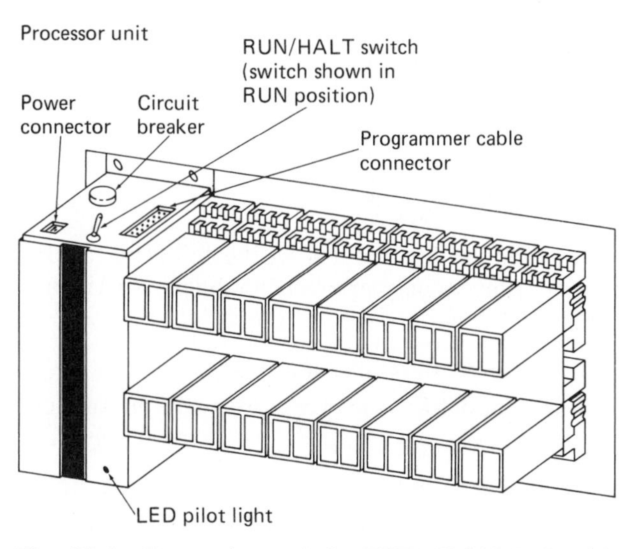

Fig. 16-4 Processing unit for EPTAK 200 P.C. with cover removed. *(Eagle Signal.)*

Power-up and Normal Operation Sequence. Upon application of service power, the processor initially turns all outputs off. If the user program has not changed during the period of power off, the LED pilot light on the front of the processor is turned on. If the program is not identical to the one that existed at power-down, the LED pilot light flashes and the program must be reloaded from the programmer or the recorder/loader.

Program execution begins when memory has been verified. At the end of the first scan after memory verification, execution of outputs begins. During normal operation inputs are read; CRs, timers, shift registers, etc., are updated; and outputs are executed at the end of each program scan.

Power-down Sequence. When the processor detects that service power has fallen below 102 V ac, program scan is halted and all outputs are turned off.

EXPANDER UNIT

The I/O expander unit interfaces additional I/O tracks with the processor. The physical size and mounting are identical to the processor. The unit contains a resettable circuit breaker. Power connections are made to a three-terminal screw connector block on the end of the unit. The I/O track connects to the expander unit with a P.C. edge connector. Input/output tracks are not designated for use with the processor or expander units since any I/O track can be used with either unit. The expander unit connects to the processor with a ribbon cable. Expander units can be connected together to give additional I/O.

PROGRAMMER

The EPTAK 200 programmer, as illustrated in Fig. 16-5, is a compact, self-contained portable unit. The programmer is housed in a formed metal enclosure approximately 7½ by 7¾ by 2½ in. and weighs approximately 3 lb. The unit's top surface is slanted for desktop use and for ease of reading the LED displays. The keyboard layout simplifies program entry and is divided into the following four function-related groups: command, control, data/address/numeric, and diagnostic/edit. The LED display contains two numeric displays for statement addresses and values, and the LED annunciators indicate selected functions.

Fig. 16-5 Programmer for EPTAK 200 Programmable Controller. *(Eagle Signal.)*

Figure 16-6 on the next page is a sample relay circuit that has been numbered for programming. The program to place this circuit in the Eagle Signal EPTAK 200 P.C. is at the right of circuit. Steps 1 to 7 are used to clear the memory of the P.C. Step 8 starts at the left side of line 1.

Every time switch IN 101 is closed, OUT 106 and OUT 107 will take turns energizing.

Programming the EPTAK 200 controller may be accomplished either at the processor site or remote

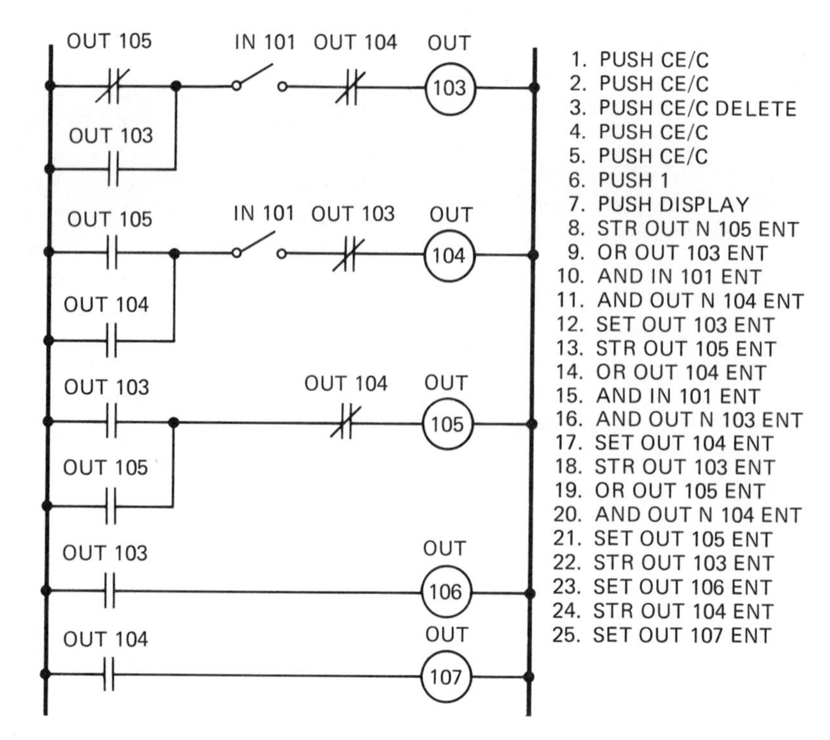

Fig. 16-6 ABCD reference circuit and program for Eagle Signal EPTAK 200 P.C.

from the processor. A *Program mode* switch on the front of the programmer allows the user to select: *Internal*—to create programs contained in the programmer for later transfer to the processor unit (an internal battery maintains the program in memory for 30 days); *External*—to create programs to be placed directly into the processor memory; and *Both*—to create programs to be placed into both the processor and the programmer memory depending on the position of the program mode switch (the program Load key causes the program to be transferred bidirectionally between processor and programmer; the programmer cable must be connected to the processor for any communication between these devices). In *Both* and *External* modes, the processor RUN/HALT switch must be in HALT position before the processor will accept programs or program changes.

INPUT/OUTPUT MODULES

The EPTAK 200 P.C. input/output modules are relay-type square-base plug-in elements. Each I/O module contains two address points. Each I/O track will house up to 16 modules or 32 addressable input or output points. Solid-state input and output modules are available in 120 or 240 V ac and 10 to 55 or 102 to 132 V dc. A 10-A, 120 V ac relay output module is also available.

ACCESSORY MODULES

Accessory modules include a simulator module, Watchdog timer module, and Load circuit test module.

All modules are molded in the following colors, from translucent polycarbonate for quick visual identification:

Yellow—Ac input module

Smoke or black—Ac output, relay output module

Clear—Dc input module

Red—Dc output module

Blue—Simulator, watchdog, or load circuit test modules

The simulator module has two manual single-pole toggle switches on top of the module. The switches allow the user to manually simulate an input for diagnostic or program-testing purposes. Two LED indicators, visible through the translucent module housing, annunciate the status of the I/O. The simulator module may be used in any of the 16 positions on the I/O track.

Diagnostics are performed through the Edit and Diagnostic function keys. Diagnostic and edit functions include: Search; Insert; Delete; Status (displays status of designated I/O and CR addresses); Load (loads program to or from the processor); and Display (displays contents of a program statement). A power flow (PWRFLOW) annunciator in the display panel indicates the accumulator status for each program statement.

A Watchdog module is used to monitor the scan of the user program. The module is basically a free-running, time-delay timer with a normally open–normally closed solid-state output. If the program scan fails, the Watchdog timer times out, causing the outputs to change state; that is, the normally on triac turns off, and the normally off triac turns on. The output may be used to signal an alarm or to shut down the system.

The load circuit test module is used to check load wiring before the load module is installed. The module contains two protector circuits, each with a resettable breaker and a manual ON/OFF toggle switch. With the switch in the ON position, the load will turn on if properly wired. If the load circuit has a fault, the protector will instantaneously trip. The load circuit test module is used for a load wiring check only and cannot be used during system operation.

The data access and display module (DAD/M) is a peripheral unit that allows visual access to user data to monitor or change the actual and set-point values of timers and counters, data registers, and auxiliary ports, and to change the values while the controller and machine being controlled are in operation. The DAD/M is housed in a formed-metal enclosure and can be panel-mounted at the machine site. The keypads and displays are the sealed-membrane type, which make the unit highly resistant to moisture and oil. The unit measures approximately 4½ by 7 by 4 in. (Fig. 16-7).

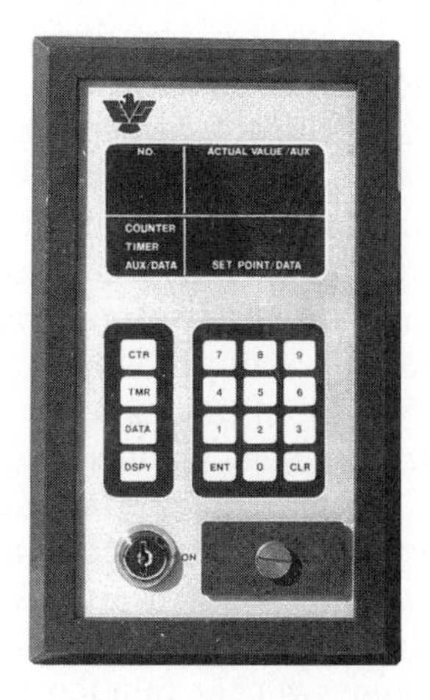

Fig. 16-7 Data access-and-display module for EPTAK 200 P.C. *(Eagle Signal.)*

Two 4-digit LED displays on the DAD/M indicate the timer/counter or auxiliary data set-point and actual readings. An LED two-digit display indicates the timer/counter or auxiliary data number being displayed.

A keylock switch on the front of the unit prevents unauthorized personnel from changing set points in the processor. Any set point or actual value can be displayed even with the switch in the LOCK position, but set-point values cannot be changed.

A single cable on the rear of the DAD/M connects the unit to the EPTAK 200 processor. An access panel on the front of the unit covers the DAD/M-to-programmer connector. This connector allows direct access between the processor and the programmer without disconnecting the DAD/M. Program changes other than timer/counter, data register, or auxiliary input set points are programmed into the processor through the connector on the front of the DAD/M.

The recorder/loader is a peripheral device supported by the EPTAK 200 programmer. Bidirectional communication between these two devices allows programs to be recorded from the programmer or loaded into the programmer for subsequent transfer to the processor. The actual program storage is on magnetic cards the size of a credit card. Each card can hold an entire 1K program (420 program statements).

The Program printer is a peripheral device supported by the programmer. The communication link is unidirectional and allows any program contained in the programmer memory to be printed out on a cash register–type paper strip. Each printed program statement consists of the statement number, instruction, and address.

16-8 MODICON MICRO 84 AND 584 PROGRAMMABLE CONTROLLERS

Gould manufactures a small, easy-to-program, low-cost P.C. called the *Modicon Micro 84* (Fig. 16-8). This solid-state system is a very cost-effective replace-

Fig. 16-8 Modicon Micro 84 P.C. *(Gould.)*

ment relay control panel. It is designed for both relay and drum switch control applications. As a replacement for conventional control, it is ideal for jobs requiring 6 to 60 relays. The Modicon Micro 84 performs all relay switching, timing, counting, arithmetic, drum, and sequencing functions. It costs less to install than a panel of 10 relays.

An electrician or technician with a sound knowledge of ladder diagrams will have no trouble replacing relay control circuits with programs for the Modicon Micro 84. To make the changeovers from conventional relay control to P.C. control requires no special training. Anyone who can work with ordinary relay ladder diagrams is ready to put the Modicon Micro 84 to work. Programming uses standard ladder logic symbols and is done directly from regular ladder diagrams.

Figure 16-9 represents a small control circuit that is numbered for control change over to the Modicon-Micro 84 P.C. Each of the elements in the motor control circuit is represented by an identical symbol on the hand-held programming panel. The ladder diagram with the symbol numbers will be placed into the programmer by pushing the appropriate keys. The programmer with its liquid crystal display (LCD) is illustrated in Fig. 16-10. The display lets the programmer see the exact information programmed into the P.C. With the Modicon Micro 84 programmer it takes only a few seconds of pressing a few keys to make complex control circuit changes. These circuit changes would take hours or even days of complicated panel rewiring.

Notice again the circuit of Fig. 16-9 and the LCD of Fig. 16-10. The display that is being indicated is the

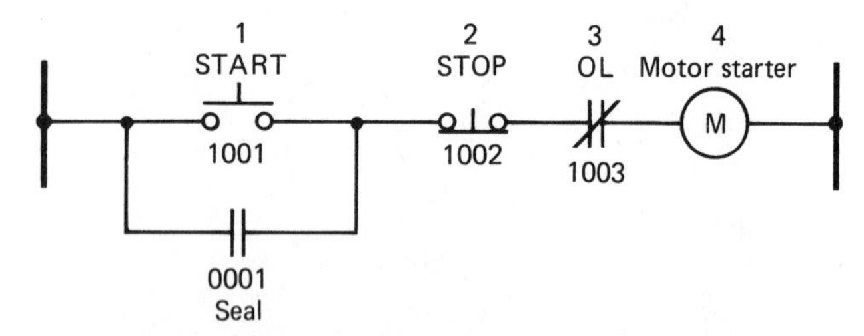

Fig. 16-9 Circuit numbered for programming into Modicon Micro 84 P.C.

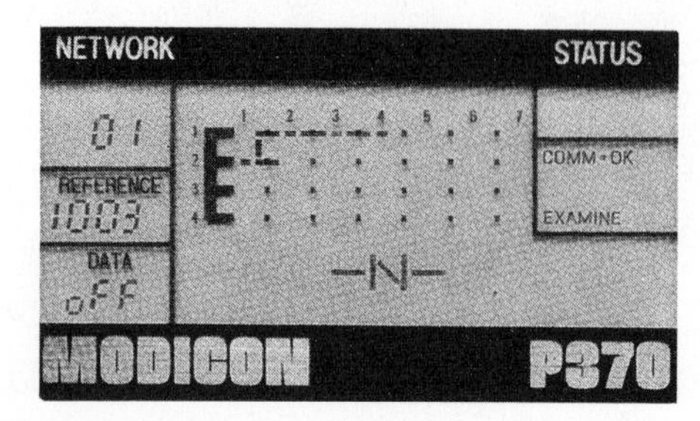

Fig. 16-10 Liquid crystal display of OL 1003 being entered into P.C. *(Gould.)*

overload relay contact. It has been numbered 1003 and it is being programmed for location 3 in the circuit display.

With a little practice the Modicon Micro 84 programming panel can be read with ease. The programmer has a unique full-network LCD that displays all vital information. Element reference numbers and network and data status are all instantly accessible. Power-flow direction is always indicated while the programmer is in use. Convenient user-oriented system status keys facilitate the entire programming process.

The Micro 84 P.C. system is ideal for all types of small machinery. Machine control that requires as few as 4 relays or as many as 60 relays can be replaced by the P.C. Whether the Micro 84 is used to replace relays, drum switches, stepping relays, or sequencers in existing machinery, or incorporated in new machinery or new machine designs, the result is greater reliability and greatly reduced costs. It offers P.C. flexibility to applications of all kinds including control for such things as transfer lines, machine tools, small assembly machines, batch sequencing, materials handling, energy management systems, and the like. Programmable controllers are especially cost effective in small control application.

Programmable controllers such as the Modicon Micro 84 are engineered for years of dependable operation in typical industrial environments. In contrast to conventional relay control systems, the P.C. is completely solid-state. It is highly resistant to ambient electrical noise and vibration. These P.C.s are built to tolerate temperatures ranging from 0 to 60°C.

Troubleshooting a P.C. system is much less a task than troubleshooting a conventional control relay system. Generally, a lot of time is consumed in testing, probing, and tracing relay circuits before the trouble can be located. Troubleshooting large control systems using stepping relays, drum switches, and timers can often be a very complicated process. Troubleshooting an equivalent control system using a P.C. is quick and simple. Since limit switches, motor starters, and other devices are wired directly to the I/O points on the P.C., status indicators let the maintenance technician or electrician isolate problems almost immediately. The programmer can be used to determine the status of all circuits and to isolate the circuits giving trouble in a fraction of the time required to isolate trouble in a conventional relay circuit.

The ease of maintenance of P.C. machine control systems keeps machine downtime to an absolute minimum.

When a control system is to be expanded, there is generally enough reserve control capacity in a P.C. that new modules do not need to be purchased. If timing, counting, and sequencing capabilities are to be added to an existing control system, the P.C. already contains these capabilities. The Micro 84 also features arithmetic capability for value addition and subtraction. It also has a full register capacity, which means you can collect, monitor, and manipulate data, and can handle up to 64 I/O points. The P.C. can be reprogrammed to take advantage of all its capabilities.

Industrial machines frequently require changes in their sequence of operation. As little downtime as possible in making the change is desired. This is an easy problem for the P.C. The Micro 84 P.C. has a memory feature which makes it possible to develop and store standard programs tailored to specific machines and/or operations and to simply load them into the system when needed. This can be done with a minimum of downtime, with little loss in production.

The cost of such P.C.s as the Modicon Micro 84 has made them very cost effective for every type of small control applications. Since 1978, when the change from relay systems to P.C.s began, the work of industrial control electricians and technicians has changed considerably. The installation of an industrial control system is a faster and simpler job now. Hours of relay and panel wiring and construction are eliminated by the P.C. The most costly part of a conventional control system is in the interwiring of the circuit components. The wiring of timers, drum switches, and relay switching circuits can become a very complicated and demanding job for the construction electrician. Now printed circuits are used for the wiring of the logic circuits in P.C.s. System buses are used for making the connections between logic sections within the P.C. Now, printed circuits and ICs eliminate all the internal wiring of a control system, which would normally take hours or days on conventional relay control. Installing a P.C. system is a simple process. All that is required of the construction electrician is to connect the P.C. input terminals to pilot devices and the output terminals to power devices such as motor starters and solenoids.

The Modicon 584 P.C. is a solid-state device designed to perform logic, timing, sequencing, and various calculations for industrial control applications. It is a general-purpose controller and is used as a direct replacement for relays or solid-state electronics in an industrial environment. Its design makes it applicable for a wide variety of industries. These include

Machine tools
Food processing
Pipelines
Transfer lines
Energy management
Pulp and paper
Plastics
Refineries
Test stands
Petrochemicals
Forest products

Modicon's 584 P.C. is widely used to replace minicomputers in industry. When compared to these devices, P.C.s offer the following advantages:

- Solid-state construction for maximum reliability
- Operability in "hostile" industrial environments (i.e., heat, electrical transients, EMI, vibration, etc.), without fans, air conditioning, or electrical filtering
- Programmability from simple ladder diagrams with no need to learn a new language
- A cathode-ray tube programming panel for easy programming
- Reusable control if no longer required on original application
- Indicator lights at major diagnostic points to simplify troubleshooting
- Simple maintenance, based upon module replacement, with minimum downtime and maximum productivity

The 584 controller offers unparalleled flexibility, expanding from a 50-relay system to a system with over 2000. Users can select either a basic relay replacement system; a sophisticated diagnostic monitoring, data collection/storage, and report-generating system; or anything in between. These widely differing capabilities are provided with the same hardware to minimize the quantity of spare parts in large multiple-controller installations. In addition the Modicon 584 offers the following benefits:

- Low-cost hardware (less than installed relays)
- Flexibility (start small and add data management later)
- Expandable memory (from 4K to 32K 16-bit words)
- Fast scan rate (between 10 and 60 ms)
- Adaptability (uses existing Modicon I/O devices, fewer I/O spares required)
- Easy installation of field wiring (intermixing any type of I/O)
- Retentive memory (for logic and numerical storage)
- Connectability (programming and peripheral devices plug directly into controller without effecting scan)
- Real-time, complete on-line programming

A typical P.C. can be divided into three sections. These are the processor, power supply, and input/output sections. In the 584 P.C., both the processor and the power supply are contained within the same cabinet. The user's manual provides all instructions for programming, wiring, and applying the P.C.

16-9 THE TI 510 PROGRAMMABLE CONTROLLER

The Texas Instruments Model 510 Programmable Controller is designed for use in small industrial applications.

Texas Instruments Model 510 Programmable Controller is priced and sized to replace control systems using from 4 to 20 relays, timers, and counters or a single drum timer. The die-cast aluminum housing and compact size make the 510 ideal for small industrial applications. Equipped with 20 I/O points and 256-word RAM for user program storage, a 510 program can include 64 internal control relays, 16 timers and counters, and 4 drum timers. Options include expanded I/O capabilities, to 40 I/Os, an EPROM programmer/downloader, and 24-V dc output levels.

The TI 510's power supply, controller, and I/O circuits are all contained in the same unit, thus replacing the mounting hardware, large enclosure size, and panel wiring needed for installation of relay banks. Field wiring is connected to a removable terminal strip for easy servicing.

Eight ac output points are standard on the 510-1101, and 24-V dc output points are available on the 510-1102. Load side indicators display the state of each output. Output power is switched by triacs on the 120-V ac version and transistors on the 24-V dc version that are socket-mounted for easy replacement.

The 12 inputs operate at ac or dc levels in groups of six and have load side LEDs to display the state of each input. The logic circuits are optically isolated from the input circuits up to 2500 V, thus protecting the internal circuitry from the damage caused by improper input signals.

A battery supplied with the 510 will maintain the user memory in RAM for 6 months with no ac power supplied.

Troubleshooting the 510 is easy with the built-in indicators and diagnostic routines. In addition to indicator lights on each input and output, the 510 has LEDs to show that the power, processor, and battery are good. Internal malfunctions are detected by internal diagnostics performed every half-second and flagged by these LEDs. User-initiated diagnostic routines on the P.C. and programmer I/O tests are accessed through auxiliary functions of the programmer and indicate problems in these areas.

Several simple-to-use functions are available to check out and debug the user program. The Status operation indicates the logical status (ON-OFF) of any logical element, input, output, and control relay. The Read operation indicates the current or preset values of counters and timers as well as the data values associated with drum instructions. The power flow indicator checks the condition of the logical element being addressed as well as all preceding elements in the line. The Force instruction checks out the operation of electromechanical devices connected to the P.C. as well as the user program by forcing an output or control relay on or off.

The Single Scan instruction allows the user to initiate a single scan of the program in order to follow the sequential logic of timers, counters, and other logical elements.

Logic available in the user program includes: 64 control relays (32 retentive, 32 nonretentive) with contacts that can be combined in series, parallel, or complex arrangements and used as many times as needed; 16 timers (with a range from 16 ms to 9 min); and counters (which can count up or down between 1 and 32,768). Timers and counters can be cascaded to give larger timing or counting values.

Jump instructions can be programmed to conditionally maintain a section of the user program in the same state as at Jump initiation. Master control relays can be programmed to conditionally maintain a section of the user program in the OFF state.

The TI Model 510's most powerful instruction is the drum timer. Four drum timers are available and each can replace up to 200 ladder logic elements. The drum can control up to 15 outputs through 16 timer steps of varying duration. The drum replaces electromechanical drum timers and their ladder logic equivalents.

16-10 PROGRAMMING THE TI 510 PROGRAMMABLE CONTROLLER

The TI Model 510 is programmed with a hand-held programmer. Its keyboard consists of 38 keys used in various functions. Figure 16-11 illustrates the TI Model 510 hand-held programmer. The *mode control* section consists of the following two keys:

P/R (program/run)—To switch the programmer from the Program mode to the Run mode and vice versa

SSN (single scan)—In the program mode, to initiate a single scan of the user program

The Program/Edit section consists of 28 keys:

STR (store)—To initiate a line of ladder logic

AND (AND)—To combine logic elements in series

OR (OR)—To combine logic elements in parallel

OUT (out)—To end a line of ladder logic with an output or control relay coil

NOT (NOT)—To invert the normal operation of ladder logic I/O elements as in normally closed contacts

X—To designate an input to the P.C.

Y—To designate an output to the P.C.

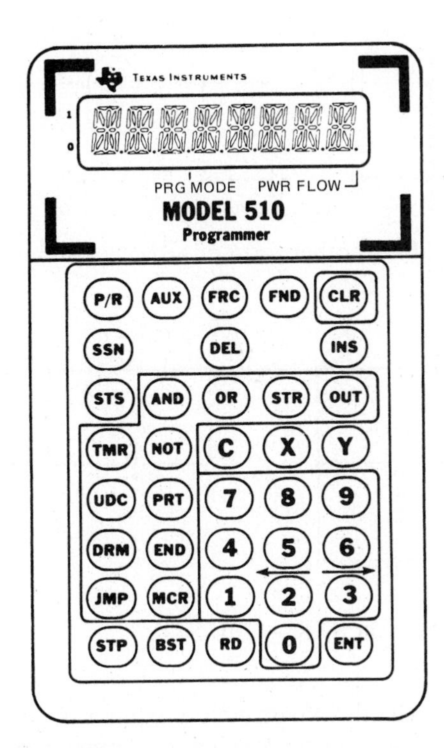

Fig. 16-11 Model 510 programmer. *(Texas Instruments.)*

C—To designate an internal control relay (CR)

TMR (timer)—To enter a timer in the user program

UDC (up-down counter)—To enter a counter in the program

PRT (protect)—To protect timers and counters from change by operator interface devices

DRM (drum)—To enter the drum timer into the program

MCR (master control relay)—To enter a master control relay into the user program

JMP (jump)—To enter a jump statement in the user program

END (end)—To terminate the MCR's field of control, JMP's field of control, and to specify the end of the scan field

Numeric keys (0–9)—In conjunction with other keys, to identify logic elements and enter preset values

INS (insert)—To insert new instructions into an existing user program

DEL (delete)—To delete a single element of the user program

ENT (enter)—To transfer instructions from the programmer to the controller and to enter masks for drum timers

The following four keys update the display for editing or program troubleshooting:

FND (find)—To call up logic elements and storage elements to the display

STP (step)—To move forward in the program display

BST (backstep)—To move backward in the program display

RD (read)—To display values associated with various logic elements

The following three keys in the monitor and troubleshoot section:

STS (status)—To read the status of internal or external elements

FRC (force)—To force on or force off logic elements

AUX (auxiliary)—To access several diagnostics and start-clear functions

The Clear key (CLR) is used to abort operations before the ENT key is pressed.

REPLACING AN EXISTING PROGRAM

If there is a program already in the TI 510 and a new program is to replace the existing program, the existing program must first be removed. The program is removed by using the AUX (auxiliary) and function 13 keys in the Program mode.

The steps for replacing an existing program are:

Step	Display
1. Press AUX key. Enter 13 from numeric key pad.	AUX 13
2. Press RD key.	CLR MEM?
3. Press ENT key.	P000 OK
4. Press CLR key.	END

After the Clear key has been pressed, the new program can be entered and the first instruction will be stored in memory location.

We will now write programs for some sample circuits using the TI 510.

PROGRAM CODE DEVELOPMENT FOR THE TI 510

Inputs and outputs can be either external devices (limit switches, pushbuttons, solenoids, indicators) or internal logic devices (control relays, timers, counters). Hard-wired I/O devices are designated by X's for inputs and Y's for outputs. Internal control relays are designated by C's.

The basic logical configurations available on the 510 are series, parallel, and complex combinations of logic elements. With the hand-held programmer, the number of series or parallel contacts that can be used in a given rung on the ladder diagram is unlimited.

Circuit Conversion. Series contacts are programmed using the AND instruction. For example, this rung of ladder logic would be programmed by the keystroke sequence that follows. The boolean algebra equation for the circuit in Fig. 16-12 is X9 · X10 = Y1. The TI 510 program for circuit is

STR X9 ENT	Begins the line with X9
AND X10 ENT	Adds X10 in series
OUT Y1 ENT	Ends the line with Y1

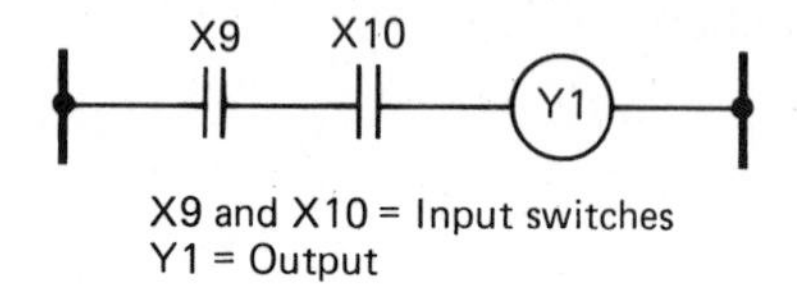

Fig. 16-12 AND Circuit.

Parallel contacts are programmed using the OR instruction. The boolean algebra equation for the circuit in Fig. 16-13 is (X11 + X12) · X13 = Y2. The TI 510 program is

STR X11 ENT	Begins the line with X11
OR X12 ENT	Adds X12 in parallel
AND X13 ENT	Adds X13 in series
OUT Y2 ENT	Ends the line with Y2

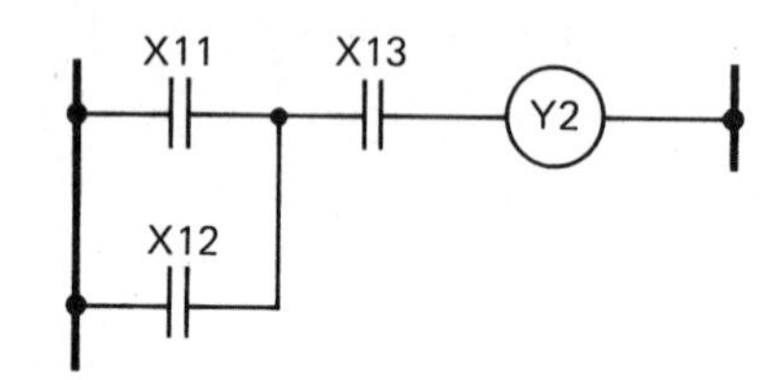

Fig. 16-13 OR/AND Circuit.

The circuit of Fig. 16-14 would appear to have the same program as that of Fig. 16-13, or vice versa. To avoid programming problems, use AND STR and OR STR instructions. The program for Figs. 16-13 and 16-14 would then be

TI 510 PROGRAM FOR FIG. 16-13	TI 510 PROGRAM FOR FIG. 16-14
STR X11 ENT	STR X11 ENT
OR X12 ENT	STR X12 ENT
STR X13 ENT	AND X13 ENT
AND STR ENT	OR STR ENT
OUT Y2 ENT	OUT Y2 ENT

When using OR STR and AND STR instructions, the STR instructions, for example, STR X9 ENT, STR X10 AND X11 ENT, STR X12 AND X13 ENT, OR STR will not work. The X9 instruction would not appear in the program. The OR STR instruction would only OR the last two STR; the AND STR instruction will only AND the last two STR (Fig. 16-15).

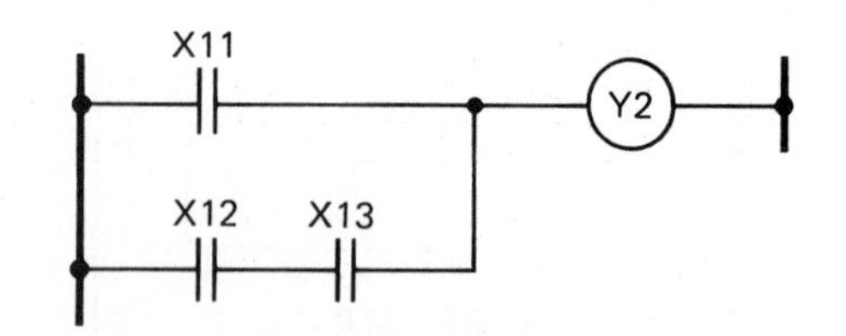

Fig. 16-14 Another form of AND/OR Circuit.

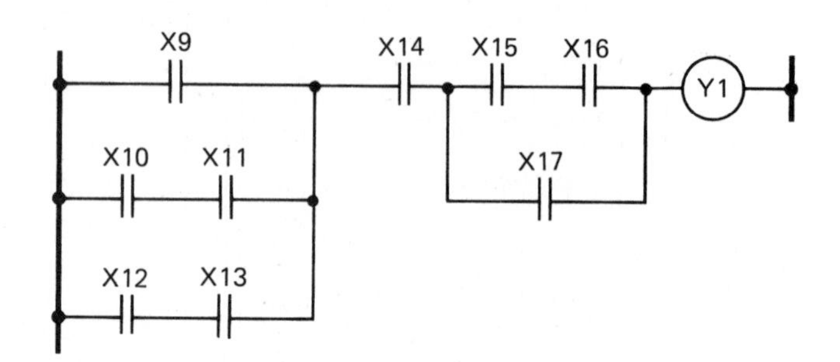

Fig. 16-15 OR STR—AND STR.

Timers. The timer instruction requires two words of user memory. The top line into the timer is the initiate line. All contacts on this line must be closed for the timer to operate. Opening the contacts on this line stops the countdown but does not reset the timer.

The bottom line into the timer is the reset/enable line. All contacts on this line must be closed for the timer to operate. Opening the contacts on this line stops the timer and resets it to the preset value.

When both the initiate and the reset-enable lines are closed, the timer counts down from the preset value to zero, at which time the output is energized.

The time base for the timer is 16.67 ms, and so a preset of 00030 will give a time delay of 0.5 s (30 × 16.67 ms) before Y1 is energized.

Programming the Timer Circuit. The timer in Fig. 16-16 is programmed with the following keystroke sequence:

STR X9 ENT	Initiate line
AND X10 ENT	
STR X9 ENT	Reset/enable line
AND X11 ENT	
TMR 1 PRT ENT	
30 ENT	
OUT Y1 ENT	

Labels. The next step in programming the circuit is to label each input, output, and control relay with letters the TI 510 understands. Notice that input switches are X's. The inputs can be from X9 to X20. The outputs are identified with the letter Y. The TI 510 has the eight outputs Y1 to Y8. Relays are identified by the letter C. Notice the numbering of the input/output controls and control relays in Table 16-1.

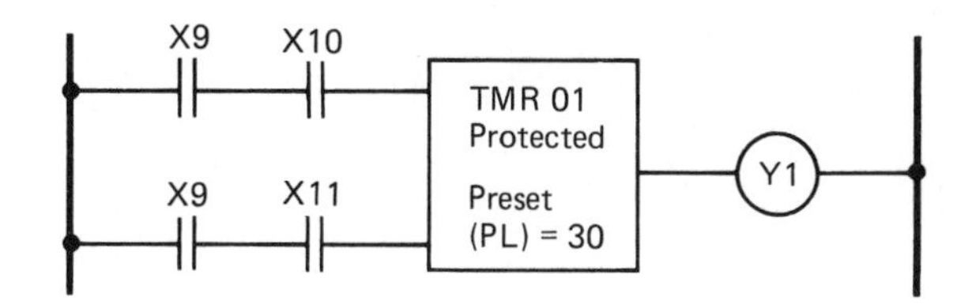

Fig. 16-16 Symbol for timer.

TABLE 16-1
I/O Symbols

I/O Control	Symbol	Notes
X9	PB1	Cycle start
X10	PB2	Manual feed
X11	LS1	N.O.
X12	LS2	N.O.
X13	LS3	N.C.
X14	LS4	N.C.
X15	ES1	Stop
Y1	B	Start-up bell
Y2	MS1	Forward-motor starter
Y3	MS2	Reverse-motor starter
C1	CR1	Control relay 1
C2	CR2	Control relay 2
C3	CR3	Control relay 3
C4	CR4	Control relay 4
TMR1	T1	T1 delay on
C5	CR5	Timer contacts

Each I/O point is assigned to a corresponding input or output device. Table 16-1 gives the input/output controls and their symbols and explains their significance.

Final Program. The ladder logic diagram is converted to a TI Model 510 program code and keyed into the hand-held programmer as follows:

STR X9 ENT
OR C1 ENT
AND NOT X15 ENT
OUT C1 ENT
STR C1 ENT
TMR 1 ENT
300 ENT (Bell rings 5 s)
OUT C5 ENT
STR NOT C5 ENT
AND C1 ENT
OUT Y1 ENT
STR C1 ENT
AND C5 ENT
MCR 1 ENT
STR C3 ENT
AND C2 ENT
OUT Y2 ENT
STR C3 ENT
AND NOT C2 ENT
OUT Y3 ENT
STR X11 ENT
OR C4 ENT
OUT C3 ENT
STR X12 ENT
AND NOT X13 ENT
OR C4 ENT
OUT C2 ENT
STR NOT X14 ENT
AND X10 ENT
OUT C4 ENT
END MCR 1 ENT

When placed into the TI 510 P.C., the program will provide the same control as a hard-wire relay circuit. It takes minutes to program the P.C. and hours to wire the relay circuit.

The conversion of a relay circuit, circuit specifications, or a logic diagram to a TI 510 ladder logic diagram is the hard part of programming the P.C. Once the ladder logic diagram is obtained, converting each coil circuit to a boolean algebra equation will make the programming of the circuit relatively easy.

The Model 510 is programmed with standard relay ladder diagram techniques. The first step is to define the control system, using relay techniques. Figure 16-17 on the next page illustrates a relay ladder diagram. This circuit uses conventional relay and electrical control symbols.

By analyzing the control system's circuitry it can be seen that it provides machine operation where a forward and a reverse motor are used. When the cycle start button (PB1) is pressed, CR1 latches on. Timer 1 begins timing and the warning bell (B) sounds. After the timer times out, the bell stops ringing and power is supplied to the rest of the circuit. If limit switch 1 (LS1) is closed, energizing CR3, and either LS2 or LS3 is open, then MS2 is energized and the reverse motor operates. If, while LS1 is closed, LS2 and LS3 are closed, MS1 is energized and the forward motor operates. If LS4 is closed and the feed forward button (PB2) is pressed, MS1 is energized and the forward motor operates.

Counters. The counter instruction requires three words of user memory (Fig. 16-18 on the next page). The bottom line into the counter is the reset/enable line. All contacts on this line must be closed for the counter to operate. Opening the contacts on this line stops the counter and resets it to zero.

Closing the contacts on the middle line into the counter decrements the counter. Closing the contacts on the top line into the counter increments the counter.

When the counter is at zero, the contact designated

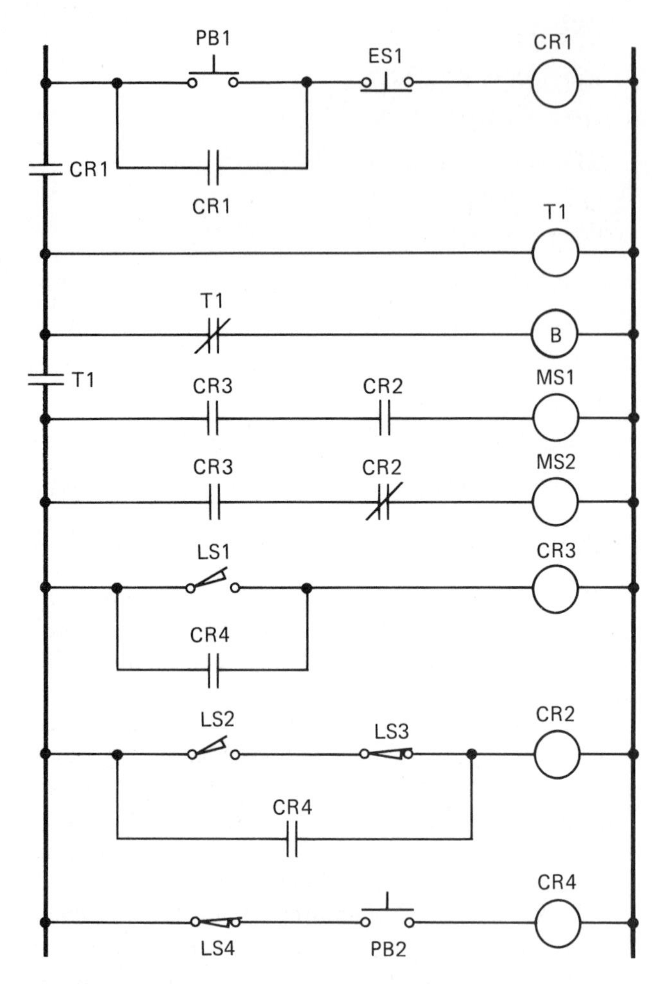

Fig. 16-17 A sample relay ladder diagram.

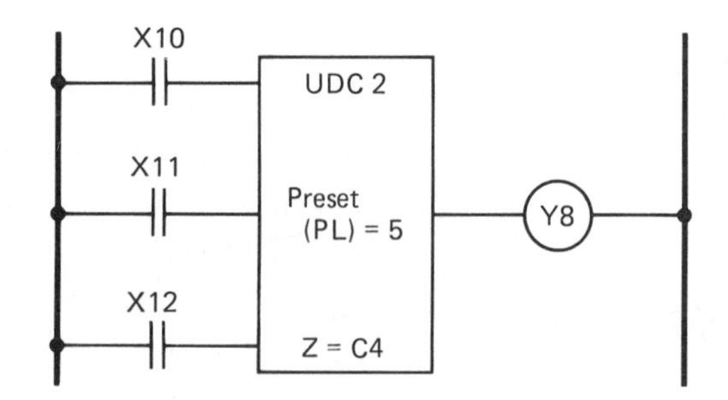

Fig. 16-18 Up-down counter.

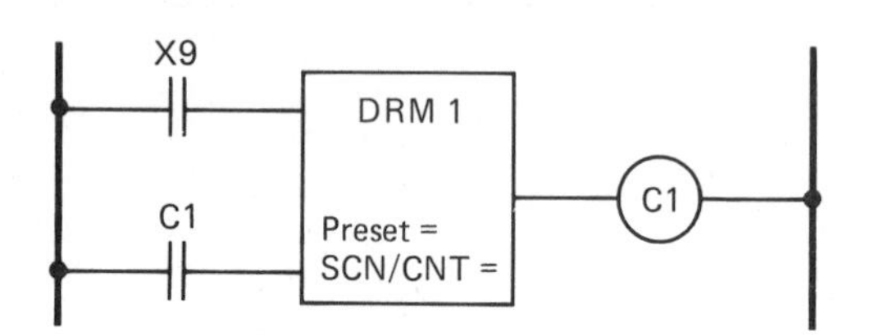

Fig. 16-19 TI 510 symbol for a drum timer.

as the zero flag (C4 in this example) and the output (Y8 in this example) are both energized. When the counter reaches the preset value, only the output (Y8) is energized, so the user can utilize the up-down counter in many configurations.

To program the counter the following keystrokes are needed:

STR X10 ENT	Enter up-count line
STR X11 ENT	Enters down-count line
STR X12 ENT	Enters reset/enable line
UCD 2 ENT	Names counter no. 2
5 ENT	Sets preset value to 5
C4 ENT	Designates C4 as zero flag
OUT Y8 ENT	Designates Y8 as output

ADDITIONAL INSTRUCTIONS

Drum Timers. The drum timer is a special instruction requiring only 50 words of user memory to achieve a series of timed outputs which are turned on and off sequentially as the 16 drum steps are executed. A single drum timer can replace up to 200 ladder logic elements. See Fig.16-19 for the drum timer symbol.

The bottom line into the drum is the reset/enable line, which must be closed for the drum to operate. Opening the contacts in this line resets the drum to the user-defined preset step.

The top line into the drum is the initiate line, which must be closed for the drum to operate. Opening the contacts in this line stops the drum operation without resetting it. With this feature, external logic "events" (interlocks, limit switches) can control the drum operating progress.

The output line is energized when the drum timer's programmed sequence has been executed.

If normally closed contacts from the output coil are used in the reset/enable line, the drum can be made to reset automatically upon completion. This configuration is useful in applications requiring continuous operation.

The drum's operation can be visualized as a cylinder model. This is illustrated in Fig. 16-20.

In this model, the drum can be imagined as a rotating cylinder with 15 outputs. At each step in the cylinder's rotation, the output mask gives every output a state (on or off). The cylinder can rotate through 16 steps before halting. The measure of time the drum remains on a given step is programmed using the counts-per-step parameter. The length of one count is specified in the scans-per-counts parameter, where each scan is 16.67 ms.

Drum programming can be simplified by entering all necessary information onto a drum timer programming form. This is illustrated in Fig. 16-21. A later example will show how the drum timer is used to program industrial-type applications.

Master Control Relay and Jump. The master control relay (MCR) and jump (JMP) instructions require two

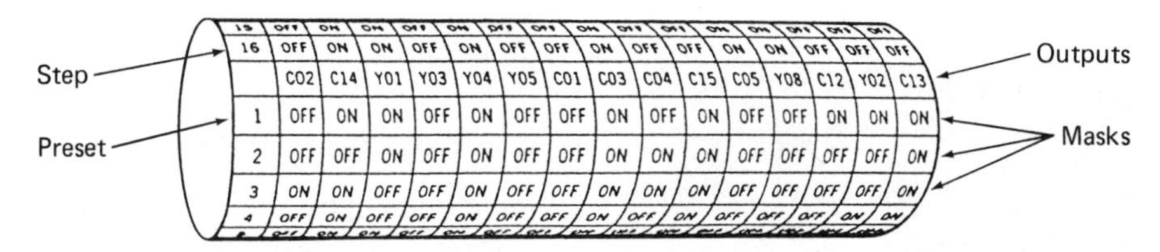

Fig. 16-20 Drum timer cylinder model. *(Texas Instruments.)*

DRUM 01 PRESET = 02		C	C	Y	Y	Y	Y	C	C	C	C	C	Y	C	Y	C
SCN/CNT = 32000		0	1	0	0	0	0	0	0	0	1	0	0	1	0	1
STEP CNT/STP*		2	4	1	3	4	5	1	3	4	5	5	8	2	2	3
1	00002	0	1	1	0	1	0	0	1	0	1	0	0	1	1	1
2	00001	0	0	1	0	1	0	0	1	1	1	0	0	0	0	1
3	02000	1	1	0	0	1	0	0	1	1	1	0	0	0	0	1
4	00000	0	1	0	0	1	0	0	1	0	1	0	0	0	1	1
5	00202	0	1	1	0	1	0	0	1	0	1	0	0	0	0	0
6	10000	0	1	1	0	1	0	0	1	1	0	0	0	1	1	0
7	00003	0	1	1	0	1	0	0	0	1	0	0	0	0	0	0
8	00010	0	0	1	0	1	0	0	1	1	1	0	0	1	0	0
9	00001	0	1	0	0	1	0	0	0	1	1	0	0	0	1	0
10	02000	1	1	0	0	1	0	0	1	1	1	0	0	1	0	0
11	01230	0	1	1	0	1	0	0	1	1	1	0	0	0	0	1
12	00300	0	0	1	0	1	0	0	1	1	1	0	0	0	0	1
13	00001	1	1	1	1	1	0	0	0	0	0	1	1	1	0	1
14	00001	0	1	1	0	1	0	0	1	1	1	0	0	1	0	0
15	00002	0	1	1	0	1	0	0	1	0	0	1	1	0	0	0
16	00002	0	1	1	0	1	0	0	1	0	0	1	1	0	0	0

*Counts/step indicates length of time outputs are in the prescribed state for each step.
1 – indicates output will be ON during step.
0 – indicates output will be OFF.

Fig. 16-21 Drum timer programming form. *(Texas Instruments.)*

lines in the program—one to begin the operation and one to end it.

The MCR and JMP instruction are programmed at the desired location and identified by a number from 1 to 8. Later in the program the END of MCR or JMP instruction is inserted to terminate the effect of the instruction on the program.

When the MCR is activated by opening contacts in its input, all output elements between the MCR and the corresponding END of MCR are turned off and kept off regardless of their input conditions.

Whenever the MCR or JMP is deactivated by energizing, its input or outputs are affected by that MCR or JMP.

The TI 510 program for the instructions in Fig. 16-22 would be the following: When the MCR is activated by opening X12, the output Y3 is turned off and kept off regardless of the status of X13. When the MCR is deactivated by closing X12, control of Y3 is returned to X13.

When JMP is activated by opening the contacts in its input, all output elements between the JMP and the corresponding END of JMP are held in the same state as when the JMP was activated, regardless of their input condition.

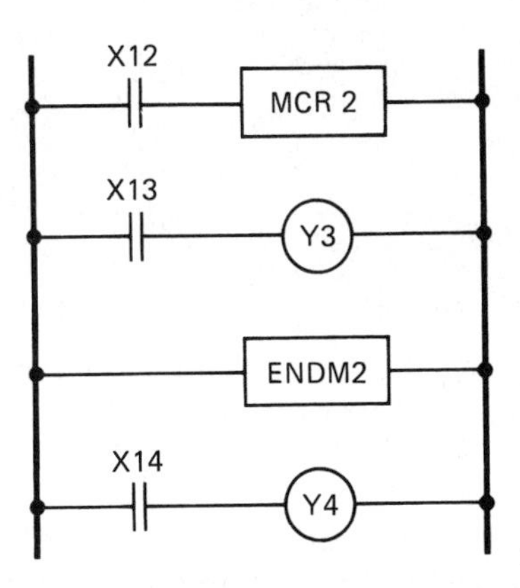

Fig. 16-22 Master control relay for TI 510 P.C.

In the JMP example in Fig. 16-23, if Y1 is on when the JMP is initiated by opening X9, then Y1 will stay on whether X10 is on or off. If Y1 is off when the JMP is initiated, then Y1 will stay off whether X10 is on or off. When the JMP is deactivated by closing X9, the status of Y1 is controlled by X10 as usual. Since Y2 is outside the field of the JMP's control, it is not affected by the JMP and operates normally.

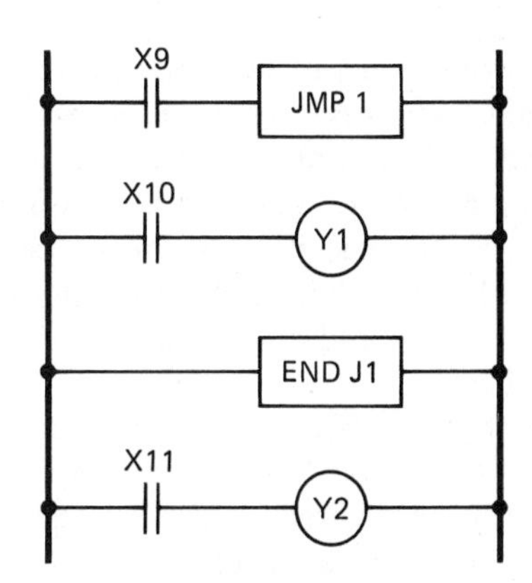

Fig. 16-23 Jump instructions for TI 510 P.C.

INSTALLATION

Installation of TI 510 (Fig. 16-24 on the next page) is simple because it is completely self-contained: The power supply, sequencer, and I/O circuits are all on board. Of course, to avoid injury to personnel or damage to equipment the installation instructions in the user's manual should be read and understood com-

Fig. 16-24 A TI 510 programmable controller with I/O simulator. *(Texas Instruments.)*

pletely before attempting to install, operate, or program the Model 510 Programmable Controller.

Figure 16-25 shows the wiring diagram for the connections of the inputs and outputs to a TI 510 programmable controller.

Planning for installation of the 510 should include enclosure selection, grounding, and master control relay considerations.

The 510 should be mounted in a NEMA type 12 enclosure to protect the controller from atmospheric contaminants, including oils, moisture, conductive dust and particles, and corrosive agents. Inside the enclosure the controller must be mounted with its cooling fins vertically oriented so that airflow will be even.

Proper grounding not only minimizes personnel hazard but also reduces electrical noise caused by electromagnetic inductive coupling from other wires, cables, and machinery. The 510, its enclosure, and all associated controlled devices must be properly grounded. Recommended practices found in the *IEEE Ground Book* and *ANSI Standard C1141* should be observed.

As a safety precaution, a hard-wired master control relay should be installed as part of every P.C. In an emergency this hard-wired MCR provides for removal of power from all output devices in the controlled system. All emergency OFF switches must be wired in series with the MCR to shut down the controlled system by removing power from the I/O points without affecting controller operation. When the STOP button is pressed, the MCR de-energizes, removing power from the I/O points. Note that power to the controller is not affected.

Figure 16-26 on page 302 illustrates the conversion of small relay circuit in the key stroke sequence to place the circuits in the TI 510 programmable controller. We will use this circuit to show how easy it is to troubleshoot P.C. programs.

ALTERING THE PROGRAM

Sometimes upon the completion of programming a controller, it is necessary to review the program, add another ladder logic instruction, delete a ladder logic instruction, or to change a ladder logic instruction that was previously programmed. This can be accomplished with the use of the following four keys:

FND Key. The FND (find) key is used to search forward in the program and display the next occurrence of one of the following:

- I/O element (X,Y,C. Example: FND X09)
- Ladder logic storage location (Example: FND 052 —find 52nd word of user memory)
- Ladder elements (TMR, UDC, DRM, MCR, JMP, ENDM, ENDJ, END).

STP Key. The STP (step) key is used to advance the display forward to the next ladder logic instruction and from one function to another in the program.

BST Key. The BST (backstep) key is used to step backward through the program and display the previous ladder logic instruction.

RD Key. The RD (read) key is used with the STP and BST keys to display values associated with program instructions that require more than one word (UDC, DRM, TMR).

Problem 1. If the program of Fig. 16-26 had been placed into the TI 510 P.C., step 14 would have had to be corrected for the program to work. After punching in the program, review the program for any errors by pushing the FND key then X9 ENT. This will bring you back to line 1 of the program. Push the STP (step) key, and this will advance you to line 2 of the program. Continue pushing the STP key to examine the rest of the program. Push the BST (backstep) key if it is desired to step backward through the program. When the line with the error is found, correct step 14 by changing the instruction to read STR X13 ENT.

Problem 2. After verifying that the program is correctly entered into the P.C., it is now desired to add switch X15 in series with X13 to energize C3. To insert switch X15 without redoing the whole program, push FND X13 ENT. This brings you to the 14th line of the program. Next push the STP (step) key and you should be on line 15 (C3 ENT). Now push the INS (insert) key and punch in AND X15 ENT. The instruction AND X15 has been inserted between line 14 and 15 of the program.

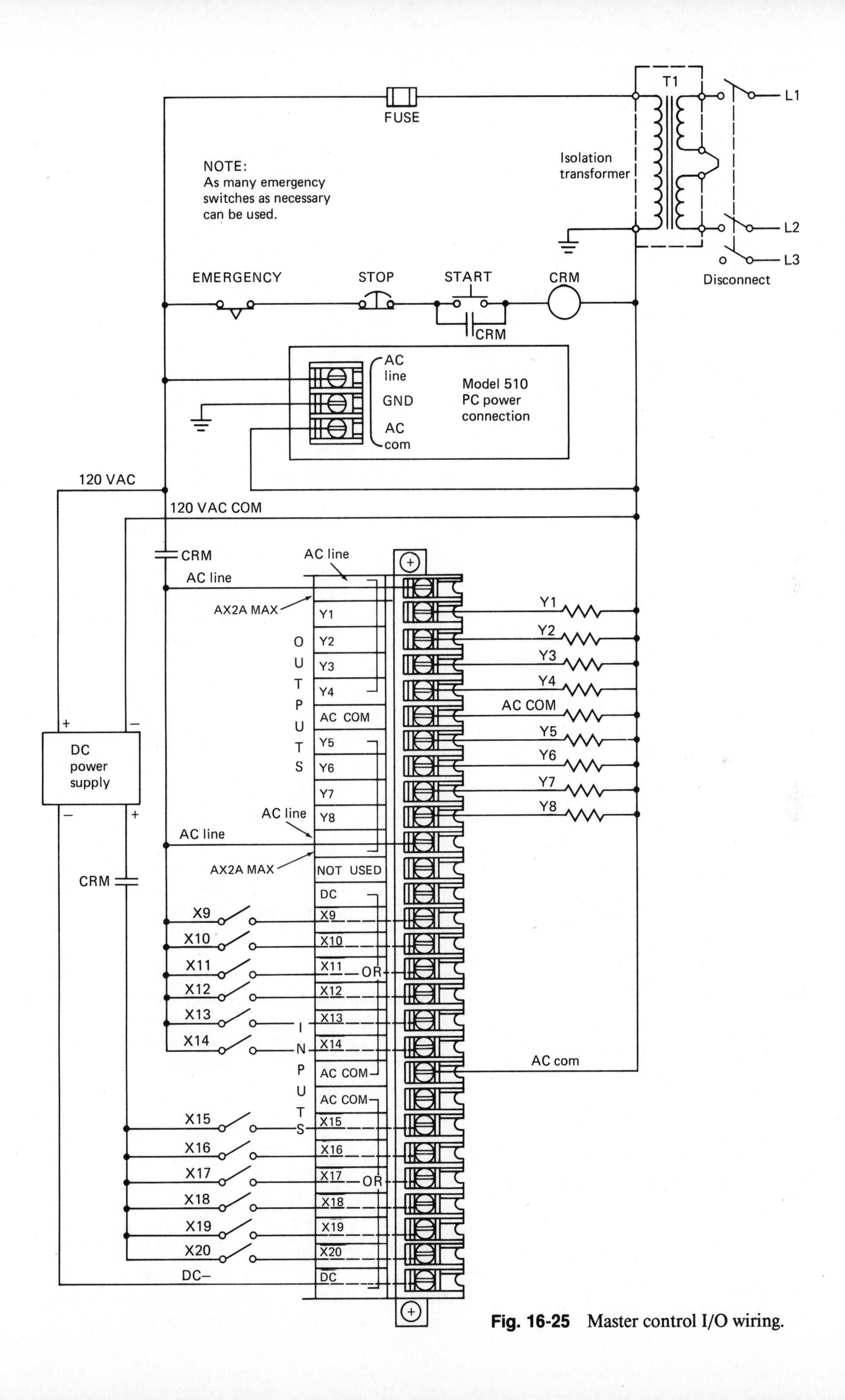

Fig. 16-25 Master control I/O wiring.

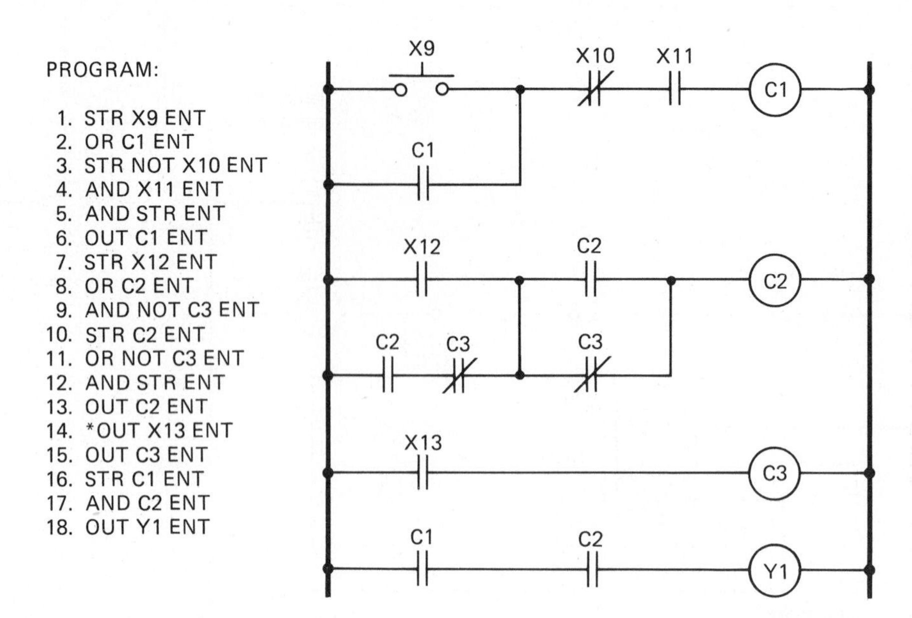

Fig. 16-26 Small relay circuit converted to keystroke sequence—Step 14 illustrates error in programming.

Problem 3. It is now desired to delete switch X15 that was just put into the program. Push FND X15 ENT. This should bring you to the ladder logic instruction just added. Now push the DEL (delete) key and enter. You should have just deleted the instruction that was added above.

FORCING

This instruction is used in troubleshooting. The FRC (force) key is used in the program on the Run mode to do the following:

- Force and hold selected ladder logic I/O elements on or off
- Unforce (release) selected ladder logic elements
- Read the forced state of selected ladder logic I/O elements
- Locate and read the next forced element

Forcing I/O Points. Once forced, an I/O element remains forced until unforced in the Program or Run modes.

Problem 4. Enter the program of Fig. 16-27 into the TI 510 Programmable Controller.

The following keystrokes illustrate how to force Y1 in Fig. 16-27 on and off using the FRC key.

Key	Display	
FRC	FRC	
Y	FRC Y	
1	FRC Y 01	
ENT	Y1 = ON	
ENT	Y1 = OFF	Each time ENT is pushed it changes Y1 to the on and off state.

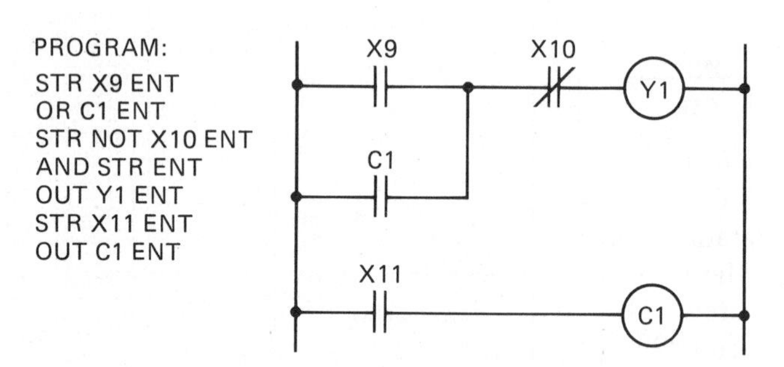

Fig. 16-27 Circuit to demonstrate forcing.

No matter what contacts or switches open or close, Y1 remains in the forced state until it is unforced.

Now to get Y1 from the forced state and back to normal operation use the following keystrokes:

Key	Display
CLR	END
END	END
FRC	ENDF ALL
Y	ENDF Y
1	ENDF Y 01
ENT	UNFORCED

Problem 5. The following keystrokes illustrate how to force X11 closed and opened using the FRC key:

Key	Display	
FRC	FRC	
X	FRC X	
1	FRC X 01	
1	FRC X 11	Pushing the ENT key changes the forced state
ENT	X11 = ON	
ENT	X11 = OFF	

With X11 forced on, C1 energizes and C1 contacts close, energizing Y1. Opening X10 will open the circuit to Y1, but X11 is forced closed, keeping C1 energized.

The following keystrokes illustrate how to unforce X11 and return operation back to normal:

Key	Display
CLR	END
END	END
FRC	ENDF A11
X	ENDF X
1	ENDF X 01
1	ENDF X 11
ENT	UNFORCED

MONITORING A TIME-DELAY PROGRAM FOR THE TI 510 P.C.

After the program instructions are entered into the P.C., the user may monitor the instruction in the Program mode to ensure the accuracy of the program. The preset times of timers and counters may be monitored and changed if desired. Counters, timers, and drum timers can be accessed in the Run mode to monitor the execution of the program.

Problem 6. Enter the program that follows (Fig. 16-28) into the TI 510 P.C. The following keystrokes illustrate accessing a timer to monitor or change present values:

Key	Display
TMR	TMR
1	TMR 01
RD	CV = 00100

Now close switch X9, and you will be able to watch the time countdown. Open switch X9, and it resets the timer. Press the RD key again, and the display should show PV = 00100. This is the preset value of the programmed timer. To change this value, simply punch in the new time desired.

If it is desired to see the state of relay C1, enter the following keystrokes:

Key	Display
CLR	TMR1
STS	STAT
C	STAT C
1	STAT C 01
RD	C1-OFF

This tells us that C1 is not energized. Now close switch X9, and after the time delay, C1 will be energized and the display will now show C1 as on.

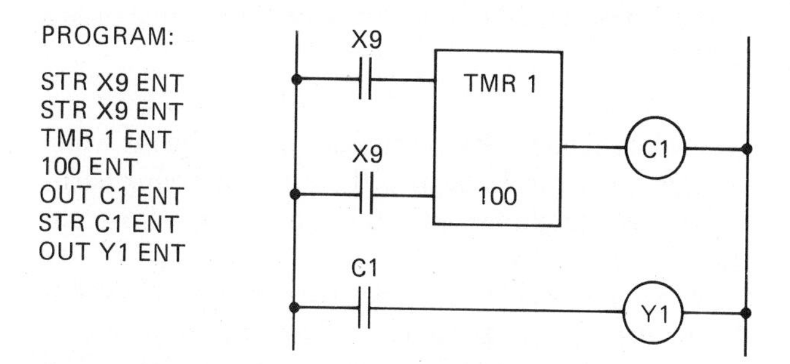

Fig. 16-28 Circuit to demonstrate monitoring the TI 510 time-delay.

STS KEY

The STS (status) key is used in the Program or Run mode to display the status (on or off) of a specific ladder logic I/O element.

16-11 THE PC 700 AND PC 900—WESTINGHOUSE'S PROGRAMMABLE CONTROLLERS

Westinghouse manufactures two P.C.s—the PC 700 and the PC 900—which can be broken down into three sections. Section one we will call the *input* section, which converts real-world signal levels from input devices such as switches, contacts, pushbuttons, pilot devices, and photocells to low voltages that are suitable for use by the P.C. Section two consists of the solid-state P.C., which electronically processes all data necessary to make the system operate. Part of this section is the memory, which receives instructions from the program loader. The third section of the P.C. is the *output.* In this section the low-level processor signals are converted to voltage levels required to operate output devices. For example a 5-V dc processor signal is raised to 120 V ac to drive a control relay.

The P.C. uses the reliability of solid-state control with the flexibility of a microprocessor to solve both the input and the output problems of control system. It does this by performing the necessary logic to process all input data as the instructions programmed into the memory call for the responses that are to be made.

TABLE 16-2
Comparison of the Westinghouse PC 700 and PC 900

Function	PC 700	PC 900
Discrete input modules	Up to 256	Up to 128
Analog input modules*	Up to 32*	Up to 16
Register input modules*	Up to 32*	Up to 16
Discrete output modules	Up to 256	Up to 128
Analog output modules†	Up to 32	Up to 16†
Register output†	Up to 32	Up to 16†

* Analog and register I/O share the registers so that the total of the two must not exceed 32.

† Analog and register I/O share the registers so that the two must not exceed 8 inputs and 8 outputs.

The Westinghouse P.C.s solve control problems while using relay control design as their programming basis.

The Westinghouse PC 700 and PC 900 are compared in Table 16-2.

Westinghouse P.C.s can be programmed by two different types of programmers. The smaller programmer is the hand-held Mini loader NLPL 789. The other is the NLPL 780, which has a CRT to display the circuits as they are developed from ladder diagrams. Once the reference diagram has been derived from a conventional relay diagram and is displayed on the screen of the CRT, the following procedure should be followed before it is loaded into the P.C.:

1. Ensure that circuit modification is not required so that the reference diagram is compatible with the scanning process used in the P.C.
2. Rearrange contacts so that interconnections are more easily recognizable.
3. If space is a problem, the circuits may be modified to fit within the allowable contact area.
4. All contacts and coils must be assigned appropriate reference numbers.

Figure 16-29*a* is a relay ladder diagram. A reference diagram for the PC 700/900 family is shown in Fig. 16-29*b*. The reference diagram is loaded into the P.C. controller via the keyboard on the programmer loader.

CRT PROGRAMMING PANELS

Two programming panels, the NLW 440 and the NLD-441, are available to program and monitor the PC 430 and PC 431 P.C.s. These panels display relay ladder diagram symbols on a CRT screen.

NLW 440 Programming Panel. The NLW 440 is lightweight and rugged and utilizes pushbuttons for entering N.O. and N.C. contacts, branch circuits, and output coils. Function buttons are used to enter time settings, shift registers, and counter data. Up to eight parallel branch circuits containing 10 elements per line may be entered at one time. A cassette tape deck (NL-442) can be used with this unit.

Each element is visually displayed on the CRT as it is entered, verifying each entry by IN for input, CR for output, TN for on delay, and TF for off delay. Activated contacts are highlighted to simplify troubleshooting.

Ladder diagrams are entered left to right on the CRT. A position indicator indicates where in the circuit the next entered element will be shown.

The CRT provides contact-by-contact editing of circuits for easy modification. Since the memory is nonformatted (each circuit uses only what is required), any circuit may be modified without re-entering the total circuit. Modifications within the 10-element-line by eight-branch circuit grid require no change in output line number, which simplifies drawing changes.

Inputs and outputs may be forced to the ON or OFF state, providing a simple means of disabling or bypassing inputs or outputs.

Hard-copy listings of the complete control program in ladder diagram format may be generated by teletype or printer via the CRT program panel.

NLD-441 Programming Panel. As an alternative to the NLW 440 programmer, a standard Datapoint 11000 CRT terminal with built-in magnetic tape cassette drives may be used.

When the unit is in operation, the program is loaded into the terminal's processor from the system tape located in the rear tape drive. This program permits the user to enter and display relay format control sequences on to the tape and to store the contents of the tape in the P.C.'s memory, or to recall the controller's contents on to the magnetic tape for program storage. Up to six parallel branch circuits containing 10 elements per line can be displayed at one time.

This CRT terminal is very versatile and can be used as a calculator, computer terminal, etc., or, by using special diagnostic tapes, provides detailed troubleshooting information concerning the P.C. or detailed diagnostic information about the machine being controlled.

Figure 16-30 on page 306 illustrates the typical CRT display for a programmed circuit. The relay circuit shown in Figs. 16-31 and 16-32, on page 307, appears in a hard copy printout as programmed on the PC 430 or PC 431.

16-12 WESTINGHOUSE PROGRAM LOADER NLPL 789

Westinghouse's Mini Loader NLPL 789 (Fig. 16-33) can be held in a person's hand. It's not much larger than a calculator. It's compact enough to fit in

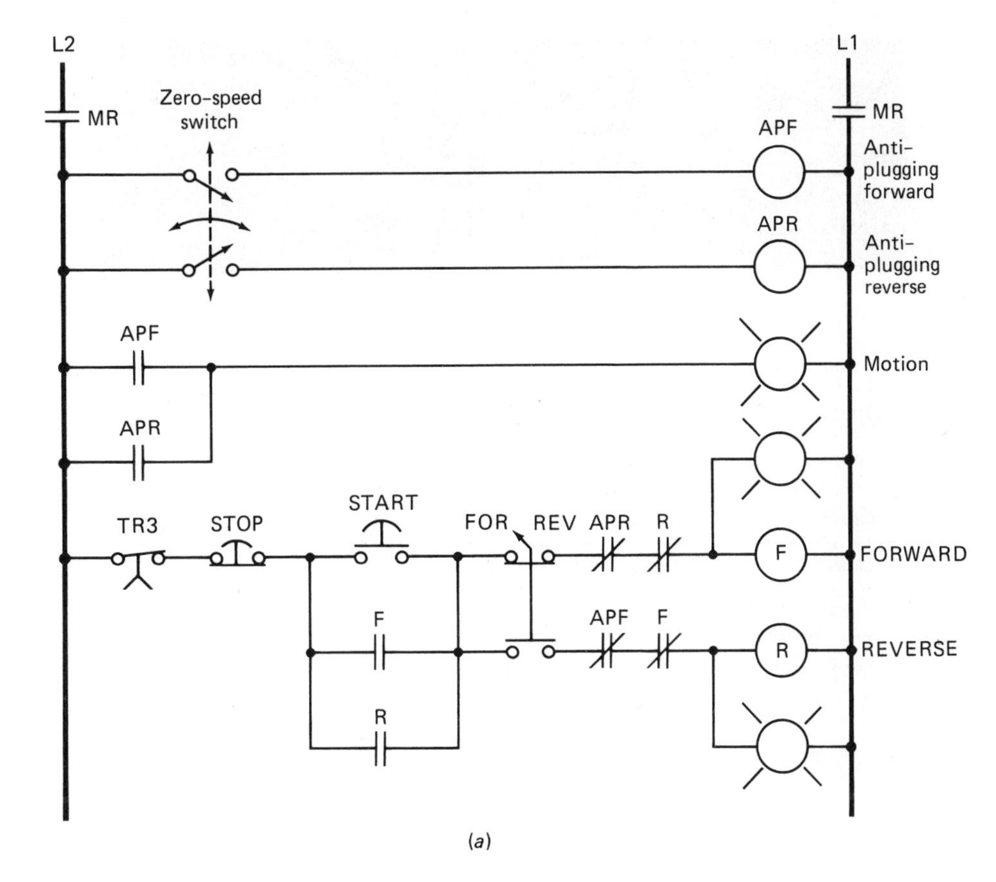

(a)

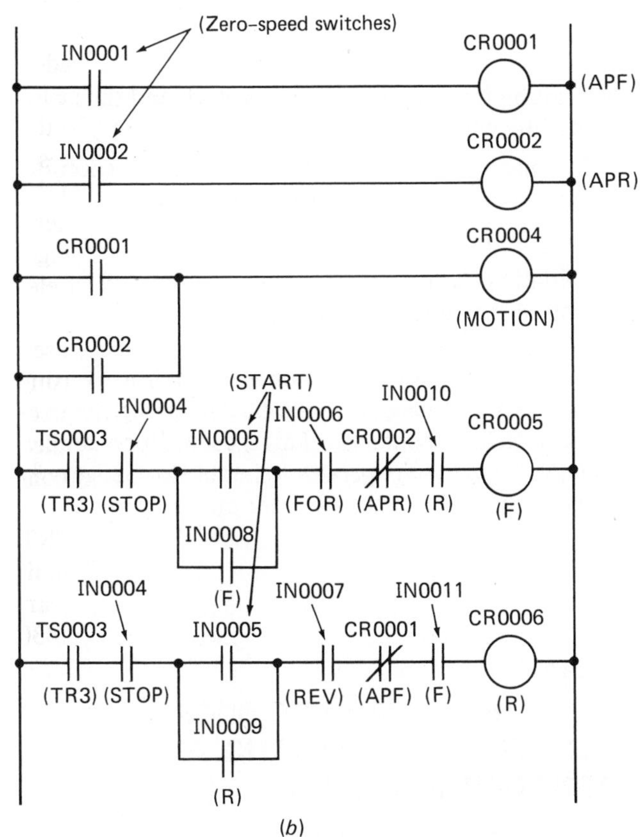

(b)

Fig. 16-29 (a) Relay ladder diagram. (b) The 700/900 P.C. reference diagram made from relay circuit.

your hand but versatile enough to mount on your machine.

The Mini Loader allows you to program and monitor ladder diagrams with the ease of expensive CRT models. It can also display diagnostic messages, process alarms, or prompt data entry. Complete with sealed, tactile feedback keyboard and computer-tested, the loader is engineered for a variety of industrial applications. It is designed for total compatibility with either the Westinghouse PC 700 or PC 900 programmable controllers.

As can be seen from Fig. 16-33, the display is relatively small. When the loader is placed in the Message mode, diagnostic or prompting messages will appear on the display to prompt operators, maintenance personnel, or operating engineers.

The number of messages and the length of each message are limited only by the memory size of the processor. If a message does not fit on the 31-character display, the loader automatically displays consecutive portions of the message at short intervals until the whole message is displayed.

Characters are stored in hold registers in the P.C. Each register can store two alphanumeric characters, providing a storage capacity of thousands of characters per P.C. Characters are loaded into the holding regis-

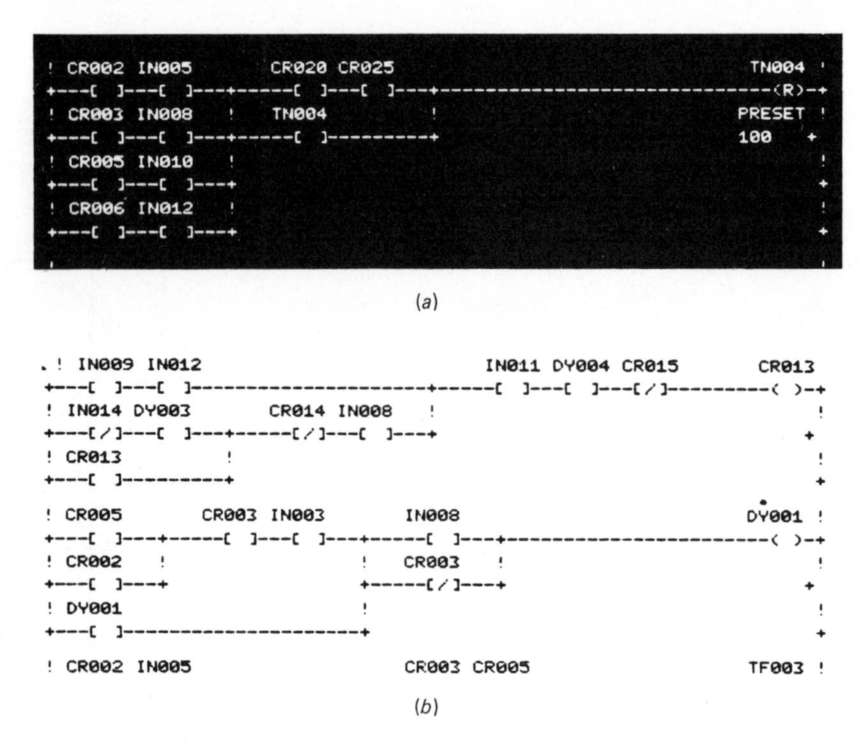

Fig. 16-30 (a) Typical CRT display. (b) Hardcopy printout. *(Westinghouse.)*

ters by using the Register mode of either the Numa-Logic CRT loader or the Mini Loader. The Westinghouse family of programmable controllers and loader is illustrated in Fig. 16-34 on page 308.

The Westinghouse P.C. can use special codes that can be used to maximize memory utilization. For example, subroutined messages can be used to store common words or phrases to be used later as part of longer messages. Other codes can be used to handle real-time decimal data. Thus codes could be used to display the time of day, date, or process data.

Figure 16-35 on page 309 illustrates a program that would be used to convert the circuit of Fig. 16-31 to a program for the Westinghouse PC 430 or 431 programmer.

Westinghouse manufactures its P.C.s and loaders with five *assignable keys.* These five keys can be programmed to each represent a sequence of up to 200 keystrokes. Assignment keys can expand your applications by providing an easily understood, customized operator interface. These assignable keys are the letters K1 to K5 on the Mini Loader.

The assignable keys can be used for such functions as

- Calling a message to request numerical data from the operator
- Locating the proper register destination for that data
- Waiting for the operator to respond, and then proceeding with other requests/responses
- Forcing specified contacts or changes to specified registers while blocking all other changes
- Programming repetitive ladder diagram structures
- Initiating sequences or individual modes of operation in machines or processes

An assignable key instruction is available to override the keylock, if needed, to make changes in register contents, or to force contacts. This is useful, for example, when you want your operator to be able to change only one specific item, such as a timer preset.

16-13 THE MINI LOADER PROVIDES CRT-TYPE PROGRAMMING AND MONITORING

The Mini Loader NLPL 789 from Westinghouse incorporates free-format programming. It programs 7 rows of 10 elements, one row at a time. It uses ladder dia-

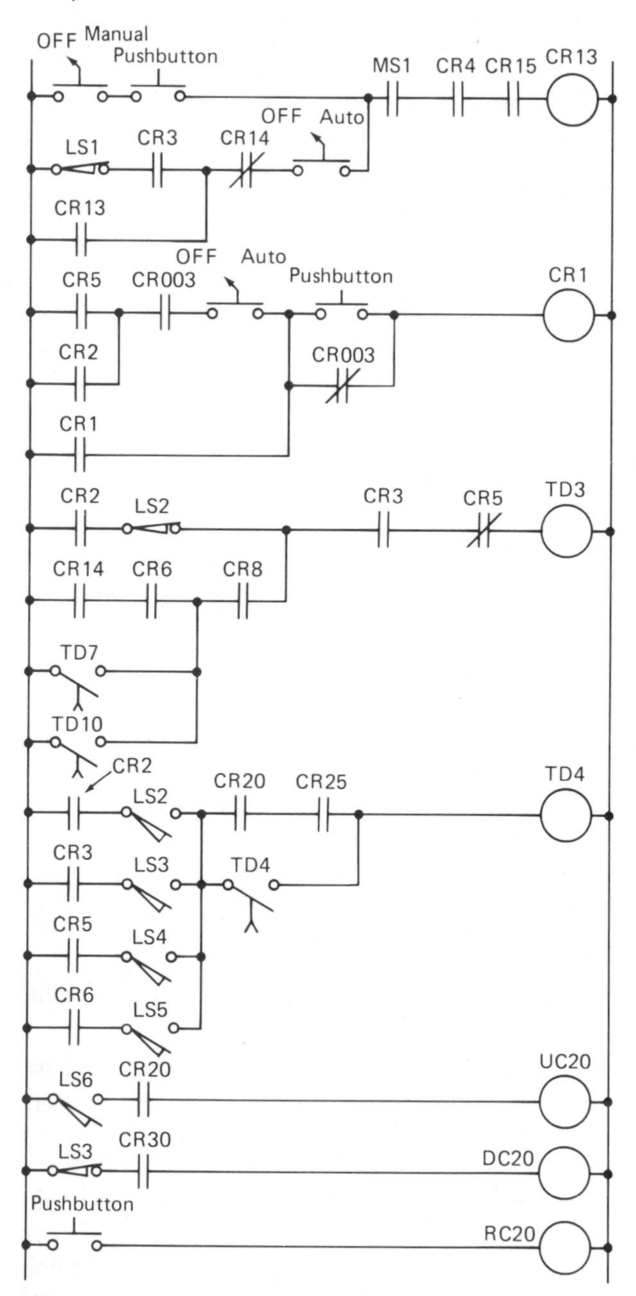

Fig. 16-31 Relay diagram.

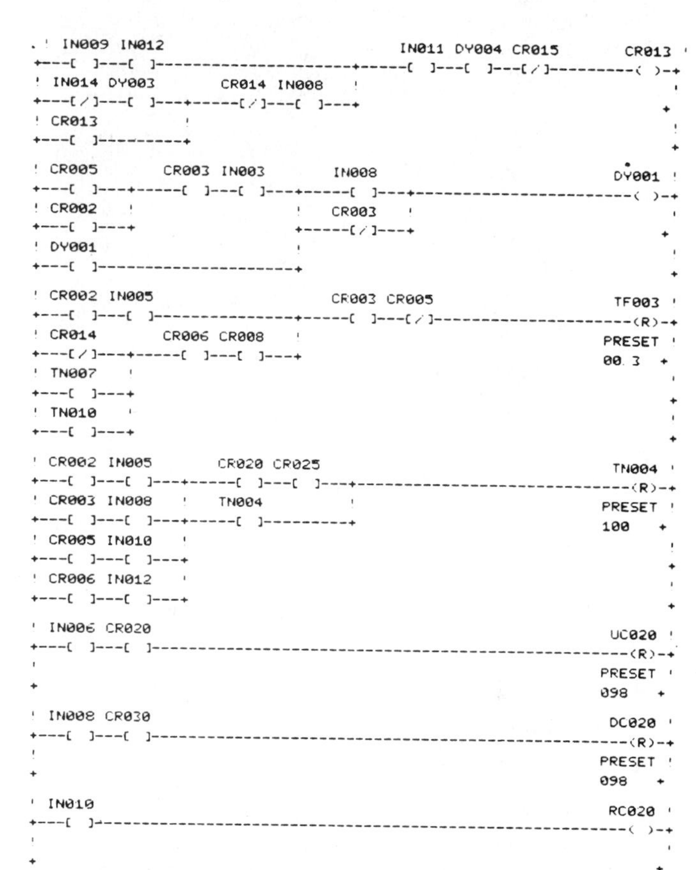

Fig. 16-32 The relay diagram as it appears on the PC 430 or PC 431 Programmer. *(Westinghouse.)*

Fig. 16-33 Westinghouse Mini Loader 789. *(Westinghouse.)*

gram symbols and continuously indicates the ON, OFF, and forced status of all displayed contacts. The Mini Loader can insert, delete, or alter any complete ladder rung within the P.C. memory, either on or off line. It can also monitor and force any row of contacts while monitoring register data.

The Mini Loader has extensive search capabilities. It can search for

- Coils
- Contacts

Fig. 16-34 The Westinghouse family of P.C.s, I/Os, and loaders. *(Westinghouse.)*

- Registers
- Forced contacts and coils
- Next ladder rung
- Previous ladder rung

During a search, the Mini Loader remembers where you initiated the search and allows you to return to the same point upon request. It can even find the location of a requested contact within a ladder rung.

LOW-COST TAPE MODE

Low-cost audio cassette decks can be used for program storage. When used with the Mini Loader, a low-cost audio cassette deck approaches the performance level of tape decks costing thousands of dollars. Westinghouse's new analog bar graph display, coupled with an Increase and Decrease key, provides a flexible process interface. You can customize each variable with its own four-character label. The unique Micro Zoom capability of the Westinghouse Mini Loader gives the bar graph an unbelievable ± 0.2 percent resolution. In addition, the continuous bump feature of the Mini Loader permits you to insert or delete steps within a drum controller sequence or characters within a message even while the P.C. is on line.

The keylock on the Mini Loader locks out forcing, register alternations, key assignments, and ladder diagram changes. It can be disconnected inside the loader for monitor-only operation. Westinghouse's security system is even more extensive. Internal jumpers can be used to disable ladder programming, key Y assignment, and Force modes. This security system offers much more flexibility than a simple keylock.

The optional Mini Loader tape capability permits you to record, load, or verify ladder diagrams from standard audio cassettes. You can also record, load, or verify data on line without disturbing the ladder diagram, which means it can be used for tasks such as

1. Process recipe selection
2. Palletizer stacking pattern generation
3. Data acquisition
4. Paper tape reader applications
5. Message libraries
6. Assignable key libraries

The powerful Register mode of the Mini Loader NLPL 789 allows the user to monitor and change register contents in five common formats. These are decimal, binary, ASCII, hexadecimal, and analog bar graph. The display can be updated as rapidly as 5 times a second.

The registers can be monitored simultaneously in any format. A single stroke of the key advances the display from one register to the next for simplified monitoring and loading of large tables of data.

16-14 THEORY AND APPLICATIONS OF ALLEN-BRADLEY'S PLC-2 PROGRAMMABLE CONTROLLERS

Allen-Bradley's PLC-2 family of programmable controllers are efficient and effective in providing industrial control. Allen-Bradley was one of the first companies to manufacture P.C.s that have proved capability of enhancing plant productivity. The PLC-2 is illustrated in Fig. 16-36 on page 310.

Allen-Bradley defines its P.C. as a solid-state logic control device for industrial applications. As the term *programmable* implies, its memory can be readily changed without the extensive rewiring that is necessary to make the same changes in a hard-wired relay control system. The programmable controller is composed of two parts: the processor and the interface. The processor is the heart of the P.C. It makes control decisions based upon the application program stored in its memory. Input/output modules form the interface between the processor and the controlled operation. Sensor devices wired to input modules provide machine or process status information to the processor, which then decides the appropriate status for output devices, such as motor starters, solenoids, and control valves.

The first application of Allen-Bradley P.C.s was to replace large panels of hard-wired relays. Changes in the operation could be made with relative ease by replacing the program stored in the P.C.'s memory. This is done by inputting ladder diagram programs into the industrial terminal CRT with a modular keyboard. Whereas hours or days of relay wiring would be re-

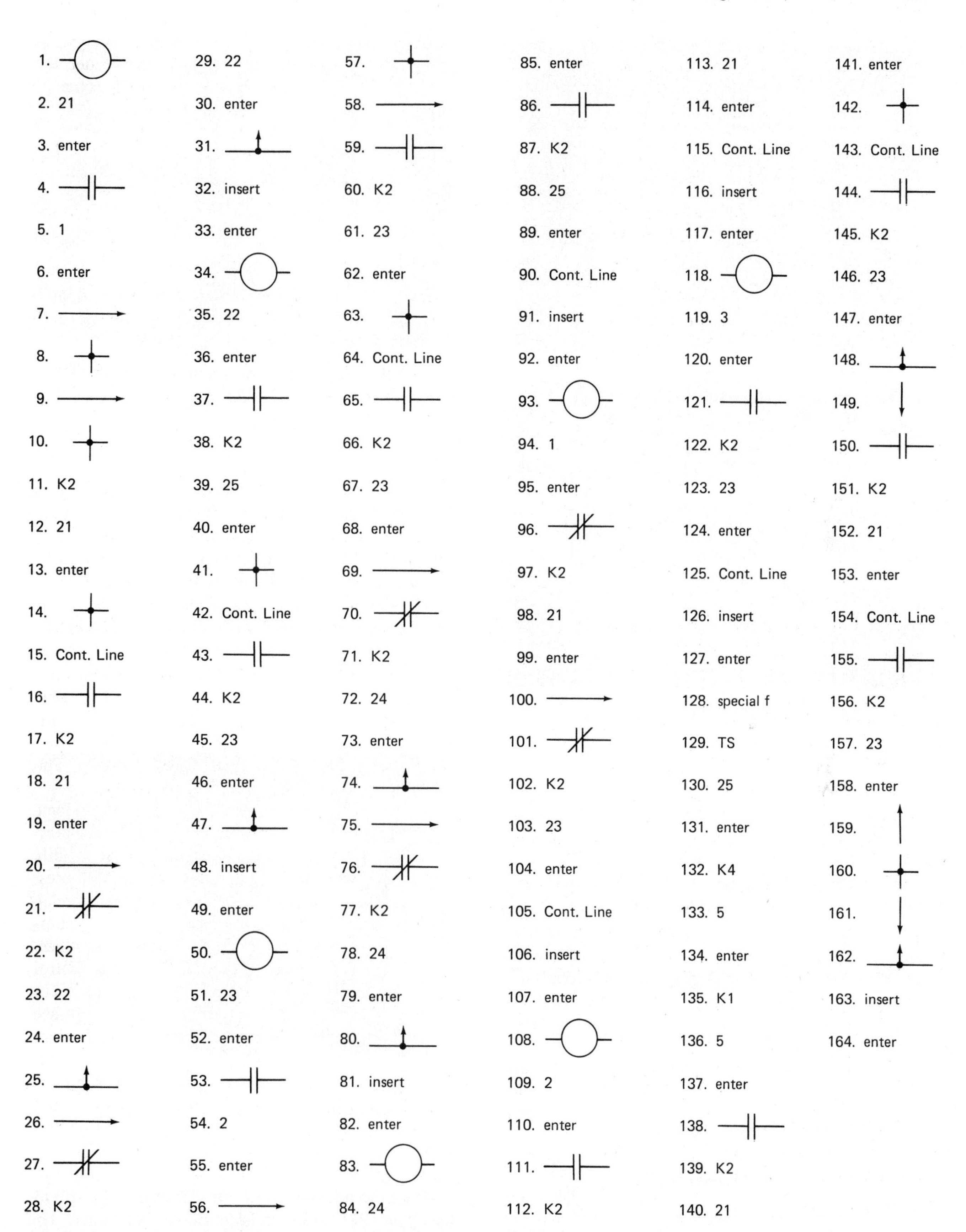

Fig. 16-35 Steps in putting program into Westinghouse P.C. 403 or PC 431. *(Westinghouse.)*

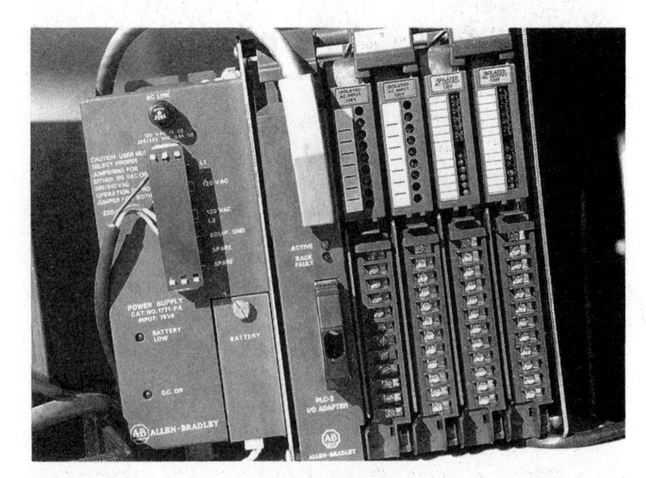

Fig. 16-37 Mini PLC2 programmable controller with I/O adapter. *(Allen-Bradley.)*

Fig. 16-36 Mini PLC-2. *(Allen-Bradley.)*

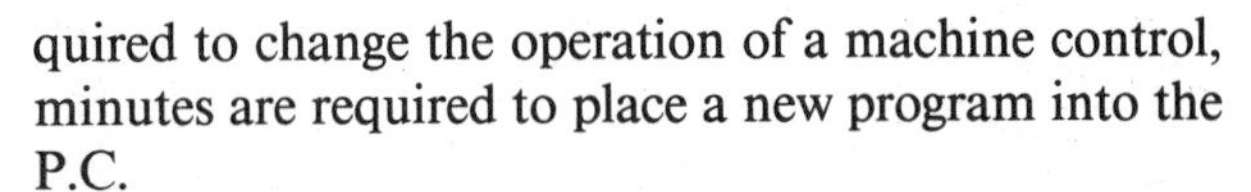

quired to change the operation of a machine control, minutes are required to place a new program into the P.C.

For limited applications, the small, PLC-2/15 (Fig. 16-37) P.C. is recommended by Allen-Bradley. It has a CMOS RAM with a 2K capacity and can monitor and control up to 128 I/O devices. It is supported by an EPROM, (or erasable programmable read-only memory) for more permanent memory storage. The EPROM can be erased with an external ultraviolet light source and reprogrammed.

The PLC 2/15 processor module has the following features:

1. Fully expandable data table
2. On-line programming
3. Up to 488 timers and counters
4. Complete I/O forcing
5. Optional external battery backup when PLC-2/15 processor module is removed from I/O rack
6. Expanded instruction set, including Jump subroutines, sequencers, and file transfers

The PLC-2/20 controller (Fig. 16-38) is the smallest member of the PLC-family. It can accommodate up to 128 I/O points, has up to 40 timers/counters in three time bases (1.0, 0.1, and 0.01), and is available with a maximum 1K CMOS RAM, with battery backup.

The design of the processor allows it to be moved easily in and out of an I/O rack slot, for simple replacement, if needed. Its size and performance features also make it ideal for a distributed control network.

Fig. 16-38 Programmable controller PLC 2/20 with terminal and I/O unit. *(Allen-Bradley.)*

The most advanced and versatile of the PLC-2 family is the PLC-2/30 Programmable Controller. The PLC-2/30 controller (Fig. 16-39) is a low-cost, versatile, and simple P.C. that offers an 8K RAM or core memory, and either 512 or 896 I/O points. The 2/30 is recommended for such standard P.C. operations as

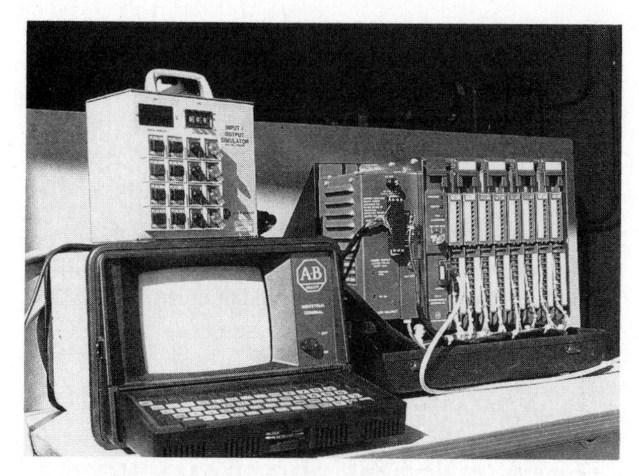

Fig. 16-39 Industrial terminal programmable and I/O simulator. *(Allen-Bradley.)*

Fig. 16-40 Industrial terminal with PLC top overlay. *(Allen-Bradley.)*

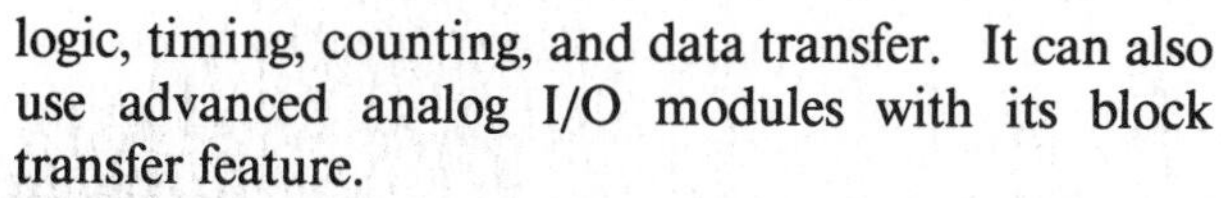

logic, timing, counting, and data transfer. It can also use advanced analog I/O modules with its block transfer feature.

The standard items on the PLC-2/30 controller include:

1. Word memory segments of 512 or 2048. This gives over 10 different memory configurations.
2. A 1000-timer/counter capacity. The number of timers and counters for an application or system can be determined and the remaining memory can be used for application programs.
3. Three timer time bases. These allow you to select the time base resolution that best fits the application.
4. A 512 or 896 I/O capacity. This affords greater flexibility and makes later expansion easier.

Figure 16-40 illustrates an industrial terminal with PLC key-top overlay. This terminal is used to program the PLC-2/20 and the PLC-2/30 Programmable Controllers.

To complement the extensive memory and I/O capabilities of the PLC-2/30 controller, it has a new, remarkable powerful instruction set. Many functions that used to require multiple keystrokes on the industrial terminal and many words of memory are now simpler to enter, leaving more memory available for other functions and, in some cases, increasing the speed of the program scan.

In addition to standard relay-type functions, this instruction set includes Jump instructions to allow unneeded parts of the program to be skipped over and to increase program execution speed. If a part of your application program must be repeated at different locations in memory, you can enter a Jump to Subroutine instruction. This tells the processor to jump directly to your separate subroutine. No need to take up memory over and over by entering the subroutine. Program it once and you can direct the processor to repeat only that section of memory.

The interfacing capabilities of the I/O modules of the Allen-Bradley P.C.s range from the control and sensing of relay signals to sensing the ON/OFF status of input control devices to indicating the status of the outputs and what is energized and de-energized. The I/Os are also capable of sensing analog signals, thermocouple voltages, encoders, and a host of other signals that would be normally difficult to interface with a relay system.

Some of the features of the I/O modules can be summarized as follows:

1. Electrical-optical isolation between logic circuitry and I/O circuitry
2. Filtering on input module circuitry to guard against transients caused by contact bounce
3. Keying of individual module types, so that only the correct type of module is inserted into an I/O slot

In addition to new instructions, the PLC-2/30 controller also has other useful features. It can be used with remote I/O. Previously available only in Allen-Bradley's larger P.C.s, remote I/O lets the I/O rack be located near the controlled machine or process, and the processor near the operator. That means added convenience in changing programs and checking the system's operation, as well as savings on field wiring. The PLC-2/30's shift register instructions increase the efficiency of shift register programming and reduce the time needed to program them. Binary-to-decimal BCD (binary-coded decimal) and BCD binary conversion allows use of the full resolution of analog modules and maximum memory use. Also of special interest is the PLC-2/30's complete I/O forcing and on-line programming.

Complete I/O forcing lets any input or output be "forced" on or off, regardless of the program's command, to test the program's command and to test the system for faulty I/O devices such as limit switches or solenoids. This is a real help in troubleshooting.

Sequencer instructions automatically emulate the functions of drum sequencers and allow repetitive operations to be programmed in one group to maximize usable memory and make programming simpler.

On-line programming enables changes in the ladder diagram program to be entered into memory without interrupting the operation of the machine or process.

The PLC-2/30 controller also has an expandable data table, whose size is limited only by the program length. That gives more control over the setting up of your program.

16-15 CRT PROGRAMMERS AND THEIR APPLICATIONS IN PROGRAMMING, EDITING, MONITORING, AND DIAGNOSTIC OPERATIONS

The industrial terminals used to program Allen-Bradley P.C.s are illustrated in Figs 16-39 and 16-40. The difference between the terminals is in their key-top overlays.

These industrial terminal systems are intelligent CRT terminals that will interface with all Allen-Bradley P.C.s. Each functions as the programming terminal for its PLC-2 family product and as a general-purpose ASCII (alphanumeric) data terminal.

The P.C's. program instructions for loading, editing, and monitoring a ladder diagram program into memory are entered through the keyboard and monitored on the industrial terminal's CRT. These program instructions are in the form of symbols similar to the familiar relay characters and functional block displays. ASCII report messages can also be generated from the keyboard, as well as received and displayed on the screen.

STANDARD FEATURES

Rugged, portable, and suitable for most industrial environments

Self-diagnostics generate audio and visual fault prompts to alert you that a problem with the industrial terminal has been detected.

Multiple ladder diagram display easily viewed on a large 9-in (diagonal) CRT.

Color-coded key-top overlays provide quick, easy reference for selecting the correct key-top overlay for the desired mode of operation.

Detachable touch-pad keyboard for use with the entire PLC-2 family or other Allen-Bradley processors. Only the keyboard need be changed. The basic CRT is compatible with all-detachable keyboards.

Graphic characters provide the capability to draw flow diagrams, production control reports, and/or machine diagnostic messages.

Protective keyboard/display cover and container snap onto the front of the industrial terminal to protect the keyboard and display screen.

16-16 LADDER DIAGRAM PROGRAMMING OF THE PROGRAMMABLE CONTROLLER

The industrial terminal system uses a simple programming language, the ladder diagram, to program Allen-Bradley P.C.s. The ladder diagram has the following distinct benefits:

- It's easy to use.
- It can be used with existing Allen-Bradley P.C. programming knowledge.
- It has quickly recognizable symbols.
- It parallels relay logic format.

The user enters the program from the sealed, touch-pad keyboard by pressing the key tops in the proper sequence. As each instruction is entered, it is added to the visually displayed ladder diagram. Editing capabilities allow any rung in the program to be corrected without having to re-enter all the rungs. Also, new instructions can be inserted any place within the ladder diagram program. The editing function can be performed anytime the P.C. processor is in the Program or Remote Program Load mode. Once programmed, any rung(s) can be displayed. Most of the rung instructions shown on the screen will be highlighted when "true." This feature allows the user to view a rung(s) and observe the conditions that exist on both input and output devices.

Programming is simple. Its flexibility allows the user to program a P.C. in ladder diagram language. An example of a relay ladder diagram and a corresponding P.C. ladder diagram is illustrated in Fig. 16-41.

The five digits above each P.C. instruction represent the instruction's address in memory as well as the terminal location of I/O devices connected to the controller.

Allen-Bradley also has a hand-held terminal that is not much bigger than a large calculator. The 1770 TII hand-held terminal enables examining, altering, and I/O forcing words or bits in the data table of a PLC-2 family processor. It's small and light and easily transported and operated.

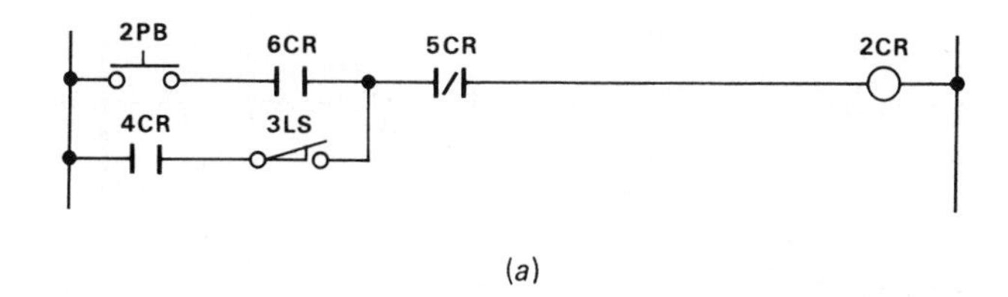

(a)

```
 |  11304     01017      01002                          01004  |
 +---] [-------] [---+-----]/[-----------------------------( )---+
 |                   |
 |  01010     11315  |
 +---] [-------] [---+
 |                                                             |
```

(b)

Fig. 16-41 Relay ladder diagram and P.C. ladder diagram for Allen-Bradley's P.C.s. *(Allen-Bradley.)*

Primarily, the hand-held terminal has the following features:

Three user-selectable functions

Permits word and bit examination only

Permits word and bit examination and alteration

Permits word and bit examination and alteration and I/O bit forcing

ADDITIONAL HIGHLIGHTS OF 1770 TII HAND-HELD TERMINAL

Power supply option—either internal battery pack (standard) or external ac converters (optional); panel mount assembly also optional.

Sealed membrane keyboard—with positive tactile feedback

ON-OFF switch—also serves as a contrast control;

Liquid crystal display—two rows, dot-matrix, eight digits each

16-17 EXAMPLE OF ALLEN-BRADLEY PROGRAMMABLE CONTROL PROGRAMMING

The circuit of Fig. 16-42 is the electric relay circuit for the sequencing of a hydraulic ram. When the CYCLE START button is pressed, the solenoid valve energizes. The ram moves off ILS and travels until it hits the 2LS limit switch. It stops the valve, de-energizes, and the ram returns to ILS. The solenoid valve again energizes and the ram advances past 2LS to 3LS, where it stops. It again returns to ILS. The solenoid valve again energizes, causing the ram to advance. It passes 2LS, 3LS,

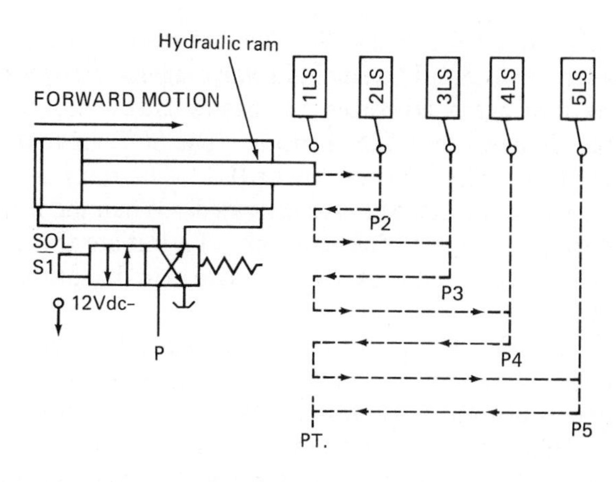

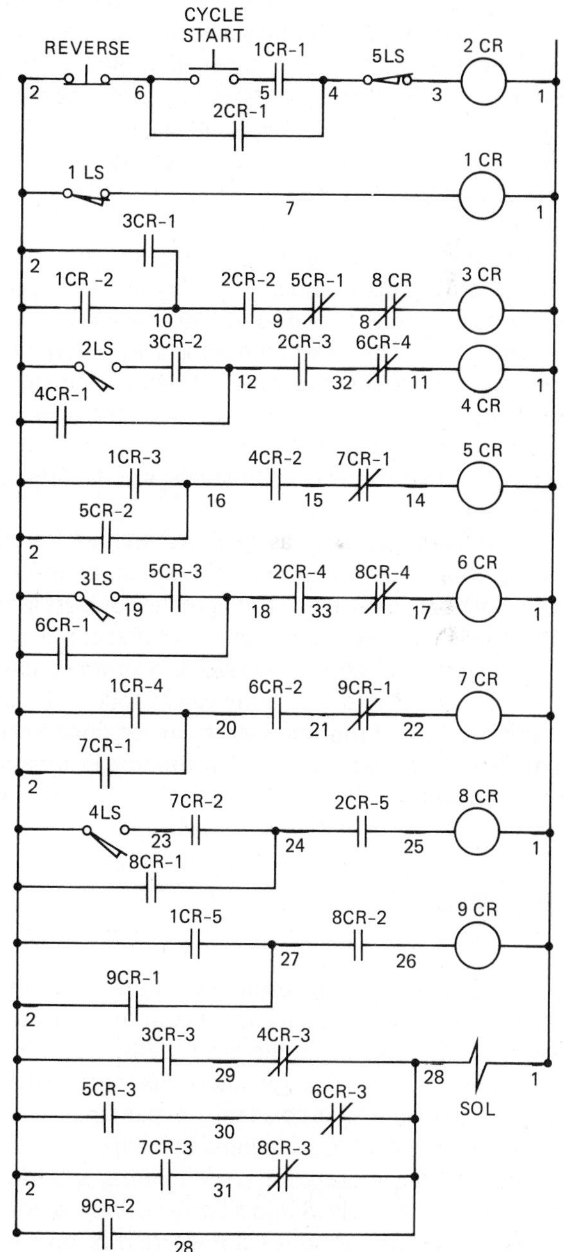

Fig. 16-42 Electric relay circuit for ram-sequence problem for Allen-Bradley's PLC-2.

and stops when it hits 4LS. The ram stops and again returns to ILS. The solenoid valve again energizes. The ram again advances. It passes 2LS, 3LS, and 4LS. When it hits 5LS, it stops. The SOL valve de-energizes. The ram returns to ILS. The relay 2CR de-energized when 5LS was activated. When the ram returns to ILS, nothing happens. To initiate the sequence again requires CYCLE START to be pushed again.

Study Fig. 16-42 and see if you can follow the sequence of operations that takes place when the CYCLE START pushbutton is pressed. When the operation of the electric circuit is clear, the next step is to develop a ladder diagram with Allen-Bradley P.C. numbering. Figure 16-43 is the same circuit shown in Fig. 16-42 but is numbered with five-digit numbers that the P.C. understands. Notice that if a coil has a number, the contacts it operates have the same number.

When the ladder diagram is complete and the numbering of contacts, switches, and coils is completed, we are ready to write a program.

Figure 16-44 shows the keys that are pushed on the keyboard of the programmer. The figure also shows the five-digit numbers used to identify switches, contacts, and coils. The keys and numbers are entered in steps 1 to 133 for this program. When all the steps have been entered into programmer, the key is switched from PROGRAM to RUN, and the P.C. will provide the same control as all the relays and contacts of Fig. 16-42.

The ladder diagram will be displayed on the CRT of the programmer terminal. The circuit can be monitored by observing the highlighting of the contacts and coils. A highlighted contact is on; a dimmer contact is off. A highlighted coil is energized, a dimmer coil is off. The sequence of operation can be followed easily on the CRT. Troubleshooting can be done very rapidly and with ease on the Allen-Bradley Programmable Controller.

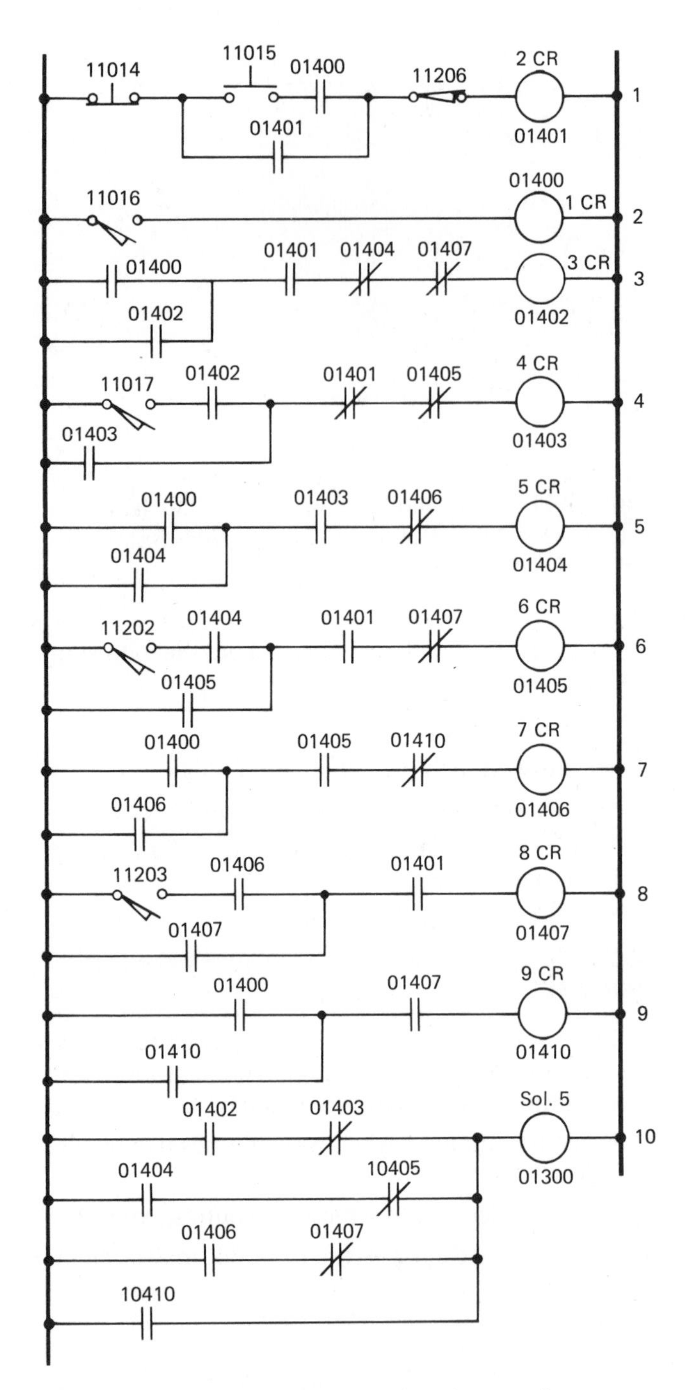

Fig. 16-43 Programmable ladder diagram for ram-sequence problem in Fig. 16-42.

SUMMARY

The world's first programmable controller—the Modicon P.C.—was installed in an automotive transmission plant in Detroit in the early 1970s. A lot has happened since then, and the enormous potential of programmable control is constantly expanding.

Manufacturers of P.C.s are making it happen. The P.C. has become more than an individual machine controller. It has evolved into a hard-working factory automation system. And it's still where it belongs—right down on the factory floor, delivering sustained operation in rigorous plant environments. Individual components now constitute the P.C. network and are installed, operated, modified, and maintained by the control specialists and electricians in the plant.

Communication, documentation, monitoring, programming, and direct control are all capabilities of the sophisticated P.C.s systems automating today's factories. Applications opportunities are virtually limitless. The continued advancement of P.C. technology and the partnership of P.C.s with computers and other digital devices are providing industry with control systems that are doing unbelievable jobs.

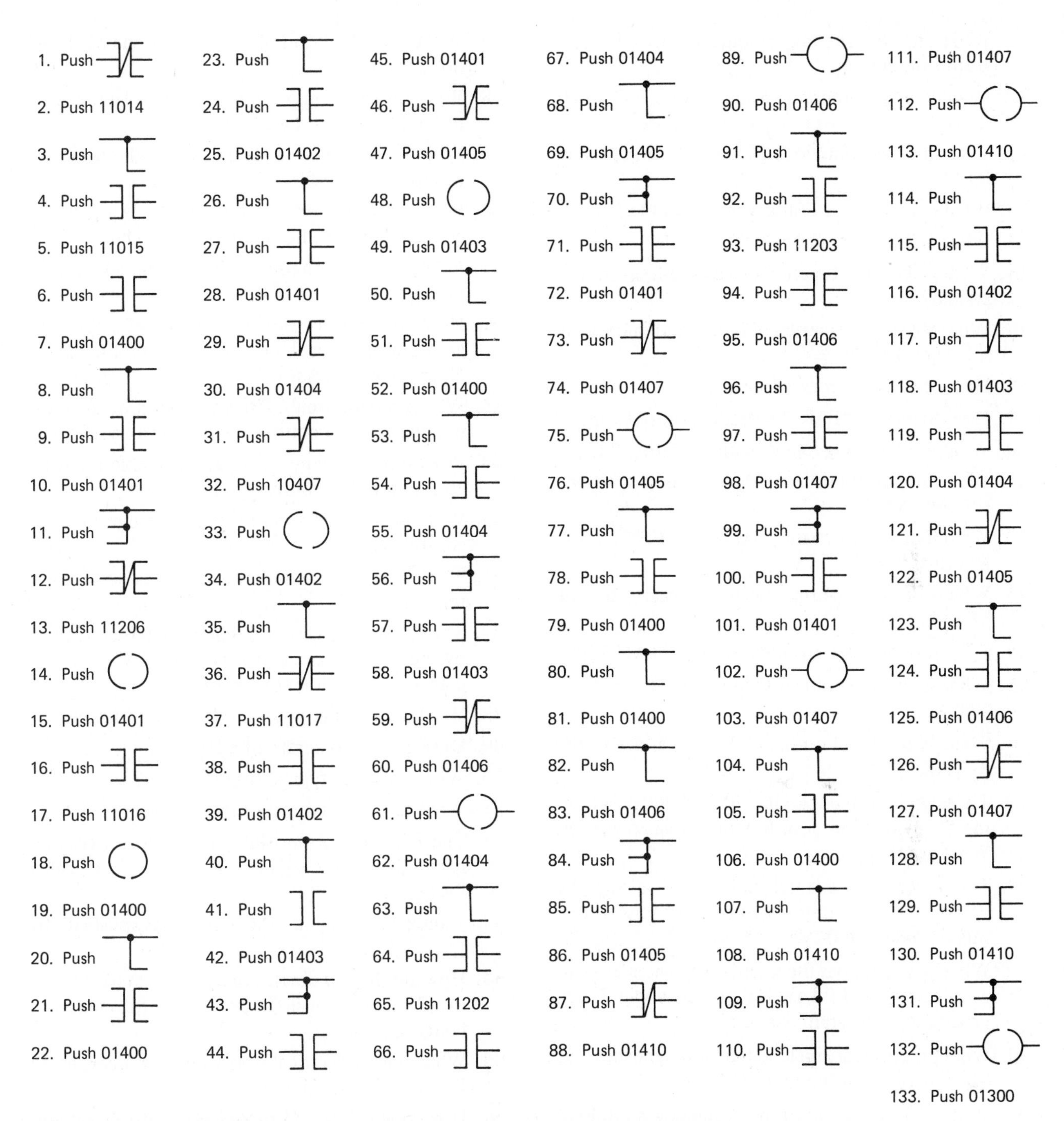

Fig. 16-44 Steps to program PLC-2 with ram-sequence problem in Figs. 16-42 and 16-43.

REVIEW QUESTIONS

1. What types of control do modern P.C.s provide?
2. How is a P.C. reprogrammed?
3. What features of P.C.s have contributed to their proliferation?
4. What are the main elements of a P.C.?
5. What tells the CPU in a P.C. what to do?
6. How long does it take the CPU to scan all the I/Os?
7. Compare a P.C. with a minicomputer.
8. The sequence of design of a P.C. is intended to emulate what?
9. Why is a P.C. slower than a general-purpose computer or a minicomputer?
10. How do the programming languages for P.C.s and general-purpose computers differ?

11. What are the four major languages used in programming P.C.s?
12. How does the program structure for a P.C. differ from that of a general-purpose computer?
13. What are the functions of the I/O modules in P.C.s?
14. What is a "smart" I/O and what are its advantages?
15. What is the function of a PID module?
16. What is the basic function of a closed-loop control system?
17. What are some examples of inputs and outputs on the Allen-Bradley P.C.?
18. How does a P.C. make decisions?
19. How does the EPTAK 200 produce output signals?
20. What are the units found in an operating P.C.?
21. What are the peripheral devices used with the EPTAK 200?
22. What happens when the power to the EPTAK 200 is turned on?
23. When does program execution begin in the EPTAK 200?
24. When are outputs executed in the EPTAK 200?
25. What is the powerdown sequence in the EPTAK 200?
26. What is the function of the I/O expander in the EPTAK 200?
27. The keyboard layout for the EPTAK 200 programmer is divided into four functions related groups. What are they?
28. The Program mode selector switch allows the EPTAK 200 programmer to be programmed two ways. What are they?
29. What position does the RUN/HALT switch have to be in before the EPTAK 200 processor will accept programs or program changes?
30. What functions are included in the diagnostics and edit functions on the EPTAK 200?
31. What is the function of the Watchdog module of the EPTAK 200?
32. What is the function of the load circuit test module of the EPTAK 200?
33. What is the function of the data access and display module in the EPTAK 200 P.C. system?
34. How are program changes other than the timer/counter, data register, or auxiliary input set points programmed into the EPTAK 200 processor?
35. What is the function of the recorder/loader of the EPTAK 200?
36. What does an electrician need to know to program the Micro 84?
37. What is the function of the LCD on the Micro 84 programmer?
38. What information is made accessible on the LCD of the Micro 84 programmer?
39. What industrial control devices can the Modicon Micro 84 replace?
40. What effect do electrical noise and temperature changes have on the Modicon Micro 84?
41. How is troubleshooting the Micro 84 accomplished?
42. How does the electrical installation of a new programmable control system compare to that of older relay systems?
43. Why is programmable control becoming the most popular form of control for industrial systems?
44. Programmable controllers are available for what general types of control applications?
45. What is the NEMA definition of a programmable controller?
46. The TI 510 P.C. contains a ________ control system.
47. The P.C. allows a control scheme to be determined by the ________, which is entered via the programmer in a form similar to ________.
48. What is the function of the P.C.'s I/O devices?
49. Draw a block diagram of a P.C.
50. What are some typical input devices?
51. What are some typical output devices?
52. The P.C. checks the status of input devices and responds to these signals by ________ as called for by ________.
53. Troubleshooting a P.C. is accomplished by ________.
54. How are the I/O devices of a control system wired to a P.C., and how is the processor protected from faulty I/O signals?
55. What causes electrical noise and how does the TI P.C. model 510 handle it?
56. How long will the TI model 510 controller maintain the user memory in RAM?
57. Troubleshooting a control system with the TI model 510 is accomplished with ________.
58. Logic available in the user program includes ________ control relays, ________ which are nonretentive, ________ which are retentive. The contacts can be combined in ________, ________, ________ arrangements and used as many times as needed ________ timers with a range of ________ to ________. Counters which can count up or down between ________. Timers and counters can be ________ to larger timing or counting values?

59. What does the Jump instruction do?

60. The TI model 510's most important instruction is the ________. ________ of these are available and each can replace up to ________ ladder logic elements. The drum can control up to ________ outputs through ________ timer steps of varying duration. It effectively replaces both ________ and their ________ equivalent.

61. A ladder logic diagram is converted to a TI model 510 program code and keyed into the hand-held programmer a rung at a time on the P.C. ladder logic diagram. Complete the following problem by drawing each rung of the P.C. ladder logic diagram that goes with each instruction.

SRT X9 ENT
OR C1 ENT
AND NOT X15 ENT
OUT C1 ENT
STR C1 ENT
STR C1 ENT
TMR1 ENT
300 ENT
OUT C5 ENT
STR NOT C5 ENT
OUT Y1 ENT
STR C1 ENT
AND C5 ENT
MCR 1 ENT
STR C3 ENT
AND C2 ENT
OUT Y2 ENT
STR C3 ENT
AND NOT C2 ENT
OUT Y3 ENT
STR X11 ENT
OR C4 ENT
OUT C3 ENT
STR X12 ENT
AND NOT X13 ENT
OR C4 ENT
OUT C2 ENT
STR NOT X14 ENT
AND X10 ENT
OUT C4 ENT
END MCR 1 ENT

62. The basic logical configurations available on the TI510 are ________. With the hand-held programmer, the number of series or parallel contacts that can be used in a rung on a ladder diagram is ________.

63. Series contacts are programmed by using the ________ instruction.

64. What is the keystroke sequence used to program the following circuit?

X9 X10 Y1

65. Parallel contacts are programmed by using the ________ instructions.

66. What is the keystroke sequence used to program the following rung of ladder logic?

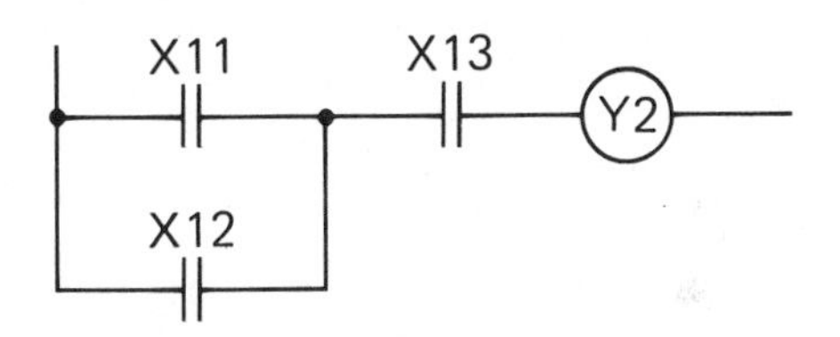

67. What instruction is used to create normally closed contacts or outputs?

68. How are parallel outputs or inverted inputs created?

69. Draw the P.C. circuit for a counter.

70. What is the purpose of the bottom line of the counter?

71. What does closing the contact on the middle line of the counter do?

72. What does closing the contact on the top line of the counter accomplish?

73. When the counter is at zero the contact designated as the zero flag will be ________ and the Y output will be ________.

74. When the counter reaches the preset value, what is the condition of the zero flag and the Y output?

75. The drum timer is a special instruction requiring only ________ works of user memory to achieve a ________ which are turned on and off ________ as the 16 drum steps are executed. A single drum timer can replace up to ________ ladder logic elements.

76. Draw the P.C. logic circuit for a drum timer.

77. What is the bottom line into the drum timer used for?

78. What does opening the contacts in the bottom line of the drum timer do?

79. The to line to a drum timer is the ________ line.

80. What does opening the contacts in the top line to the drum timer do?

81. How are external logic "events" such as interlocks and limit switches used to control the drum operating process?

82. The output line of the drum timer is energized when the drum timer ________ has been ________.

83. How can the drum timer be made to reset automatically at the completion of the programmed sequence?

84. Draw a ladder logic diagram of a circuit using an MCR.

85. When the MCR is activated by opening ________ the output Y3 is ________ and ________ regardless of the status of ________. When the MCR is deactivated by closing ________, control of ________ is returned to ________.

86. Draw a P.C. ladder diagram for a circuit with JMP.

87. How is the JMP activated?

88. What happens when the JMP is activated?

89. If Y1 is on when JMP is initiated by ________ X9, then Y1 will ________. If Y1 is off when JMP is deactivated by closing ________ the status of Y1 is controlled by ________.

90. How does the JMP instruction affect Y2?

91. What are some of the special features of the Allen-Bradley 1771 I/O modules?

92. What are some of the useful features of the PLC-2/30?

93. What does I/O forcing do?

94. What do sequencing instructions allow the P.C. to do?

95. What does on-line programming do?

17

Microprocessors and Programmable Logic for Motor Control

The microprocessor, microcomputer, and programmable logic have become the "brains" of many industrial control systems. Decision making in large industrial control systems is often delegated to the microprocessor. For dedicated control systems the microprocessor is hard to beat. Programmable logic also has some cost-saving features that make it desirable for dedicated control.

17-1 DESCRIPTION OF A MICROPROCESSOR

A *microprocessor* is a large-scale integrated circuit that contains most of the logic circuitry usually associated with digital computers. Many digital computers use the microprocessor as the central processing unit (CPU). A motor control system that includes a microprocessor is said to be *microprocessor-based control system.*

A microprocessor's job in the control of motors is to serve as the central processing unit, or "brain," of the system. It looks at the input information from sensors and pilot devices. It also takes a look at instructions that have been programmed into the memory. With the memory instructions and help from other parts of the system, the microprocessor makes the decision as to what control action is required next. The output section feeds output control signals to the power control devices in the system.

Microprocessors are not appropriate for all types of motor control. For example, a group of machines whose sequence of operation is always the same with the same time delays in start-up and shutdown may not require a microprocessor for control. Conventional control may be adequate. If the control required for a group of machines requires continuous changes in the sequence of operation, if the time delays require continual adjustment, if rapid decisions in control action are required, or if continual monitoring of the process is needed, then microprocessor control is the route to go. Because it's easy to program, the microprocessor offers many advantages in controlling machines with complex operations.

In the control of motors, the microprocessor is often used to provide what is called *dedicated control.* Dedicated control is designed to do limited tasks repeatedly. Sequence, or programmed, control is an example of dedicated control. The microprocessor is a logic device that can provide the same control that drum switches and stepping relays do in the dedicated control provided by conventional electric motor control systems. The microprocessor can replace relay systems or hard-wired logic systems for automatic control of large electrical systems.

The microprocessor is used with digital electronic control devices to make up the decision-making portion of industrial control systems. Figure 17-1 on the next page indicates a block diagram of an automatic motor control system.

Figure 17-1 illustrates the use of the microprocessor in dedicated control. As indicated, the microprocessor is used to make decisions concerning how the power control should react to various input information. The microprocessor unit is a complex logic element that is capable of performing arithmetic, logic, and control operations. The computational functions of the microprocessor also imply that it has many uses in data processing as well. The microprocessor unit (MPU) is an integrated circuit (IC). When combined with read-only memory (ROM), it is used as a hard-wired logic replacement. As a logic replacement element, it has simplified and broadened the scope of motor control.

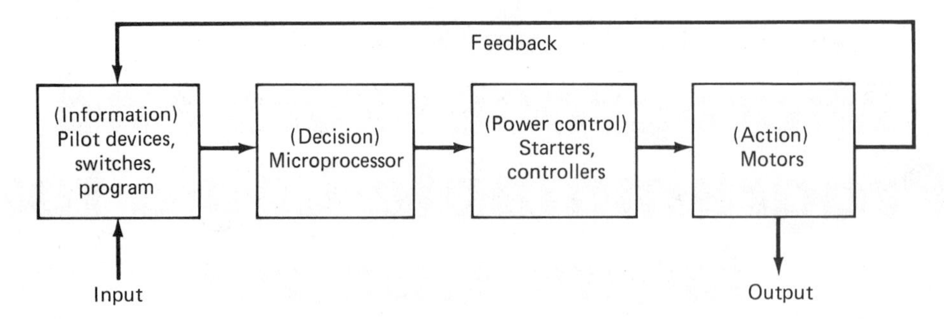

Fig. 17-1 Block diagram of automatic motor control.

17-2 MICROPROCESSORS VERSUS MICROCOMPUTERS

Technically, there is a difference between a microprocessor and a microcomputer. A microcomputer contains a microprocessor and other circuits such as memory devices to store information, interface adapters to connect it to the outside world, and a clock to act as a master timer for the system. Figure 17-2*a* illustrates a block diagram of a basic microcomputer using a microprocessor unit. Figure 17-2*b* is a block diagram of an elementary microprocessor.

As indicated in Fig. 17-2, a microprocessor must be combined with other devices to become a computer. A microprocessor with ROM can supply a wide range of the same random logic applications as found in dedicated control. Remember, a dedicated control application is one where a piece of electrical or electronic hardware provides a limited task repeatedly and on demand. When power is applied and the start signal is initiated, the dedicated control system does its job. There is no need to process volumes of data as a computer would do. Dedicated control does jobs that are unique. They are often one of a kind. The job is precisely defined and it must be easy to design and change. No set programs are used as with computers. The ease with which microprocessors can be programmed make them ideal control systems for industrial automatic control systems.

Microprocessors offer many advantages over relay-based systems, such as a wider range of logic functions, lower cost, more flexibility, lower power consumption, smaller size, higher reliability, easier replacement, and easier interfacing with I/O devices. While microprocessors make poor computer replacements, they are finding many applications in the high-volume product-producing market that uses dedicated control.

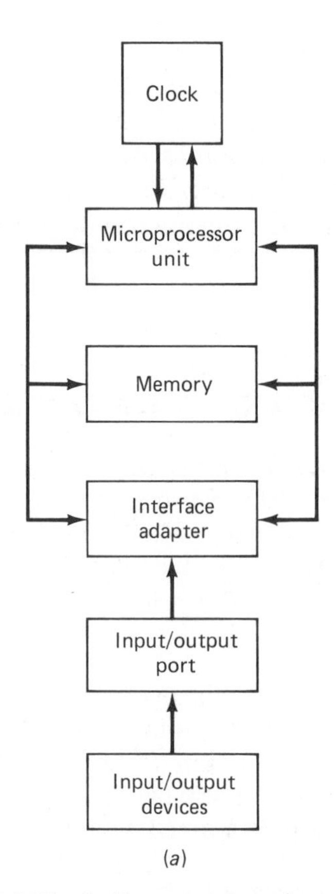

Fig. 17-2 (a) Block diagram of a microcomputer.

17-3 MICROPROCESSOR PROGRAMS

A group of instructions that enable the microprocessor to perform a specific job is called a *program.* To place a new microprocessor into operation or to replace a program in an old microprocessor, the technician must understand the principles of programming. The length of a program for the microprocessor is related to the complexity of the task to be performed. The program for a machine that continuously does the same task over and over again may be relatively short, with just a few instructions. A program for a process line in a large manufacturing industry may be complex enough to require many program instructions.

Data means numbers, facts, information, or anything that the microprocessor must know before it can solve a dedicated control problem. A sequence of instructions must be developed and programmed into the microprocessor before it can perform its decision-making role in a motor control system.

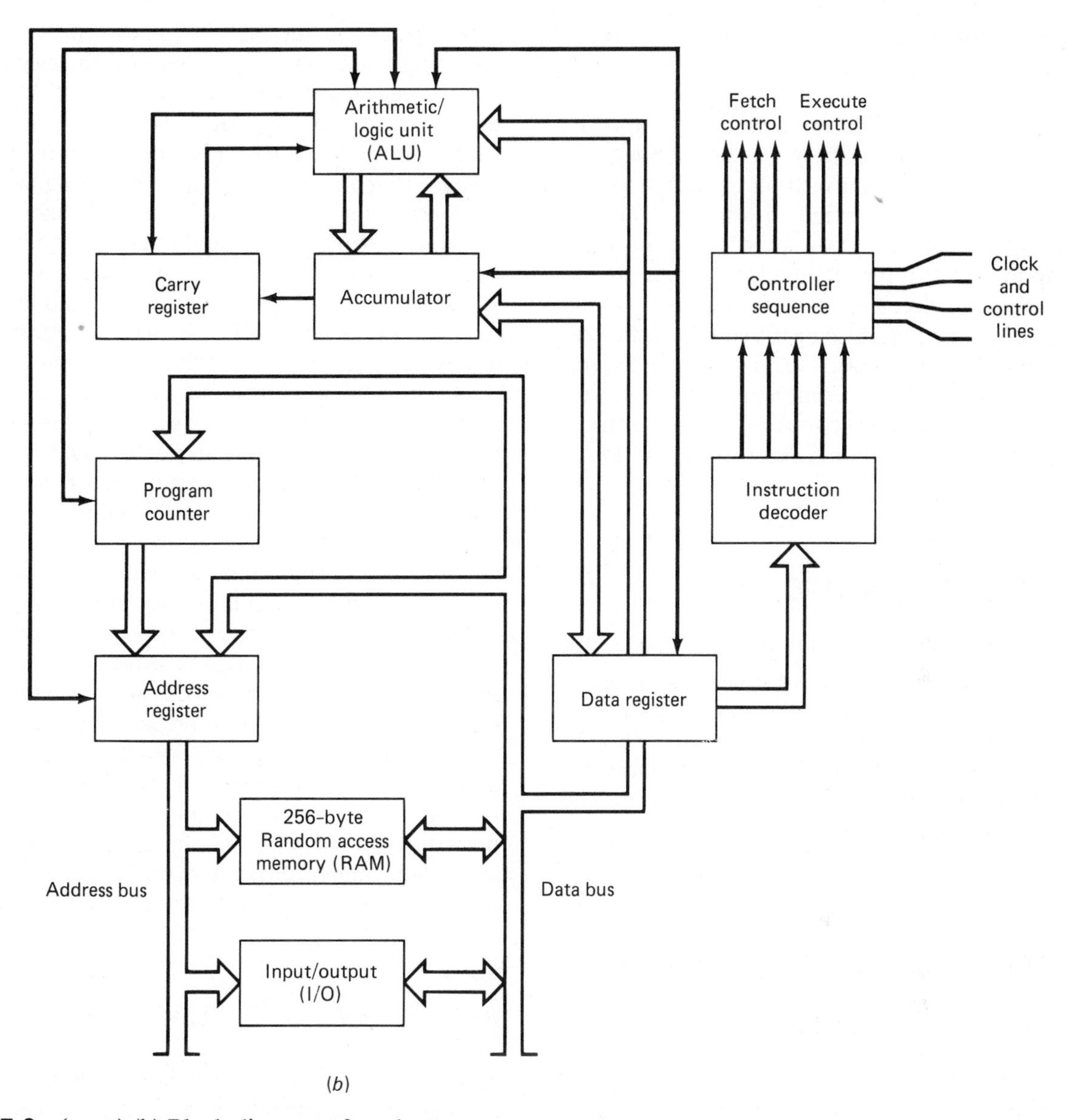

Fig. 17-2 (*cont.*) (b) Block diagram of a microprocessor.

A program is a step-by-step procedure that solves a problem, performs some control operation, or otherwise manipulates data. Programs are stored in memory along with the data they process.

Instructions and data are stored in or retrieved from memory during processing. The data stored in the memory is said to be *written* into the memory. When data is retrieved from the memory, we call this process *reading from the memory.*

When a program or set of instructions is stored in the memory of the microprocessor, it can be stored for a long period of time. A program in memory can be cleared and a new program can be entered into memory at any time. For example, suppose a set of instructions for a machine that is making 1-qt milk cartons needed to be changed to a set of instructions so the machine would make 2-qt cartons, or containers. All that would be required is to clear the memory in the microprocessor and program into memory the new set of instructions for the machine control. A program change such as this may be a 5-min operation. If a control system using relays required a change such as that just described, it may take days to rewire the control to obtain the desired control action. One of the advantages of using a microprocessor is its ability to provide rapid control changes. A system using a microprocessor takes only minutes to reprogram for a change in machine operation.

17-4 MICROPROCESSOR CONTROL SECTION

As mentioned, a memory-write operation and a memory-read operation respectively store or retrieve data or instructions.

When the microprocessor is in operation, the control section causes the program to be executed automatically. The control section sequentially examines the instructions in the program, performs some control operation, or manipulates data, and then initiates control signals for carrying out the programmed operations.

The control section of the microprocessor automatically fetches instructions from memory, interprets them, and then executes the instructions one at a time. When an instruction is executed, this may involve sending signals to other sections in the microprocessor that help to carry out the programmed operation.

As each instruction is examined or decoded, timing and control signals generated by the control section set up the control action to carry out the operation specified by the instruction. Completing an instruction may require the combined control action of the arithmetic, logic, input and output, and memory section of the microprocessor.

17-5 THE ARITHMETIC/LOGIC UNIT

The arithmetic/logic unit (ALU) in the microprocessor is that section that is responsible for such operations as adding and subtracting and performing logic manipulations such as ANDing and ORing. If a number is to be added to a number and the result stored in memory, the ALU would be responsible for this operation. It would also be responsible for any logic decisions that are made.

17-6 THE PROGRAMMING PROCESS

There are five steps in the programming process:

1. Define a problem.
2. Develop a solution for the problem.
3. Code and program the problem.
4. Test and debug the program.
5. Document the program.

To define a problem requires that the problem be stated as clearly as possible. Sometimes a problem will be easier to define if it is broken up and blocked out. Charts, diagrams, and logic expression also help in defining a problem. A solution to a problem will be much easier to see if the problem is thoroughly defined.

Once a problem is defined, the next stop is to develop a solution. If a person is using a microprocessor to run a program, he or she will require a thorough knowledge of the codes and instructions of the programming language. In many cases, a program for a dedicated control system is not altered often enough to justify reprogramming the entire control operation. The microprocessor manufacturer may even "burn in" a program in memory that will make the microprocessor respond to the instructions found in a table. Whatever is placed in the table becomes instructions for operating a machine or process. All the electrician, maintenance technician, or control system repair personnel have to do is to write new instructions for the table. The program which is burned into memory is called a *state machine program.* For example, if the microprocessor were to be used for traffic light control at an intersection, all that would have to be prepared is a table indicating the sequence in which the red, amber, and green lights of the intersection would turn on. Also, the time delays involved in switching lights would be needed in the table. For this table *hex numbers* can be developed that become the instructions for the microprocessor. The instructions are programmed into the state machine program table only. A microprocessor programmed as a state machine becomes a simple solution to writing a program for industrial applications that use a dedicated control.

An industrial application that requires a number of motors to start and stop it in some predetermined sequence can be controlled very easily by a state machine program with instructions for a sequence of operation in the table. All the programmer would do to change the sequence of control is to prepare a new table.

Solving any problem is usually a step-by-step procedure. Each step in the procedure will produce some part of the solution for the problem. A step-by-step procedure for solving a specific problem is called an *algorithm. Algorithm* is a term used to describe a set of procedures by which a given result is obtained.

17-7 ALGORITHMS AND FLOWCHARTS

Algorithms are useful in developing programs. When a program is prepared for a microprocessor, algorithms, flowcharts, and tables all help in its development. The flowchart is often used to provide solutions to designing a program for the microprocessor. Symbols used with flowcharts represent operations, data, flow, equipment, etc. Programmers often use flowcharts to provide a visual, or graphical, representation of the solution of a program.

Flowcharts use a number of standard symbols. These are shown in Fig. 17-3.

The rectangle represents some type of operation. To represent a decision point on the flowchart the diamond is used. A flowchart for the multiplication process is shown in Fig. 17-4. Here the problem is to write a program that could be used to find the product of multiplying two numbers together by repeated addition. The flow diagram is applicable for any microprocessor but the program instructions or codes would be different for each type of microprocessor. Figures 17-5 and 17-6 on page 324 illustrate flow diagrams for two different industrial problems.

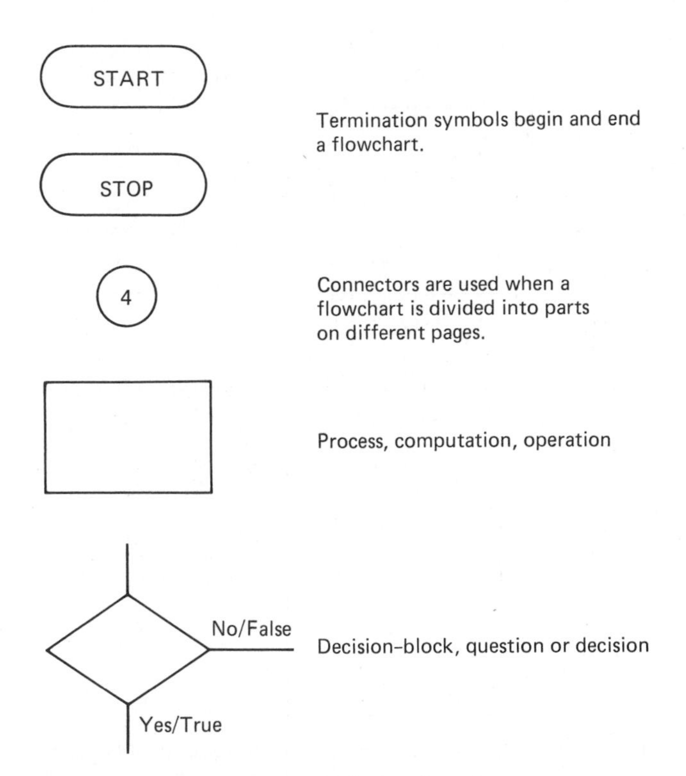

Fig. 17-3 Standard symbols for flowcharts.

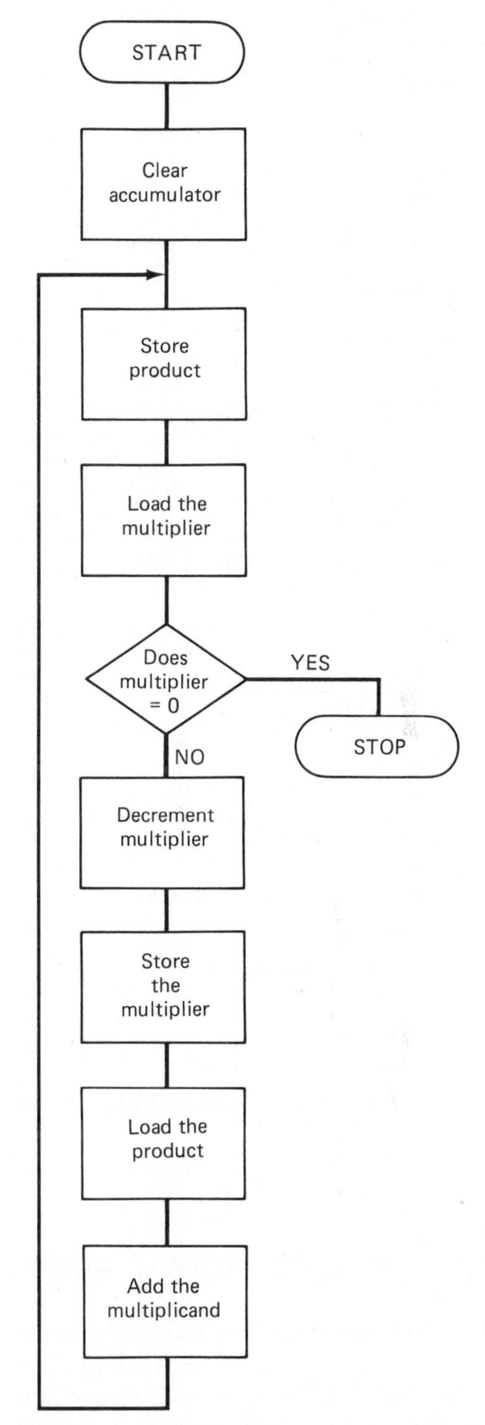

Fig. 17-4 Flowchart for multiplying by repeated addition.

17-8 MICROPROCESSOR APPLICATIONS

The objective of this section is to prepare electricians, maintenance technicians, and control systems repair personnel to work with microprocessors. In many large industries downtime is calculated in dollars per minute. Downtime on a machine may be $50 to $100 a minute. Production line downtime may be several thousand dollars a minute. When trouble occurs because of a faulty microprocessor, the cost of downtime dictates that the processor be replaced and not repaired. A maintenance electrician does not really have to know too much about the electronics of microprocessors to replace them. If a microprocessor costs $15 and downtime is $2000 a minute, you can see that 20, 30, or 60 min to repair a microprocessor circuit would not be economically feasible. A printed circuit board with many microprocessors may cost thousands of dollars, but even these are cheaper to replace than repair on site.

When printed circuit boards are replaced to minimize downtime what becomes of them? In many large industries a special electronics shop is maintained with personnel who are trained and equipped to repair any electronic equipment. If an industry has this type of facility, a replaced board will end up there for troubleshooting and perhaps repair. In many smaller industries that do not have the maintenance personnel trained in electronics, the faulty board will be returned to the manufacturer or an outside service shop for repairs. In many cases, the cost of a board is such that repairs are not justified. New ones are always used. In some cases, industry expects the maintenance department to do all repairs. They also can tolerate a little downtime. Generally, the less complex boards can be easily repaired because the usual trouble is in the ICs. All that is required to repair such boards is to replace the defective ICs with a known good IC.

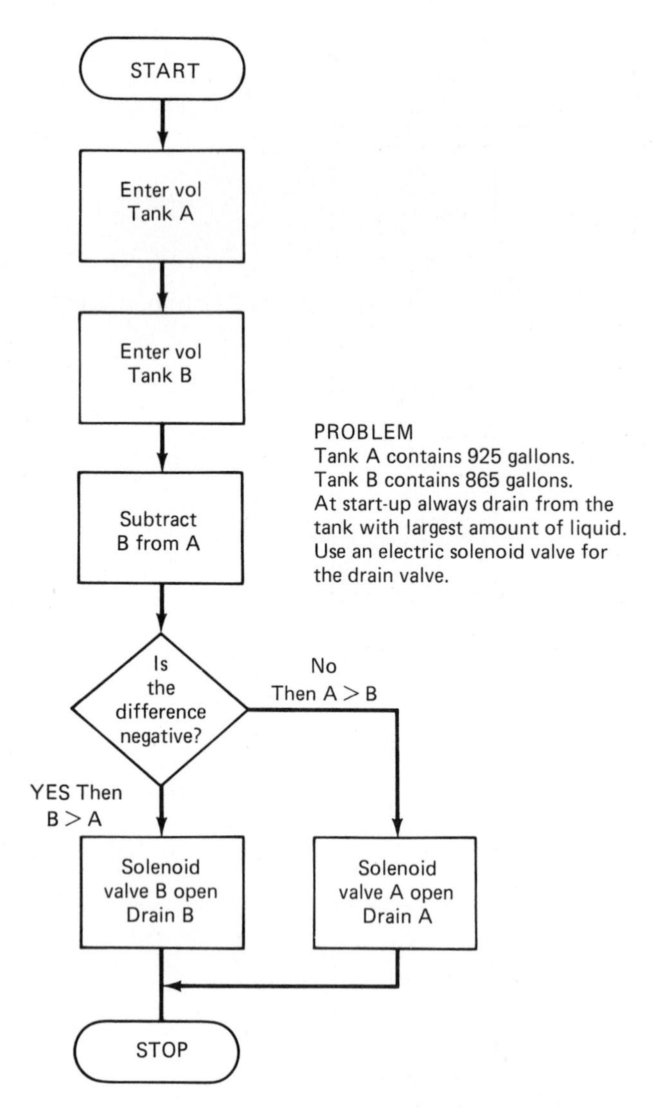

Fig. 17-5 Flowchart for a tank drainage problem.

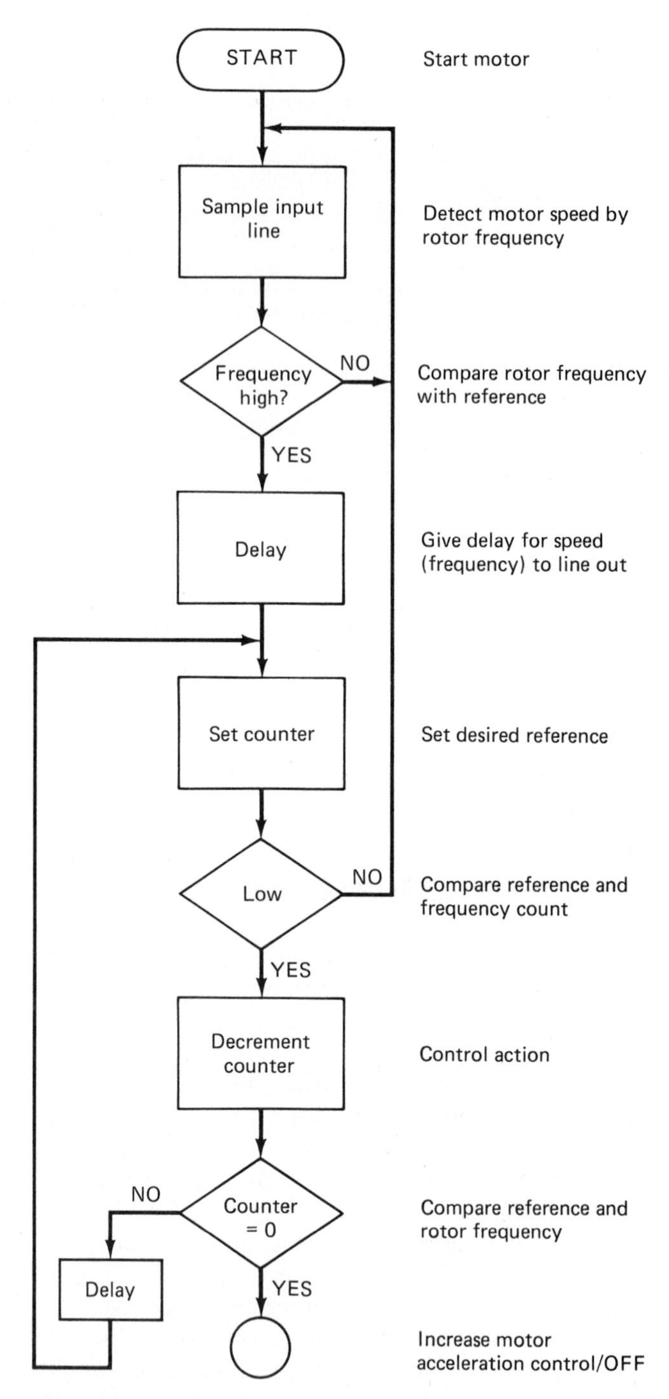

Fig. 17-6 Flowchart for speed control of a wound-rotor motor.

17-9 BINARY AND HEXADECIMAL NUMBERS

We will now show some practical examples to illustrate how numbering systems are used to program microprocessors and put them to use. *Binary* and *hexadecimal* numbers are two of the common numbering systems used in programming microprocessors. Look at Table 17-1 for the relationship between decimal, binary, and hexadecimal numbers.

The decimal system is based on the number 10. Each position in the decimal number represents a power of 10. The first digit to the left of the decimal point is the *units* digit; to its left is the *tens* digit; to its left is the *hundreds* digit; followed by the *thousands, ten thousands, hundred-thousands* and so on. For example, the number 273 can be represented by the sum of the values in each position.

Position value	10^2	10^1	10^0
(Decimal equivalent)	(100)	(10)	(1)
Decimal digits	2	7	3
Decimal number	200 +	70 +	3 = 273

Binary numbers can be represented in the same way, except that each position is a power of 2 (for the sake of this explanation the decimal symbol is used; the binary system does not use the symbol 2). The binary system

Position value	2^8	2^7	2^6	2^5	2^4	2^3	2^2	2^1	2^0	
(Decimal equivalent)	(256)	(128)	(64)	(32)	(16)	(8)	(4)	(2)	(1)	
Binary digits	1	0	0	0	1	0	0	0	1	
Decimal number	256 +	0 +	0 +	0 +	16 +	0 +	0 +	0 +	1	= 273

Fig. 17-7 Finding the decimal equivalent of a binary number.

has only two digits, 0 and 1. If we wanted to represent 2 in binary it would be 10 (read as "one-zero"). In the binary system, the point is called a *binary point.* The digit to the left of the binary point is the units digit, as before. To the left of the units digit is the *twos* place; in turn each position to the left is multiplied by 2. Thus reading left from the units digit we have twos, fours, eights, sixteens, and so on. The binary number 100010001 can be represented by the sum of the values in each position. Using the decimal 2 to denote these positions, we can find the decimal equivalent of the binary number as shown in Fig. 17-7.

Therefore $100010001_2 = 273_{10}$. The subscripts are used to indicate the base of the numbering system being used: 2 stands for the binary system, 10 stands for the decimal system. The routine used above also gives us an easy way to convert a binary number to a decimal number. A decimal number can be converted to a binary number by finding the highest power of 2 less than the given decimal number, and then subtracting this number from the original number. The process is repeated with each difference until the final subtraction is 0.

TABLE 17-1
Hexadecimal to Decimal and Binary Conversion

Hexadecimal Digit	Decimal Value	Binary Value
0	0	0000
1	1	0001
2	2	0010
3	3	0011
4	4	0100
5	5	0101
6	6	0110
7	7	0111
8	8	1000
9	9	1001
A	10	1010
B	11	1011
C	12	1100
D	13	1101
E	14	1110
F	15	1111
10	16	10000
11	17	10001
12	18	10010
13	19	10011
14	20	10100
15	21	10101
16	22	10110
17	23	10111
18	24	11000
19	25	11001
1A	26	11010
1B	27	11011
1C	28	11100
1D	29	11101
1E	30	11110
1F	31	11111
20	32	100000

Example: Convert binary number 01011001 to its decimal equivalent.

Solution: Add the values of each of the positions of the binary digits using the powers of 2:

2^7	2^6	2^5	2^4	2^3	2^2	2^1	2^0	
(128)	(64)	(32)	(16)	(8)	(4)	(2)	(1)	
0	1	0	1	1	0	0	1	
0 +	64 +	0 +	16 +	8 +	0 +	0 +	1	= 89

Therefore binary 01011001 = decimal 89.

Example: Convert decimal 106 to its equivalent binary number.

Solution: The highest power of 2 less than or equal to 106 is 64. Subtract 64 from 106:

$$106 - 64 = 42$$

The highest power of 2 less than or equal to 42 is 32. Subtract 32 from 42:

$$42 - 32 = 10$$

The highest power of 2 less than or equal to 10 is 8. Subtract 8 from 10:

$$10 - 8 = 2$$

The highest power of 2 less than or equal to 2 is 2. Subtract 2 from 2:

$$2 - 2 = 0$$

We then put a 1 in the 64, 32, 8, and 2 positions; all other positions get a 0:

64	32	16	8	4	2	1
1	1	0	1	0	1	0

Therefore decimal 106 = binary 1101010.

Another method for finding the binary equivalent of a decimal number is to take successive divisions of 2, noting only the remainders in each case. The above example will be done using this method.

$\frac{106}{2} = 53$ remainder 0

$\frac{53}{2} = 26$ remainder 1

$\frac{26}{2} = 13$ remainder 0

$\frac{13}{2} = 6$ remainder 1

$\frac{6}{2} = 3$ remainder 0

$\frac{3}{2} = 1$ remainder 1

$\frac{1}{2} = 0$ remainder 1

The binary equivalent will be the remainders reading from the bottom up:

1101010

This is the same as the number found using the first method.

The hexadecimal (or simply *hex*) number system is based on the number 16. Thus, there are 16 symbols denoting the 16 digits of the hexadecimal system. The symbols 0 through 9 denote the first 10 digits of the system. For the next 6 symbols (equivalent to decimal 10 through 15) the letters A through F are used. Thus decimal 10 = hex A; decimal 11 = hex B; 12_{10} = hex C; 13_{10} = hex D; 14_{10} = hex E; 15_{10} = hex F. To summarize, the 16 digits of the hex system are:

0, 1, 2, 3, 4, 5, 6, 7, 8, 9, A, B, C, D, E, F

In a manner similar to the binary and decimal systems, each position in the hex numbering system is equal to a power of 16.

16^4	16^3	16^2	16^1	16^0
65536	4096	256	16	1

Thus each digit in the 16^3 position must be multiplied by 4096 to find its decimal equivalent; each digit in the 16^4 position must be multiplied by 65536, and so on.

The hexadecimal equivalent of a binary number can be found by separating the binary number into units of four digits starting from the right. (If the last group has fewer than four digits, zeroes are added just for the sake of uniformity.) Thus the binary number for decimal 22 is 10110. To find the hex equivalent of this binary number, we separate the binary digits into groups of four starting from the right 1 0110; but for the sake of uniformity we add three zeros to the end of the left-hand group: 0001 0110. A group of four binary digits can represent the hex numbers from 0 through F (decimal equivalent 0 through 15). The groups of four will therefore represent the digits of a hex number.

Example: Find the hex equivalent of the binary number 0001 0110.

Solution: Binary number 0001 0110
hex number 1 6
Therefore binary 10110 = hex 16.

The reverse process can be used to find the binary equivalent of a hex number. Each digit of the hex number can be converted to a four digit binary number and combining the four digit groups we will have the equivalent binary number.

Example: Find the binary equivalent of hex FB5.

Solution: The binary equivalent of F is 1111
The binary equivalent of B is 1011
The binary equivalent of 5 is 0101
Combining these groups gives us 1111 1011 0101.
Therefore hex FB5 = binary 111110110101.

Example: Find the decimal equivalent of FB5.

Solution:

16^2	16^1	16^0
(256)	(16)	(1)
F	B	5

Hex F = decimal 15; $15 \times 256 = 3840$
Hex B = decimal 11; $11 \times 16 = 176$
Hex 5 = decimal 5; $5 \times 1 = 5$
Total 4021

Therefore hex FB5 = decimal 4021.

In a previous example we found the binary equivalent to FB5. We will now find the decimal equivalent of the binary number, 111110110101.

The positions of the powers of 2 are shown in Fig. 17-8.

2048	1024	512	256	128	64	32	16	8	4	2	1
1	1	1	1	1	0	1	1	0	1	0	1

$2048 + 1024 + 512 + 256 + 128 + 0 + 32 + 16 + 0 + 4 + 0 + 1 = 4021$

Fig. 17-8 Verifying the hex-to-binary conversion by converting from binary to decimal.

This is the same answer we obtained when converting the hex number directly to its decimal equivalent.

Try completing the following examples yourself.

Example 1: Convert the following hex numbers to their decimal and binary equivalents.

Hex	Decimal equivalent	Binary equivalent
A3	A = (10 × 16) + 3 = 163	A = 1010 and 3 = 0011 10100011
27		
7A		
FE		
DA		
FF		
68		
F7		
74		

Example 2: Convert the following binary numbers to their hexadecimal equivalents:

Binary	Hexadecimal equivalent
11111010	1111 = F and 1010 = A 11111010 = FA
11000110	
01001001	
10110011	
10011101	
01111101	
00111111	
01011010	
00101110	
11011011	

17-10 STEPPING RELAYS AND ROTARY DRUM SWITCHES

A rotary stepping relay or drum switch is an electromagnetic device that, as the name implies, is used to open, close, and transfer circuits. A rotary stepping switch or drum controller can do any of a long list of control functions. It can select, indicate, count, and provide time intervals. It can also provide pulse and digital, start and stop, and sequence control. It can perform decimal-to-binary conversion. It can produce binary numbers. It can be used to program the operation and sequence of electrical machinery; it can be used to monitor and test. In large industry, programming the operation sequence of machines and equipment is probably the rotary stepping relay's or drum switch's main application (Fig. 17-9).

A microprocessor can also do all of the above-mentioned control functions. As we progress, it should become evident that microprocessors can replace drum switches or stepping relays that have been programmed to provide dedicated control for conventional electric control systems.

ROTARY STEPPING SWITCH

Rotary stepping switches at one time were some of the most widely used components for industrial sequence control. They were used to switch, select, count, indi-

Fig. 17-9 Rotary stepping switches (stepping relays). Microprocessors have replaced stepping switches in modern equipment.

cate, monitor, time, control, test, and program in a wide variety of industrial products and applications.

What is a rotary stepping switch? It is precisely what its name implies: it has a rotating arm which steps across a number of contacts, making and breaking circuits (switching) in the process. More fully, a rotary stepping switch is a mechanism consisting essentially of one or more wiping springs fixed on a shaft moved by a pawl and ratchet which is actuated by an electromagnet in response to momentary pulses of current. At each pulse, the pawl engages the ratchet, moving the wipers (to which circuits are connected) one step forward into contact with stationary terminals to which a selected circuit is connected. These stationary terminals are assembled in one unit in the form of a semicircular (or arc) *bank,* and may number up to 250 or more individual contacts. Bank levels, equal in number to pairs of wipers, are separated from each other by an insulating material.

A rotary stepping switch performs the same basic switching functions as a "switch." That is, it can "make" or "break" a circuit. It can transfer a circuit with a "break-before-make" action. In rotary stepping switch nomenclature, this is called *nonbridging* and designated *NB.* If the transfer action is "make-before-break," it is called *bridging* and designated *Br.*

In the make-before-break operation the tip of the rotatable member (the wiper) is of sufficient length to touch the contact moved to before leaving the contact last rested upon. This additional length of wiper tip is the physical component that gives bridging, or make-before-break, action to rotary stepping switch operation.

The descriptions used in designating rotary stepping switch parts, for reordering or for explaining circuit function, are confusing to many people. The rotated group of wipers and supporting mechanism, including the ratchet wheel, is called a *wiper assembly* (some call it a *rotor*).

The wipers are of two basic kinds—those which close ("make") the circuit and those which open ("break") the circuit. The former is by far the most common.

There are two kinds of wiper tips: bridging and nonbridging, both of which are often supplied on the same wiper assembly. Bridging wipers have long flat tips which, during rotation, permit the wiper to engage the next bank contact before breaking away from the previous contact. Bridging wipers are used when the circuit through them must be continuous and unbroken. Nonbridging wipers have shorter tips that leave one bank contact before engaging the next, avoiding electrical interconnections between circuits of adjoining bank contacts.

The arrangement of contacts in a single arc, each electrically insulated, to which the wiper will be sequentially connected as it rotates is called a *physical bank level* or, more simply, a *level.* The total accumulation of levels in an assembly is called a *bank.*

There are two types of driving mechanisms used in rotary stepping switches: indirect and direct. When the armature-pawl combination acts directly on the ratchet, under the magnetic attraction generated by the electromagnet, the rotary stepping switch is said to be *directly driven* (see Fig. 17-10*a*).

When the pawl is caused to act on the ratchet wheel from mechanical force resulting from potential energy stored mechanically in some form of a drive spring, the mechanism is said to be an *indirectly driven* rotary stepping switch (see Fig. 17-10*b*).

Rotary stepping switches are classified either as *unidirectional* or *dual-directional.* Dual-directional is sometimes designated as "around-and-return" or "bidirectional." Unidirectional rotary stepping switches step in one direction only.

Schematically, the dual-directional switch can be pictured as operating in the following manner. In Fig. 17-10*a*, the switch moves one tooth notch each time the rotary magnet is energized. It is held in this position by the detent of the release magnet armature when the rotary armature repositions at the end of the energized period. Successive steps take place in the same manner. When the release magnet is energized, the detent is removed from the holding position and the rotor returns to its home position (against the step) from the force of the restoring spring.

The same kind of description can be made for the unidirectional rotary stepping switch. Figure 17-10*b* is a schematic diagram of the pulse-controlled operation of a unidirectional rotary stepping switch. Each time the magnet is energized, the armature is attracted against the tension of the driving spring, and the pawl slips over the adjacent tooth. When the magnet is de-energized, the driving spring pushes the armature pawl against the ratchet tooth it rests upon and moves the wiper assembly the prescribed distance, carrying the wipers forward to the next bank contact. The detent prevents movement in the unwanted direction while the armature is being attracted to the "cocked" position. The wiper is always moved in the same rotary direction in successive steps.

17-11 STEPPING RELAY ILLUSTRATIVE EXAMPLE

Figure 17-11 on page 330 represents the decks of a stepping relay that energize relays CR1 to CR8 in some predetermined sequence. The energizing coil circuit for the stepping relay is not shown. Assume that the switches advance every 30 s. The rotating contact of the switch for deck A through deck H would all advance together after the time interval. Can you determine

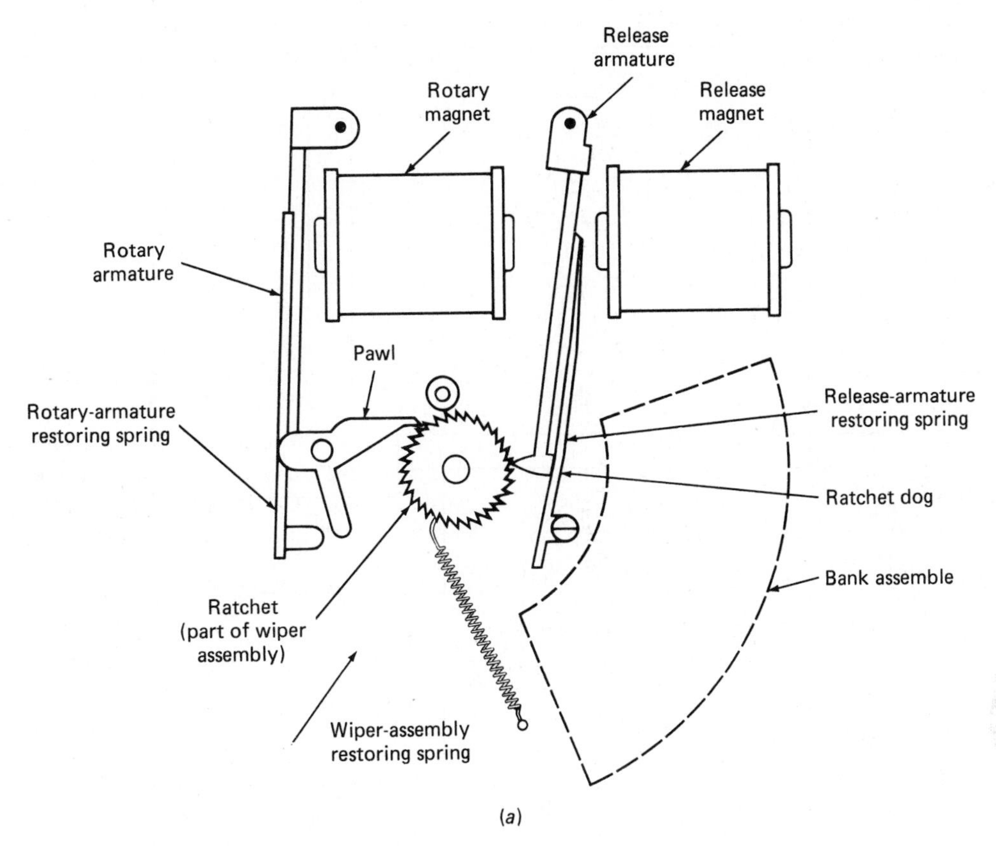

Fig. 17-10*a* Direct-drive mechanism.

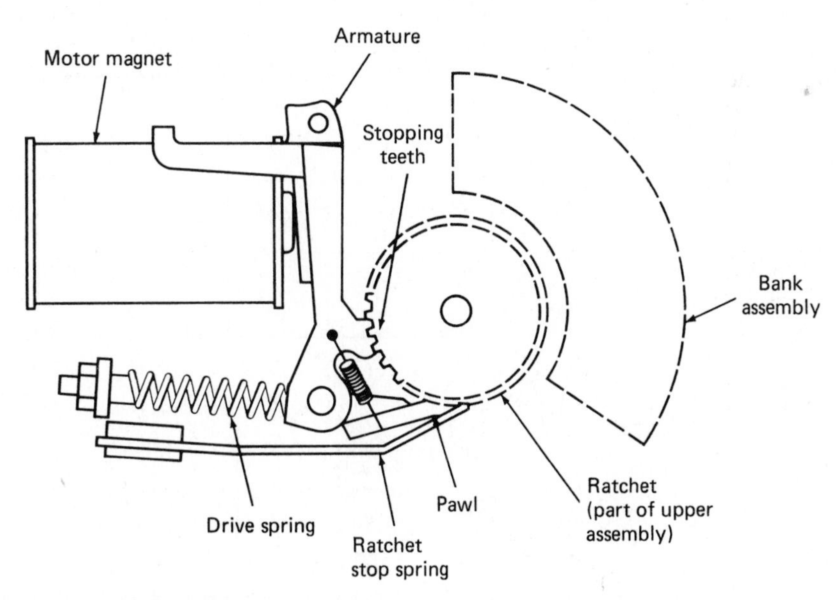

Fig. 17-10*b* Indirect-drive mechanism.

which relays are energized in each of the different switch positions through 11 for each deck of the stepping relay?

The table in Fig. 17-11 is used to develop a wiring chart for the decks of a stepping relay. When completed, the table will represent the eight decks for a stepping relay wired as in Fig. 17-11. If a wire connection is made to switch 1 on deck A, an X is placed on the line under A. To indicate a connection, a 1 is placed in the space preceding line A. Notice in Fig. 17-11 that no wiring connection is made to switch 1 from deck C. With no connection at switch 1 from deck C, we would put a 0 by it. As we look at each deck of the stepping relay in the different switch positions, we can determine if the switches are wired. If we mark an X for a connection, we would have the beginning of a *wiring*

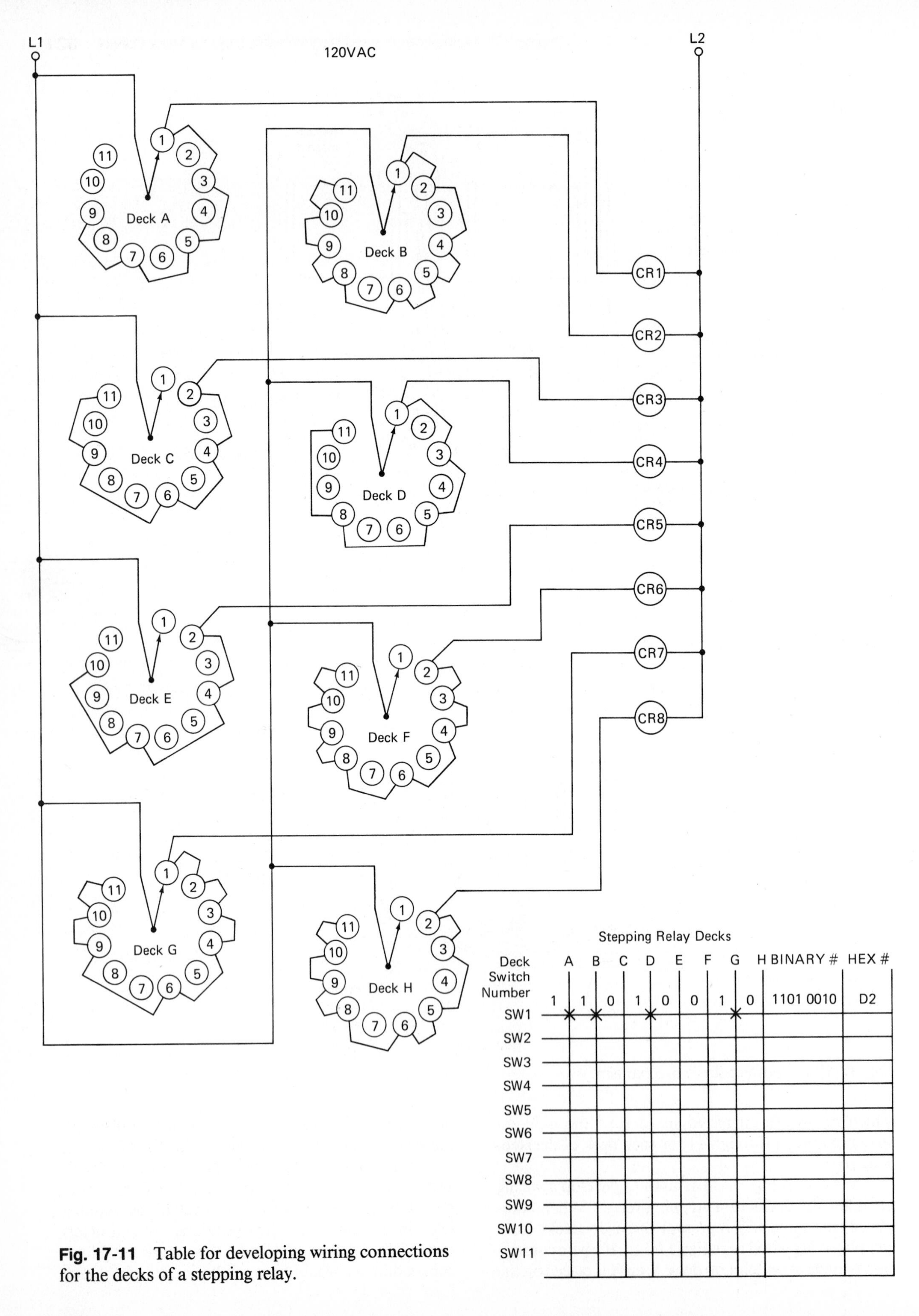

Deck Switch Number	A	B	C	D	E	F	G	H	BINARY #	HEX #
SW1	1	1	0	1	0	0	1	0	1101 0010	D2
SW2										
SW3										
SW4										
SW5										
SW6										
SW7										
SW8										
SW9										
SW10										
SW11										

Fig. 17-11 Table for developing wiring connections for the decks of a stepping relay.

table. An X indicates a connection; a blank, no connection. By putting a 1 by a connection and a 0 by no connection, we also obtain a binary number (1101 0010 for switch position 1, deck A, B, C, D, E, F, G, and H). This is converted to the hex number D2.

If we were to replace the stepping relay with a microprocessor and connect the microprocessor through a suitable coupling device to relays 1 through 8, the instruction D2 in the table of the program for the microprocessor would turn on the same relay as the wiring on the stepping relay.

It can be seen then as you complete the wiring table for the stepping relay and convert the connections to 1s and 0s that you will end up with binary numbers. These binary numbers, when converted to hex numbers, become the program instructions for the table of the microprocessor's state machine program. The only thing missing now is the time delay between which successive relays would be energized. Suppose a 25-s time delay were desired. The binary number 0001 1001 equals 25. This binary number equals 19 in hexadecimal form. The hex number 19 placed between each hex number that indicates the switch-closing pattern will complete the program for the microprocessor. The hexadecimal numbers will cause the relays to energize in the same sequence that the stepping relay did. Hexadecimal numbers placed in between the hex numbers developed in the table become the time delay for the system. A microprocessor then can become a replacement for a stepping relay.

17-12 HEX NUMBERS AS INSTRUCTIONS FOR MICROPROCESSORS

A stepping relay with 12 steps and 8 decks is wired according to the wiring chart of Fig. 17-12. This is another form of the table shown in Fig. 17-11. An X indicates a connection; a blank, no connection. If we convert this wiring table to hexadecimal numbers, we can replace the stepping relay with a microprocessor. The hex numbers will become the instructions for the table of the microprocessor program. The wiring of the decks of a stepping relay takes hours. The programming of the microprocessor takes minutes.

In Fig. 17-12 convert each step, from 1 to 12, on the stepping relay to binary numbers. Then convert each binary number to a hex number. When finished, you have instruction for a microprocessor table.

Study the electric control circuit of Fig. 17-13 on the next page. This circuit uses a stepping relay to control the sequence in which hydraulically operated gates open to let material transfer from a storage area to a conveyor belt. Complete the table in Fig. 17-14 on page 333 to show the wiring connections on the stepping relays. First convert the rows of the connection table to binary numbers, indicating a connection with a 1 and no connection with a 0. Then convert these binary numbers to hexadecimal numbers. The hex numbers become the instructions for the microprocessor table if the control of the gates were to use a microprocessor-based control system instead of a stepping-relay system. If each gate were to be opened for 35 s, what hex number would provide a 35-s delay? The binary number 00100011 equals the decimal number 35. The binary number 00100011 = hexadecimal 23. Thus 23 is the hex code for 35-s time delay.

If the conveyor belt were reversible so that it could send material to two different storage areas, the solenoid valves in Fig. 17-13 should open in reverse sequence. The valve SOL V8 should open first, SOL V7 next, etc. Notice that the table for the operation of the hydraulically operated valves in reverse sequence would look like Fig. 17-15 on page 334. Provide the hex numbers for Fig. 17-15. These hex numbers will

	Deck or banks on stepping relay								Binary number	Hexadecimal number
Step	A	B	C	D	E	F	G	H		
1		X		X		X		X	01010101	55
2	X	X	X	X	X	X	X	X		
3	X	X	X	X	X	X				
4	X	X	X	X						
5	X	X								
6	X				X			X		
7	X	X					X	X		
8	X	X	X			X	X	X		
9	X	X		X	X		X	X		
10			X	X	X	X				
11		X	X	X	X	X	X			
12	X		X		X		X			

Fig. 17-12 A wiring chart used to prepare microprocessor instructions.

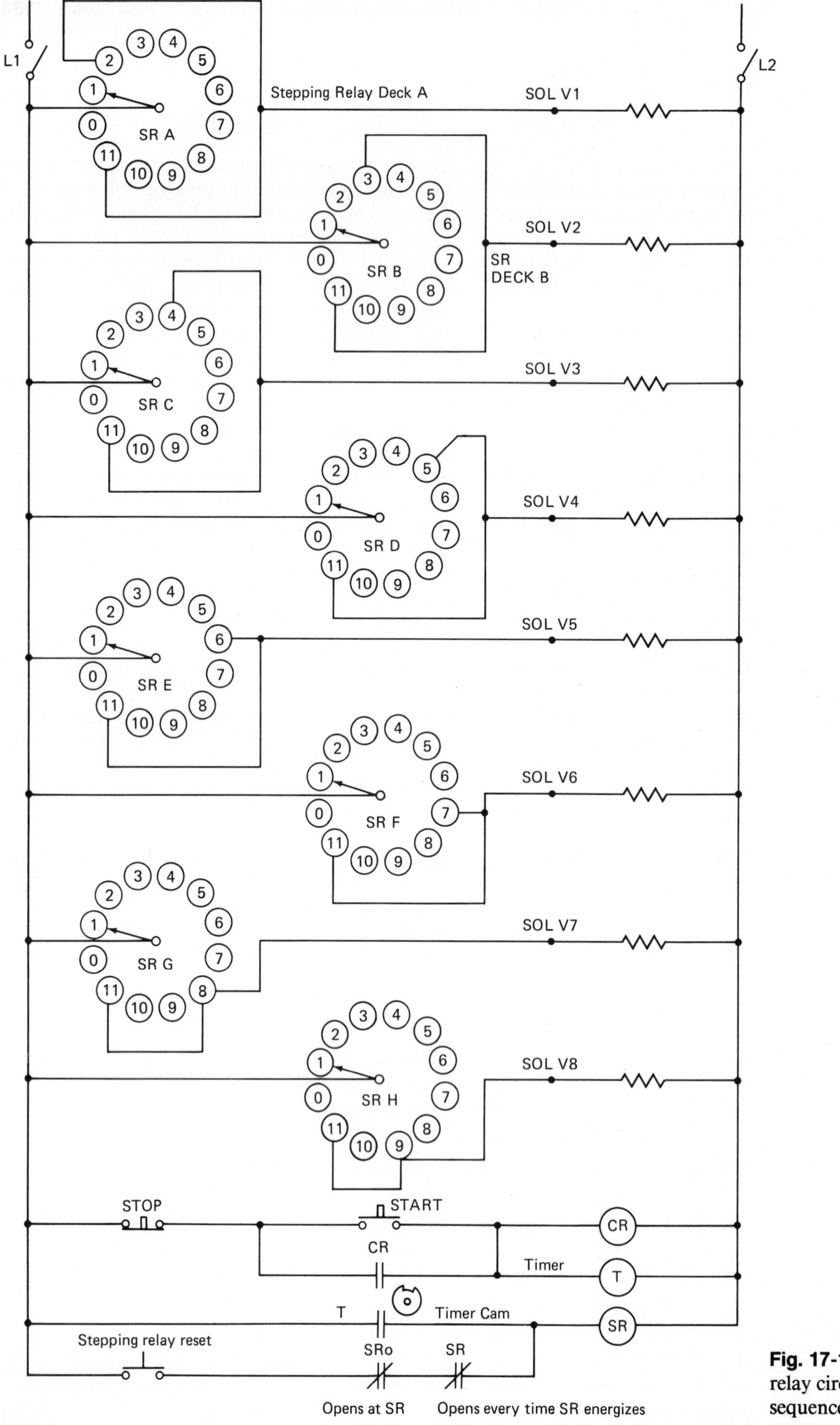

Fig. 17-13 Stepping relay circuit to give sequence control.

1. Represents deck on stepping relays or drum switch sections
2. Represents solenoid valves or relays that are energized
3. Represents binary number for conversion to hex numbers

Instruction:	(1) A (2) 1 (3) D′	B 2 C′	C 3 B′	D 4 A′	E 5 D	F 6 C	G 7 B	H 8 A	Binary numbers	Hexadecimal numbers
1. Start SW 1 on	0	0	0	0	0	0	0	0	0000 0000	00
2. Time Delay 35 s		35 Seconds							0010 0011	23
3. SW 2 on	1	0	0	0	0	0	0	0	1000 0000	80
4. Time Delay 35 s		35 Seconds							0010 0011	23
5. SW 3 on										
6. Time Delay										
7.										
8.										
9.										
10.										
11.										
12.										
13.										
14.										
15.										
16.										
17.										
18.										
19.										
20.										
21.										
22.										
23.										
24.										
25.										
26.										
27.										
28.										
29.										
30.										
31.										
32.										
33.										
34.										
35.										

Fig. 17-14 Microprocessor worksheet.

become the instructions for the microprocessor program giving the valves reverse-sequence operation. Suppose the time delay for the solenoid valves to be open is 45 s. What is the hex number for a 45-s time delay?

Something should now be obvious about the problems depicted in Fig. 17-13 through 17-15. If the conveyor belt were required to be reversed every 2 or 3 h, with each reversal the solenoid valves opening the gates energize in reverse sequence, it would be an impossible problem with stepping relays or drum switches. The time to rewire the sequence on the relays would be considerable. If the cycle were repeated several times a day, the stepping relays would be rewired so often that they would soon be destroyed. How about a microprocessor? To change the sequence of energizing would require only that the program in the memory addresses for the table be cleared and the new hex in-

Solenoid valves								
1	2	3	4	5	6	7	8	Hex
D′	C′	B′	A′	D	C	B	A	numbers
0	0	0	0	0	0	0	1	______
0	0	0	0	0	0	1	0	______
0	0	0	0	0	1	0	0	______
0	0	0	0	1	0	0	0	______
0	0	0	1	0	0	0	0	______
0	0	1	0	0	0	0	0	______
0	1	0	0	0	0	0	0	______
1	0	0	0	0	0	0	0	______
0	0	0	0	0	0	0	0	______
1	1	1	1	1	1	1	1	______
0	0	0	0	0	0	0	0	______

Fig. 17-15 Valve sequence problem.

structions be programmed back into the memory addresses that were cleared. How long would this take? Less than 5 min. Would any new wiring be required? No. It is obvious that a dedicated control problem that requires occasional changes in its cycle would be simple for a microprocessor.

To give you a variety of problems, we will make a few more changes to the program for you to work out. Suppose that the gates were to open in the sequence indicated in Fig. 17-16. What would be the new program for the table in the microprocessor? Allow the gates to open for 40 s each.

Solenoid valves								
1	2	3	4	5	6	7	8	Hex numbers
0	0	0	0	0	0	0	1	______
0	0	0	0	0	0	1	1	______
0	0	0	0	0	1	1	1	______
0	0	0	0	1	1	1	1	______
0	0	0	1	1	1	1	1	______
0	0	1	1	1	1	1	1	______
0	1	1	1	1	1	1	1	______
1	1	1	1	1	1	1	1	______
0	0	0	0	0	0	0	0	______
1	1	1	1	1	1	1	1	______
0	0	0	0	0	0	0	0	______

Fig. 17-16 Another valve sequence problem.

17-13 SDK-85 MICROPROCESSOR BOARD

One of the microprocessor boards that can be used to program our problems on in this chapter is the SDK-85. This development board uses the Intel 8085 microprocessor chip. The program for a state machine is already prepared for you in Fig. 17-17. It begins at address 2000 and ends at address 202E. This program will process whatever instructions are in the table, which starts at address 2040.

The table in Fig. 17-18 on page 336 is included in the state machine program to provide sequence control. The microprocessor does whatever the table tells it to do. In Fig. 17-18, Program 1 provides the control depicted in Fig. 17-14. Program 2 is the instructions for the control depicted in Fig. 17-15. A 1-s time delay is given to the table of Fig. 17-18.

17-14 STATE MACHINE CONTROL FOR MICROPROCESSORS

State machine motor control for eight motors is given in Fig. 17-19 on page 336. The program operates LEDs on the microprocessor. These can be coupled into solid-state optical couplers or solid-state devices such as triacs, which energize relays or contactors when the LEDs are on. The relays can be used to energize motor starters of any size.

Develop a microprocessor program for the table of the state machine that will give the operation indicated by Fig. 17-19. Use a microprocessor worksheet to develop the program. If you have an Intel SDK-85 microprocessor board, you can verify that your program works by placing it into the SDK-85. Execute the program and watch to see if the motors operate in the sequence indicated by the program.

Design a program that will control the traffic lights at an intersection. Use LEDs 1, 2, and 3 to represent red, amber, and green lights for the north and south lights. Use LEDs 6, 7, and 8 for red, amber, and green for the east-west lights. Use a time delay for switching the lights that is comparable to the time delay of the traffic lights in your area. Place the sequence on a microprocessor worksheet and develop the instructions for the table of the state machine program. Load your program into the SDK-85, and verify that the program is correct.

To the traffic light problem above, add turn signals. Let LED 4 represent the turn signal lights for the north-south traffic. Let LED 5 represent the turn signal lights for the east-west turn signals. Watch an intersection with turn signals to develop the operation sequence of the lights. Write a program for the table on a worksheet. Load the program in the microprocessor and verify it is correct.

17-15 PROGRAMMABLE LOGIC DEVICES

There are quite a number of devices that belong to the programmable logic family. Field programmable logic arrays (FPLAs) with their AND/OR/INVERTER

STATE MACHINE (Programmed sequence control)

Memory address	Instructions (mnemonic)	Memory contents (HEX)
2000	MVI ADFFA	3E
2001		FF
2002	OUT 3	D3
2003		03
2004	LXI H2040 H	21
2005		40
2006		20
2007	MOV A, M	7E
2008	ORA	B7
2009	JZ	CA
200A		04
200B		20
200C	MOV E,M	5E
200D	INX H	23
200E	MOV M.A. (A.M.)	7E
200F	CMA	2F
2010	OUT 01	D3
2011		01
2012	Wait: MVIB, 04 (FF-0) (Time delay adjust)	06
2013		04 for one sec. TD
2014	MVI C FA	0E
2015		FA
2016	MVI D, D6 (FF-01) (Time delay adjust)	16
2017		D6
2018	DCR D	15
2019	JVZ	C2
201A		18
201B		20
201C	DCR C	0D
201D	JNZ	C2
201E		16
201F		20
2020	DCR,B	05
2021	JNZ	C2
2022		14
2023		20
2024	DCR,E	1D
2025	JNZ	C2
2026		12
2027		20
2028	SUB A	97
2029	OUT 01	D3
202A		01
202B	INX H	23
202C	JMP	C3
202D		07
202E		20

PROGRAM TABLE STARTING AT ADDRESS 2040 INTO MICROPROCESSOR

EXEC

GO INSTRUCTIONS TO RUN PROGRAM

2000

EXEC

The table provides the sequence and time delay in the operation of the outputs of the microprocessor.

Fig. 17-17 SDK-85 (Intel 8085) microprocessor state machine program used for sequence control (must have a table such as is developed in the previous switching problem).

Address	Instruction	Memory Content	
		Prog. 1	Prog. 2
2040	Time 2 s	02	02
2041	LED's OFF	00	00
2042	Time 1 s	01	01
2043	#1 (left LED) ON	80	01
2044	Time 1 s	01	01
2045	#2 LED ONE	40	02
2046	Time 1 s	01	02
2047	#3 LED ON	20	04
2048	Time 1 s	01	01
2049	#4 LED ON	10	08
204A	Time 1 s	01	01
204B	#5 LED ON	08	10
204C	Time 1 s	01	01
204D	#6 LED ON	04	20
204E	Time 1 s	01	01
204F	#7 LED ON	02	40
2050	Time 1 s	01	01
2051	#8 LED ON	01	80
2052	Time 1 s	01	01
2053	All LED ON	FF	FF
2054		00	00
2055	Repeat	00	00

Fig. 17-18 Table for state machine. Memory Content is the program. Table shows two programs for two different sequences of control. Only one program is placed in the microprocessor at a time.

			Motor								
Steps	Hex #	Time	1	2	3	4	5	6	7	8	Instructions
1		Start	off	off	off	off	off	off	off	on	Motor 1 starts
2		10 s	off	off	off	off	off	off	on	on	10 s later, 2 starts
3		5 s	off	off	off	off	off	on	on	on	5 s later, 3 starts
4		10 s	off	off	off	off	on	on	on	on	10 s later, 4 starts
5		25 s	off	off	off	on	on	on	on	on	25 s later, 5 starts
6		3 s	off	off	on	on	on	on	on	on	3 s later, 6 starts
7		5 s	off	on	on	on	on	on	on	on	5 s later, 7 starts
8		10 s	on	on	on	on	on	on	on	on	10 s later, 8 starts
9		60 s	off	on	on	on	on	on	on	on	60 s later, 8 stops
10		5 s	off	off	on	on	on	on	on	on	5 s later, 7 stops
11		5 s	off	off	off	on	on	on	on	on	5 s later, 6 stops
12		20 s	off	off	off	off	on	on	on	on	20 s later, 5 stops
13		10 s	off	off	off	off	off	on	on	on	10 s later, 4 stops
14		10 s	off	off	off	off	off	off	on	on	10 s later, 3 stops
15		5 s	off	off	off	off	off	off	off	on	5 s later, 2 stops
16		10 s	off	off	off	off	off	off	off	off	10 s later, 1 stops

Fig. 17-19 Table for starting eight motors in sequence with time delay. Motors are stopped in reverse sequence. Hex numbers become the program.

functions are the programmable logic products that electricians and control system repair personnel will be most interested in. Custom programming in the field can be applied to an FPLA so that any logic expression, including sum or product equations, can be placed into a 28-pin, specially designed large-scale integrated circuit (LSIC). The programming of an FPLA requires a special programmer. The custom implementing of logic circuits into the gate array is done by "burning in," or "fuse linking," the programmable connections within the FPLA. This method is similar to the programming techniques used in placing programs in programmable read-only memories (PROMs).

Figure 17-20 illustrates a sample programmable logic array with user-programmable connections. All internal interconnections are programmable via fuse links. The number of AND, OR, and INVERTER gates used in the chip and how they are interconnected is determined by user programming. User programming burns in the fuse links in such a way as to leave the desired logic circuit interconnected on the LSIC.

Programmable logic devices (PLDs) are capable of logic synthesis. As illustrated in Fig. 17-21 on the next page, two logic circuits are converted to boolean algebra equations. Each term of the logic equation becomes a direct entry into the logic program table. Notice the logic program table in Fig. 17-22 on the next page. The terms of the logic equation for the logic circuit are entered as H and L. A NOT (inverted) term is entered as an L. A term that is not inverted is entered as an H. Example: $P1 = \overline{C}D$. It can be seen that large logic circuits using the ANDS, ORS, and INVERTERS could be entered on this table. Circuits with a large number of logic packages or modules could be replaced with PLDs.

SUMMARY

A microprocessor is a large-scale integrated circuit that contains most of the circuitry associated with digital computers. The microprocessor is used in motor control as the central processing unit, or brain, of the control system. It is ideal for dedicated control. The microprocessor is programmed to perform the task it does. It looks at the programmed instructions in memory and instructions or input signals from pilot devices. The microprocessor analyzes the input information and makes the decision as to what control action is required. It can perform arithmetic, logic, and

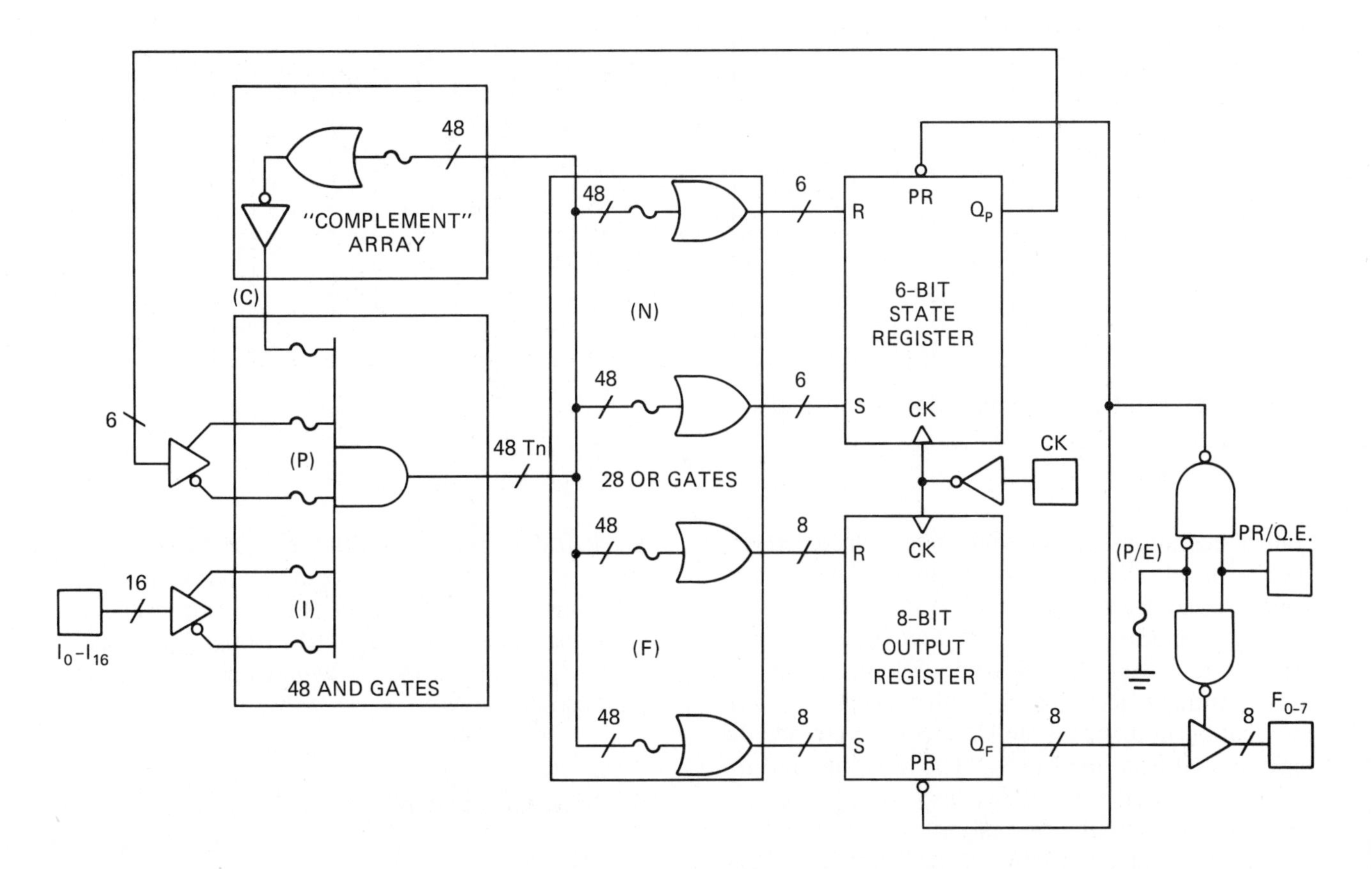

NOTE:
I, P, C, N, F and P/E are user-programmable connections.

Fig. 17-20 Sample programmable logic array showing fuse linking in the FPLAs.

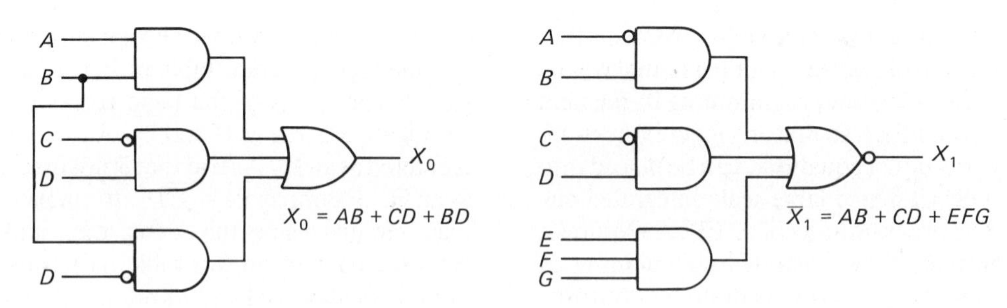

Fig. 17-21 Equivalent fixed-logic diagram.

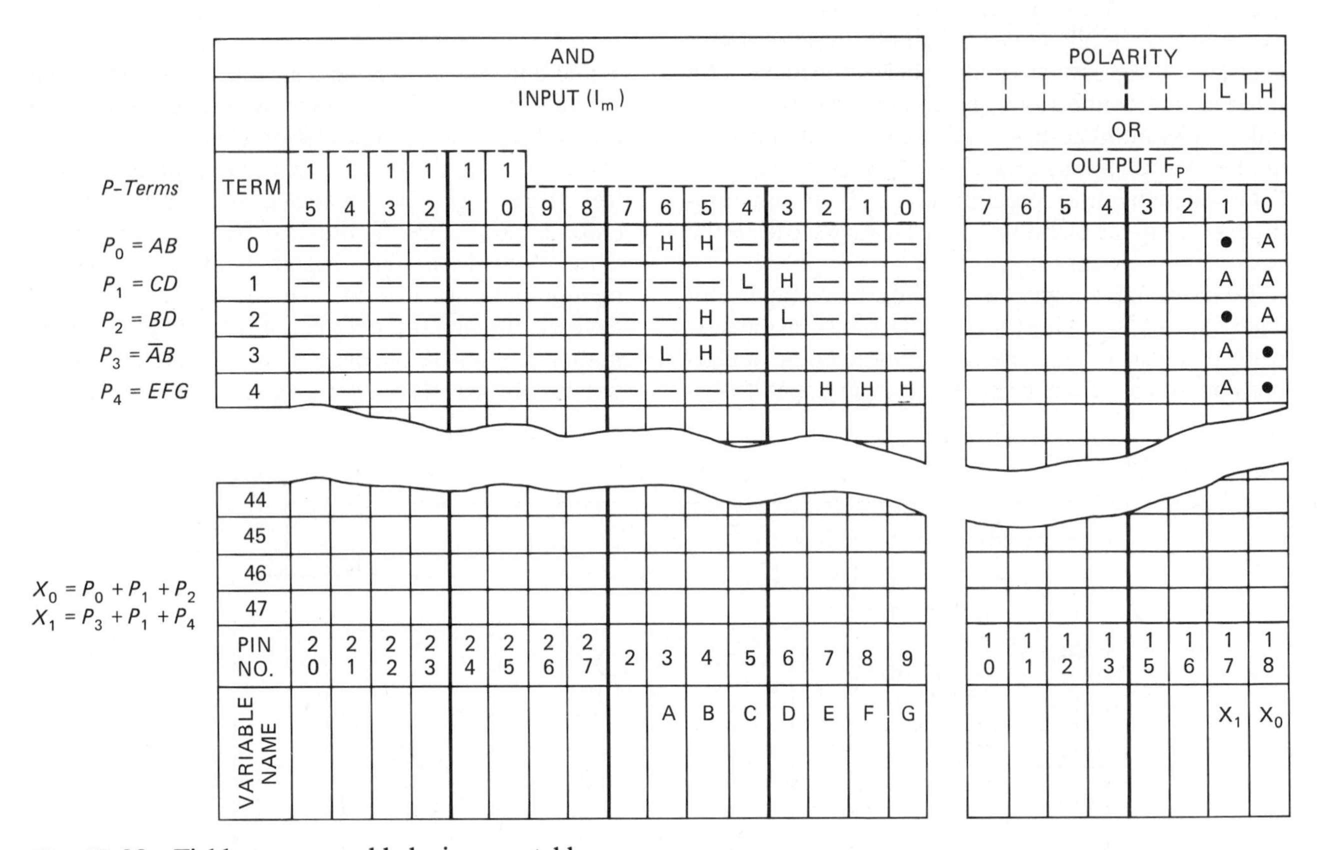

P-Terms	TERM	AND INPUT (I_m) 15	14	13	12	11	10	9	8	7	6	5	4	3	2	1	0	POLARITY / OR OUTPUT F_P 7	6	5	4	3	2	1	0
	Polarity																							L	H
$P_0 = AB$	0	—	—	—	—	—	—	—	—	—	H	H	—	—	—	—	—							•	A
$P_1 = CD$	1	—	—	—	—	—	—	—	—	—	—	—	L	H	—	—	—							A	A
$P_2 = BD$	2	—	—	—	—	—	—	—	—	—	—	H	—	L	—	—	—							•	A
$P_3 = \bar{A}B$	3	—	—	—	—	—	—	—	—	—	L	H	—	—	—	—	—							A	•
$P_4 = EFG$	4	—	—	—	—	—	—	—	—	—	—	—	—	—	H	H	H							A	•
	44																								
	45																								
$X_0 = P_0 + P_1 + P_2$	46																								
$X_1 = P_3 + P_1 + P_4$	47																								
	PIN NO.	20	21	22	23	24	25	26	27	2	3	4	5	6	7	8	9	10	11	12	13	15	16	17	18
	VARIABLE NAME										A	B	C	D	E	F	G							X_1	X_0

Fig. 17-22 Field-programmable logic-array table.

control operations. It is a solid-state replacement for such control elements as stepping relays, rotary switches, and drum timers.

Field programmable logic arrays are large-scale integrated circuits capable of being field-programmed. Programming is accomplished with a special programmer that burns in, or fuse-links, circuits to provide the logic desired from the FPLA. The AND, OR, and INVERTER functions are capable of logic synthesis. Conventional logic circuits are analyzed and put into boolean algebra equations form. A programmable logic array table is prepared from these equations. The table is sent to the manufacturer of the PLD logic chip with the desired logic burned into it. The PLD logic chips can be field programmed with the desired logic circuits if the customer has his or her own programmer. Field-programmed logic arrays offer the advantages of simplified logic boards and reduced cost per logic circuit. Field programmable logic arrays are opening a new era in industrial control.

REVIEW QUESTIONS

1. Define a microprocessor.
2. What is the microprocessor's job in the control of motors?

3. What part does the microprocessor play in a control system?
4. What determines if microprocessor-based control is justified for a control system?
5. Define dedicated control.
6. What types of control systems can a microprocessor replace?
7. What types of operations is a microprocessor capable of performing?
8. What do the abbreviations CPU and ROM stand for?
9. What is the function of interface adapters?
10. What device programs a microprocessor?
11. What takes place during "processing" in a microprocessor?
12. What determines when instructions are retrieved from the memory of microprocessors?
13. The control section of the microprocessor automatically fetches each instruction from memory and does what with it?
14. What happens when an instruction is executed?
15. What does decoding refer to?
16. What is the function of the arithmetic/logic unit in a microprocessor?
17. What are the five steps in the programming process?
18. What are the applications for state machine programs?
19. What is an algorithm?
20. What do the different symbols in flowcharts represent?
21. What is a mnemonic symbol used for?
22. What are the steps a maintenance electrician would take in troubleshooting a microprocessor-based control system?
23. What is the biggest source of trouble in a microprocessor?
24. Where is the best place to look for aid in troubleshooting a microprocessor?
25. What is the binary number for the decimal number 164?
26. What is the decimal number for 10111011?
27. What is the hex number for 10111011?
28. What is a rotary stepping switch used for in industrial control work?
29. Describe the mechanism of a rotary stepping relay.
30. What is meant by nonbridging in stepping relay nomenclature?
31. What does BR mean in stepping relay nomenclature?
32. How are stepping relays classified?
33. How does a microprocessor give the same control action as a stepping relay?
34. How is the time-delay instruction determined for a program in a microprocessor table?
35. When continuous daily changes are required in a control system, why is a microprocessor more desirable than a stepping relay?
36. At what address does the table for a state machine program start for the SDK-85 microprocessor board?
37. What microprocessor is used in the SDK-85 microprocessors board?
38. What do programmable logic devices consist of?
39. Define the abbreviations FPLA and PLD.
40. How is a logic program placed into a LSIC?
41. How are internal connections placed into a programmable logic array?
42. How is the required logic placed in a LSIC?
43. How are PLDs capable of logic synthesis?
44. What is the purpose of a logic program table?
45. In Fig. 17-11 determine the sequence in which the relays energize. Complete the table in Fig. 17-11. Use binary and hex numbers in your answers.

18
Industry Outlook

This chapter touches briefly on the use of computers for motor control, microprocessors for numerical control, and robotics in control systems. It discusses the opportunities for employment in an otherwise highly automated field. One of the largest manufacturers of industrial control valves in the West uses automation to the limit. The production of a valve starts with design engineers creating a drawing for the valve parts by using computerized drafting and design (CADD). The computer stores the prints and specifications in its memory. When the production people are ready to start the manufacturing process, the plans are retrieved from the computer memory. The computer processes programs for the machining process using numerical control.

18-1 NUMERICAL CONTROL

Numerical control is often defined as the use of coded numerical instructions for automatically controlling machines. The coded numerical instructions can originate in a computer.

Numerical control can also be defined as a method of controlling movements of a machine tool with the aid of a numerical language. Numerical control directs the feed movements of a machine through the use of numbers.

18-2 HOW NUMERICAL CONTROL IS USED IN INDUSTRY

Numerical control machines can direct the movement of 25 or more tools during a manufacturing process. For example, a machine may drill a hole in a part to a certain depth. This would require a drill bit as the tool. After it is finished, the tool is automatically returned to its storage place. The next operation might call for a tap to thread the hole just drilled by the machine. This process may continue until up to 25 different tools have been used in the manufacturing process. A part can be precisely duplicated by retrieving the program from the computer. The numbers, remember, are symbolic codes for the instructions that operate the machine tools. This is the way a machined product is created.

A number of special motors move the operating elements on a numerically controlled machine. The motors get their instructions from electronic control units. The control units tell the motor when to start, whether a slow start or a fast start is required, what direction to turn, how fast to turn, when to stop, whether a soft stop or fast stop is required, how long to stay stopped, etc. These electronic control units contain many IC devices presented in earlier chapters.

Some numerical control machines are controlled by tapes. The program to run the numerical control machine is placed on a tape. Some tapes are magnetic; some are perforated tapes. Magnetic tapes have magnetic codes representing instructions for the machine. Perforated tapes have holes punched in them that are codes for a program. Each time a machine is to manufacture a part, a tape is loaded into a tape reader. The information for the control and sequencing of the motors on the machine is derived from decoding the information on the tapes.

The simplest type of numerical control is *point-to-point control.* In this type of control, the part or tool is moved first in one direction until a certain programmed position is reached. The machine then moves the part in another direction until a second position in a X or Y axis is reached.

A more complicated form of numerical control is called *three-axis machine control.* With three-axis control, all motors may be running at the same time and machining or positioning the part in three dimensions. The motors may all be running at different speeds to place the part and tool at the proper location when the axis motors stop.

Figure 18-1 illustrates a small numerical control machine.

18-3 SKILLS NEEDED BY INDUSTRIAL TECHNICIANS

The industrial maintenance technician must work with robots and robotic systems. The valve manufacturing company cited earlier uses *robots* in their production line. Until the robots were installed, there were three people involved in the manufacture of a certain part. The three people worked a 40-h week to produce the necessary number of parts required by the production schedule. When a robot was placed in the manufacturing system, it produced in 1 day what three people had been producing in 5 days.

The original process involved moving a part through six different machines. Each machine performed an operation on the manufactured part. The robot could have several parts moving through the machines at the same time. Mechanically, its job was to move parts from machine to machine.

Robot control was not much different than numerical control. The robot's main job was to move the part from one machine tool to another as each machine completed its operation. When ready for the next operation, the part was picked up by the robot and moved to the next machine in the manufacturing sequence. The robot did not move the part again until the next operation had been completed on the part. Instead of taking just one part at a time through the machine, the robot kept as many parts as possible in the machine at a time. Where the machine operators followed one part through the machine at a time and manually moved the part from machine to machine, the robot took the part through many operations simultaneously. If necessary, it could operate continuously for an entire 24-h day. An example of a highly automated factory is shown in Fig. 18-2 on the next page.

It should be obvious that the computer control for a numerical control machine and that for robot are very similar. It should also be obvious that with all the sophisticated machines in modern industry, the need for highly skilled industrial electricians and technicians is growing. There is a need for more high-tech training programs to provide technically qualified people for industry. However, industry's needs require workers with a wide range of education.

"Displacement" of workers does not necessarily

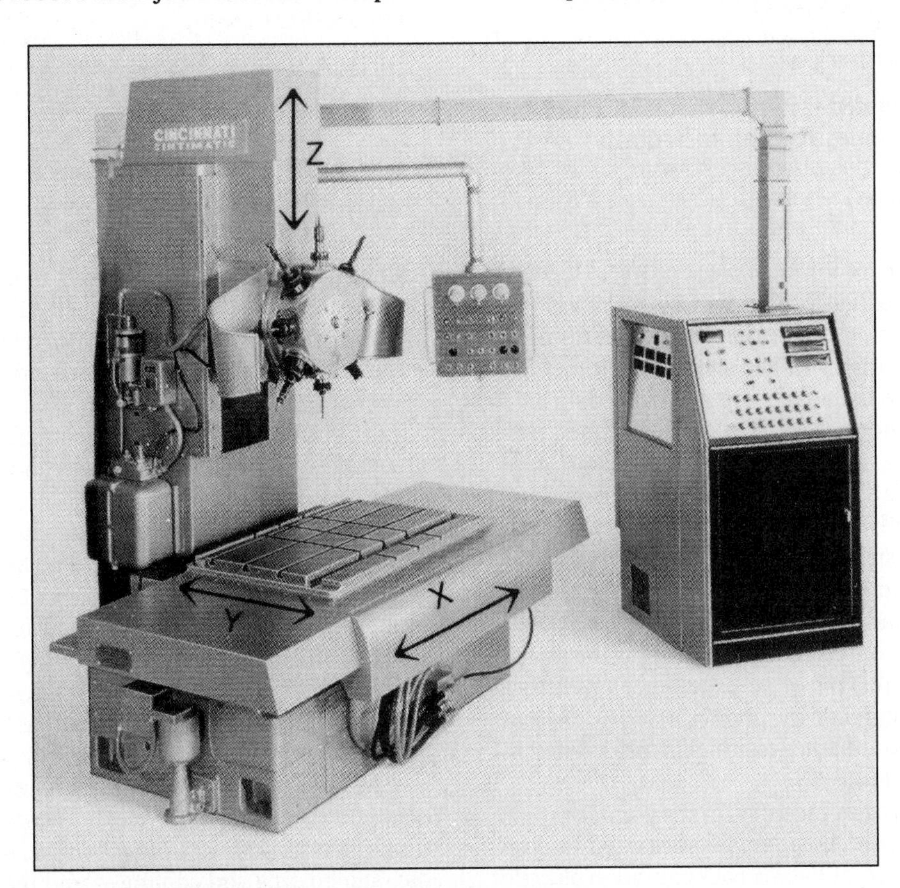

Fig. 18-1 A numerically controlled single-spindle drilling machine showing the *X, Y,* and *Z* axes and the machine tool it controls. (*Cincinnati-Milacron, Inc.*)

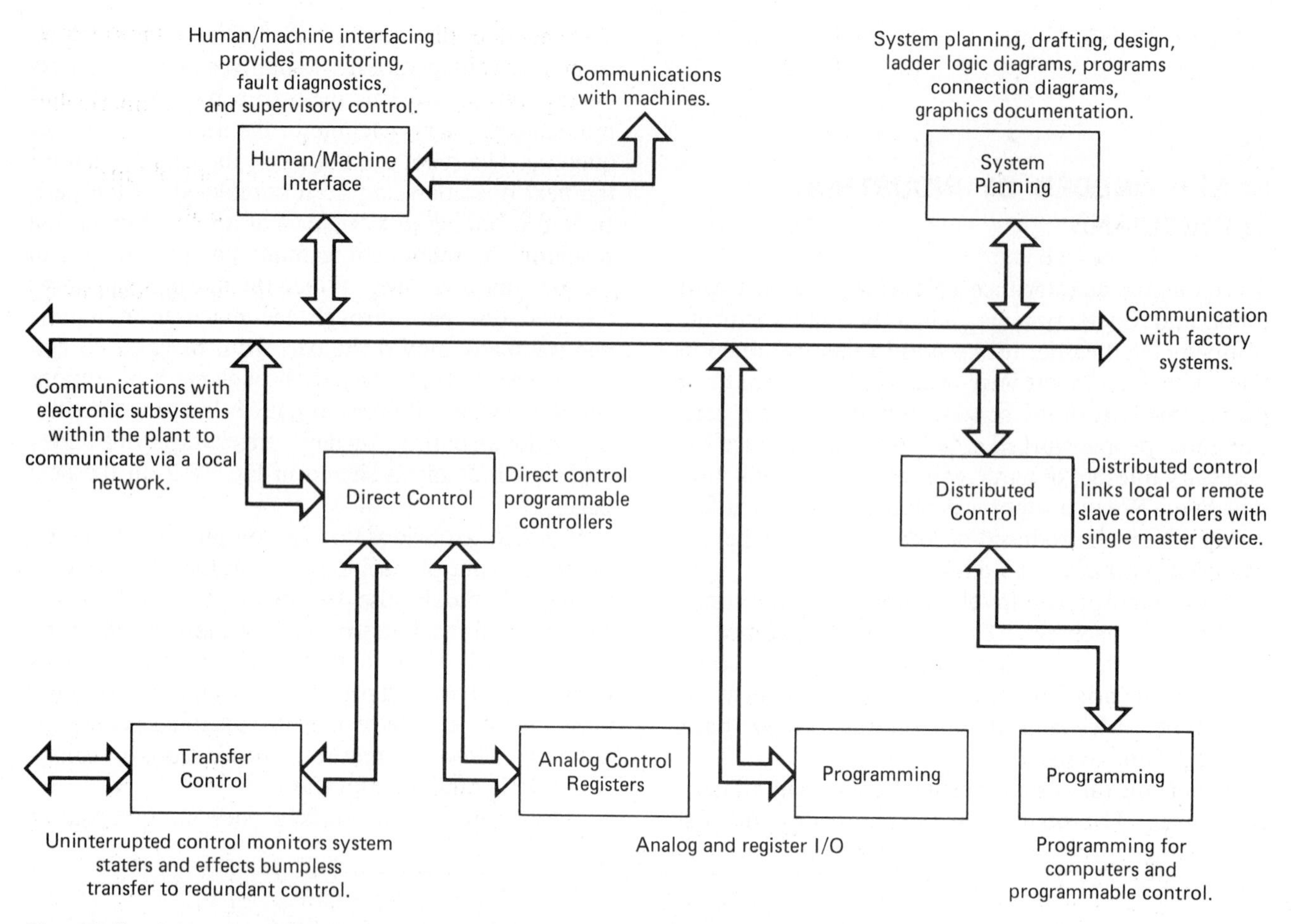

Fig. 18-2 Automated factory systems. Computers and programmable controllers communicating with each other.

mean unemployment for most plant people. It generally means that skilled people will be retrained (frequently to work on or with one or more automated systems), or they simply will be shifted to another part of the plant where their skills are still needed.

Industry does not need a lot of skilled robotic technicians. It needs skilled welders, electricians, computer specialists, lab technicians, and other people with the same basic skills that schools have been teaching for years. Strong basic electrical, electronic, mechanical, hydraulic, and pneumatic technical skills must first be acquired before industrial skills can be applied effectively. It requires only a short time to apply the knowledge and technical skills of an experienced industrial electrician or technician to robotics systems. Repair personnel and maintenance technicians are needed instead of machine tenders.

Robotics training in industry, in the plant, or in the robot supplier's own classrooms is measured in days, not months or years. Given a background in electricity, electronics, hydraulics, or pneumatics (depending on the type of robot to be serviced), a robotics repair technician can attain the necessary skills in a short time. That training is specific to the robot that the trainee will service. The payback to industry is immediate. Schools planning generalized robotics training are competing with their own best customers instead of serving them. The message to educators is clear: America needs more of what we have been doing all along—turning out skilled electricians, technicians, mechanics, and other people with good basic technical skills.

Many industrial suppliers and users agree that the education of a technician should be broadly based and multidisciplinary. Related careers that are this comprehensive are instrumentation technology and electromechanical technology. Training in these areas requires a strong basis in physics, chemistry, and electrical fundamentals, all supported by mathematics, graphics, and shop skills.

In addition, students in these technologies must understand six classes of devices. These devices are electrical, electronics, electromechanical, mechanical, fluidic, and heating and cooling—and the computers and microprocessors used in these devices. With all these basics, only a short period of specialized application is required to make an effective technician in an automated industrial plant.

REVIEW QUESTIONS

1. What does a computer basically do in a control system?
2. What is numerical control?
3. How many tool changes may be required by a numerically controlled machine in making one part?
4. How does the numerical control machine make a part over and over again?
5. What does the control units "tell" the motors in a numerically controlled machine?
6. What type of tapes are used to control numerically controlled machines?
7. What is the simplest numerical control tape?
8. What is meant by three-axis control?
9. Define robotics.
10. What is the purpose of robots in the manufacturing process?

Index